CONVERSION FACTORS FOR MASS

multiply number of → to obtain ↓ by ↘	(lb-sec²/in)	Grams (gr)	Kilograms (kg)	Pounds mass (lbm)	Slugs (lb-sec²/ft)
(lb-sec²/in)	1	$5.711(10^{-6})$	$5.711(10^{-3})$	$2.591(10^{-3})$	$8.333(10^{-2})$
Grams (gr)	$1.751(10^{5})$	1	$1(10^{3})$	453.6	$1.459(10^{4})$
Kilograms (kg)	175.1	$1(10^{-3})$	1	0.4536	14.59
Pounds mass (lbm)	386.0	$2.205(10^{-2})$	2.205	1	32.17
Slugs (lb-sec²/ft)	12	$6.853(10^{-5})$	$6.853(10^{-2})$	$3.108(10^{-2})$	1

CONVERSION FACTORS FOR FORCE

multiply number of → to obtain ↓ by ↘	Dynes (gr-cm/sec²)	Newtons (kg-m/sec²)	Pounds force (lb)	Poundals (lbm-ft/sec²)
Dynes (gr-cm/sec²)	1	$1(10^{5})$	$4.448(10^{5})$	$1.383(10^{4})$
Newtons (kg-m/sec²)	$1(10^{-5})$	1	4.448	0.1383
Pounds force (lb)	$2.248(10^{-6})$	0.2248	1	$3.108(10^{-2})$
Poundals (lbm-ft/sec²)	$7.233(10^{-5})$	7.233	32.17	1

MECHANICS OF MACHINES

Oxford University Press is a department of the University of Oxford.
It furthers the University's objective of excellence in research,
scholarship, and education by publishing worldwide.

Oxford   New York
Auckland   Cape Town   Dar es Salaam   Hong Kong   Karachi
Kuala Lumpur   Madrid   Melbourne   Mexico City   Nairobi
New Delhi   Shanghai   Taipei   Toronto

With offices in
Argentina   Austria   Brazil   Chile   Czech Republic   France   Greece
Guatemala   Hungary   Italy   Japan   Poland   Portugal   Singapore
South Korea   Switzerland   Thailand   Turkey   Ukraine   Vietnam

Published in the United States of America by
Oxford University Press

198 Madison Avenue, New York, NY 10016
http://www.oup.com

Library of Congress Cataloging-in-Publication Data
Cleghorn, W. L. (William L.), author.
Mechanics of machines / William L. Cleghorn and Nikolai Dechev. — 2 [edition].
  pages cm
ISBN 978-0-19-538408-6
1. Mechanical engineering.   I. Dechev, Nikolai, author.   II. Title.
TJ170.C58 2015
621.8—dc23
2014015235

The front cover depicts an image of a Chalmers suspension.

Wallace G. Chalmers (1923–1987) was an engineer who combined creative design with entrepreneurial skills. In 1971, he founded *Chalmers Suspension International* to manufacture his design of a tandem axle suspension for trucks. The Chalmers suspension is now prominent in the heavy-duty trucking industry.

Wallace G. Chalmers, and his spouse Dr. Clarice Chalmers, provided substantial support to the University of Toronto that is continuing to be used for promoting design in engineering education. William L. Cleghorn had the honor of serving as the *Wallace G. Chalmers Professor of Engineering Design* and subsequently the *Clarice Chalmers Chair of Engineering Design*. Nikolai Dechev received a *Wallace G. Chalmers Engineering Design Award* while studying Mechanical Engineering.

Printing number: 9 8 7 6 5 4 3 2 1

Printed in the United States of America
on acid-free paper

MECHANICS OF
MACHINES

WILLIAM L. CLEGHORN and NIKOLAI DECHEV

SECOND EDITION

OXFORD
UNIVERSITY PRESS

CONTENTS

4. ANALYTICAL KINEMATIC ANALYSIS OF PLANAR MECHANISMS 145

5. GEARS 187

6. GEAR TRAINS 238

7.   CAMS   310

8.   GRAPHICAL FORCE ANALYSIS OF PLANAR MECHANISMS   355

APPENDIX B: SCALARS AND VECTORS   579

APPENDIX C: MECHANICS: STATICS AND DYNAMICS   583

PREFACE

This book is intended mainly for use as an undergraduate text. It contains topics and material sufficient for a one-half-year course. It may also be useful to practicing engineers.

Conventional notation is employed throughout, primarily using the International System set of units. Approximately one-quarter of the examples employ the U.S. Customary System.

Many examples included in each chapter relate to situations of practical significance. Problems of varying length and difficulty are included at the end of each chapter. In addition, twenty-six computer design projects are supplied in Appendix A. It is suggested that these be completed using the software Working Model 2D.

The companion website for this textbook (www.oup.com/us/cleghorn) has numerous files of 3D animations, mechanisms, video clips of real mechanical systems in operation. These images have been stored in the MPEG(*.mp4) format and may be viewed through the web browser with appropriate movie plug-ins installed (Windows Media Player or Quicktime player). Within the text, each of these files is identified either in boldface in square brackets (e.g., **[Video1.1]**) or within a textbox located in the margin as shown here.

> **Video 1.1**
> Single Cylinder
> Piston Engine

The companion website also contains interactive 2D models based on Working Model 2D, and requiring a file player of that software. To open and use these models, Working Model 2D must be downloaded and installed on the computer. A link to acquire this software is provided on the website. Within the text, each of these files is identified either in boldface in square brackets (e.g., **[Model 1.14A]**) or within a textbox located in the margin as shown here. Appendix D lists all models and animations and video segments.

> **Video 1.14A, 1.14B**
> **Model 1.14A, 1.14B**
> Four-Bar with
> Coupler Point

The companion website also contains more than ten Mathcad files for use with a select set of problems dealing with the analysis and synthesis of mechanisms. The problems for which the Mathcad files may be applied are identified in the statements associated with end-of-chapter problems.

Graphical and analytical methods of analysis are included for the kinematic and kinetic analyses of mechanisms. Graphical methods may be applied in situations where analysis for only one position of a mechanism is required. They often require a scaled drawing in the configuration for which an analysis is required. For several of the end-of-chapter problems, a scaled drawing may be obtained from the companion website. Such problems are identified by the icon shown in the margin. Analytical solutions are useful in instances when solutions are required for a series of mechanism configurations.

Chapter 1 covers basic concepts, including linkage classification by motion characteristics, as well as degrees of freedom of the planar joints of mechanisms. Some common examples of mechanisms, along with their associated animations, are provided.

Chapter 2 provides the background material required to carry out static and dynamic analyses of planar mechanical systems. Expressions for relative velocity and acceleration in the radial-transverse coordinate system are covered. The instantaneous center of velocity of a body is presented. Equations for the kinetics of a rigid body and associated commonly employed sets of units are also provided. The concept of mechanical advantage is covered along with its determination using the instantaneous centers of velocity of the mechanism.

Chapter 3 covers traditional graphical analysis of planar mechanisms. Both velocity and acceleration analyses are presented using vector polygons for one position of the mechanism. Velocity analyses implementing the method of instantaneous centers of velocity are also covered.

Chapter 4 presents an analytical method based on complex numbers for the kinematic analysis of a planar mechanism. The equations generated using this technique may be programmed on a computer for completing an analysis in a series of positions.

Chapter 5 gives a comprehensive synopsis of gears. This includes many common types of gears and related animations of the gears in meshing action. This chapter also includes some of the common methods of gear manufacture.

Chapter 6 presents an analysis of gear trains. For analysis of planetary gear trains, an algorithm suited for computer implementation is provided.

Chapter 7 presents design procedures of cam mechanisms. Both graphical and analytical methods are covered. In addition, the computer program Cam Design is included on the companion website. This program can synthesize a wide variety of disc cams. For a given set of input parameters and prescribed motion of the follower, all pertinent kinematic parameters are provided as a function of input motion.

Chapter 8 covers graphical force analyses of planar mechanisms. Each graphical analysis may be applied for one configuration of a mechanism—for either static or dynamic conditions.

Chapter 9 covers analytical force analysis of planar mechanisms. The governing equations of motion are derived for an arbitrary configuration of a mechanism and may be programmed on a computer to determine results for multiple configurations. Means to balance a four-bar mechanism and a slider crank mechanism are also provided.

Chapter 10 covers the analysis and design of flywheels. This includes determining the size of a flywheel required to keep speed fluctuations within a desired tolerance.

Chapter 11 presents some common methods for the synthesis of mechanisms. This includes graphical and analytical techniques for function synthesis and rigid-body guidance synthesis of four-bar and slider crank mechanisms.

Chapter 12 introduces and describes the design process methodology as it applies to mechanisms and machines. The complete design process is described, including problem formulation, conceptualization, preliminary design, detailed design, embodiment, testing, and documentation/reporting. Various examples are provided throughout the chapter for each stage of the process.

Chapter 13 presents an extensive set of case studies in the design of mechanisms and machines. The case studies allow the reader to follow along in the problem formulation stage, whereby a structured set of design goals and design objectives is developed. Next, one or more possible design solutions that satisfy the design goal and objectives are described. These are solutions that are commonly used in the field; description of their operation is included as well. Finally, a design summary and a set of recommendations highlight the strength and limitations of the proposed design solutions.

Appendix A includes a set of design projects. They can be ideally solved using the Working Model 2D software. Appendix B provides background reference material related to scalars and vectors. Appendix C gives a brief review of mechanics, which is the basis of much of the material presented in this textbook. Appendix D lists the files included on the companion website.

William L. Cleghorn
Nikolai Dechev

Acknowledgments

We express our sincere appreciation to the many colleagues and faculties from the University of Toronto, the University of Victoria, and the University of Manitoba who played a significant part in the preparation of this book. We are most grateful to the following individuals for reviewing the text, improving figures, developing software, and offering their comments and suggestions: Mr. Sean Voskamp, Professor Mina Hoorfar, Professor Homayoun Najjaran, Professor Kenneth C. Smith, Ms. Laura Fujino, Professor James K. Mills, Dr. Henry Chu, Professor Petru-Aurelian Simionescu, Dr. Leif E. Becker, Mr. Joshua Coutts, Mr. Hoi Sum (Sam) Iu, Mr. Peter Bahoudian, Mr. Martin Côté, Mr. Andy K. L. Sun, the late Professor Robert G. Fenton, Professor John Van de Vegte, Professor Ron P. Podhorodeski, Dr. George Tyc, Dr. Masoud Alimardani, Mr. Daniel Ohlsen, and Mr. K. K. (John) Mak. We acknowledge the assistance of numerous undergraduate students who reviewed portions of drafts and provided useful feedback.

We thank the following for providing extremely useful comments while carrying out assessments of the chapters and accuracy checks of the solutions to problems at the ends of the chapters: Professor Hamid Nayeb-Hashemi, Northeastern University; Professor Jeremy Daily, University of Tulsa; Dr. Ali Gordon, University of Central Florida; Professor Thomas Winthrow, Vanderbilt University; Mr. Andrew Mosedale, Texas Tech. University - Lubbock; Professor Peyman Honarmandi, City College of New York; Professor Su-Seng Pang, Louisiana State University; Professor Robert Gerlick, Pittsburgh State University; Professor Janak Dave, University of Cincinnati; Professor Mohamed B. Trabia, University of Nevada; Professor Madhu Madhukar, University of Tennessee, Knoxville; Professor Jizhou Song, University of Miami; Professor David Rocheleau, University of South Carolina–Columbia; Professor Christopher J. Massey, Widener University; Professor Daejong Kim, University of Texas at Arlington; Professor Tonya Stone, Mississippi State University; Professor Peter Goldsmith, University of Calgary; Professor Teong Tan, University of Memphis; Professor Masoud Mojtahed, Purdue University; Professor Calumet James L. Glancey, University of Delaware; Professor Christopher Niezrecki, University of Massachusetts, Lowell; Professor Chad B. O'Neal, Louisiana Tech University; Professor Robert Rogers, University of New Brunswick.

Valued help and contributions were received from the following individuals in industry: Mr. Andrew Kowalski, Magna Powertrain; Mr. Richard Houghton, General Gear Limited; the late Mr. Hagop Artinian, Swissway Machining Limited; Mr. John Augerman, Mr. Will J. Bachewich, Mr. Michael W. Borowitz, Mr. Steven J. Grave, and Mr. Bruce M. Kretz, General Motors of Canada Limited; Mr. William Frey, HD Systems, Inc; and Mr. Andy Eaton, AE Auto Plus Inc.

The Department of Mechanical and Industrial Engineering and the Information Technology Courseware Development Fund at the University of Toronto are gratefully acknowledged for providing funding for the development of much of the software.

A special expression of my appreciation is extended to the late Dr. Clarice Chalmers and the late Mr. Wallace G. Chalmers. Completion of this book would not have been possible without their supportive sponsorship and encouragement.

William L. Cleghorn
Nikolai Dechev

CHAPTER 1

INTRODUCTION

1.1 PRELIMINARY REMARKS

Mechanics is a science that predicts the conditions of a system either at rest or in motion, when under the action of forces and moments. Within mechanics, the study can involve stationary systems at rest (statics) or systems in motion (dynamics). In statics, the motion of the parts in a system is zero (or insignificant), so the analysis is about the forces that act between the parts. In dynamics, the analysis is concerned with the forces that arise in a system of moving parts.

A *machine* is defined as an apparatus that transmits energy through its parts to perform desired tasks. This definition does not restrict the form of energy transmitted, nor does it restrict the size and rigidity of the parts. It therefore encompasses a vast array of mechanical and electrical devices, many of which have a profound effect on our lives. One of the simplest types of machines is the lever. The lever works by transforming a small force with large motion into a high force with short motion. In this way, the small force available from human effort can be magnified into high force for various farming or construction activities. An example of a more complex and recent machine is the automobile. An automobile is a machine that transfers energy from the engine into the drive shaft and onto the wheels, to provide motion. In fact, we can also say the engine is a machine that transfers the energy from the fuel (gasoline) into the drive shaft.

This textbook, entitled *Mechanics of Machines*, deals with the statics and dynamics of various mechanical systems that transmit energy through their parts. The wheel and the lever were among the earliest inventions of machines. Engineers have since designed numerous ingenious machines for the advancement of society. Machines are widely used in all aspects of our society, such as transportation, farming, manufacturing, medicine, war, sports, and many others. For example, doctors employ machines to treat patients, or they can prescribe machines such as prosthetic hands, arms and legs to replace natural limbs. In recent years, robotic machines have had a significant influence in manufacturing methods. Some are able to rapidly perform repetitive tasks in a reliable and accurate manner and can be used to work in hostile and dangerous environments.

Some machines are a combination of mechanical, electrical, and hydraulic components. An automobile is composed of mechanical parts of the drive train, suspension, and steering systems, along with electrical parts used to ignite the air–fuel mixture in the engine and to provide lighting and sensory feedback for the driver.

Machines having electrical components usually incorporate at least some mechanical parts. The hard drive within a computer has electric motors with rotating shafts for spinning the disc that holds the data. It also has a mechanical arm carrying a magnetic sensor head that swings above the disc to store and access data. The keyboard may be considered a machine with mechanical parts. Printers and scanners have machine components such as gears and lever arms to handle the paper. Many electronic watches have hands operated by miniature ratchet wheels that are pushed around one tooth at a time.

Figure 1.1 Single-cylinder piston engine [Model 1.1].

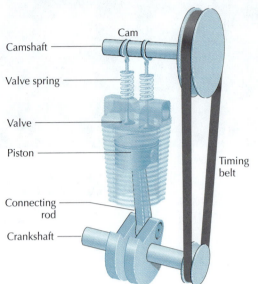

Camshaft

Cam

Valve spring

Valve

Piston

Connecting rod

Crankshaft

Timing belt

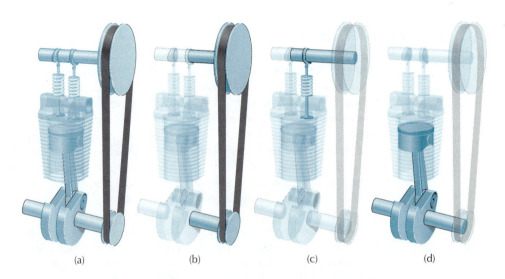

(a)   (b)   (c)   (d)

Figure 1.2 Mechanisms in a single-cylinder piston engine: (a) Engine. (b) Timing belt drive. (c) Cam mechanism. (d) Slider crank mechanism.

This textbook will consider only the mechanical parts of machines, along with their analysis. To conduct such analyses, it is usually convenient to divide a machine into subsystems, referred to as *mechanisms* (or *linkages*), rather than attempting to analyze all parts of a machine simultaneously. They may be either moving or stationary. Even with this restriction, there are countless types of machine.

A well-known example of a machine is the single-cylinder internal combustion piston engine, as illustrated in Figure 1.1. This machine serves as a good example, since it contains mechanisms within itself, which are described later in this text. In particular, it highlights three mechanisms such as a *timing belt drive*, a *cam mechanism*, and a *slider crank mechanism*, which are illustrated in Figures 1.2(b), 1.2(c), and 1.2(d), respectively. An animation of this machine is provided by [**Video 1.1: Single-Cylinder Piston Engine**], included on the companion website. A detailed description of the operation of this machine is given in Chapter 13, Section 13.2.1.

A simple machine may also be considered as a single mechanism. For instance, the tongs shown in Figure 1.3(a) can be considered either as a machine or as a mechanism. Figure 1.3(b) shows a free-body diagram of the system used to analyze the manual force required to generate sufficient gripping force.

Mechanisms are widely used in applications where precise relative movement and transmission of force are required. Motions may be continuous or intermittent, linear, and/or angular. Examples

Video 1.1
Single-Cylinder
Piston Engine

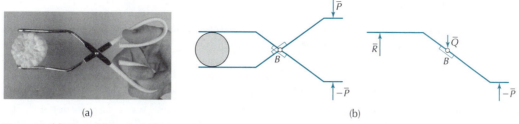

(a)   (b)

Figure 1.3 (a) Tongs. (b) Free-body diagram.

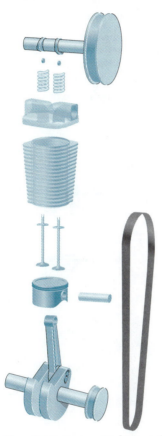

Figure 1.4 Engine assembly [Video 1.4].

of several types of machines and mechanisms are presented in Section 1.2 and Chapter 13. Engineers, and those particularly concerned with design, should ensure that they are aware of the wide range of machines and mechanisms and be familiar with their behavior. There is a good reason for this; it is often possible to apply or slightly modify an existing machine design to meet the needs of new problems.

Individual parts of a machine or mechanism are also referred to as *links*. They may be nonrigid, such as cables or belts. Alternatively, they may be rigid bodies such as cranks, levers, wheels, bars, or gears. Figure 1.4 shows an exploded view of an engine. [**Video 1.4**] provides an animation of assembling this engine, link by link. Most links of this mechanism may be considered rigid. The timing belt and the valve springs are examples of flexible links.

Video 1.4 Engine Assembly

1.2 COMMONLY EMPLOYED MECHANISMS

This section provides a small sampling of common mechanisms. Many more examples of machines and mechanisms and their descriptions are contained in Chapter 13.

1.2.1 Slider Crank Mechanism

A *slider crank mechanism* is illustrated in Figure 1.5. This mechanism incorporates a stationary *base link*, designated as link 1. All portions of the base link are depicted with hatched lines. The other links can move relative to the base link. The *crank*, designated as link 2, rotates about a *base pivot*, denoted as O_2. At the other end of link 2 is point B where a pivot point or bearing allows relative rotation between links 2 and 3. Link 3 is referred to as the *coupler* or *connecting rod*. The coupler is connected to link 4, through a bearing, denoted as point D. Link 4 is called a *slider* or *piston*. The slider moves with respect to the *slide*, which is part of the base link. The straight-line path of the center of the slider is referred to as the *line of action*.

When analyzing the motions of mechanisms, it is often convenient to employ highly simplified drawings, called *skeleton diagrams*. The dimensions of a skeleton diagram are the ones critical for determining motions. Figure 1.6(a) shows a slider crank mechanism as it would appear in an engine. The skeleton form of this mechanism is given in Figure 1.6(b). Bearings in skeleton diagrams are represented by either small circles or dots. The crank and connecting rod are each drawn simply as a straight line. The piston is shown as a rectangle. Hatched lines indicate the base link.

Another slider crank mechanism is shown in Figure 1.7. In this instance, base pivot O_2 is not on the line of action of point D. Dimension r_1 is referred to as the *offset*. Dimensions r_2 and r_3 are lengths of the crank and coupler, respectively.

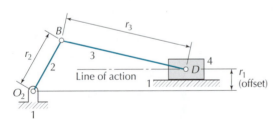

Figure 1.5 Slider crank mechanism [Video 1.5].

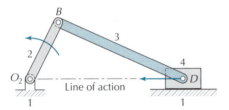

Figure 1.7 Slider crank mechanism with offset [Video 1.7].

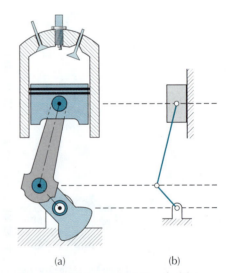

(a)   (b)

Figure 1.6 Skeleton representation of a slider crank mechanism: (a) Mechanism. (b) Skeleton representation.

1.2.2 Four-Bar Mechanism

Four-bar mechanisms are among the most common and useful mechanisms. A typical four-bar mechanism is illustrated in Figure 1.8(a). It has four *bars* or links: the stationary base link and three moving links. Links 2 and 4 are connected to the base link through base pivots O_2 and O_4, respectively. The coupler, link 3, is attached to links 2 and 4 through moving pivots, designated as points B and D. For the mechanism shown, if link 2 is driven to rotate about its base pivot, then link 4 will in turn be forced to also rotate about its base pivot.

For the mechanism shown in Figure 1.8(a), angular displacements of links 2 and 4 are designated as θ_2^* and θ_4^*. Figure 1.8(b) shows the angular displacement of link 4 as a function of the angular displacement of link 2. This mechanism is also known as an *angular function-generating mechanism*.

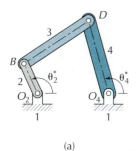

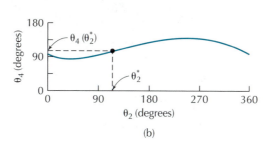

Figure 1.8 (a) Four-bar mechanism [Video 1.8]. (b) Function graph.

(a)

(b)

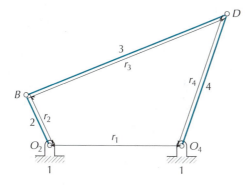

Figure 1.9 Dimensions of a four-bar mechanism.

Figure 1.9 shows a skeleton diagram of a four-bar mechanism. Also shown are the lengths of the links. These lengths dictate the type and extent of motion that may be achieved. The types of motion are presented in Section 1.7.

Many mechanisms can be developed from a single skeleton form. Examine the skeleton diagram of the four-bar mechanism shown in Figure 1.10(a). All other mechanisms illustrated in Figure 1.10 are essentially the same four-bar mechanism and are therefore considered to be *equivalent mechanisms*. As shown through the animation provided using [**Video 1.10**], all of these mechanisms transmit the same motions. When analyzing a motion, without consideration of the associated forces, it is not necessary to know the widths of links and sizes of bearings between links. The critical information in determining motions are the *center-to-center distances*. Figures 1.10(a) and 1.10(c) illustrate a typical distance r that is the same for two of the equivalent mechanisms.

Mechanisms can be built up by starting with a simpler mechanism and then adding links to create more complicated mechanisms. Figure 1.11 shows a four-bar mechanism used in a domestic washing machine. Figure 1.12(a) is a drawing of a similar four-bar mechanism. Figure 1.12(b) shows the four-bar mechanism combined with additional links that generate the required motions for the washing machine. Constant input motion is supplied through a gear, link 2, which drives a larger gear, link 3. Both gears are represented as circles. Links 1, 3, 4, and 5 constitute the four links of the four-bar mechanism. Link 5 is a gear sector that meshes with link 6. The proportions of the links are such that when link 2 is driven at a constant rotational speed, link 6 will have oscillatory rotational motion. In operation, link 6 drives an agitator to provide the washing action.

Various paths of motions may be obtained from a selected point on the coupler of a four-bar mechanism, also called the *coupler point*. These mechanisms are referred to as *path-generating mechanisms*. The shape of the path traced by the coupler point is a function of the dimensions shown in Figure 1.13. These consist of distances r_1, r_2, r_3 and r_4 between pivots. In addition, dimensions r_5 and r_6 define the location of the coupler point, C, with respect to the pivot between links 2 and 3. For instance, Figure 1.14 shows a four-bar mechanism for which

$$r_1 = 7.2 \text{ cm}; \qquad r_2 = 2.5 \text{ cm}; \qquad r_3 = 6.0 \text{ cm}; \qquad r_4 = 7.2 \text{ cm}$$

Also shown are three different *coupler curves* that were obtained by employing distinct values of r_5 and r_6. Using [**Model 1.14A**] and [**Model 1.14B**], it is possible to specify the position of the coupler point with respect to the coupler and also view the corresponding coupler curve.

Video 1.10
Equivalent Four-Bar Mechanisms

Video 1.12
Washing Machine Mechanism

Video 1.14A, 1.14B
Model 1.14A, 1.14B
Four-Bar with Coupler Point

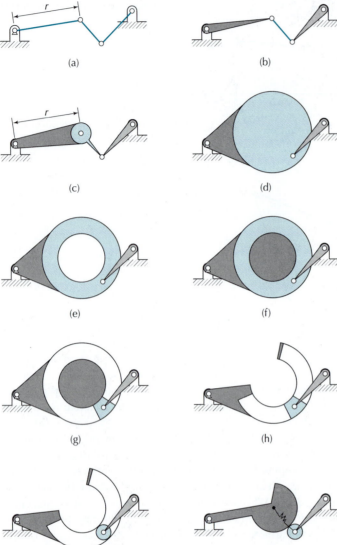

Figure 1.10 Equivalent four-bar mechanisms [Video 1.10].

(a)

(b)

(c)

(d)

(e)

(f)

(g)

(h)

(i)

(j)

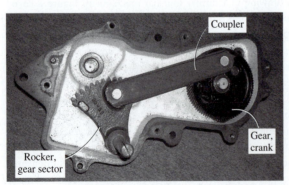

Coupler

Gear, crank

Rocker, gear sector

Figure 1.11 Washing machine mechanism.

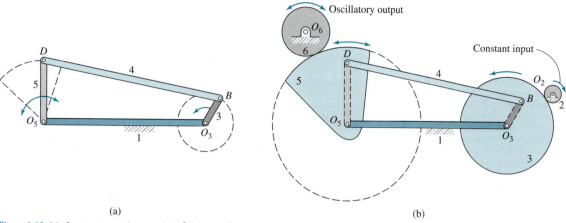

(a)

(b)

Figure 1.12 Mechanism in a washing machine [Video 1.12].

Figure 1.13 Dimensions of a four-bar mechanism with a coupler point.

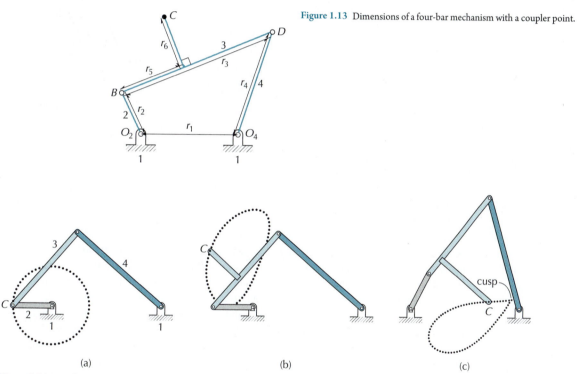

(a)

(b)

(c)

Figure 1.14 Four-bar mechanism with coupler point [Model 1.14A, Model 1.14B].

Video 1.15 A
Chebyshev Straight-Line Mechanism

Video 1.15 B
Roberts Straight-Line Mechanism

Video 1.15 C
Watt Straight-Line Mechanism

Video 1.16 Film Transport Mechanism

Some four-bar mechanisms with a properly selected coupler point can generate a path that is very close to a straight line, for a portion of the cycle. Such mechanisms are referred to as *straight-line mechanisms*. Figure 1.15 shows three types of such mechanisms, called the *Chebyshev straight-line mechanism*, the *Roberts straight-line mechanism*, and the *Watt straight-line mechanism*.

Figure 1.16 illustrates an application of a four-bar mechanism incorporating a coupler point. This mechanism is employed to intermittently advance film in a movie projector. During each cycle, the coupler point enters a perforation (Figure 1.16(a)) in the film and advances it to the next picture frame. When the end of the arm is not engaged in a perforation (Figure 1.16(b)), the film remains stationary, and a shutter (not shown) is then opened to allow light to pass through to project an image momentarily.

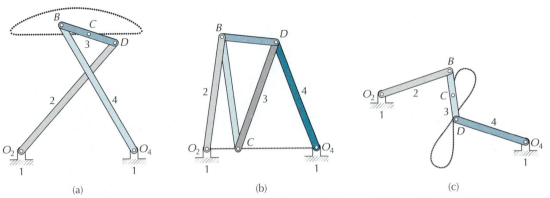

Figure 1.15 Straight-line mechanisms: (a) Chebyshev [Video 1.15A]. (b) Roberts [Video 1.15B]. (c) Watt [Video 1.15C].

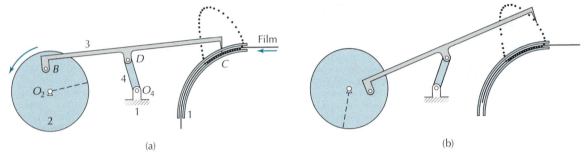

Figure 1.16 Film transport mechanism [Video 1.16].

1.2.3 Belt Drive, Chain Drive, and Gearing

A common requirement in the design of machinery is to smoothly transmit rotational motion from one shaft to another. When the shafts are parallel, this may be accomplished by mounting pulleys on the shafts and wrapping a continuous belt around both pulleys under tension, as shown in Figure 1.17. For this system, known as an *open-loop friction drive*, both pulleys rotate in the same direction. Figure 1.18 shows another arrangement of an open-loop friction drive. The *idler pulley* enables the input and output pulleys to turn in different planes. Still another arrangement is achieved by crossing the belt between the pulleys, as shown in Figure 1.19. Now the pulleys will rotate in opposite directions. This arrangement is called a *cross belt friction drive*.

Figure 1.20 shows a system similar to that of the open-loop friction drive, known as a *chain drive*. In the chain drive, a continuous chain replaces the belt, and *sprockets* replace the pulleys. The chain consists of a series of discretely spaced links, usually made of a metal material. Each link is connected to the next link by a pin, forming a rotational joint. Chain drives have a few key differences compared

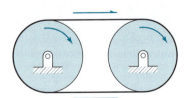

Figure 1.17 Open-loop friction drive.

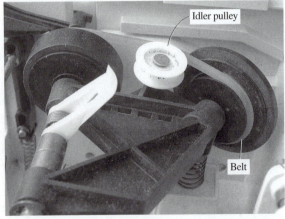

Figure 1.18 Open-loop friction drive.

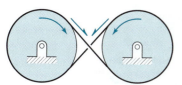

Figure 1.19 Cross belt friction drive.

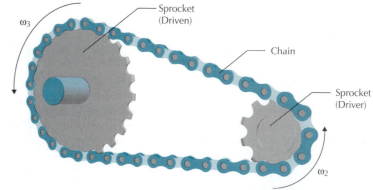

Figure 1.20 Chain drive [Video 1.20].

to belt–pulley systems. The links in the chain are in mesh with the teeth on the sprockets, and hence the chain cannot slip relative to the sprockets during normal operation. This allows for the transmission of high amounts of torque from one sprocket to the other, via the chain. Unlike belt–pulley systems, the sprockets must have discrete diameters that correspond to the link-to-link distances in the chain. Additionally, the center-to-center spacing between the sprockets must correspond to the chain link spacing. Examples of chain drive systems can be found in bicycles (foot pedal motion to wheel motion), motorcycles (transmission output to wheel motion), factory conveyor systems, and industrial equipment, among other uses. Special (extreme) examples of chain drives include (a) chainsaws and (b) tread systems on excavators or armoured tanks.

Figure 1.21 shows an alternative means of transferring rotational motion through a pair of rolling cylinders in physical contact. Transfer of motion in this instance relies on friction between the cylinders. This mechanism is known as a pair of *friction gears*. Smooth transmission is achieved as long as the friction capacity between the friction gears is not exceeded. If the friction capacity is exceeded, slippage occurs, and the motion of the follower will not have a constant ratio relative to that of the driver.

Slippage is avoided if the gears have interlocking teeth. These are called *toothed gears*. Figure 1.22 shows three meshing toothed gears that form a *gear train*. The gear in the middle, being the smallest of the set, is referred to as the *pinion*. It is also referred to as an *idler gear*. In the gear set shown in Figure 1.23, one of the gears, known as an *internal gear*, has its teeth on the inside of a ring. When an internal gear meshes with a gear having teeth on its periphery (referred to as an *external gear*), both gears turn in the same direction. Internal gears are commonly employed in planetary gear trains, which are discussed in Chapter 6.

Figure 1.24 shows a *rack and pinion* gear set. The *rack* provides for straight-line motion and is equivalent to a portion of a gear of infinite radius. Figure 1.25(a) illustrates a rack and pinion gear set employed in the steering mechanisms of automobiles. Figure 1.25(b) shows a close-up view of this gear set.

Gearing and gear trains are discussed in more detail in Chapters 5 and 6.

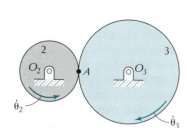

Figure 1.21 Friction gears.

Figure 1.22 Toothed gears [Video 1.22].

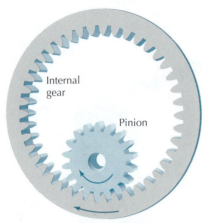

Figure 1.23 Internal gear [Video 1.23].

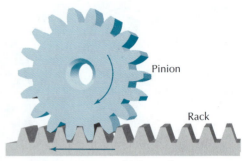

Figure 1.24 Rack and pinion [Video 1.24].

Figure 1.25 Rack and pinion steering [Video 1.25].

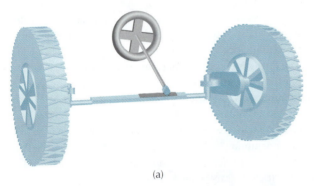

(a)

(b)

1.2.4 Cam Mechanism

Video 1.26 Disc
Cam Mechanism

Video 1.27 Disc
Cams

A *cam* is a link of a mechanism used to transmit motion to another link, known as a *follower*, by direct contact. One type of cam mechanism, known as a *disc cam mechanism*, is illustrated in Figure 1.26. Starting from the position shown in Figure 1.26(a), rotation of the disc cam, link 2, causes the follower, link 4, to translate. Figure 1.26(b) illustrates a configuration of the mechanism after the follower has undergone a translation. If the roller remains in contact with the cam, and the cam is driven at constant speed, then motion of the follower is periodic. An application of a disc cam mechanism is in the valve train of a piston engine. A typical valve train, including the intake and exhaust valves, is given in Figure 1.27. This illustration is a close-up view of a portion of Figure 1.1.

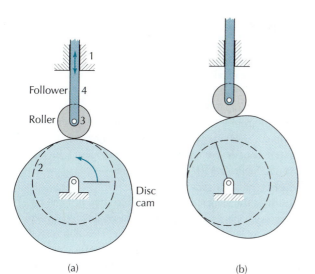

Figure 1.27 Disc cams [Video 1.27].

Figure 1.26 Disc cam mechanism [Video 1.26].

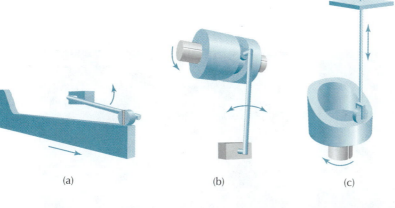

Figure 1.28 Types of cams:
(a) Wedge cam [Video 1.28A].
(b) Cylindrical cam [Video 1.28B].
(c) End cam [Video 1.28C].

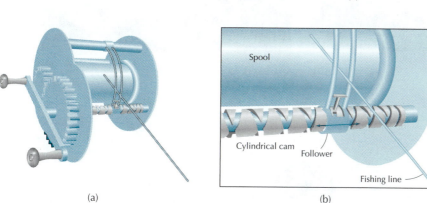

Figure 1.29 Fishing reel [Video 1.29].

Video 1.28A
Wedge Cam
Mechanism

Video 1.28B
Cylindrical Cam
Mechanism

Video 1.28C End
Cam Mechanism

Video 1.29
Fishing Reel

Cam mechanisms are simple and inexpensive, have fewer moving parts, and can be designed to occupy a smaller space compared to other mechanisms that provide similar output motions. Furthermore, the shape of a cam may be determined in a straightforward manner to produce a wide variety of follower motions. For these reasons, cam mechanisms are used extensively in modern machinery. The disadvantages of cam mechanisms are poor wear resistance and the noise generated by impacts between the cam and follower when operated at high speed.

Figure 1.28 illustrates additional types of cam mechanisms: the *wedge cam mechanism*, the *cylindrical cam mechanism*, and the *end cam mechanism*. An application of a cylindrical cam is in a fishing reel, as illustrated in Figure 1.29. As the cylindrical cam rotates, the follower translates back and forth, causing the fishing line to evenly wind onto the *spool*. A wedge cam is employed in a key lock as shown in Chapter 13, Section 13.16.1.

The analysis and design of disc cam mechanisms are presented in Chapter 7.

1.3 PLANAR AND SPATIAL MECHANISMS

Planar motion is restricted to a plane. For a *planar mechanism*, the motions of all of its links must take place either in the same plane or in planes that are parallel to one another. These motions include translation and rotation, within the plane. The slider crank mechanism and four-bar mechanism illustrated in Figures 1.5 and 1.8 are examples of planar mechanisms.

In a *spatial mechanism*, links translate and rotate in all three dimensions. For example, Figure 1.30 illustrates a prosthetic hand [1]. The individual fingers are planar mechanisms, however, the overall hand is a spatial mechanism. The fingers start from the configuration shown in Figure 1.30(a), and then they wrap around an object as shown in Figure 1.30(b). The thumb moves in a plane that is not parallel to the planes of motion of the other four fingers.

Another example of a spatial mechanism is the vehicle suspension illustrated in Figure 1.31. The suspension shown is called a *Chalmers suspension* and is employed in heavy duty vehicles such as cement trucks. Often, road surfaces can be uneven, as well as driveways and loading areas. The purpose of this mechanism is to better distribute the weight of the vehicle on all of the tires shown. To create good contact between the tires and the road, each tire must have the ability to move up and down relative to the base link. Yet, the mechanism must allow the differential (via the axles) to provide rotational power to the wheels. During operation, motion of the frame, lower tie rods, axles, and upper tie rods are all in different planes.

Video 1.30
Prosthetic Hand

Video 1.31
Chalmers
Suspension

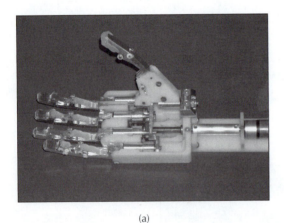
(a)

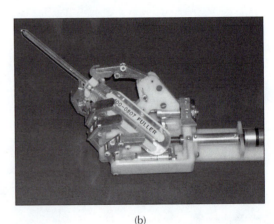
(b)

Figure 1.30 Prosthetic hand [Video 1.30].

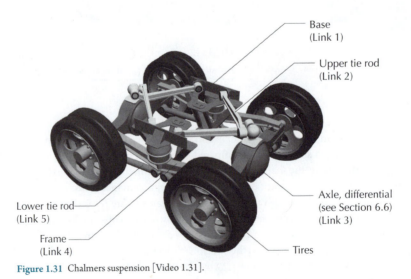

Base
(Link 1)

Upper tie rod
(Link 2)

Axle, differential
(see Section 6.6)
(Link 3)

Lower tie rod
(Link 5)

Frame
(Link 4)

Tires

Figure 1.31 Chalmers suspension [Video 1.31].

1.4 KINEMATIC CHAINS AND KINEMATIC PAIRS

A *kinematic chain* is an assembly of links connected together without specifying the base link. Figure 1.32 illustrates a four-bar kinematic chain. All links are connected by pivots. Figure 1.32(b) shows the skeleton diagram representation of this kinematic chain.

A series of alternative mechanisms may be produced, in turn, by holding one of the links of the kinematic chain in a fixed position to become the base link. Usually, the type and amplitude of absolute motions (i.e., with respect to the base link) depend on the choice of the base link. Section 1.6 presents examples of various types of motion that may be produced from a single kinematic chain.

The links of a mechanism are connected together by *kinematic pairs* or *joints*. Three common types of kinematic pairs in planar mechanisms are as follows:

- *Turning pairs* allow relative turning motion between two links. Such pairs are also called *bearings, pivots*, or *pin joints*. A four-bar mechanism (Figure 1.8) has a total of four turning pairs. Figure 1.33 illustrates other examples of links connected by turning pairs, along with their skeleton representations.
- *Sliding pairs* allow relative sliding motion between two links. For example, in Figure 1.5, link 4 is permitted to undergo sliding motion with respect to link 1. Figure 1.34 illustrates other examples of links connected by sliding pairs.
- *Rolling pairs* allow relative rolling motion between two links, such as employed in a pair of friction gears (Figure 1.21). Another example of a rolling pair is illustrated in Figure 1.35. In this mechanism, ball bearings roll along a rail track. By using the rolling balls sandwiched between the outer base rails and the inner slider rails, the friction between the base link and the slider link can be reduced. In this way, there is relative linear sliding motion between the two links. Linear bearing sliding stages are typically used in applications that require precise linear motion, such as milling machine beds, or prismatic links on precision machines. In such applications, transverse motion (motion in axes not along desired linear path) is unwanted. Bearing sliding stages are also used in applications where heavy loads require linear translation, such as cabinet drawers on toolboxes. Figure 1.36 illustrates other examples of links connected by rolling pairs. For a rolling pair, it is assumed that there is no slippage between the links.

Video 1.35 Linear Bearing Sliding Stage

Each turning pair, sliding pair, and rolling pair permits one relative motion between adjacent links. All of these kinematic pairs are referred to as *one degree of freedom pairs*.

Another type of kinematic pair is referred to as a *two degree of freedom pair*. It allows two relative motions between the adjacent links. Examples are shown in Figure 1.37. Figure 1.37(a) demonstrates two degrees of freedom between links 2 and 3. One of the degrees of freedom is the relative turning between the links. The other degree of freedom is the translational motion of the pin at one end of link 3 in the slot of link 2. As shown in Figure 1.37(a), this two degree of freedom pair may be represented in skeleton form using one sliding pair and one turning pair.

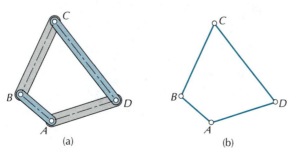

(a)   (b)

Figure 1.32 (a) Four-bar kinematic chain. (b) Skeleton representation.

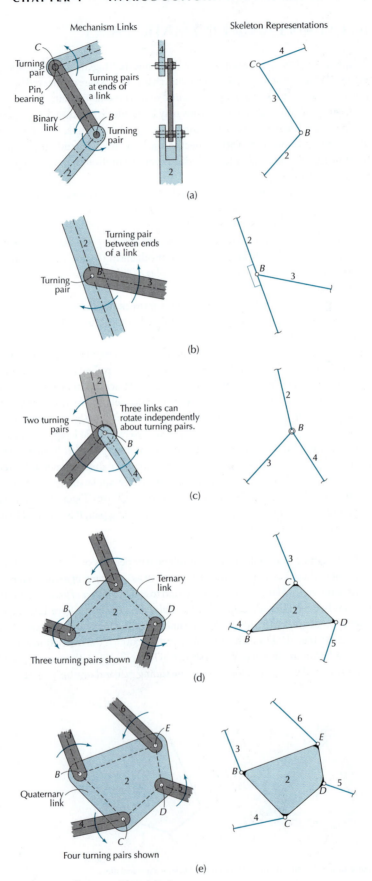

Mechanism Links   Skeleton Representations

Figure 1.33 Illustrations of turning pairs.

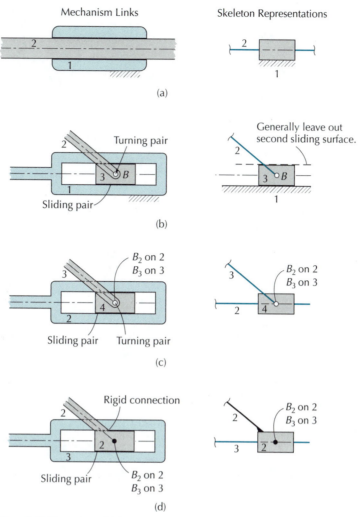

Figure 1.34 Illustrations of sliding pairs.

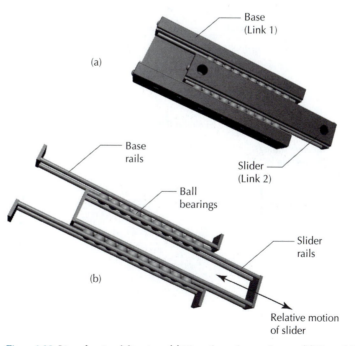

Figure 1.35 Linear bearing sliding stage. (a) View of complete mechanism. (b) View of slider rails and ball bearings only [Video 1.35].

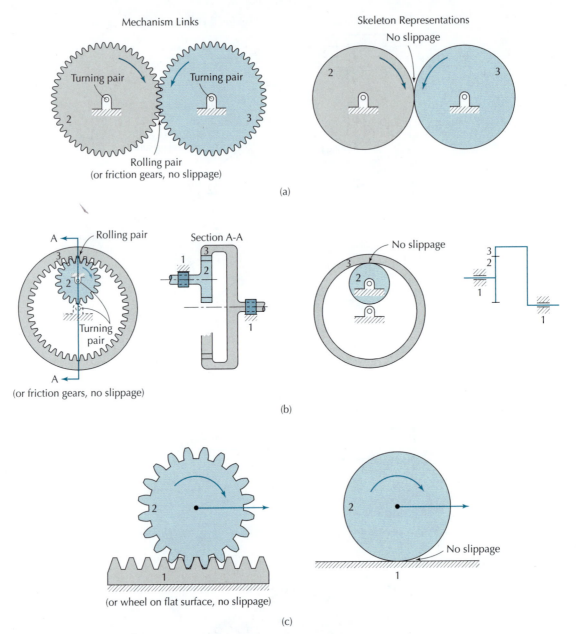

Figure 1.36 Illustrations of rolling pairs.

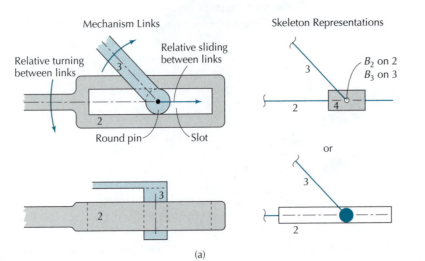

Figure 1.37 Illustrations of two degree of freedom pairs.

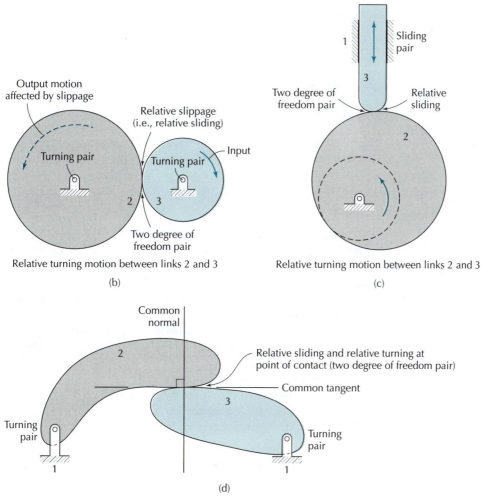

Relative turning motion between links 2 and 3

(b)

Relative turning motion between links 2 and 3

(c)

(d)

Figure 1.37 *(Continued).*

1.5 MECHANISM MOBILITY

The *mobility* of a mechanism is defined as the number of independent parameters required to specify the position of all links of the mechanism. This section is restricted to considering the mobility of planar mechanisms.

The location of a rigid body executing planar motion can be described by specifying three independent parameters. A planar mechanism with n links, has $n - 1$ links that can move. Thus, $3(n - 1)$ parameters would need to be specified if the links moved independently. Figure 1.38(a) shows a four-bar mechanism. In Figure 1.38(b), one of the links has been isolated from the rest of the mechanism. The location of this link may be described by specifying the x coordinate of point B, the y coordinate of point B, and angle θ. Similar specifications may be applied for the other movable links.

Links of a mechanism do not move independently but instead are coupled by kinematic pairs. Each kinematic pair between links provides one or more constraints that reduce the number of independent motions of the mechanism. Each one degree of freedom pair (i.e., turning pair, sliding pair, or rolling pair) allows just one relative motion between links, while two relative motions are constrained. For a turning pair (Figure 1.33), only relative rotation is permitted. The constrained relative motions are the linear translations at the turning pair in the horizontal and vertical directions. For a sliding pair (Figure 1.34), the only relative motion permitted is translation of the slider, tangent to the direction of the slide. The constrained motions consist of translation perpendicular to the direction of the slide, and relative rotation between the links. For a rolling pair (Figure 1.36),

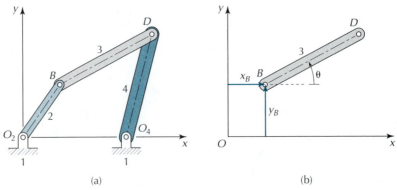

Figure 1.38 Specification of position of a mechanism link in a plane.

the links remain in contact. Therefore, one constraint is that the center of rotation of one link must remain a fixed distance from the surface of the other link forming the pair. Also, for a rolling pair, no slippage is allowed between the links. Therefore, the second constraint arises due to restricted relative rotations that occur between the links. Each two degree of freedom pair (Figure 1.37) allows two relative motions between links, and just one relative motion is constrained. For the two degree of freedom kinematic pair shown in Figure 1.37(d), there can be both relative turning and sliding between links. Relative sliding takes place along the *common tangent* to the two surfaces at the point of contact. The only constraint when considering these two links is that at the point of contact there cannot be any relative motion along the *common normal.*

Implementing the above examination, it is possible to derive at an expression for the mobility, *m*, of a planar mechanism in terms of the number of links and the numbers and types of kinematic pairs. In the instance that each kinematic pair of a mechanism imposes either one or two constraints that are not applied by any other kinematic pair of the mechanism, then the mobility may be expressed as

$$m = 3(n-1) - 2j_1 - j_2 \qquad (1.5\text{-}1)$$

where

n = number of links in the mechanism
j_1 = number of one degree of freedom pairs
j_2 = number of two degree of freedom pairs

The coefficient 2, which multiplies j_1, corresponds to the two constraints associated with each one degree of freedom pair. The minus signs in front of the terms involving the numbers of kinematic pairs relate to a reduction of the mobility brought about by the constraints that are introduced.

If more than one kinematic pair of a mechanism impose the *same* constraint, then it is said to possess a *redundant constraint.* In this case, Equation (1.5-1) is no longer valid, and instead we employ

$$m = 3(n-1) - 2j_1 - j_2 + N_R \qquad (1.5\text{-}2)$$

where

N_R = number of redundant constraints

The presence of redundant constraints are found by inspection of the mechanism. Examples are provided below.

1.5.1 Examples of Mechanism Mobility with No Redundant Constraints

For all of the examples presented in this section, there are no redundant constraints. Therefore, the mobility may be calculated by either using Equation (1.5-1) or using Equation (1.5-2) with $N_R = 0$.

The four-bar mechanism shown in Figure 1.39(a) has four links, consisting of the base link and three moving links, and four kinematic pairs, all of which are turning pairs. There are no two degree of freedom pairs for this mechanism. In summary

$$n = 4; \qquad j_1 = 4; \qquad j_2 = 0$$

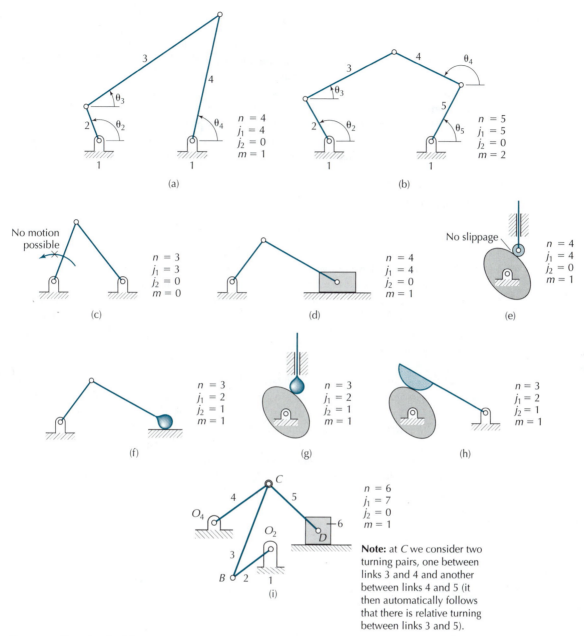

Figure 1.39 Examples of mobility.

Substituting the above values in Equation (1.5-1) yields

$$m = 3\,(n-1) - 2j_1 - j_2 = 3(4-1) - 2 \times 4 - 0 = 1$$

That is, one input value needs to be specified to allow the position of all links to be determined. If θ_2 is specified, then the other variable quantities (i.e., θ_3 and θ_4) of the mechanism may be calculated.

As a second example, consider the five-bar mechanism shown in Figure 1.39(b). In this instance, there are five links $(n = 5)$ and a total of five one degree of freedom pairs, all of which are turning pairs $(j_1 = 5)$. There are no two degree of freedom pairs $(j_2 = 0)$. Using Equation (1.5-1), we find $m = 2$. Thus, in order to determine the geometry of the mechanism, it is necessary to specify two independent parameters. As shown in Figure 1.39(b), this may be accomplished by specifying the values of both θ_2 and θ_5.

Seven more examples are provided in Figure 1.39. The number of links, the numbers of one and two degree of freedom pairs, and the calculated values of mobility are listed for each.

1.5.2 Examples of Mechanism Mobility with at Least One Redundant Constraint

For mechanisms with at least one redundant constraint, the mobility may be calculated using Equation (1.5-2) with the appropriate value of N_R.

Figure 1.40(a) shows a pair of friction gears. There are a total of three links. Links 2 and 3 are each connected to the base link through a turning pair. Links 2 and 3 contact at A. If there is no slippage at A, then links 2 and 3 are connected by a rolling pair. A specified rotation of gear 2 leads to a definite rotation of gear 3. The positions of all links can be determined for the input motion of gear 2, and by inspection the mobility is equal to one. For this mechanism

$$n = 3; \qquad j_1 = 3; \qquad j_2 = 0$$

If Equation (1.5-1) was employed, then we obtain

$$m = 3(n-1) - 2j_1 - j_2 = 3(3-1) - 2 \times 3 - 0 = 0$$

indicating that the mechanism cannot move, which is an incorrect result. This is because in the above calculation we have not accounted for the existence of one redundant constraint in the mechanism. The redundant constraint is the fixed distance, c, between points O_2 and O_3. This constraint is applied *twice* by more than one kinematic pair. It is applied by the two turning pairs (Figure 1.40(b)); however, the *same* constraint is applied by considering the rolling pair (Figure 1.40(c)). For this mechanism, therefore, $N_R = 1$. Substituting the values of the numbers links and kinematic pairs, along with the value of N_R, in Equation (1.5-2) gives the correct value of $m = 1$.

Additional examples are shown in Figure 1.41, where in each case the number of links and kinematic pairs are given. Each mechanism generates the same relative motion between the input and output, and for each mechanism we have $m = 1$. For the mechanisms shown in Figures 1.41(a) and 1.41(b), there are no redundant constraints. However, for the mechanism shown in Figure 1.41(c), there is a redundant constraint of the relative rotation between links 2 and 3. This constraint is applied once by the sliding pair connecting links 2 and 3. The same constraint is applied by a combination of the sliding pairs between links 1 and 2 and between links 1 and 3.

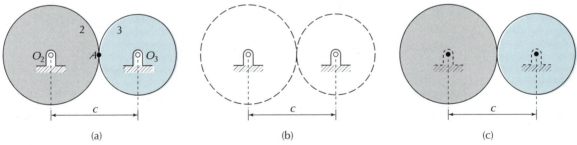

(a)   (b)   (c)

Figure 1.40 Friction gears.

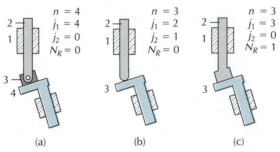

(a)   (b)   (c)

Figure 1.41 Mobility examples.

1.5.3 Example of Using Calculation of Mobility to Ensure a Stationary System

Video 1.42
Model 1.42
Front-end Loader
Mechanism

Consider the example of a front-end loader given in Figure 1.42. This mechanism is driven by means of two actuators, components 6 and 7. If the lengths of the two actuators are kept constant, then they can be considered as links, and in this case $n = 9$. The total number of kinematic pairs (all turning pairs) is 12. There are no redundant constraints. Therefore

$$m = 3(n-1) - 2j_1 - j_2 = 3(9-1) - 2 \times 12 - 0 = 0$$

This result indicates that if the lengths of two actuators are kept constant, then the front-end loader is not permitted to move.

The mechanism in fact has a mobility equal to two by allowing the operator to adjust the lengths of the two actuators. Typical positions are shown in Figure 1.42. The starting position is shown in Figure 1.42(a). Then the bucket is rolled back (Figure 1.42(b)) by reducing the length of component 6, also known as the *dump actuator*. This is followed by extending component 7, also known as the *lift actuator*, to raise the load (Figure 1.42(c)). Finally, the dump actuator is extended to empty the bucket (Figure 1.42(d)).

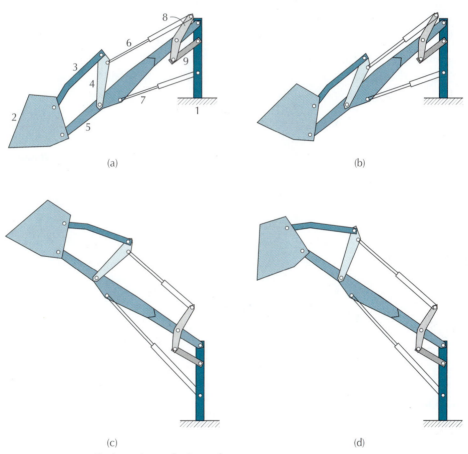

(a)   (b)

(c)   (d)

Figure 1.42 Front-end loader mechanism [Video 1.42].

1.6 MECHANISM INVERSION

Every mechanism has one stationary base link. All other links may move relative to the fixed base link.

An *inverted mechanism* is obtained by making the originally fixed link into a moving link and selecting an originally moving link to be the fixed link. As an illustration of the inversion of a mechanism, Figure 1.43 shows four mechanisms that were created from the same kinematic chain.

In each case, a different link of the kinematic chain is held fixed. The slider crank mechanism is illustrated in Figure 1.43(a), where link 1 is the base link. Figures 1.43(b), 1.43(c), and 1.43(d), respectively, correspond to the cases where links 2, 3, and 4 are held fixed, and link 1 can move. These three mechanisms are inversions of that given in Figure 1.43(a). All four mechanisms shown in Figure 1.43 have a driving motor between links 1 and 2. Sets of accumulated images throughout a cycle of motion are given in Figure 1.44. In each mechanism, *absolute* motions of the links with respect to the base link are distinct. However, since all mechanisms have the same link dimensions, and all have a driving motor located at the same position in the kinematic chain, the *relative* motions between links are identical for all of these mechanisms. In general, a mechanism having n links can have $n - 1$ inversions.

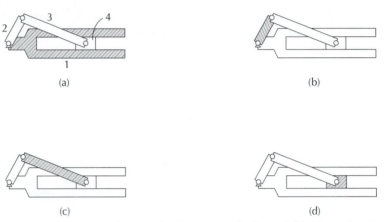

(a)   (b)

(c)   (d)

Figure 1.43 Slider crank mechanism and its three inversions [Video 1.43]: (a) Slider crank mechanism. (b) Inversion #1 (link 2 fixed). (c) Inversion #2 (link 3 fixed). (d) Inversion #3 (link 4 fixed).

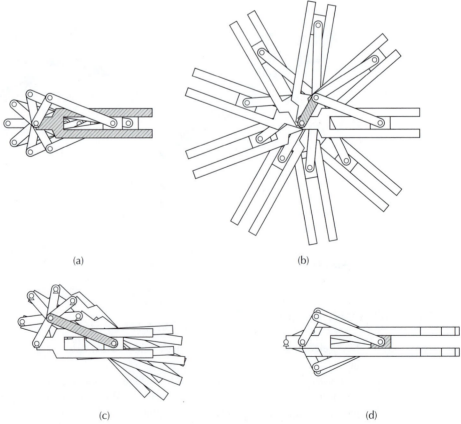

(a)   (b)

(c)   (d)

Figure 1.44 Accumulated images of a slider crank mechanism and its three inversions [Video 1.43].

Another example of the inversion of a mechanism is shown in Figure 1.45. We start with the four-bar mechanism shown in Figure 1.45(a), in which link 1 is held fixed. The three inversions of this mechanism are shown in Figures 1.45(b), 1.45(c), and 1.41(d), in which links 2, 3, and 4, respectively, are held fixed. Sets of accumulated images throughout a cycle of motion are given in Figure 1.46.

Video 1.45
Four-Bar
Mechanism and Its
Three Inversions

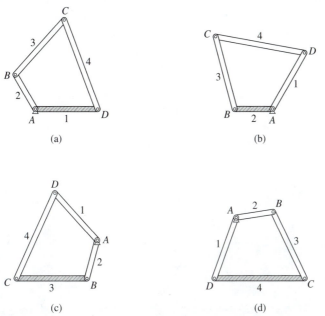

(a)

(b)

(c)

(d)

Figure 1.45 Four-bar mechanism and its three inversions [Video 1.45]: (a) Slider crank mechanism. (b) Inversion #1 (link 2 fixed). (c) Inversion #2 (link 3 fixed). (d) Inversion #3 (link 4 fixed).

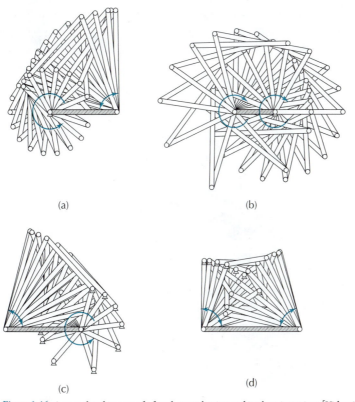

(a)

(b)

(c)

(d)

Figure 1.46 Accumulated images of a four-bar mechanism and its three inversions [Video 1.45]: (a) Crank rocker. (b) Drag link. (c) Crank rocker. (d) Rocker–rocker.

A *speed governor* is a mechanism employed to regulate (control) the operation of another mechanism. A speed governor incorporates an inverted slider mechanism. Figure 1.47(a) shows an old-fashioned steam-powered locomotive, whose engine power is controlled by the use of a speed governor, as shown in Figure 1.47(b). The speed governor operates on the principle of converting rotational input motion into linear output motion. For the governor shown here, rotational input motion from the locomotive wheels (corresponding to the locomotive speed) is input by a belt-and-pulley system. The pulley rotates the bevel gear set, which in turn rotates the two ball masses. As the masses rotate, they experience centripetal force, which causes them to swing radially outward. If the masses rotate fast, this radial outward motion (part of the inverted slider) lifts the control rod. If the masses rotate more slowly, the control rod will drop. This linear motion of the control rod is used to throttle/regulate the amount of water entering the heat exchanger of the engine, and hence regulate the output power of the engine. If the locomotive speed is too high, the control rod automatically lifts and then attenuates the engine power. If the locomotive speed becomes too low, the control rod automatically drops and then increases the engine power. In this sense, a speed governor is an example of a purely mechanical-based closed-loop control system. An illustration of a different speed governor (with a different configuration) is shown in Figure 1.48.

Video 1.48 Speed Governor

(a)   (b)

Figure 1.47 (a) Steam powered locomotive. (b) Speed governor.

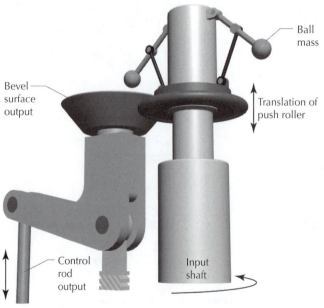

Figure 1.48 Speed governor [Video 1.48].

1.7 TYPES OF FOUR-BAR AND SLIDER CRANK MECHANISMS

Four-bar and slider crank mechanisms may be classified according to *type*. A type is characterized by the number of links that are able to undergo full rotation, and which are connected to the base link at base pivots. This section provides the method of determining the type of mechanism based on the lengths of the links.

1.7.1 Four-Bar Mechanism—Grashof's Criterion

Figures 1.45 and 1.46 illustrate three different types of four-bar mechanisms. For Figures 1.45(a) and 1.45(c), link 2, the *crank,* is able to make a full rotation, whereas link 4, the *rocker,* has a limited oscillatory motion. Such four-bar mechanisms are called *crank rockers.* In Figure 1.45(b), both links 1 and 3 are able to make a full rotation, and it is called a *drag link.* For Figure 1.45(d), neither link 1 nor link 3 is able to make a full rotation, and it is referred to as a *rocker–rocker.* Another illustration of these three types of four-bar mechanism is given in Figure 1.49. For the crank rocker mechanism shown in Figure 1.49(a), if link 2 makes complete rotations, then link 4 rocks between two *limit positions* illustrated as dashed lines. The amplitude of the rocking motion is $\Delta\theta_4$.

Four-bar mechanisms may be studied by distinguishing the link lengths as follows:

- *s*: the length of the shortest link
- *l*: the length of the longest link
- *p, q*: the lengths of the other two links

To assemble the kinematic chain, it is necessary that

$$s + p + q \geq l$$

When the two sides of the above expression are equal, all links are constrained to remain collinear, and no motion is permitted. Only when the sides of the expression are not equal can there be rotations of the links.

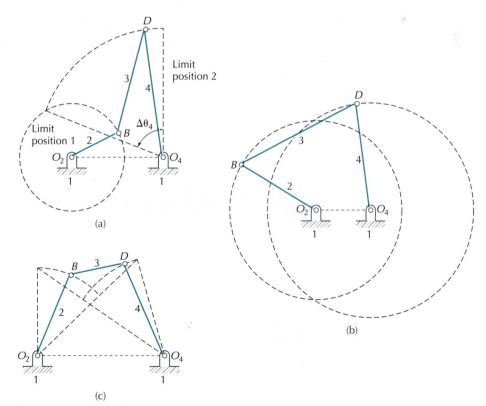

Figure 1.49 Types of four-bar mechanisms.

TABLE 1.1 CLASSIFICATION OF FOUR-BAR MECHANISMS

Class I Kinematic Chain $s + l < p + q$	Class II Kinematic Chain $s + l > p + q$
If s is the input link, then the mechanism is a crank rocker. If s is the base link, then the mechanism is a drag link. If otherwise, then mechanism is a rocker–rocker.	The mechanism is a rocker–rocker.

The type of a four-bar mechanism may be determined using *Grashof's Criterion* [2]. Using the above designations of link lengths, Grashof's Criterion is given in Table 1.1.

In Table 1.1, two classes of kinematic chains are identified. Only for a Class I kinematic chain is it possible to obtain all three types of a four-bar mechanism.

As an illustration, consider the four-bar mechanisms shown in Figure 1.45. In this instance

$$s = 1.9 \text{ cm}; \qquad p = 3.2 \text{ cm}; \qquad q = 3.8 \text{ cm}; \qquad l = 4.8 \text{ cm}$$

Since

$$s + l = 6.7 \text{ cm} < p + q = 7.0 \text{ cm}$$

the mechanisms are made from a Class I kinematic chain, and thus through inversion it is possible to obtain the different types of four-bar mechanisms, as illustrated in Figure 1.46.

The case where

$$s + l = p + q$$

is not covered in Table 1.1. Such a mechanism, referred to as a *change point mechanism,* may be brought into a geometry for which all links are collinear. Figure 1.50 shows an example of a change point mechanism.

A special case of a change point mechanism occurs when

$$r_1 = r_3 \qquad \text{and} \qquad r_2 = r_4$$

This is referred to as a *parallelogram four-bar mechanism.* Figure 1.51 shows a skeleton representation of such a mechanism. For this mechanism, link 1 always remains parallel to link 3, and link 2 always remains parallel to link 4. Figure 1.52 shows an automobile lift that incorporates a parallelogram four-bar mechanism. The automobile will always remain parallel to the ground (base link) while being raised.

Video 1.51
Parallelogram
Four-Bar
Mechanism

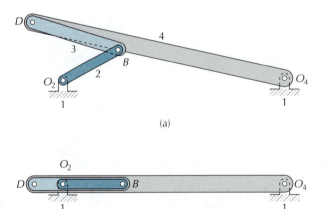

(a)

(b)

Figure 1.50 Change point mechanism.

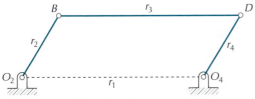

Figure 1.51 Parallelogram four-bar mechanism [Video 1.51].

Figure 1.52 Example application of a parallelogram four-bar mechanism.

With [**Model 1.53**] it is possible to adjust the length of the base link, and view the resulting motion. The lengths of the three moving links are

$$r_2 = 2.0 \text{ cm}; \qquad r_3 = 4.0 \text{ cm}; \qquad r_4 = 5.0 \text{ cm}$$

Figure 1.53 shows two cases where the length of the base link takes on values

$$r_1 = 6.0 \text{ cm} \qquad \text{and} \qquad r_1 = 1.5 \text{ cm}$$

which correspond to crank rocker and rocker–rocker mechanisms, respectively.

Corresponding to the mechanism shown in Figure 1.53, Table 1.2 lists the ranges of values of r_1 and the corresponding types of four-bar mechanisms. The range of link lengths represents the theoretical range of permissible values. However, the practical limits in a design are somewhat restrictive, and it may sometimes be difficult to transmit motions between links. This is described further through examples presented in Chapter 2.

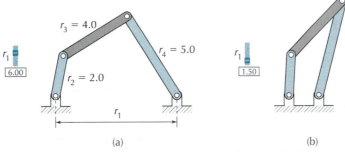

(a)   (b)

Figure 1.53 Four-bar mechanism with variable base length [Video 1.53]: (a) Crank rocker. (b) Rocker–rocker.

TABLE 1.2 EXAMPLE OF TYPES OF FOUR-BAR MECHANISMS

r_1 (cm)	Type of Four-Bar Mechanism
$0.0 < r_1 < 1.0$	Drag link
$r_1 = 1.0$	Change point
$1.0 < r_1 < 3.0$	Rocker–rocker
$r_1 = 3.0$	Change point
$3.0 < r_1 < 7.0$	Crank rocker
$r_1 = 7.0$	Change point
$7.0 < r_1 < 11.0$	Rocker–rocker
$r_1 = 11.0$	All links collinear, no motion possible
$r_1 > 11.0 = r_2 + r_3 + r_4$	Impossible to assemble mechanism

Note: $r_2 = 2.0$ cm; $r_3 = 4.0$ cm; $r_4 = 5.0$ cm.

1.7.2 Slider Crank Mechanism

Consider the slider crank mechanism shown in Figure 1.7. The offset is designated by r_1, and the lengths of the crank and coupler are r_2 and r_3, respectively. In order for the crank to have full rotation, it is necessary that

$$r_2 < r_3 \quad \text{and} \quad |r_1| \le r_3 - r_2$$

When

$$|r_1| > r_2 + r_3$$

it is impossible to assemble the mechanism.

As an illustration of the above, Figure 1.54 shows a slider crank mechanism for which

$$r_2 = 1.5 \text{ cm}; \qquad r_3 = 2.5 \text{ cm}$$

Figures 1.54(a), 1.54(b), and 1.54(c) show offsets of 0.0, –0.9, and –1.4 cm, respectively. For the mechanisms shown in Figures 1.54(a) and 1.54(b), the crank is able to make full rotation. However, for the mechanism shown in Figure 1.54(c), since

$$|r_1| > r_3 - r_2$$

the crank has a limited range of permissible motion. With $\boxed{\textbf{Model 1.54}}$ it is possible to adjust the offset and view the resulting motion.

For the mechanisms shown in Figures 1.54(a) and 1.54(b), the slider moves between two limit positions. The distance between the two limit positions of the slider is referred to as the *stroke* as illustrated in Figure 1.54(d). The offset influences the value of the stroke. When the offset is zero (Figure 1.54(a)), we have

$$\text{stroke} = 2\, r_2 = 3.0 \text{ cm}$$

and for the mechanism shown in Figure 1.54(b) with a nonzero offset we have

$$\text{stroke} > 2\, r_2$$

Video 1.54
Model 1.54
Variable-Offset
Slider Crank
Mechanism

1.8 COGNATES OF MECHANISMS[†]

Associated with every four-bar mechanism and slider crank mechanism incorporating a coupler point is at least one other mechanism that will generate the identical path of the coupler point. These associated mechanisms are referred to as *cognates.* In a case where a mechanism has been designed to generate a suitable path of the coupler point, but the links of the mechanism have inappropriate locations, it may be possible to implement instead a cognate. This section presents the method of constructing cognate mechanisms of four-bar and slider crank mechanisms.

1.8.1 Cognates of a Four-Bar Mechanism

A four-bar mechanism with a coupler point has two associated cognates. The four-bar mechanism shown in Figure 1.55 has two base pivots O_2 and O_4. It is designated as the primary mechanism. The moving pivots are designated as B and D, and the coupler point is labeled as C.

Figure 1.56(a) shows two additional four-bar mechanisms added to the primary mechanism. Their four-bar kinematic chains include points $O_{10}HGO_8$ and O_5EFO_7. As indicated, the following points coincide: O_2 and O_{10}; O_4 and O_5; O_7 and O_8. All three mechanisms share the same coupler point C.

For the three four-bar mechanisms illustrated in Figure 1.56(a) it is necessary that

- The triangles formed by the following points are similar: BDC, HCG, and CEF.
- The following points must form parallelograms: O_2BCH, DO_4EC and CFO_7G.
- The base pivots, O_2, O_4 and O_7 form a triangle that is similar to that formed by points BDC.

[†]The material presented in this section is often not covered in a first course on the mechanics of machines. It does not specifically relate to any of the remaining topics presented in this textbook.

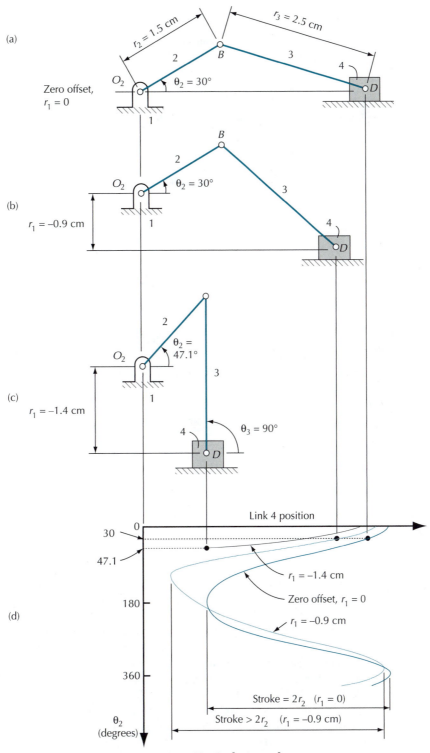

Figure 1.54 Slider crank mechanism with variable offset [Video 1.54].

Figures 1.56(b), 1.56(c), and 1.56(d) show the same three four-bar mechanisms separated from one another. Figure 1.57 shows three four-bar mechanisms that can all move together while attached, along with three separated mechanisms. All mechanisms can move in unison to trace out the same coupler point path.

A straightforward means to determine the lengths of the links of the cognates of a four-bar mechanism, given the links of the primary mechanism, is to construct *Cayley's triangles.* To do

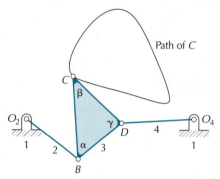

Figure 1.55 Four-bar mechanism with a coupler point.

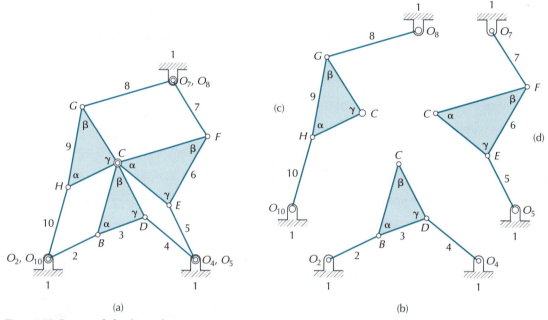

Figure 1.56 Cognates of a four-bar mechanism.

this for the mechanism shown in Figure 1.58(a), the base pivots are disconnected, and the links are repositioned with O_2, B, D, and O_4 on a straight line as shown in Figure 1.58(b). Points B, C, and D form a triangle with internal angles α, β, and γ. A line is drawn through base pivot O_2 and parallel to BC of the repositioned links, and another line is drawn through base pivot O_4 and parallel to DC. The two lines intersect at O_7, which is the location of the third base pivot for the cognates (Figure 1.58(c)). A line is drawn through O_2 on the repositioned links and parallel to BC. Another line is drawn through O_4 and parallel to DC, and a third line is drawn through C and parallel to BD (Figure 1.58(d)). Triangles GHC and FCE are similar to the triangle CBD. The lines are darkened to reveal Cayley's triangles, which contain the geometries of the links of the cognates (Figure 1.58(e)). The links of the three mechanisms may be separated and connected with the base pivots (Figure 1.58(f)).

[**Video 1.58**] provides an animated sequence of the construction of the two cognates illustrated in Figure 1.58.

Video 1.58
Cayley's Triangles

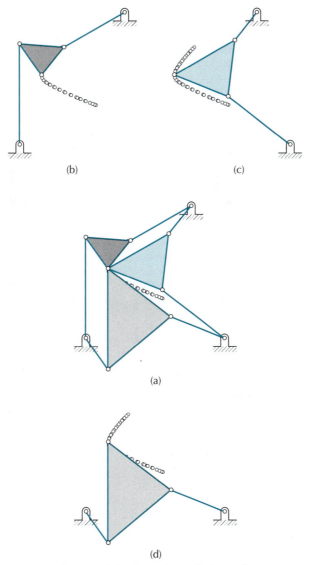

Figure 1.57   Cognates of a four-bar mechanism [Video 1.57].

1.8.2 Cognate of a Slider Crank Mechanism

A slider crank mechanism has one associated cognate. Consider the slider crank mechanism shown in Figure 1.59. As the crank rotates, coupler point C traces out the path indicated. Construction of the cognate mechanism is shown in Figure 1.60(a). For the dashed line indicated, O_2BCE forms a parallelogram. Also, BDC and ECF form similar triangles. The cognate is illustrated in Figure 1.60(b). Figure 1.61 shows two slider crank mechanisms that can move together while attached, along with two separated mechanisms. All mechanisms can trace out the same coupler point path.

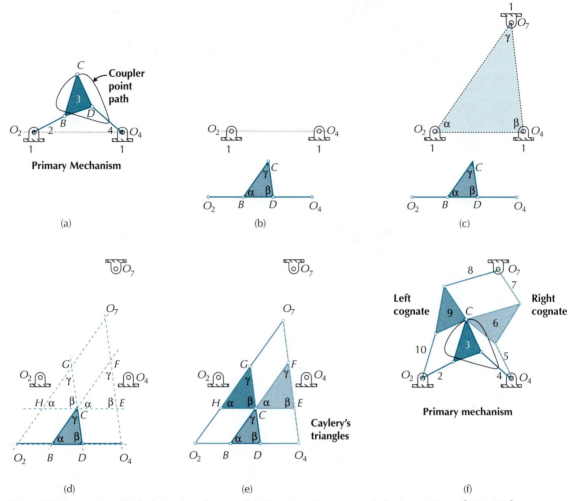

Figure 1.58 Construction of Cayley's triangles to determine link geometries of the cognates of a four-bar mechanism [Video 1.58].

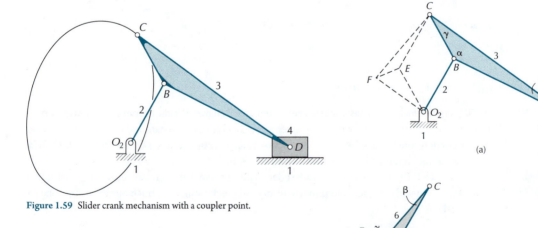

Figure 1.59 Slider crank mechanism with a coupler point.

Figure 1.60 Cognate of a slider crank mechanism.

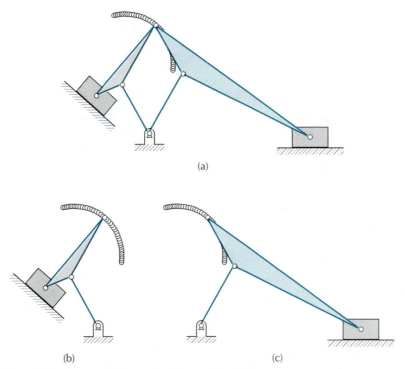

Figure 1.61 Cognate of a slider crank mechanism [Video 1.61].

PROBLEMS

Four-Bar and Slider Crank Mechanisms

P1.1 For each set of link lengths given in Table P1.1 (see Figure 1.9)

(i) determine whether or not the links can actually form a mechanism;

(ii) if a mechanism exists, determine the type of four-bar mechanism.

TABLE P1.1

	r_1 (cm)	r_2 (cm)	r_3 (cm)	r_4 (cm)
(a)	2	6.5	3	7
(b)	2	8	3	9
(c)	2.5	1	2.5	2
(d)	2.5	3	1	2
(e)	2	4.5	1.5	9
(f)	1.5	3	2.5	6

P1.2 Given the prescribed lengths of three links of a four-bar mechanism (see Figure 1.9), determine the range of values of the length of link 2 so that the mechanism will become a

(a) crank rocker mechanism

(b) drag link mechanism

(c) change point mechanism

(d) rocker–rocker mechanism

$$r_1 = 1.0 \text{ cm}; \qquad r_3 = 2.5 \text{ cm}; \qquad r_4 = 2.0 \text{ cm}$$

P1.3 Given the prescribed lengths of three links of a four-bar mechanism (see Figure 1.9), determine the range of values of the length of link 4 so that the mechanism will become a

(a) crank rocker mechanism

(b) drag link mechanism

(c) change point mechanism

(d) rocker–rocker mechanism

$$r_1 = 1.0 \text{ cm}; \qquad r_2 = 3.0 \text{ cm}; \qquad r_3 = 2.5 \text{ cm}$$

P1.4 Given the dimensions of a slider crank mechanism (see Figure 1.7), determine the range of values of the length of link 2 so that

(a) link 2 can make a full rotation

(b) the mechanism can be assembled

$$r_1 = 1.0 \text{ cm}; \qquad r_3 = 2.5 \text{ cm}$$

P1.5 Given the prescribed link dimensions of a slider crank mechanism (see Figure 1.7), determine the range of values of the length of link 3 so that

(a) link 2 can make a full rotation

(b) the mechanism can be assembled

$$r_1 = 1.0 \text{ cm}; \qquad r_2 = 2.5 \text{ cm}$$

P1.6 For the four-bar mechanism shown in Figure P1.6, the following link dimensions are given:

$r_1 = 7.0$ cm   (length of the base link)

$r_2 = 2.0$ cm   (length of the input link)

$r_4 = 6.0$ cm   (length of the output link)

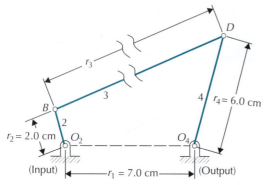

Figure P1.6

Determine the type of four-bar mechanism when the length of the coupler, link 3, is

(a) $r_3 = 3.5$ cm

(b) $r_3 = 11.0$ cm

(c) $r_3 = 11.5$ cm

P1.7 Consider the washing machine mechanism shown in Figure 1.12. To help your understanding, you should watch [**Video 1.12**] on the companion website. Is the four-bar mechanism (a) a crank rocker, (b) a drag link, (c) a change point, or (d) a rocker–rocker? Justify your answer by explaining which link serves as the crank, drag, or rocker.

P1.8 Consider the film transport mechanism shown in Figure 1.16. To help your understanding, you should watch [**Video 1.16**] on the companion website. Is the four-bar mechanism (a) a crank rocker, (b) a drag link, (c) a change point, or (d) a rocker–rocker? Justify your answer by explaining which link serves as the crank, drag link, or rocker.

Kinematic Pairs, Number of Links, and Mobility

P1.9 For each of the mechanisms shown in Figure P1.9, specify the number of links. List the types of kinematic pairs present in each of the mechanisms, and the number of each type. Calculate the mobility.

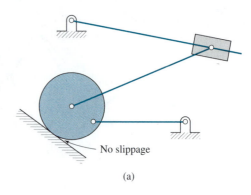

No slippage

(a)

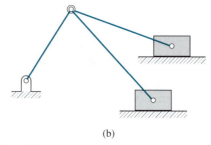

(b)

Figure P1.9

P1.10 For each of the mechanisms shown in Figure P1.10, determine the mobility.

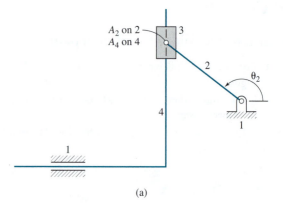

(a)

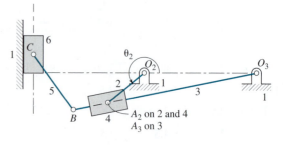

10.5 cm

(b)

Figure P1.10

P1.11 For each of the mechanisms shown in Figure P1.11, list the types of kinematic pairs present and the number of each type. Calculate the mobility.

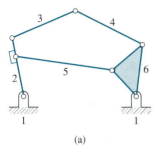

(a)

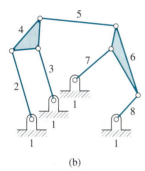

(b)

Figure P1.11

P1.12 For each of the mechanisms shown in Figure P1.12, list the types of kinematic pairs present and the number of each type. Calculate the mobility.

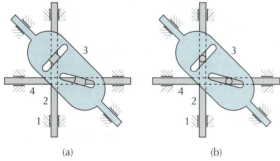

(a)   (b)

Figure P1.12

P1.13 For each of the mechanisms shown in Figure P1.13, list the types of kinematic pairs present and the number of each type. Calculate the mobility.

P1.14 Consider the Roberts straight-line mechanisms shown in Figure 1.15(b). (a) Specify the number of links, (b) list the types of kinematic pairs, and (c) calculate the mechanism mobility.

P1.15 Consider the film transport mechanism shown in Figure 1.16. (a) Specify the number of links, (b) list the types of kinematic pairs, and (c) calculate the mechanism mobility.

P1.16 Consider the rack and pinion steering mechanism shown in Figure 1.25. To help your understanding,

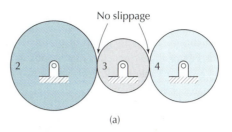

No slippage

(a)

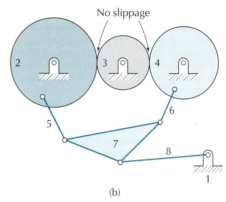

No slippage

(b)

Figure P1.13

you should watch [**Video 1.25**] on the companion website. (a) Specify the number of links, (b) list the types of kinematic pairs, and (c) calculate the mechanism mobility.

P1.17 Consider the disc cam mechanism shown in Figure 1.26. To help your understanding, you should watch [**Video 1.26**] on the companion website. (a) Specify the number of links, (b) list the types of kinematic pairs, and (c) calculate the mechanism mobility.

P1.18 For each of the mechanisms shown on Figure P1.18, calculate the mobility.

$$r_{O_2O_4} = r_{O_4O_5} = r_{O_5O_6} = r_{BD} = r_{DE} = r_{EF}$$

$$r_{O_2B} = r_{O_4D} = r_{O_5E} = r_{O_6F}$$

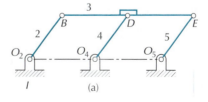

(a)

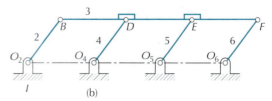

(b)

Figure P1.18

Cognates

P1.19 The nomenclature for this group of problems is given in Figure P1.19, and the dimensions and data are given in Table P1.19. For each, draw the two cognates of the mechanism.

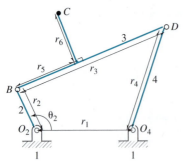

Figure P1.19

TABLE P1.19

	r_1 (cm)	r_2 (cm)	r_3 (cm)	r_4 (cm)	r_5 (cm)	r_6 (cm)	θ_2 (degrees)
(a)	1	0.2	0.4	1	0.2	0.5	30
(b)	5	1	4	3	-2	0	60
(c)	8	2	10	6	6	-3	120
(d)	0.2	1	0.4	0.9	0.2	0.4	210

P1.20 The nomenclature for this group of problems is given in Figure P1.20, and the dimensions and data are given in Table P1.20. For each, draw the cognate of the mechanism.

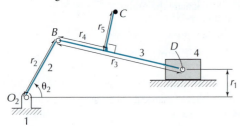

Figure P1.20

TABLE P1.20

	r_1 (cm)	r_2 (cm)	r_3 (cm)	r_4 (cm)	r_5 (cm)	θ_2 (degrees)
(a)	0	5	10	-4	0	30
(b)	0	2	6	2	4	60
(c)	2	4	8	12	0	120
(d)	-2	4	7	5	-3	315

P1.21 For the mechanism shown on Figure P1.21, construct Cayley's triangles and determine the geometries of the links of the two cognates.

$$r_{O_2O_4} = 13.0 \text{ cm}; \quad r_{O_2B} = 5.0 \text{ cm};$$

$$r_{BD} = 9.0 \text{ cm}; \quad r_{O_4D} = 6.0 \text{ cm}$$

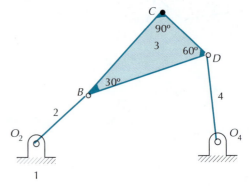

Figure P1.21

P1.22 For the mechanism shown on Figure P1.22, construct Cayley's triangles and determine the geometries of the links of the two cognates.

$$r_{O_2O_4} = 12.0 \text{ cm}; \quad r_{O_2B} = 5.0 \text{ cm};$$

$$r_{BD} = 9.0 \text{ cm}; \quad r_{O_4D} = 6.0 \text{ cm}$$

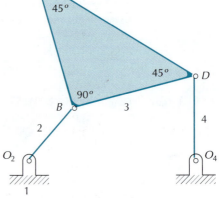

Figure P1.22

CHAPTER 2

MECHANICS OF RIGID BODIES AND PLANAR MECHANISMS

2.1 INTRODUCTION

This chapter begins by covering the analysis of the positions and displacements of points and the links of planar mechanisms. After that, expressions are presented for the relative velocities and relative accelerations between two moving points in a plane. This material is the background for kinematic analyses of planar mechanisms, which is studied in Chapters 3 and 4.

The concept of *transmission angle* is introduced. It is a parameter that should be considered in the design of planar mechanisms. The *instantaneous center of velocity* is then defined followed by the presentation of a method to locate all of the instantaneous centers of velocity for a planar mechanism. They are used in this chapter to determine the *mechanical advantage* of mechanisms and will also be employed in Chapter 3 to carry out their velocity analyses.

The *limit positions* and *time ratio* of a mechanism are presented from which it is possible to determine an average speed of the output moving between its limit positions when the input is driven at a constant speed.

The governing equations for the static and dynamic force analyses of rigid bodies are reviewed. In Chapters 8 and 9 they will be used for the force analyses of planar mechanisms. Common sets of units of quantities employed in these equations are tabulated.

2.2 POSITION ANALYSIS

2.2.1 Position Analysis of Points

Figure 2.1(a) shows points A and B located in a plane. The positions of the points may be described by the *position vectors* $\bar{r}_{AO}$ and $\bar{r}_{BO}$ with respect to the origin of the stationary coordinate system.

As illustrated in Figure 2.1(a), the position of point B may also be described as the sum of two vectors, that is,

$$\bar{r}_{BO} = \bar{r}_{AO} + \bar{r}_{BA} \qquad (2.2\text{-}1)$$

where $\bar{r}_{BA}$ is the *relative position vector* of point B with respect to point A.

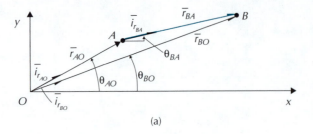

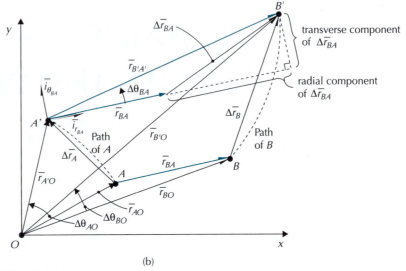

Figure 2.1 (a) Positions of points. (b) Displacements of points.

With the vectors drawn in a set of Cartesian axes, each vector may be expressed in terms of its x and y components. Alternatively, a vector may be defined by its magnitude multiplied by a vector of unit length that points the same direction as the vector. For the relative position vector shown in Figure 2.1, we have

$$\bar{r}_{BA} = r_{BA}\,\bar{i}_{r_{BA}} \qquad (2.2\text{-}2)$$

where $\bar{i}_{r_{BA}}$ is the *unit relative radial vector*. Note that the orientation of the relative position vector and the unit relative radial vector are defined by angle θ_{BA} measured at the tail end of the vectors from the horizontal in the counterclockwise direction.

In the analysis of a mechanism, it is often necessary to monitor the positions of points on the mechanism. The position of any point on a mechanism may be described by a position vector. For the mechanism shown in Figure 2.2(a), the origin of the coordinate system is placed at the base pivot O_2. The positions of points B, C and D are described by position vectors. Any point on the mechanism may also be described as the sum of a position vector and a relative position vector or vectors. For instance, the position of point C may be expressed as

$$\bar{r}_{CO_2} = \bar{r}_{BO_2} + \bar{r}_{CB} \qquad (2.2\text{-}3)$$

2.2.2 Position Analysis of Rigid Bodies and Mechanisms

The position of a rigid body or link of a mechanism cannot always be defined with a position vector. For the slider crank mechanism shown in Figure 2.2(a), the crank, link 2, is restricted to rotate about the base pivot. Its position may therefore be defined by angle θ_2 measured counterclockwise from the horizontal to a line on the link. The slider, link 4, at most undergoes pure translation and is constrained to move horizontally. Therefore, the position of link 4 is

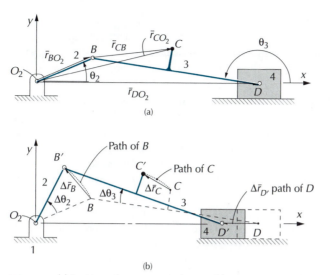

Figure 2.2 (a) Positions of points on a mechanism. (b) Displacement of a mechanism.

defined using the position vector of point D. The coupler, link 3, can undergo a more complicated motion. Therefore, its position cannot be defined with either a single position vector or just an angular position. The position of link 3 may be specified by a position vector to a point on the link, say $\bar{r}_D$, and angle θ_3.

The graphical position analysis of a planar mechanism is a straightforward task. As an example, we are provided with the geometries of all of the links of a slider crank mechanism (see Figure 2.3(a))

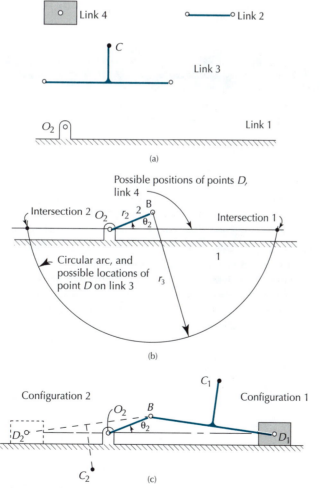

Figure 2.3

along with the prescribed angular position of link 2, θ_2, and are required to determine the configuration of the assembled mechanism. Since the mechanism has a mobility (see Section 1.5) equal to one, if we are given θ_2, then the entire mechanism geometry may be determined. We start by drawing links 1 and 2 in their prescribed positions as illustrated in Figure 2.3(b) and identify point B, the turning pair between links 2 and 3. We now seek the position of point D, the turning pair between links 3 and 4. Since point D is on link 3, we draw an arc, centered at B, with radius equal to the length of link 3. Points on the arc provide the possible locations of point D on link 3. Point D is also on link 4, which is constrained translate horizontally with respect to the base link. We therefore draw a straight horizontal line that corresponds to the possible locations of point D on link 4. Since for this example the slider has zero offset, the line is drawn through O_2. The intersection between the arc and the straight line gives the acceptable position of point D for the given position of link 2. Note that there are actually two points of intersection. That is, for the given position of link 2, the mechanism can be assembled in either one of two configurations as illustrated in Figure 2.3(c). It is often possible to assemble a mechanism in two configurations for a prescribed position(s) of the input link(s).

The above example determined graphically the configuration of the mechanism. The scaled diagram may be subsequently used for the graphical velocity, acceleration and force analyses of the mechanism. The procedures are presented in Chapters 3 and 8. For each configuration of the mechanism where an analysis is required, a separate graphical construction of the mechanism is drawn. Alternatively, an analytical position analysis of planar mechanisms is presented in Chapter 4. Here, analytical expressions are derived that may be used repeatedly to calculate the values of the position variables for all needed configurations of the mechanism. The results may then be employed for the analytical velocity, acceleration, and force analyses. These procedures are presented in Chapters 4 and 9.

EXAMPLE 2.1 GRAPHICAL CONSTRUCTION OF A FOUR-BAR MECHANISM

For the given link lengths of a four-bar mechanism and the position of link 2, construct the permissible geometries of the mechanism

$$r_1 = 10.0 \text{ cm}; \quad r_2 = 4.0 \text{ cm}; \quad r_3 = 8.0 \text{ cm}; \quad r_4 = 5.0 \text{ cm}; \quad \theta_2 = 30°$$

SOLUTION

We first draw links 1 and 2 in their specified configurations as shown in Figure 2.4(a). For this configuration, the turning pair joining links 3 and 4 must be located a distance equal to the length of the coupler, r_3, from the turning pair at B. The same turning pair must also be located distance r_4 from the base pivot O_4. The two location requirements may be represented graphically by drawing circular arcs of radii r_3 and r_4 centered at B and O_4, respectively. In order to satisfy both requirements, the turning pair must be located at an intersection of the arcs. The two intersection points indicate it is possible to assemble the mechanism links in either one of the two geometries as illustrated in Figure 2.4(b).

(Continued)

EXAMPLE 2.1 Continued

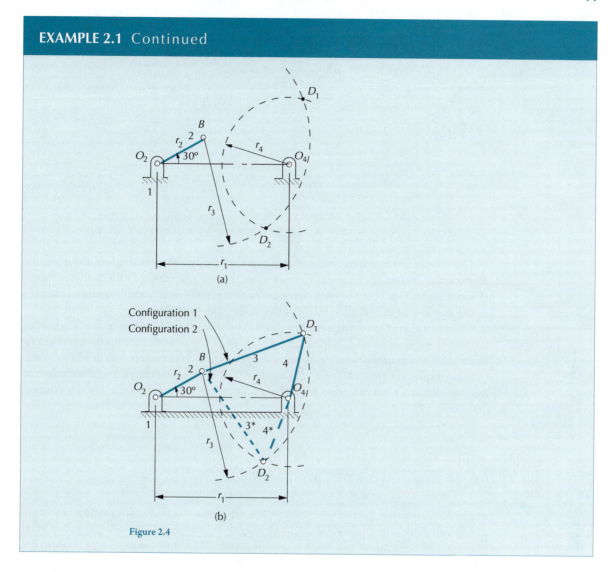

Figure 2.4

2.3 DISPLACEMENT ANALYSIS

2.3.1 Displacement Analysis of Points

Displacement of a point is the change of its position. It may be represented by a *displacement vector*. Figure 2.1(b) shows points A and B that have moved to points A' and B', respectively, along paths that are depicted as dotted lines. The displacement vectors $\Delta \bar{r}_A$ and $\Delta \bar{r}_B$ extend between the points before and after the motions have taken place. This figure illustrates that a path of motion and a displacement vector need not coincide. The relative position vector after the motions is shown as

$$\bar{r}_{B'A'} = \bar{r}_{BA} + \Delta \bar{r}_{BA} \qquad (2.3\text{-}1)$$

where $\Delta \bar{r}_{BA}$ is the *relative displacement vector* of point B with respect to point A. In this illustration $\bar{r}_{B'A'}$ is not parallel to $\bar{r}_{BA}$. We therefore introduce $\bar{i}_{\theta_{BA}}$, the *unit relative transverse vector*, which points 90° counterclockwise from the direction of $\bar{i}_{r_{BA}}$. Vectors $\bar{i}_{r_{BA}}$ and $\bar{i}_{\theta_{BA}}$ are the unit vectors of the radial-transverse coordinate system associated with $\bar{r}_{BA}$. The radial and transverse components of the relative displacement vector are illustrated in Figure 2.1(b). Also, as a result of the motions, the relative position vector has changed its orientation by $\Delta \theta_{BA}$.

Consider the slider crank mechanism shown in Figure 2.2(b). The crank is rotated $\Delta \theta_2$ and the mechanism takes up a new configuration. Points B, C, and D move to B', C', and D', respectively. Dotted lines illustrate the paths of motions. Three displacement vectors are also illustrated. For point D, its path and displacement vector coincide. However, the paths of points B and C are distinct from their displacement vectors.

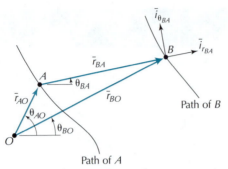

Figure 2.5 Relative position vector between two moving points.

2.3.2 Displacement Analysis of Rigid Bodies and Mechanisms

The description of the displacement of a rigid body is usually more involved than a point. The displacement analysis of a mechanism requires that the displacements for all of its links be determined. This may be accomplished by drawing two configurations of the mechanism and identifying the displacements that take place. For the slider crank mechanism shown in Figure 2.2(b), link 2 is constrained to undergo pure rotation about the base pivot O_2. The displacement of this link is therefore defined by the changes of its angular position about the base pivot, $\Delta\theta_2$. Since link 4 is constrained to undergo linear translation, its displacement is defined by the displacement vector of point D. The specification of the displacement of link 3 requires values of the linear displacement of a point on the link, say point D, *and* the change of angular position of the link, $\Delta\theta_3$.

2.4 RELATIVE VELOCITY BETWEEN TWO POINTS UNDERGOING PLANAR MOTION

Appendix C presents expressions for the position, velocity, and acceleration of a moving point in a plane with respect to a fixed point. In this section, the more general case of having two moving points in a plane is considered. An expression is developed for the relative velocity between the points. Figure 2.5 shows two moving points, A and B, and their paths of motion. The relative position vector is illustrated along with the associated unit vectors of the radial-transverse coordinate system.

The relative velocity between the two points is found by differentiating Equation (2.2-2) with respect to time, yielding

$$\overline{v}_{BA} = \frac{d}{dt}(\overline{r}_{BA}) = \frac{d}{dt}(r_{BA}\,\overline{i}_{r_{BA}}) = \dot{r}_{BA}\,\overline{i}_{r_{BA}} + r_{BA}\,\frac{d}{dt}(\overline{i}_{r_{BA}}) \tag{2.4-1}$$

An expression for the time derivative of the unit relative radial vector may be found using a procedure similar to that presented in Appendix C. The result is

$$\frac{d}{dt} = (\overline{i}_{r_{BA}}) = \dot{\theta}_{BA}\,\overline{i}_{\theta_{BA}} \tag{2.4-2}$$

where $\dot{\theta}_{BA}$ is the *angular velocity* of the line segment joining A and B.

Substituting Equation (2.4-2) in Equation (2.4-1) yields

$$\overline{v}_{BA} = \dot{r}_{BA}\,\overline{i}_{r_{BA}} + r_{BA}\,\dot{\theta}_{BA}\,\overline{i}_{\theta_{BA}} \tag{2.4-3}$$

Components of Equation (2.4-3) are illustrated in Figure 2.6. Special cases of Equation (2.4-3) are presented in Section 2.5.

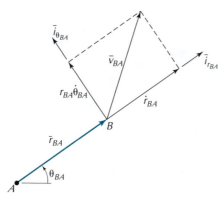

Figure 2.6 Relative velocity components between two points.

2.5 SPECIAL CASES OF RELATIVE VELOCITY EXPRESSION

2.5.1 Two Points Separated by a Fixed Distance

If points A and B are separated by a fixed distance, then

$$\dot{r}_{BA} = 0 \tag{2.5-1}$$

and the relative velocity expressed by Equation (2.4-3) reduces to

$$\overline{v}_{BA} = r_{BA}\,\dot{\theta}_{BA}\,\overline{i}_{\theta_{BA}} \tag{2.5-2}$$

For Equation (2.5-2), the relative velocity is in the transverse direction as defined by the line segment joining A and B. It is therefore perpendicular to the relative position vector.

In summary, for two points separated by a fixed distance, the relative velocity must always be in a direction perpendicular to the line segment joining the points.

As an illustration, consider the four-bar mechanism shown in Figure 2.7(a). A separate drawing of link 2 is shown in Figure 2.7(b). Points B and O_2 are on the same link and thus are separated by a fixed distance. Therefore

$$\overline{v}_{BO_2} = r_{BO_2}\,\dot{\theta}_{BO_2}\,\overline{i}_{\theta_{BO_2}} \tag{2.5-3}$$

Since O_2 is fixed, the relative velocity given in Equation (2.5-3) also represents the absolute velocity, and

$$\overline{v}_{BO_2} = \overline{v}_B = r_{BO_2}\,\dot{\theta}_{BO_2}\,\overline{i}_{\theta_{BO_2}} \tag{2.5-4}$$

Figure 2.7(c) shows a drawing of link 3. Points B and C are on the same link and separated by a fixed distance. The relative velocity between the two points is

$$\overline{v}_{CB} = r_{CB}\,\dot{\theta}_{CB}\,\overline{i}_{\theta_{CB}} \tag{2.5-5}$$

The absolute velocity of point C may be expressed as

$$\overline{v}_C = \overline{v}_B + \overline{v}_{CB} \tag{2.5-6}$$

Substituting Equations (2.5-4) and (2.5-5) in Equation (2.5-6) yields

$$\overline{v}_C = r_{BO_2}\,\dot{\theta}_{BO_2}\,\overline{i}_{\theta_{BO_2}} + r_{CB}\,\dot{\theta}_{CB}\,\overline{i}_{\theta_{CB}} \tag{2.5-7}$$

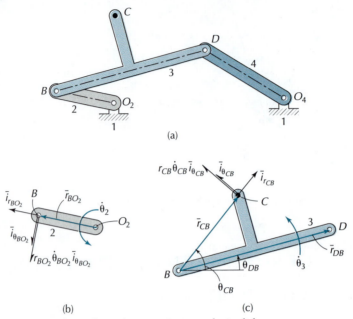

(a)

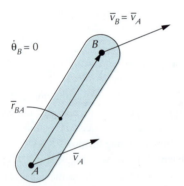

(b)   (c)

Figure 2.7 Relative velocities between points on mechanism links.

Figure 2.8 Body undergoing pure translation.

A moving link has a single value of rotational speed. Considering link 3 shown in Figure 2.7(c).

$$\dot{\theta}_{CB} = \dot{\theta}_{DB} = \dot{\theta}_3 \qquad (2.5\text{-}8)$$

Now consider the rigid body shown in Figure 2.8. Two points separated by a fixed distance on this body are A and B. For the instant shown, let us say

$$\bar{v}_B = \bar{v}_A$$

and therefore using

$$\bar{v}_B = \bar{v}_A + \bar{v}_{BA} = \bar{v}_B + r_{BA}\dot{\theta}$$

we conclude

$$\bar{v}_{BA} = \bar{0}$$

and then

$$\dot{\theta} = 0$$

In this instance the body is undergoing pure translation.

2.5.2 Slider on a Straight Slide

As a second special case of Equation (2.4-3), consider the movement of a slider on a straight slide, as shown in Figure 2.9. Point A is fixed. Point B_2 is on the slide, link 2, and point B_3 is on the slider, link 3. For convenience in this analysis, points B_2 and B_3 coincide for the instant at which the relative motion is being considered.

Since point B_2 remains a fixed distance from fixed point A, its velocity may be found by considering the special case presented in Section 2.5.1. The result is

$$\bar{v}_{B_2A} = \bar{v}_{B_2} = r_{B_2A}\dot{\theta}_{B_2A}\bar{i}_{\theta_{B_2A}} \tag{2.5-9}$$

For point B_3, the associated position vector equation is

$$\bar{r}_{B_3A} = \bar{r}_{B_2A} + \bar{r}_{B_3B_2} \tag{2.5-10}$$

and the relative velocity equation is

$$\bar{v}_{B_3A} = \bar{v}_{B_2A} + \bar{v}_{B_3B_2} \tag{2.5-11}$$

or

$$\bar{v}_{B_3} = \bar{v}_{B_2} + \bar{v}_{B_3B_2} \tag{2.5-12}$$

where, using Equation (2.4-3), we obtain

$$\bar{v}_{B_3B_2} = \dot{r}_{B_3B_2}\bar{i}_{r_{B_3B_2}} + r_{B_3B_2}\dot{\theta}_{B_3B_2}\bar{i}_{\theta_{B_3B_2A}} \tag{2.5-13}$$

However,

$$r_{B_3B_2} = 0 \tag{2.5-14}$$

Also, even though B_2 and B_3 coincide at the instant under consideration, if the slider were to move with respect to the slide, it would be along a line in the same direction as line segment B_2A, and therefore we are permitted to write

$$\bar{i}_{r_{B_3B_2}} = \bar{i}_{r_{B_2A}} \tag{2.5-15}$$

Combining Equations (2.3-9) and (2.3-12)–(2.3-15) yields

$$\bar{v}_{B_3} = \bar{v}_{B_3} + \bar{v}_{B_3B_2} = r_{B_2A}\dot{\theta}_{B_2A}\bar{i}_{\theta_{B_2A}} + \dot{r}_{B_3B_2}\bar{i}_{r_{B_2A}} \tag{2.5-16}$$

Comparing expressions for velocities of B_2 and B_3, given in Equations (2.5-9) and (2.5-16), we find the following:

- B_2 and B_3 share the *same* transverse velocity component.
- B_2 and B_3 can have *different* radial components.

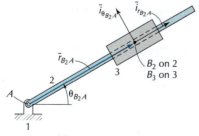

Figure 2.9 Slider on a straight slide.

EXAMPLE 2.2 DETERMINATION OF CONFIGURATIONS OF ZERO ROTATIONAL SPEED OF A MECHANISM LINK

For the slider crank mechanism shown in Figure 2.10, determine

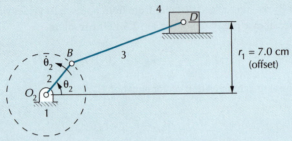

Figure 2.10

(a) the value(s) of θ_2 when the rotational speed of link 3 is zero
(b) the corresponding linear speeds of the slider

$$r_1 = 7.0 \text{ cm}; \quad r_2 = 4.0 \text{ cm}; \quad r_3 = 12.0 \text{ cm}; \quad \dot{\theta}_2 = 20 \text{ rad/sec CCW (constant)}$$

SOLUTION

(a) Two points on link 3 are B and D. The direction of velocity of B changes as the mechanism moves and for any configuration, it is perpendicular to $\bar{r}_{BO_2}$ (This is an example of the special case presented in Section 2.5.1). Point D is on the slider and is constrained to move horizontally (This is an example of the special case presented in Section 2.5.2.). Link 3 will have zero rotational velocity and undergoing pure translation when the directions of the velocities at B and D are the same. This is achieved when link 2 is perpendicular to the slide (i.e., $\theta_2 = 90°$ and $270°$), as illustrated in Figure 2.11.

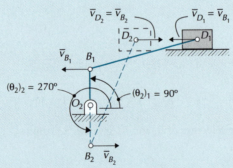

Figure 2.11

(b) The speed of the slider for both configurations shown in Figure 2.11 is

$$v_{D_1} = v_{D_2} = v_{B_1} = r_2 \dot{\theta}_2 = 80 \text{ cm/sec}$$

2.6 RELATIVE ACCELERATION BETWEEN TWO POINTS UNDERGOING PLANAR MOTION

In this section, we continue our analysis of the motions of a moving point B with respect to another moving point A, as illustrated in Figure 2.5. We now consider the relative acceleration between the two points. Using an approach similar that employed in section 2.4 for the analysis of velocity, the relative acceleration between the points is determined to be

$$\bar{a}_{BA} = \frac{d}{dt}(\bar{v}_{BA}) = (\ddot{r}_{BA} - r_{BA}\dot{\theta}_{BA}^2)\bar{i}_{r_{BA}} + (r_{BA}\ddot{\theta}_{BA} + 2\dot{r}_{BA}\dot{\theta}_{BA})\bar{i}_{\theta_{BA}} \qquad (2.6\text{-}1)$$

The above expression for acceleration is made up of two components in the radial direction, and two in the transverse direction. This is illustrated in Figure 2.12. Each of these four components is discussed separately below.

2.6.1 Sliding Acceleration

Relative *sliding acceleration*, illustrated in Figure 2.13(a), is designated as

$$a_{BA}^S = \ddot{r}_{BA} \qquad (2.6\text{-}2)$$

where point A is on the slide, and point B is on the slider.

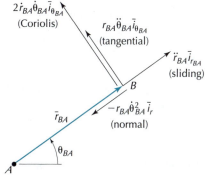

Figure 2.12 Relative acceleration components between two points.

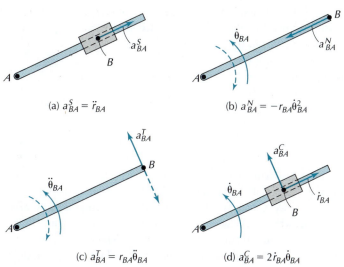

(a) $a_{BA}^S = \ddot{r}_{BA}$

(b) $a_{BA}^N = -r_{BA}\dot{\theta}_{BA}^2$

(c) $a_{BA}^T = r_{BA}\ddot{\theta}_{BA}$

(d) $a_{BA}^C = 2\dot{r}_{BA}\dot{\theta}_{BA}$

Figure 2.13 Components of relative acceleration expression.

In this case, the relative acceleration is in the radial direction defined by the locations of points A and B. The direction of sliding acceleration must be in a positive or negative radial direction.

2.6.2 Normal Acceleration

Relative *normal acceleration*, illustrated in Figure 2.13(b), is designated as

$$\boxed{a_{BA}^N = -r_{BA}\dot{\theta}_{BA}^2} \tag{2.6-3}$$

where the negative sign indicates that normal acceleration is always directed in the negative radial direction. The direction of normal acceleration is independent of the direction of angular velocity.

Normal acceleration occurs in circular motion. Consider the disc with radius r shown in Figure 2.14. It is rotating about base pivot O at a constant rate of $\dot{\theta}$. During a short time interval, point B on the periphery of the disc moves to point B'. The velocity of a point may be expressed as a vector in the complex plane (refer to Appendix B, Section B.4). The velocities at points B and B' are

$$\bar{v}_B = r\dot{\theta}e^{i\pi/2}; \quad \bar{v}_{B'} = r\dot{\theta}e^{i(\pi/2+\Delta\theta)} \tag{2.6-4}$$

Point B is moving at constant speed. However, the direction of its velocity is changing. Hence, it undergoes acceleration expressed as

$$\bar{a}_B = \lim_{\Delta t \to 0} \frac{\bar{v}_{B'} - \bar{v}_B}{\Delta t} \tag{2.6-5}$$

Substituting Equation (2.6-4) in Equation (2.6-5) and simplifying yields

$$\bar{a}_B = \lim_{\Delta t \to 0} \frac{\bar{v}_{B'} - \bar{v}_B}{\Delta t} = \lim_{\Delta t \to 0} \frac{\Delta \bar{v}_B}{\Delta t} = \lim_{\Delta t \to 0} -\frac{r\dot{\theta}\Delta\theta}{\Delta t} = -r\dot{\theta}^2 \tag{2.6-6}$$

The negative sign in Equation (2.6-6) indicates that the acceleration is in the negative radial direction.

2.6.3 Tangential Acceleration

Relative *tangential acceleration*, illustrated in Figure 2.13(c), is designated as

$$\boxed{a_{BA}^T = r_{BA}\ddot{\theta}_{BA}} \tag{2.6-7}$$

This relative acceleration is perpendicular to the line segment joining A and B. It can be in either the positive or negative transverse direction, depending on the positive or negative sign depicting the *angular acceleration*, $\ddot{\theta}_{BA}$.

2.6.4 Coriolis Acceleration

Relative *Coriolis acceleration*, illustrated in Figure 2.13(d), is designated as

$$\boxed{a_{BA}^C = 2\dot{r}_{BA}\dot{\theta}_{BA}} \tag{2.6-8}$$

where point A is on the slide and point B is on the slider.

In order for Coriolis acceleration to be present, relative sliding (i.e., $\dot{r}_{BA}$) must take place in a frame of reference that has angular velocity (i.e., $\dot{\theta}_{BA}$). This would occur when a slider block moves along a slide, while at the same time the slide has rotational motion. Figure 1.43(b) presents a typical example. The direction of the Coriolis acceleration is dependent upon both the sign of the sliding velocity and the sign of the angular velocity. In this textbook (following the usual convention for analyses of planar systems), positive angular displacements, angular velocities, and angular accelerations are typically taken to be in the counterclockwise direction. The positive value of the

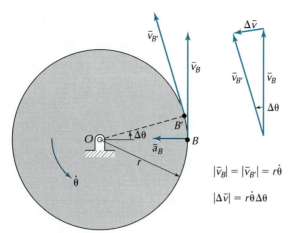

Figure 2.14 Circular motion.

sliding velocity corresponds to motion in the positive radial direction defined by the two points under consideration. Figure 2.13(d) shows the combination of positive sliding velocity and positive angular velocity. The resultant Coriolis acceleration is then in the positive transverse direction.

Figure 2.15 shows the four possible combinations of positive and negative angular velocity and sliding velocity. The relative Coriolis acceleration component has been drawn for each combination. All Coriolis acceleration components have a magnitude of

$$\left| a_{BA}^{C} \right| = \left| 2\dot{r}_{BA}\dot{\theta}_{BA} \right| \tag{2.6-9}$$

Another illustration of Coriolis acceleration is shown in Figure 2.16. A rotating disc is turning about base pivot O. A radial line is painted on the disc between O and point P on the periphery. Let us say there is an ant on the surface of the disc at point B. It moves radially along the line from point B toward point P at speed $\dot{r}$. During a short time interval, the ant moves from point B to point B', and the disc rotates $\Delta\theta$. At points B and B', the velocities are

$$\bar{v}_B = \dot{r} + r\dot{\theta}e^{i\pi/2}; \quad \bar{v}_{B'} = \dot{r}e^{i\Delta\theta} + (r+\Delta r)\dot{\theta}e^{i(\pi/2+\Delta\theta)} \tag{2.6-10}$$

Although there is a constant rate of radial movement of the ant with respect to the center of the disc, it experiences an increase in the component of velocity perpendicular to the direction of the relative motion. Therefore, the ant undergoes acceleration in that direction. The acceleration of the ant at point B is

$$\bar{a}_B = \lim_{\Delta t \to 0} \frac{\bar{v}_{B'} - \bar{v}_B}{\Delta t} \tag{2.6-11}$$

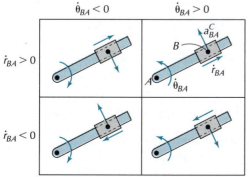

Figure 2.15 Coriolis acceleration of a slider (B) with respect to a slide (A).

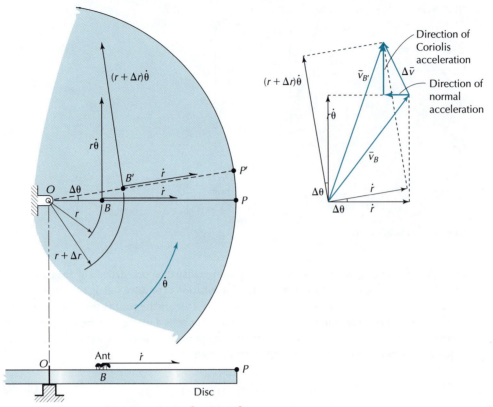

Figure 2.16 Illustration of Coriolis acceleration [Model 2.16].

Substituting Equation (2.6-10) in Equation (2.6-11) and simplifying yields

$$\bar{a}_B = -r\dot{\theta}^2 + i2\dot{r}\dot{\theta}$$

(2.6-12)

Video 2.16
Model 2.16
Coriolis
Acceleration

Equation (2.6-12) indicates there is a normal component of acceleration directed back toward the base pivot, and a Coriolis component perpendicular to the direction of relative motion. Using [**Model 2.16**], one is able to simulate the various combinations of the radial motion of the object (i.e., ant) with respect to the disc and the rotational speed of the disc.

2.7 SPECIAL CASES OF THE RELATIVE ACCELERATION EQUATION

Combining Equations (2.6-1)–(2.6-5), we may express the relative acceleration between points A and B as

$$\bar{a}_{BA} = (a_{BA}^{S} + a_{BA}^{N})\bar{i}_{r_{BA}} + (a_{BA}^{T} + a_{BA}^{C})\bar{i}_{\theta_{BA}}$$

(2.7-1)

The components are illustrated in Figure 2.12. In Subsections 2.7.1–2.7.3, we examine a few special cases of this expression.

2.7.1 Two Points Separated by a Fixed Distance

When two points are separated by a fixed distance, we obtain

$$\dot{r}_{BA} = 0; \quad \ddot{r}_{BA} = 0$$

(2.7-2)

Then from Section 2.6, we have

$$a_{BA}^{C} = 0; \quad a_{BA}^{S} = 0 \tag{2.7-3}$$

and Equation (2.7-1) reduces to

$$\boxed{\begin{aligned} \bar{a}_{BA} &= (-r_{BA}\dot{\theta}_{BA}^{2})\bar{i}_{r_{BA}} + (r_{BA}\ddot{\theta}_{BA})\bar{i}_{\theta_{BA}} \\ &= (a_{BA}^{N})\bar{i}_{r_{BA}} + (a_{BA}^{T})\bar{i}_{\theta_{BA}} \end{aligned}} \tag{2.7-4}$$

which indicates that between two points separated by a fixed distance there can only be normal and tangential relative accelerations.

As an illustration, consider the four-bar mechanism shown in Figure 2.17(a). A separate drawing of link 2 is shown in Figure 2.17(b). Points B and O_2 are on the same link and separated by a fixed distance. Therefore

$$\bar{a}_{BO_2} = (-r_{BO_2}\dot{\theta}_{BO_2}^{2})\bar{i}_{r_{BO_2}} + (r_{BO_2}\ddot{\theta}_{BO_2})\bar{i}_{\theta_{BO_2}} \tag{2.7-5}$$

Since O_2 is fixed, the relative acceleration given in Equation (2.7-5) also represents the absolute acceleration, and

$$\bar{a}_{BO_2} = \bar{a}_B = (-r_{BO_2}\dot{\theta}_{BO_2}^{2})\bar{i}_{BO_2} + (r_{BO_2}\ddot{\theta}_{BO_2})\bar{i}_{\theta_{BO_2}} \tag{2.7-6}$$

Furthermore, if link 2 of the mechanism is driven at a constant rate, then

$$\dot{\theta}_{BO_2} = \text{constant}; \quad \ddot{\theta}_{BO_2} = 0$$

Equation (2.7-6) reduces to

$$\bar{a}_B = (-r_{BO_2}\dot{\theta}_{BO_2}^{2})\bar{i}_{r_{BO_2}} \tag{2.7-7}$$

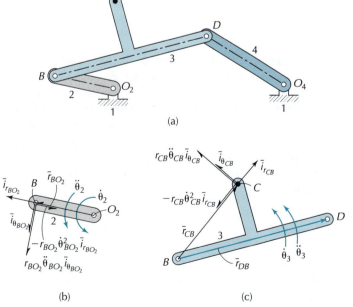

(a)

(b)   (c)

Figure 2.17 Relative accelerations between points on mechanism link: (a) Mechanism. (b) Link 2. (c) Link 3.

Figure 2.17(c) illustrates link 3. Since points B and C are on the same link, they are separated by a fixed distance. Therefore

$$\bar{a}_{CB} = (-r_{CB}\dot{\theta}_{CB}^2)\bar{i}_{r_{CB}} + (r_{CB}\ddot{\theta}_{CB})\bar{i}_{\theta_{CB}} \tag{2.7-8}$$

The absolute acceleration of point C may be expressed as

$$\bar{a}_C = \bar{a}_B + \bar{a}_{CB} \tag{2.7-9}$$

Substituting Equations (2.7-7) and (2.7-8) in Equation (2.7-9) yields

$$\bar{a}_C = (-r_{BO_2}\dot{\theta}_{BO_2}^2)\bar{i}_{r_{BO_2}} + (-r_{CB}\dot{\theta}_{CB}^2)\bar{i}_{r_{CB}} + (r_{CB}\ddot{\theta}_{CB})\bar{i}_{\theta_{CB}} \tag{2.7-10}$$

A moving link can have only one value of angular acceleration. Thus, for link 3 shown in Figure 2.17(c) we have

$$\ddot{\theta}_{CB} = \ddot{\theta}_{DB} = \ddot{\theta}_3 \tag{2.7-11}$$

2.7.2 Slider on a Straight Slide

We now consider the special case of a slider moving along a straight slide as illustrated in Figure 2.9. Considering points B_2 and B_3 for which

$$r_{B_2 B_3} = 0 \tag{2.7-12}$$

Then from Section 2.6 we obtain

$$a_{B_2 B_3}^N = 0; \quad a_{B_2 B_3}^T = 0 \tag{2.7-13}$$

and Equation (2.7-1) reduces to

$$\begin{aligned}
\bar{a}_{B_3 B_2} &= (\ddot{r}_{B_3 B_2})\bar{i}_{r_{B_2 A}} + (2\dot{r}_{B_3 B_2}\dot{\theta}_{B_2 A})\bar{i}_{\theta_{B_2 A}} \\
&= (a_{B_3 B_2}^S)\bar{i}_{r_{B_2 A}} + (a_{B_3 B_2}^C)\bar{i}_{\theta_{B_2 A}}
\end{aligned} \tag{2.7-14}$$

That is, for two points that momentarily coincide, it is possible to have relative sliding acceleration and relative Coriolis acceleration occurring between the two points.

2.7.3 Slider on a Curved Slide

Figure 2.18(a) illustrates a slider on a curved slide. The slide has a radius of curvature ρ, angular velocity $\dot{\theta}_2$, and angular acceleration $\ddot{\theta}_2$. The velocity of the slider with respect to the slide is given as $\bar{v}_{B_3 B_2}$. Given these motions, Figure 2.18(b) shows two components of acceleration of point B_2, relative to point A on the slide. These consist of normal and tangential components. Components of acceleration of the slider, represented by point B_3, relative to point B_2 on the slide, are shown in Figure 2.18(c). Similar to the case of a slider on a straight slide, there are relative sliding and Coriolis components that are tangential and perpendicular to the sliding motion. In addition, there is a component of acceleration directed toward the center of curvature of the slide. The magnitude of this component is

$$\frac{v_{B_2 B_3}^2}{\rho} \tag{2.7-15}$$

Figure 2.18 illustrates a slide in the shape of a circular arc, and hence the radius of its curvature is constant. Equation (2.7-15) is applicable in analyses involving noncircular arcs by applying the value of radius of curvature that corresponds to the position under consideration.

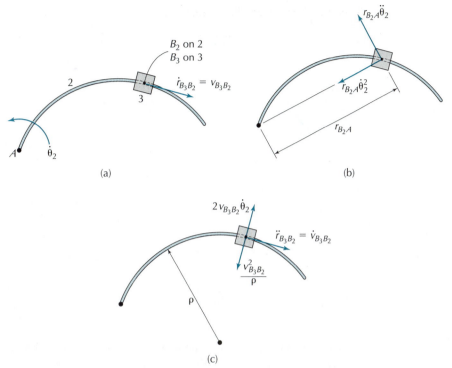

Figure 2.18 Slider on a curved slide.

2.8 LIMIT POSITIONS AND TIME RATIO OF MECHANISMS

Consider the slider crank mechanism shown in Figure 2.19. The crank is able to make full rotations, and the slider moves between two *limit positions* designated as D_1 and D_2. Limit positions define the *stroke* of the slider.

In this section, the limit positions of some common mechanisms are presented. We will restrict ourselves to conditions where the input link of the mechanism is able to make full rotation, and the output link moves between two limit positions.

2.8.1 Slider Crank Mechanism

In instances when the crank is able to make a complete rotation and is turning at a constant speed, we define the *time ratio* as the time for the slider to move in one direction between the limit positions, divided by the time it takes to move in the opposite direction between the same

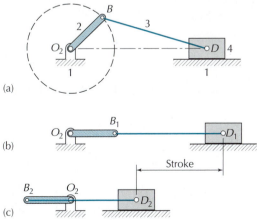

Figure 2.19 Limit positions of a slider crank mechanism: (a) Mechanism. (b) Geometry of first limit position. (c) Geometry of second limit position.

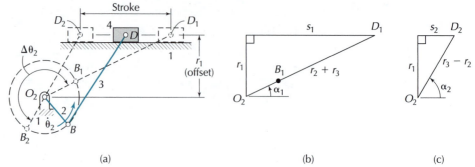

Figure 2.20 Limit positions of a slider crank mechanism: (a) Mechanism. (b) Geometry of first limit position. (c) Geometry of second limit position.

limit positions. The time ratio is dimensionless and usually arranged such that its calculated value greater than or equal to unity.

Figure 2.20(a) shows a slider crank mechanism. The crank is rotating in the counterclockwise direction. Also shown are the two limit positions. For each limit position, the crank and coupler are collinear and the speed of the slider is zero. The rotation of the crank required to move the slider from the right limit position D_1 to that on the left D_2 is designated as $\Delta\theta_2$.

Simplified outlines of the limit positions of the mechanism are shown in Figures 2.20(b) and 2.20(c). Comparing Figures 2.20(a), 2.20(b), and 2.20(c), we have

$$\Delta\theta_2 = 180° + \alpha_2 - \alpha_1 \tag{2.8-1}$$

where

$$\alpha_1 = \sin^{-1}\left(\frac{r_1}{r_2 + r_3}\right); \quad \alpha_2 = \sin^{-1}\left(\frac{r_1}{r_3 - r_2}\right) \tag{2.8-2}$$

Also

$$s_1 = [(r_2 + r_3)^2 - r_1^2]^{1/2}; \quad s_2 = [(r_3 - r_2)^2 - r_1^2]^{1/2} \tag{2.8-3}$$

and

$$\text{stroke} = s_1 - s_2 = [(r_2 + r_3)^2 - r_1^2]^{1/2} - [(r_3 - r_2)^2 - r_1^2]^{1/2} \tag{2.8-4}$$

If the crank turns at a constant rate in the counterclockwise direction, then the time taken for the slider to move to the left, from one limit position to the other, is

$$\Delta t_1 = \frac{\Delta\theta_2}{\dot{\theta}_2} \tag{2.8-5}$$

and the time taken for the slider to move to the right between the limit positions is

$$\Delta t_2 = \frac{2\pi - \Delta\theta_2}{\dot{\theta}_2} \tag{2.8-6}$$

The time ratio of the mechanism is

$$T_R = \frac{\Delta t_1}{\Delta t_2} = \frac{\Delta\theta_2}{2\pi - \Delta\theta_2} \tag{2.8-7}$$

Having established the stroke and the elapsed times required to move the slider between the limit positions, it is possible to determine the *average* velocity of the slider moving in each direction. The average velocity of the slider to the left is

$$\left(v_{4,avg}\right)_{left} = \frac{stroke}{\Delta t_1} = \frac{stroke}{\left(\dfrac{\Delta\theta_2}{\dot{\theta}_2}\right)} \tag{2.8-8}$$

and the average velocity of the slider to the right is

$$\left(v_{4,avg}\right)_{right} = \frac{stroke}{\Delta t_2} = \frac{stroke}{\left(\dfrac{2\pi - \Delta\theta_2}{\dot{\theta}_2}\right)} \tag{2.8-9}$$

For the case where the offset is zero, the time ratio is unity, and the average velocities of the slider to the left and to the right are equal.

In Chapters 3 and 4, methods are presented whereby *instantaneous* values of the speed of the slider may be determined.

2.8.2 Crank Rocker Four-Bar Mechanism

Figure 2.21(a) shows a crank rocker four-bar mechanism. The crank is rotating in the counterclockwise direction. Figures 2.21(b) and 2.21(c) show simplified outlines of the mechanism in its limit geometries. Similar to the analysis of a slider crank mechanism, the counterclockwise angular swing of the crank required to move the rocker from the first to the second limit position indicated is

$$\Delta\theta_2 = 180° + \alpha_2 - \alpha_1 \tag{2.8-10}$$

where expressions for α_1 and α_2 are determined using the cosine law (refer to Trigonometric Identities, page xxx). From Figure 2.21(b) we have

$$r_4^2 = r_1^2 + (r_2 + r_3)^2 - 2(r_2 + r_3)r_1\cos\alpha_1 \tag{2.8-11}$$

or

$$\alpha_1 = \cos^{-1}\left[\frac{r_1^2 + (r_2 + r_3)^2 - r_4^2}{2(r_2 + r_3)r_1}\right] \tag{2.8-12}$$

Similarly, the other angles shown in Figures 2.21(b) and 2.21(c) may be expressed in terms of the link lengths as

$$\alpha_2 = \cos^{-1}\left[\frac{r_1^2 + (r_3 - r_2)^2 - r_4^2}{2(r_3 - r_2)r_1}\right] \tag{2.8-13}$$

$$\beta_1 = \cos^{-1}\left[\frac{r_1^2 + (r_2 + r_3)^2 + r_4^2}{2r_1 r_4}\right] \tag{2.8-14}$$

$$\beta_2 = \cos^{-1}\left[\frac{r_1^2 - (r_3 - r_2)^2 + r_4^2}{2r_1 r_4}\right] \tag{2.8-15}$$

The amplitude of motion of the rocker may be expressed as

$$\Delta\theta_4 = \beta_1 - \beta_2 \tag{2.8-16}$$

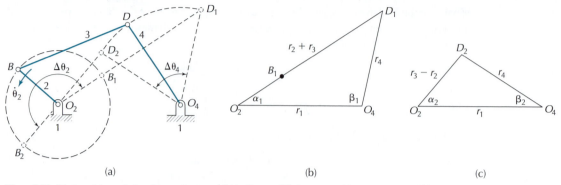

Figure 2.21 Limit positions of a four-bar mechanism: (a) Mechanism. (b) Geometry of first limit position. (c) Geometry of second limit position.

Equations (2.8-5)–(2.8-7), developed for a slider crank mechanism, may also be applied to the present case. Similar to the analysis of a slider crank mechanism, when the crank rotates in the counterclockwise direction the average rotational speeds of the rocker in the clockwise and counterclockwise directions are

$$(\dot\theta_{4,avg})_{CW} = \frac{\Delta\theta_4}{\left(\frac{2\pi - \Delta\theta_2}{\dot\theta_2}\right)} \; ; \quad (\dot\theta_{4,avg})_{CCW} = \frac{\Delta\theta_4}{\left(\frac{\Delta\theta_2}{\dot\theta_2}\right)} \tag{2.8-17}$$

2.8.3 Quick-Return Mechanism

Quick-return mechanisms are defined as those for which, when driven with a constant input speed, the time to complete the output motion in one direction differs from that in the opposite direction. Using this definition, there are a variety of mechanisms that can be classified as quick-return. For instance, a crank rocker four-bar mechanism and a slider crank mechanism can be designed as quick-return mechanisms. However, the name quick-return mechanism is often reserved for one incorporating an inverted slider crank mechanism, as illustrated in Figure 2.22. The mechanism is shown in one of its limit positions, and the geometry of the second limit position is illustrated with dashed lines. Both limit positions occur when link 2 is perpendicular to link 4. In the limit positions the velocity of link 6 is zero. We will assume that link 2 is driven at a constant rate in the counterclockwise direction. From Figure 2.22, points O_2, P_2, and O_4 form a right triangle. Therefore

$$\beta = \cos^{-1}\left(\frac{r_2}{r_1}\right) \tag{2.8-18}$$

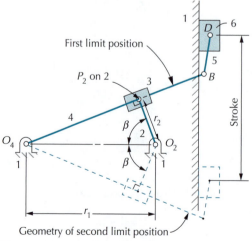

Figure 2.22 Limit positions of a quick-return mechanism.

where

$$r_1 = r_{O_2O_4}; \quad r_2 = r_{O_2P_2} \tag{2.8-19}$$

Due to symmetry of the geometry of the limit positions about the horizontal axis, the rotation of link 2 required to move link 6 downward from the first limit position to the second, is 2β. The time required to execute this motion is

$$\Delta t_1 = \frac{2\beta}{\dot\theta_2} = \frac{2\cos^{-1}\left(\dfrac{r_2}{r_1}\right)}{\dot\theta_2} \tag{2.8-20}$$

and the time required to move link 6 upward, from the second limit position to the first, is

$$\Delta t_2 = \frac{2\pi - 2\cos^{-1}\left(\dfrac{r_2}{r_1}\right)}{\dot\theta_2} \tag{2.8-21}$$

The time ratio of the motions of link 6 downward and upward is

$$T_R = \frac{\Delta t_1}{\Delta t_2} \tag{2.8-22}$$

The shaping machine illustrated in Figure 2.23 incorporates the same mechanism as shown in Figure 2.22. Link 6 is a cutting tool that reciprocates between two limit positions. Plots of the position and speed of link 6 reveal the two rotations of link 2 to move link 6 between its limit positions are not equal. Furthermore, when link 2 is driven at a constant rate, the mechanism provides a slower cutting motion, taking longer than the return motion.

Video 2.23
Shaping Machine

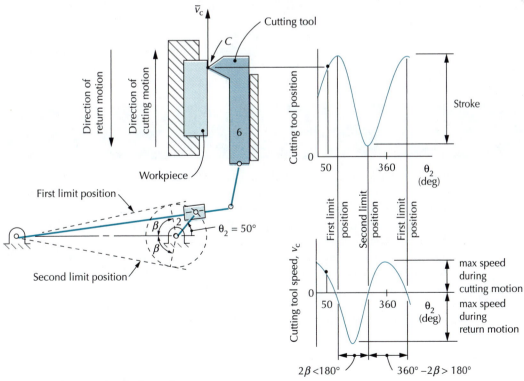

Figure 2.23 Shaping machine [Video 2.23].

EXAMPLE 2.3 AVERAGE VELOCITIES OF A SLIDER CRANK MECHANISM

For the slider crank mechanism shown in Figure 2.20(a) with dimensions and crank angular speed we have

$$r_1 = 2.0 \text{ cm}; \quad r_2 = 3.5 \text{ cm}; \quad r_3 = 10.0 \text{ cm}$$
$$\dot{\theta}_2 = 20.0 \text{ rad/sec CCW (constant)}$$

determine the average speed of the slider to the right, and to the left.

SOLUTION
From Equation (2.8-2) we obtain

$$\alpha_1 = \sin^{-1}\left(\frac{r_1}{r_2 + r_3}\right) = \sin^{-1}\left(\frac{2.0}{3.5 + 10.0}\right) = 8.52°$$
$$\alpha_2 = 17.92°$$

From Equation (2.8-1)

$$\Delta\theta_2 = 180° + \alpha_2 - \alpha_1 = 189.40° = 3.31 \text{ rad}$$

Employing Equation (2.8-4), the stroke of the slider is

$$\text{stroke} = [(r_2 + r_3)^2 - r_1^2]^{1/2} - [(r_3 - r_2) - r_1^2]^{1/2} = 7.17 \text{ cm}$$

Employing Equations (2.8-8) and (2.8-9) yields

$$(v_{4,\text{avg}})_{\text{left}} = \frac{\text{stroke}}{\left(\dfrac{\Delta\theta_2}{\dot{\theta}_2}\right)} = \frac{7.17}{\left(\dfrac{3.31}{20.0}\right)} = 43.35 \text{ cm/sec}$$

$$(v_{4,\text{avg}})_{\text{right}} = 48.14 \text{ cm/sec}$$

EXAMPLE 2.4 ANALYSIS OF LIMIT POSITIONS AND AVERAGE SPEEDS OF AN INVERTED SLIDER CRANK MECHANISM

For the inverted slider crank mechanism shown in Figure 2.24

(a) Determine the values of θ_2 when the mechanism is in its limit positions.
(b) Determine the average rotational speed of link 4 in the counterclockwise direction between its limit positions.
(c) Specify all values of θ_2 in the range $0 < \theta_2 < 360°$ for which there is no relative Coriolis acceleration between B_2 and the point on the slide, link 4, which instantaneously coincides with B_2.

$$r_{O_2 O_4} = 6.0 \text{ cm}; \quad r_{O_2 B_4} = 2.0 \text{ cm}; \quad \dot{\theta}_2 = 8.0 \text{ rad/sec CCW (constant)}$$

(Continued)

EXAMPLE 2.4 Continued

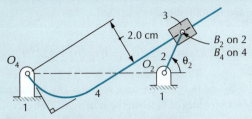

Figure 2.24 Inverted slider crank mechanism.

SOLUTION

(a) The limit positions of the mechanism occur when link 2 is perpendicular to the direction of the slide, link 4. This is because in these configurations point B_2 moves parallel to the direction of the slide and no motion is imparted to link 4. The geometries are shown in Figure 2.25. The angles of link 2 corresponding to the limit positions are

$$(\theta_2)_1 = 180° - \sin^{-1}\left(\frac{2.0}{3.0}\right) = 131.8°; \quad (\theta_2)_2 = 270°$$

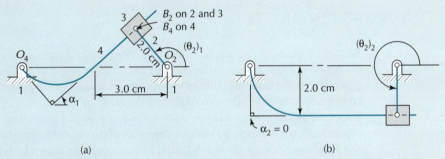

(a)   (b)

Figure 2.25 Limit positions of an inverted slider crank mechanism: (a) Geometry of the first limit position. (b) Geometry of second limit position.

(b) Angular motion of link 2 to move link 4 in the counterclockwise direction between the limit positions:

$$\Delta\theta_2 = (\theta_2)_1 + 360° - (\theta_2)_2 = 221.8°$$

Angular motion of link 4 between limit positions:

$$\Delta\theta_4 = \alpha_1 - \alpha_2 = \sin^{-1}\left(\frac{2.0}{3.0}\right) - 0 = 41.8°$$

Average rotational speed of link 4 in the counterclockwise direction:

$$(\dot{\theta}_{4,\text{avg}})_{\text{CCW}} = \frac{\Delta\theta_4}{(\Delta\theta_2/\dot{\theta}_2)} = \frac{41.8°}{\left(221.8°/8.0\dfrac{\text{rad}}{\text{sec}}\right)} = 1.51\frac{\text{rad}}{\text{sec}}$$

(c) The relative Coriolis acceleration is expressed as

$$a_{B_2B_4}^{C} = 2\dot{r}_{B_2B_4}\dot{\theta}_4$$

(Continued)

EXAMPLE 2.4 Continued

The relative Coriolis acceleration vanishes when either

$$\dot{\theta}_4 = 0 \ (\text{i.e., limit positions when } (\theta_2)_1 = 0 \text{ or } (\theta_2)_2 = 131.8°) \text{ or } \dot{r}_{B_2B_4} = 0$$

which is when the points B_2 and B_4 share the same velocity. This is illustrated in Figure 2.26. Therefore, the complete set of values of θ_2 when the relative Coriolis acceleration vanishes is 0, 131.8°, 180°, and 270°.

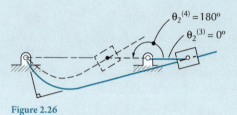

Figure 2.26

2.9 TRANSMISSION ANGLE

Section 1.7 provided the requirements to generate particular types of four-bar and slider crank mechanisms. The requirements can be used to determine ranges of the lengths of the links that generate a certain type. The limits of these ranges are theoretical. In practical designs, additional considerations must be addressed regarding the quality of transmission of motion between the links. One common consideration is the *transmission angle*, φ. For a four-bar mechanism (see Figure 2.27), φ is the angle between the centerline of the coupler and the centerline of the driven link. The transmission angle of a slider crank mechanism (see Figure 2.28) is measured between the centerline of the coupler and a line perpendicular to the line of action of the slider.

The ideal value of the transmission angle is 90°. In this instance, the line of action of the interactive force applied by the coupler on the driven link matches the line of action of motion of the kinematic pair between the coupler and the driven link. This ideal value cannot be maintained for

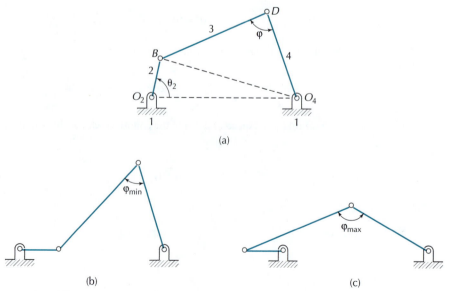

Figure 2.27 (a) Transmission angle of a four-bar mechanism. (b) $\theta_2 = 0°$. (c) $\theta_2 = 180°$.

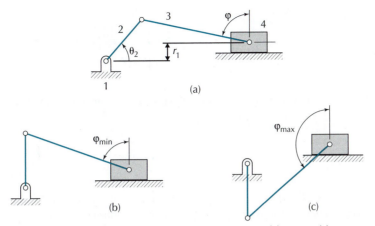

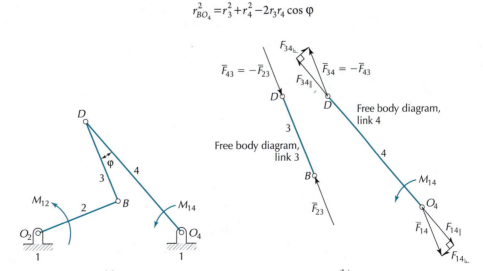

Figure 2.28 (a) Transmission angle of a slider crank mechanism: (b) $\theta_2 = 90°$. (c) $\theta_2 = 270°$.

all configurations of a mechanism. However, it is generally acceptable if transmission angles fall within the range

$$45° < \varphi < 135° \tag{2.9-1}$$

Values outside this range result in inefficient transmission of motion. Figure 2.29(a) shows the configuration of a four-bar mechanism having a small transmission angle, outside the acceptable range. Driving torque M_{12} is applied to link 2. As shown in Figure 2.29(b), the direction of force transmitted from link 3 on link 4 results in a small torque M_{14} about base pivot O_4, but a large bearing force at the same point. For transmission angles close to zero or 180°, there is a tendency for a mechanism to bind.

Using Figure 2.27(a), we may determine an expression for the transmission angle of a four-bar mechanism. Applying the cosine law for the triangle formed by points O_2O_4B, we obtain

$$r_{BO_4}^2 = r_1^2 + r_2^2 - 2r_1r_2 \cos\theta_2 \tag{2.9-2}$$

Using the cosine law again for the triangle formed by points DO_4B, we have

$$r_{BO_4}^2 = r_3^2 + r_4^2 - 2r_3r_4 \cos\varphi$$

Figure 2.29 Configuration of a four-bar mechanism with a small transmission angle.

or

$$\varphi = \cos^{-1}\left(\frac{r_3^2 + r_4^2 - r_{BO_4}^2}{2r_3 r_4}\right)$$

(2.9-3)

Combining Equations (2.9-2) and (2.9-3), we obtain the transmission angle as a function of the angular displacement of link 2:

$$\varphi = \cos^{-1}\left(\frac{r_3^2 + r_4^2 - r_1^2 - r_2^2 + 2r_1 r_2 \cos\theta_2}{2r_3 r_4}\right)$$

(2.9-4)

Figures 2.27(b) and 2.27(c) show configurations of a crank rocker four-bar mechanism where its transmission angle is extremum; these configurations occur when

$$\theta_2 = 0 \quad \text{and} \quad \theta_2 = 180°$$

(2.9-5)

Expressions for the minimum and maximum values of the transmission angle of a crank rocker four-bar mechanism are

$$\varphi_{min} = \cos^{-1}\left[\frac{r_3^2 + r_4^2 - (r_1 - r_2)^2}{2r_3 r_4}\right]$$

$$\varphi_{max} = \cos^{-1}\left[\frac{r_3^2 + r_4^2 - (r_1 - r_2)^2}{2r_3 r_4}\right]$$

(2.9-6)

EXAMPLE 2.5 TRANSMISSION ANGLES OF A FOUR-BAR MECHANISM

For a crank rocker four-bar mechanism with link lengths

$$r_1 = 8.0 \text{ cm}; \quad r_2 = 3.5 \text{ cm}; \quad r_3 = 10.0 \text{ cm}; \quad r_4 = 9.0 \text{ cm}$$

determine

(a) the transmission angle, φ, as a function of the angular displacement of link 2
(b) the minimum and maximum values of the transmission angle

SOLUTION
(a) For the typical angle

$$\theta_2 = 30°$$

we use Equation (2.9-4) to obtain

$$\varphi = \cos^{-1}\left(\frac{r_3^2 + r_4^2 - r_1^2 - r_2^2 + 2r_1 r_2 \cos\theta_2}{2r_3 r_4}\right)$$

$$= \cos^{-1}\left(\frac{10.0^2 + 9.0^2 - 8.0^2 - 3.5^2 + 2 \times 8.0 \times 3.5 \times \cos 30°}{2 \times 10.0 \times 9.0}\right) = 31.64°$$

(Continued)

EXAMPLE 2.5 Continued

Based on this and additional calculations for other values of θ_2, Figure 2.30 shows a plot of transmission angle as a function of the angular displacement of link 2.

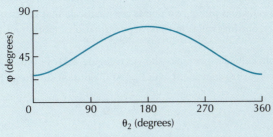

Figure 2.30 Transmission angle of a four-bar mechanism.

(b) Using Equation (2.9-6), the minimum and maximum values of the transmission angle throughout a cycle are

$$\varphi_{min} = 26.7°; \quad \varphi_{max} = 74.3°$$

The above results indicate that transmission angle is less than 45° for some positions of the mechanism, and therefore would be outside of the generally accepted range (Equation (2.9-1)) for portions of a cycle of motion.

Figure 2.28(a) illustrates a slider crank mechanism from which we obtain

$$r_3 \cos\varphi = r_2 \sin\theta_2 - r_1 \tag{2.9-7}$$

Solving for the transmission angle, we obtain

$$\varphi = \cos^{-1}\left(\frac{r_2 \sin\theta_2 - r_1}{r_3}\right) \tag{2.9-8}$$

The crank of the mechanism shown in Figure 2.28 can make a full rotation. Figures 2.28(b) and 2.28(c) illustrate the configurations of the mechanism when the transmission angle is extremum, and these configurations occur when

$$\theta_2 = 90° \quad \text{and} \quad \theta_2 = 270° \tag{2.9-9}$$

The corresponding expressions for the minimum and maximum values of the transmission angle are

$$\varphi_{min} = \cos^{-1}\left(\frac{r_2 - r_1}{r_3}\right)$$

$$\varphi_{max} = \cos^{-1}\left(\frac{-r_2 - r_1}{r_3}\right) \tag{2.9-10}$$

2.10 INSTANTANEOUS CENTER OF VELOCITY

Consider two links, designated as i and j, and at least one is undergoing planar motion. The *instantaneous center of velocity*, $P_{i,j}$, associated with these two links has the following properties:

- Two points, one on each of the two links, are coincident for at least an instant in time.
- Linear absolute velocities of the two points are identical. That is, about an instantaneous center of velocity, one link will at most have pure rotational motion relative to the other.

As an illustration, consider two links shown in Figure 2.31. Points A_2 and A_3 are on links 2 and 3, respectively. Both are coincident with $P_{2,3}$. At the instant shown we have

$$\bar{v}_{A_2} = \bar{v}_{A_3} = \bar{v}_{P_{2,3}} \qquad (2.10\text{-}1)$$

Instantaneous centers of velocity will hereafter be referred to simply as *instantaneous centers*. They may be used to determine the mechanical advantage of a mechanism that is presented later in this chapter and complete velocity analyses of planar mechanisms presented in Chapter 3.

Given the velocities of two points on a moving rigid body, it is possible to find the rigid body's instantaneous center with respect to the base link. Figure 2.32 is an illustration of link 2 undergoing planar motion. At the instant illustrated, the *directions* of velocities $\bar{v}_A$ and $\bar{v}_B$ on link 2 are defined. We now apply the condition of relative velocity between two points separated by a fixed distance presented in Section 2.5.1. From such an analysis, we can conclude that point A moves with respect to a stationary point located somewhere on the line that extends through point A and is perpendicular to $\bar{v}_A$. Likewise, point B moves about a stationary point on a line that extends through B, and it is perpendicular to $\bar{v}_B$. Therefore, the point of zero velocity on link 2, $P_{1,2}$, must be at the intersection of these two lines, where subscripts 1 and 2 indicate the base link and moving link, respectively.

Having located instantaneous center P_{12}, we can consider at that instant all points on link 2 are rotating about $P_{1,2}$. Using the analysis presented in Section 2.5.1, we have

$$|\bar{v}_A| = r_{P_{1,2}A}|\dot{\theta}_2|; \quad |\bar{v}_B| = r_{P_{1,2}B}|\dot{\theta}_2| \qquad (2.10\text{-}2)$$

or

$$|\dot{\theta}_2| = \frac{|\bar{v}_A|}{r_{P_{1,2}A}} = \frac{|\bar{v}_B|}{r_{P_{1,2}B}} \qquad (2.10\text{-}3)$$

Thus, if only $\bar{v}_A$ and the line of action of $\bar{v}_B$ are known, then a scale drawing may be used to provide $r_{P_{1,2}A}$ and $r_{P_{1,2}B}$, which in turn can be used to determine the magnitude of $\bar{v}_B$ and $\dot{\theta}_2$.

For the above case, the instantaneous center lies within the physical boundaries of link 2. This is not always the case. As illustrated in Figure 2.33(a), it is possible to find an instantaneous center that is outside the physical boundaries of a link. Here, we can consider an imaginary extension of the physical boundaries of the link as shown in Figure 2.33(b).

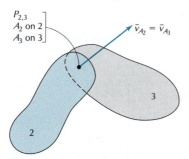

Figure 2.31 Illustration of instantaneous center.

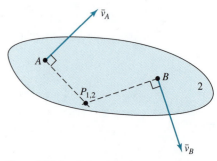

Figure 2.32 Locating an instantaneous center.

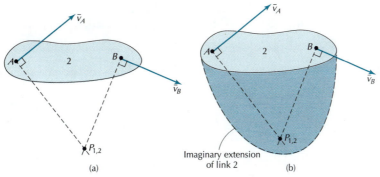

Figure 2.33

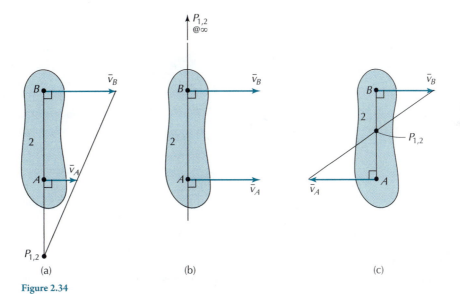

Figure 2.34

When velocities of two points on a link are parallel, as shown in Figure 2.34(a), the instantaneous center is found by applying Equation (2.10-3). However, in this case, both velocities must be known. If $\bar{v}_A$ and $\bar{v}_B$ are equal in magnitude and direction (Figure 2.34(b)), then the link is undergoing pure translation, and the instantaneous center is at an infinite distance in a direction perpendicular to the motion. If $\bar{v}_A$ and $\bar{v}_B$ are parallel and opposite in direction (Figure 2.34(c)), then the instantaneous center lies between A and B.

For two links connected through a kinematic pair, the location of the instantaneous center is summarized in Table 2.1.

It is possible that each link of a mechanism can have relative motion with respect to every other link. Consequently, the total number of instantaneous centers, p, in a mechanism having n links is

$$p = \frac{n(n-1)}{2} \qquad (2.10\text{-}4)$$

The factor ½ in Equation (2.10-4) takes into account the fact that instantaneous centers $P_{i,j}$ and $P_{j,i}$ are the same. In the four-bar mechanism illustrated in Figure 2.35(a), where $n = 4$, the number of instantaneous centers is

$$p = \frac{n(n-1)}{2} = \frac{4(4-1)}{2} = 6 \qquad (2.10\text{-}5)$$

TABLE 2.1 LOCATIONS OF INSTANTANEOUS CENTERS FOR VARIOUS TYPES OF KINEMATIC PAIRS

Type of Kinematic Pair	Location of Instantaneous Center
Turning pair	The instantaneous center coincides with the turning pair.
Sliding pair (curved slide)	The slider is in curvilinear motion with respect to the slide. The instantaneous center is at the center of curvature of the slide.
Sliding pair (straight slide)	The slider is in rectilinear motion with respect to the slide. A straight slide is equivalent to a curved slide of infinite radius. Thus, the instantaneous center is at an infinite distance in a direction perpendicular to the sliding motion.
Rolling pair	The instantaneous center is at the point of contact, this being the only point where the two links have the same velocity.
Two degree of freedom pair	Relative motion takes place along the *common tangent* between the two links at the point of contact. Consequently, the instantaneous center lies on the *common normal*. Its position on the common normal depends on the ratio of the sliding and angular velocities.

The complete set of instantaneous centers is: $P_{1,2}$, $P_{1,3}$, $P_{1,4}$, $P_{2,3}$, $P_{2,4}$, and $P_{3,4}$. Some of the instantaneous centers are found by examining the kinematic pairs and using Table 2.1. These are $P_{1,2}$, $P_{2,3}$, $P_{3,4}$, and $P_{1,4}$, which are shown in Figure 2.35(a).

The two other instantaneous centers are $P_{1,3}$ and $P_{2,4}$. Instantaneous center $P_{1,3}$ is found by an analysis similar to that illustrated in Figure 2.32. Lines are drawn perpendicular to the velocities at points B and D. The intersection of these two lines is the instantaneous center $P_{1,3}$ (Figure 2.35(b)).

The remaining instantaneous center is $P_{2,4}$. Its location may be found by inverting the mechanism and making link 2 the fixed link, as shown in Figure 2.35(c). Using the same reasoning as for the mechanism shown in Figure 2.35(b), the instantaneous center is at the intersection of the two lines that are perpendicular to the velocities at D and O_4. Since relative motions for the mechanisms shown in Figures 2.35(b) and 2.35(c) are the same (see Section 1.6), then all six instantaneous centers that have been located apply to both the original and the inverted mechanisms. The complete set of instantaneous centers is shown in Figure 2.35(d).

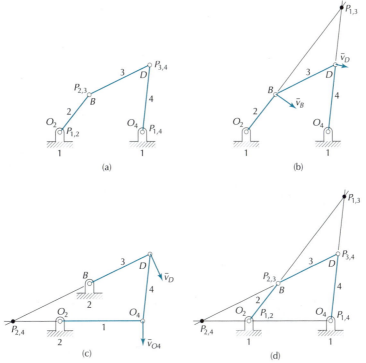

Figure 2.35 Instantaneous centers of a four-bar mechanism.

2.11 KENNEDY'S THEOREM

From the analysis of the four-bar mechanism shown in Figure 2.35, the three instantaneous centers associated with any three of the links are collinear. That is, each of the following sets of instantaneous centers lie on the same straight line:

- $P_{1,2}$, $P_{2,3}$, $P_{1,3}$

- $P_{1,4}$, $P_{3,4}$, $P_{1,3}$

- $P_{2,4}$, $P_{1,2}$, $P_{1,4}$

- $P_{2,4}$, $P_{2,3}$, $P_{3,4}$

This result conforms to *Kennedy's Theorem*, which may be applied to any planar mechanism. The theorem states that when considering three links, designated as *i*, *j*, and *k*, undergoing motion relative to one another, the three associated instantaneous centers, $P_{i,j}$, $P_{i,k}$, and $P_{j,k}$, lie on the same straight line [3].

In determining the locations of instantaneous centers of a mechanism, some can be readily located using the guidelines provided in Table 2.1. The remainder may then be found by successive applications of Kennedy's Theorem.

One may identify all instantaneous centers by drawing an *auxiliary polygon* that relates to the mechanism under study. Corresponding to the mechanism and instantaneous centers shown in Figure 2.36(a), an auxiliary polygon is illustrated in Figure 2.36(b). The number of sides and vertices in the polygon must equal the number of links in the mechanism. Each vertex of the polygon represents a link and is accordingly assigned a number.

In an auxiliary polygon, every side and diagonal is associated with two links. Each side and diagonal corresponds to an instantaneous center. Also, an auxiliary polygon illustrates the requirements of Kennedy's Theorem. Each triangle formed by sides and diagonals is associated with three links.

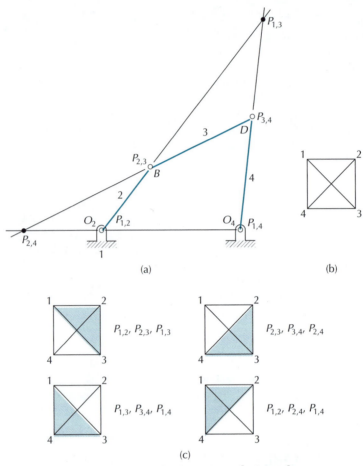

Figure 2.36 Instantaneous centers of a four-bar mechanism [Video 2.36].

The three corresponding instantaneous centers must lie on the same straight line. Figure 2.36(c) shows four auxiliary polygons, each with a shaded triangle. Beside each is listed the corresponding three instantaneous centers lying on a straight line.

[Video 2.36] provides an animated sequence of the determination of the instantaneous centers illustrated in Figure 2.36.

EXAMPLE 2.6 INSTANTANEOUS CENTERS OF A SLIDER CRANK MECHANISM

For the slider crank mechanism shown in Figure 2.37(a), determine all of the instantaneous centers for the position shown.

SOLUTION

The mechanism has four links ($n = 4$) and, using Equation (2.10-4), six instantaneous centers ($p = 6$). They are: $P_{1,2}$, $P_{1,3}$, $P_{1,4}$, $P_{2,3}$, $P_{2,4}$, and $P_{3,4}$.

The following instantaneous centers are located by examining the kinematic pairs (Table 2.1): $P_{1,2}$, $P_{2,3}$, $P_{3,4}$, and $P_{1,4}$.

(Continued)

EXAMPLE 2.6 Continued

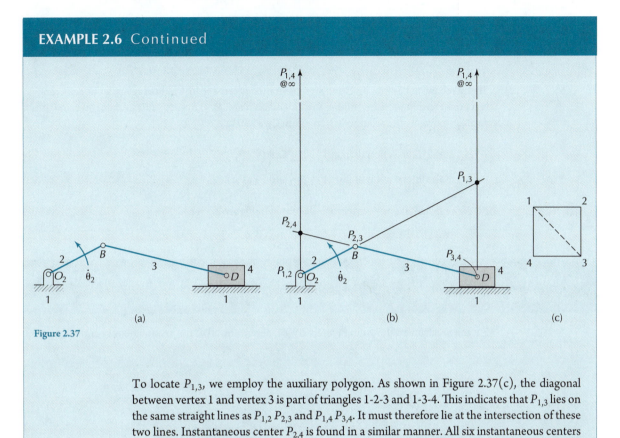

Figure 2.37

To locate $P_{1,3}$, we employ the auxiliary polygon. As shown in Figure 2.37(c), the diagonal between vertex 1 and vertex 3 is part of triangles 1-2-3 and 1-3-4. This indicates that $P_{1,3}$ lies on the same straight lines as $P_{1,2} P_{2,3}$ and $P_{1,4} P_{3,4}$. It must therefore lie at the intersection of these two lines. Instantaneous center $P_{2,4}$ is found in a similar manner. All six instantaneous centers are shown in Figure 2.37(b).

EXAMPLE 2.7 INSTANTANEOUS CENTERS OF AN INVERTED SLIDER CRANK MECHANISM

For the inverted slider crank mechanism shown in Figure 2.38(a), determine all of the instantaneous centers for the position shown.

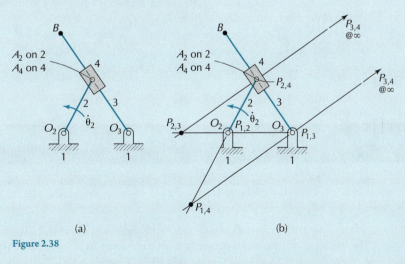

Figure 2.38

(Continued)

EXAMPLE 2.7 Continued

SOLUTION

This mechanism has four links and six instantaneous centers. The following instantaneous centers are determined by examining the kinematic pairs: $P_{1,2}$, $P_{1,3}$, $P_{2,4}$, $P_{3,4}$. Then, by using an auxiliary polygon (i.e., Kennedy's Theorem), we locate

- $P_{1,4}$ at the intersection of lines $P_{1,2} P_{2,4}$ and $P_{1,3} P_{3,4}$

- $P_{2,3}$ at the intersection of lines $P_{1,2} P_{1,3}$ and $P_{2,4} P_{3,4}$

All of the instantaneous centers are illustrated in Figure 2.38(b).

EXAMPLE 2.8 INSTANTANEOUS CENTERS OF A MECHANISM INCORPORATING A TWO DEGREE OF FREEDOM PAIR

For the mechanism shown in Figure 2.39(a), determine all of the instantaneous centers.

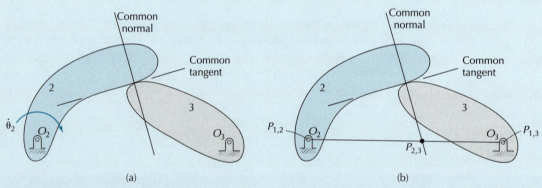

(a)   (b)

Figure 2.39

SOLUTION

This mechanism has three links. Using Equation (2.10-4), there are only three instantaneous centers. Instantaneous centers $P_{1,2}$, $P_{1,3}$ coincide with the turning pairs. Based on Kennedy's Theorem, the third instantaneous center, $P_{2,3}$, lies on the same straight line as $P_{1,2} P_{1,3}$. In addition, from Table 2.1, this instantaneous center lies on the common normal at the point of contact between links 2 and 3. Thus, instantaneous center $P_{2,3}$ is located as shown in Figure 2.39(b).

EXAMPLE 2.9 INSTANTANEOUS CENTERS OF A SIX-LINK MECHANISM

For the mechanism shown in Figure 2.40(a), determine all of the instantaneous centers.

SOLUTION

This mechanism has six links and, by employing Equation (2.10-4), 15 instantaneous centers. The following instantaneous centers are determined by examining the kinematic pairs: $P_{1,2}$, $P_{1,4}$, $P_{1,6}$, $P_{2,3}$, $P_{3,4}$, $P_{4,5}$, $P_{5,6}$. These are indicated in Figure 2.40(b). The corresponding auxiliary

(Continued)

EXAMPLE 2.9 Continued

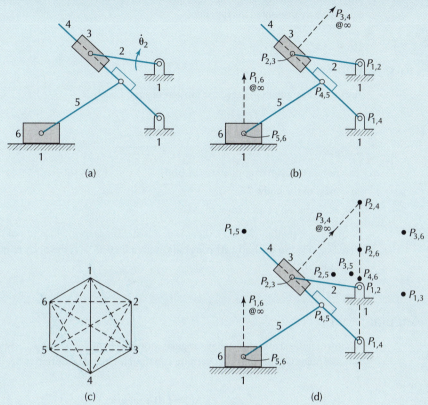

Figure 2.40

polygon is shown in Figure 2.40(c). Instantaneous centers that have been determined by examining the kinematic pairs are depicted in the auxiliary polygon as solid lines, while the remaining instantaneous centers are shown as dashed lines.

The location of all of the remaining instantaneous centers are determined by recognizing

- $P_{2,4}$ at the intersection of lines $P_{1,2} P_{1,4}$ and $P_{2,3} P_{3,4}$ (see Figure 2.40(d))

- $P_{4,6}$ at the intersection of lines $P_{1,4} P_{1,6}$ and $P_{4,5} P_{5,6}$

- $P_{1,3}$ at the intersection of lines $P_{1,2} P_{2,3}$ and $P_{1,4} P_{3,4}$

- $P_{1,5}$ at the intersection of lines $P_{1,4} P_{4,5}$ and $P_{1,6} P_{5,6}$

- $P_{3,5}$ at the intersection of lines $P_{1,3} P_{1,5}$ and $P_{3,4} P_{4,5}$

- $P_{2,5}$ at the intersection of lines $P_{1,2} P_{1,5}$ and $P_{2,4} P_{4,5}$

- $P_{2,6}$ at the intersection of lines $P_{1,2} P_{1,6}$ and $P_{2,5} P_{5,6}$

- $P_{3,6}$ at the intersection of lines $P_{2,3} P_{2,6}$ and $P_{3,5} P_{5,6}$

Figure 2.40(d) shows all of the instantaneous centers.

2.12 KINETICS

This section presents the governing equations of the kinetics of a rigid body. Derivations of these equations are presented in Appendix C.

2.12.1 Linear Motion

The governing equation for linear motion of a rigid body is

$$\overline{F} = m\overline{a}_G \tag{2.12-1}$$

where

$\overline{F}$ = net external force applied to the body

m = mass of the body

$\overline{a}_G$ = acceleration of the center of mass

In using Equation (2.12-1), there is no restriction as to the location of the point at which the force is applied to the body. Also, as indicated in Figure 2.41, the body may be subjected to additional moments.

2.12.2 Angular Motion

A body of finite size may also have angular motions. Figure 2.42 shows a rigid body with its center of mass, point G. The related governing equation for angular motion is

$$M_O = I_G\ddot{\theta} + m(x_G\ddot{y}_G - \ddot{x}_G y_G) \tag{2.12-2}$$

where

M_O = net external moment applied to the body measured about point O

$\ddot{\theta}$ = angular acceleration of the body

$x_G y_G$ = coordinates of the center of mass with respect to a reference coordinate system

$\ddot{x}_G \ddot{y}_G$ = components of acceleration of the center of mass

I_G = *polar mass moment of inertia* of the body with respect to its center of mass (see Appendix C, Section C.9)

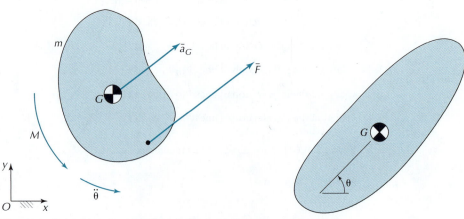

Figure 2.41 Body subjected to externally applied load and moment.   **Figure 2.42**

In this book, both the net moment applied to the body and the angular acceleration of the body are considered positive in the counterclockwise direction.

If the origin of the coordinate system is placed at the center of mass, then

$$x_G = 0 = y_G \tag{2.12-3}$$

and Equation (2.12-2) becomes

$$M_G = M_O = I_G \ddot{\theta} \tag{2.12-4}$$

Commonly employed sets of units associated with Equations (2.12-1) and (2.12-2) are presented in Section 2.13.

2.12.3 Fixed-Axis Rotation of a Rigid Body

The rigid body shown in Figure 2.43 is allowed to move about a fixed axis at point O. For this type of motion, it is convenient to place the origin of the coordinate system at point O. In this case

$$x_G = r\cos\theta; \quad y_G = r\sin\theta \tag{2.12-5}$$

where r is the radial distance from point O to the center of mass G.

Substituting Equations (2.12-5) in Equation (2.12-2) yields

$$\begin{aligned} M_O &= I_G\ddot{\theta} + m(x_G\ddot{y}_G - \ddot{x}_G y_G) \\ &= I_G\ddot{\theta} + mr\cos\theta(-r\dot{\theta}^2\sin\theta + r\ddot{\theta}\cos\theta) \\ &\quad - mr\sin\theta(-r\dot{\theta}^2\cos\theta + r\ddot{\theta}\sin\theta) \\ &= I_O\ddot{\theta} \end{aligned} \tag{2.12-6}$$

where, after simplification,

$$I_O = I_G + mr^2 \tag{2.12-7}$$

is the polar mass moment of inertia about point O. Equation (2.12-7) is referred to as the *parallel-axis theorem*.

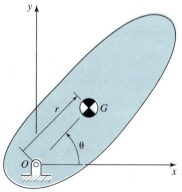

Figure 2.43

2.13 SYSTEMS OF UNITS

There are multiple systems of units employed in engineering. They are generally classified into two groups: *System International* (*SI* or *metric*) and *U.S. Customary*. Consistent sets of units for the governing equations of motion of a rigid body are presented in this section.

To accommodate these different sets, we make the following substitutions for the mass and polar mass moment of inertia:

$$m = \frac{m^*}{g_c} \; ; \quad I_G = \frac{I_G^*}{g_c} \tag{2.13-1}$$

where quantities m^* and I_G^*, like m and I_G, are referred to as the mass and polar mass moment of inertia, respectively, and g_c is a *gravitational constant*. The value and units of g_c depend on the units of the other components in the governing equations. Substituting Equations (2.13-1) into Equations (2.12-1) and (2.12-2) yields

$$\bar{F} = \frac{m^*}{g_c} \bar{a}_G \tag{2.13-2}$$

and

$$M_O = \frac{I_G^*}{g_c} \ddot{\theta} + \frac{m^*}{g_c}(x_G \ddot{y}_G - \ddot{x}_G y_G) \tag{2.13-3}$$

Tables 2.2 and 2.3 provide units for each of the components of Equations (2.13-2) and (2.13-3). For each table, a consistent set of units is obtained by employing one column in a table.

For the cases in Tables 2.2 and 2.3 for which $g_c = 1$, then from Equations (2.13-1)

$$m = m^* \; ; \quad I_G = I_G^* \tag{2.13-4}$$

Employing Equation (2.13-2) and Table 2.2, forces for the SI system may be defined as

$$\boxed{1\,\text{Newton} = 1\,\text{N} = 1\,\text{kg} \times 1\frac{\text{m}}{\text{sec}^2}}$$

TABLE 2.2 CONSISTENT SETS OF UNITS EMPLOYED FROM THE SI SYSTEM FOR EQUATIONS (2.13-2) AND (2.13-3)

	1	2
$\bar{F}$	dyne	N
M_O	dyne-cm	N-m
m^*	gr	kg
I_G^*	gr-cm^2	kg-m^2
x, y	cm	m
$\bar{a}_G, \ddot{x}, \ddot{y}$	cm/sec^2	m/sec^2
$\ddot{\theta}$	rad/sec^2	rad/sec^2
g_c	1	1

TABLE 2.3 CONSISTENT SETS OF UNITS EMPLOYED FROM THE US CUSTOMARY SYSTEM FOR EQUATIONS (2.13-2) AND (2.13-3)

	1	2	3	4
$\overline{F}$	lb	lb	lb	lb
M_O	ft-lb	ft-lb	in-lb	in-lb
m^*	lbm	1b-sec²/ft (slug)	lbm	1b-sec²/ft
I_G^*	lbm-ft²	lb-ft-sec²	lbm-in²	lb-in-sec²
x, y	ft	ft	in	in
$\overline{a}_G, \ddot{x}, \ddot{y}$	ft/sec²	ft/sec²	in/sec²	in/sec²
$\ddot{\theta}$	rad/sec²	rad/sec²	rad/sec²	rad/sec²
g_c	32.2 1bm-ft/1b-sec²	1	386 1bm-in/1b-sec²	1

and

$$1 \text{ dyne} = 1 \text{gr} \times 1\frac{\text{cm}}{\text{sec}^2}$$

A simple illustration on the use of Tables 2.2 and 2.3 is provided below.

EXAMPLE 2.10 BODY SUBJECTED TO AN EXTERNAL FORCE

For the body shown in Figure 2.44, determine the magnitude of acceleration of the center of mass if

$$|\overline{F}| = 12 \text{ lb}; \quad m^* = 21 \text{ bm}$$

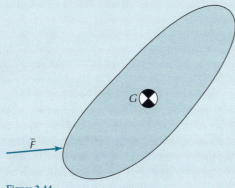

Figure 2.44

(*Continued*)

EXAMPLE 2.10 Continued

SOLUTION
Referring to the units of the given quantities, we employ a set of units listed in Table 2.3. We may use either the column numbered 1 or 3. If we use column 3, then

$$g_c = 386\frac{\text{lbm-in}}{\text{lb-sec}^2}$$

and from Equation (2.13-2)

$$|\overline{a}_G| = \frac{|\overline{F}|}{\left(\dfrac{m^*}{g_c}\right)} = \frac{12\text{ lb}}{\left[\dfrac{2\text{ lbm}}{\left(386\dfrac{\text{lbm-in}}{\text{lb-sec}^2}\right)}\right]} = 2320\frac{\text{in}}{\text{sec}^2}$$

We generally consider the *weight* of a body as the gravitational force exerted on it by the earth at its surface. Weight, being a force, is a vector quantity. The direction of this vector is toward the center of the earth.

When a body is allowed to fall freely under the influence of gravity, the magnitude of acceleration is

$$g = |\overline{a}_G| = 9.81\frac{\text{m}}{\text{sec}^2} = 32.2\frac{\text{ft}}{\text{sec}^2} = 386\frac{\text{in}}{\text{sec}^2}$$

The magnitude of this acceleration is independent of the mass of the body. Using Equations (2.13-1) and (2.13-2), the magnitude of the weight, W, of a body of mass m is

$$W = mg \tag{2.13-5}$$

A mass of $m^* = 1$ lbm weighs

$$W = mg = \frac{m^*}{g_c}g = \frac{1\text{ lbm}}{\left(32.2\dfrac{\text{lbm-ft}}{\text{lb-sec}^2}\right)}32.2\frac{\text{ft}}{\text{sec}^2} = 1\text{ lb}$$

and a mass of $m^* = 1$ kg weighs

$$W = \frac{m^*}{g_c}g = \frac{1\text{ kg}}{1} \times 9.81\frac{\text{m}}{\text{sec}^2} = 9.81\frac{\text{kg-m}}{\text{sec}^2} = 9.81\text{ N}$$

To keep an object from free-falling, an upward force must be applied of the same magnitude as the weight so that there is zero net force on the object.

In some instances, gravity may be neglected in an analysis, such as when the prescribed accelerated motions of a body require forces that are far greater than those caused by gravity. Alternatively, gravity may be neglected if it acts perpendicular to the plane of motion of the body.

EXAMPLE 2.11 PHYSICAL PENDULUM

The rigid body illustrated in Figure 2.45 is pivoted about point O at a distance l from the center of mass G. Gravity acts downward. The equilibrium position corresponds to points O and G in a vertical alignment. If the body is pivoted a small angle θ from the equilibrium position and released, show that the motion is harmonic and determine the period of one oscillation.

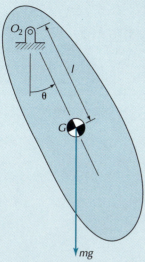

Figure 2.45 Physical pendulum.

SOLUTION
Employing Equation (2.12-6), we obtain

$$M_O = -mgl\sin\theta = I_O\ddot{\theta} \tag{2.13-6}$$

Since for small angles we have

$$\sin\theta \approx \theta \tag{2.13-7}$$

then Equation (2.13-6) becomes

$$\ddot{\theta} + \omega_n^2\theta = 0 \tag{2.13-8}$$

where the *natural frequency* of the system is defined as

$$\omega_n = \left(\frac{mgl}{I_O}\right)^{1/2} \tag{2.13-9}$$

The solution of Equation (2.13-8) is [4]

$$\theta = \Theta\cos\left(\omega_n t + \varphi\right) \tag{2.13-10}$$

where Θ and φ are arbitrary constants. Angular motion is thus harmonic in time.

Using Equations (2.13-9) and (2.13-10) the period of time, τ, for one oscillation is found by recognizing that

(Continued)

EXAMPLE 2.11 Continued

$$\omega_n \tau = \left(\frac{mgl}{I_O}\right)^{1/2} \tau = 2\pi \qquad (2.13\text{-}11)$$

from which

$$\tau = 2\pi \left(\frac{I_O}{mgl}\right)^{1/2} \qquad (2.13\text{-}12)$$

2.14 EQUATIONS OF EQUILIBRIUM

2.14.1 Static Conditions

For static conditions, where there is no motion,

$$\overline{a}_G = \overline{0}; \quad \ddot{\theta} = 0 \qquad (2.14\text{-}1)$$

Then Equations (2.12-1) and (2.12-4) reduce to

$$\boxed{\overline{F} = \overline{0}} \qquad (2.14\text{-}2)$$

$$\boxed{M_O = 0 = M_G} \qquad (2.14\text{-}3)$$

Equations (2.14-2) and (2.14-3) are the governing equations of static equilibrium of a body.

EXAMPLE 2.12 STATIC EQUILIBRIUM OF A PIVOTING LINK

The brake lever shown in Figure 2.46(a) consists of the link pivoted to the base link at point O_2. A force of 100 N is applied by the driver's foot, causing tension in the cable. Assuming conditions of static equilibrium, determine

(a) the magnitude of the tension in the cable;
(b) the reaction at O_2.

SOLUTION
(a) A free-body diagram of the link is shown in Figure 2.46(b). Summing moments about O_2 yields

$$\sum M_{O_2} = -100 \times 0.20 + |\overline{T}| \times 0.15 = 0$$

(Continued)

EXAMPLE 2.12 Continued

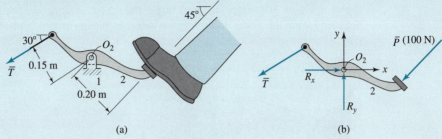

Figure 2.46

from which

$$|\overline{T}| = 133\text{N}$$

(b) Referring to the free-body diagram shown in Figure 2.46(b), we require that

$$\sum F_x = 0 = P_x + T_x + R_x = -100\cos 45° - 133\cos 30° + R_x,$$
$$R_x = 186\text{ N}$$

Also

$$\sum F_y = 0 = P_y + T_y + R_y = -100\sin 45° - 133\sin 30° + R_y,$$
$$R_y = 137\text{ N}$$

and therefore

$$\overline{R} = R_x\overline{i} + R_y\overline{j} = 186\overline{i} + 137\overline{j}\text{ N}$$

EXAMPLE 2.13 FORCE ANALYSIS OF A DOOR LATCH MECHANISM

Video 2.47
Door Latch
Mechanism

Figure 2.47(a) shows a latch mechanism mounted on a door. The latch is used to hold the door in its closed position but allow the door to open provided enough pulling force is applied to the handle. Figure 2.47(b) shows a closeup of the latch. A schematic side view of the system is illustrated in Figure 2.48. In this view, the door is restricted to move horizontally. Figure 2.48(a) shows the latch while the door is open. The spring is stretched from its free length and is under tension. The pivot arm is in contact with the stop. Figure 2.48(b) shows the door during the closing process. The striker plate on the door is in contact with the cylindrical portion of the pivot arm and the external force has moved the pivot arm off of the stop. The tension in the spring gives a clockwise moment to the pivot arm about the base pivot O_2 and counteracts the counterclockwise moment of the externally applied force. Continuing to push the door with the external force brings the door into the position illustrated in Figure 2.48(c). Now the spring does not provide a moment about the base pivot. Pushing the door a little further brings the system into the configuration shown in Figure 2.48(d). The tension in the spring gives a counterclockwise moment to the pivot arm about the base pivot and it is no longer necessary to apply an external force to close the door. The mechanism will continue to move under the action of the spring force until the door goes to the closed configuration

(Continued)

EXAMPLE 2.13 Continued

shown in Figure 2.48(e). Since the spring is still under tension, the spring will affirmatively hold the door in the closed position unless sufficient pulling force is applied on the handle.

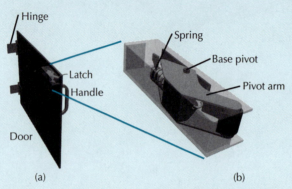

Figure 2.47 (a) Door and latch. (b) Close up of latch [Video 2.47].

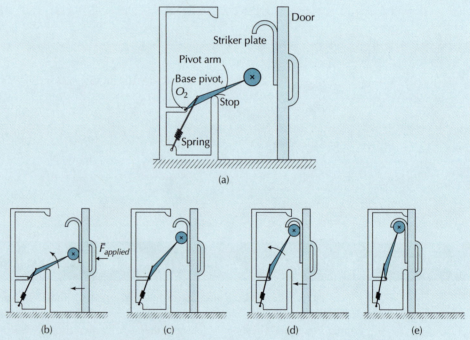

Figure 2.48 Positions of door latch.

For the parameters provided below and illustrated in Figure 2.49, determine the applied force required as a function of the horizontal position of point C to move the pivot arm, link 2, from the open position to the closed position. Neglect friction between the striker plate and the cylindrical portion of link 2. Neglect the effects of inertia and gravity.

$$r_{BO_2} = 2.0 \text{ cm}; \quad r_{CO_2} = 8.0 \text{ cm}; \quad b = 1.4 \text{ cm}; \quad c = 4.0 \text{ cm}$$

$$\textit{spring freelength} = 4.0 \text{ cm}; \quad k = 50.0 \frac{\text{N}}{\text{cm}}; \quad \alpha = 20°; \quad 25° \leq \theta \leq 75°$$

(Continued)

EXAMPLE 2.13 Continued

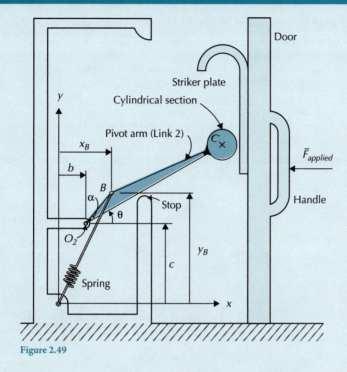

Figure 2.49

SOLUTION

Figure 2.50 shows a free-body diagram of link 2. The coordinates of point B are

$$x_B = b + r_{BO_2}\cos(\theta + \alpha); \quad y_B = c + r_{BO_2}\sin(\theta + \alpha) \tag{2.14-4}$$

The current length of the spring is then

$$spring\ length = \left(x_B^2 + y_B^2\right)^{1/2} \tag{2.14-5}$$

The force in the spring for the current amount of stretching is

$$F_{spring} = k\,(spring\ length - spring\ free\ length) \tag{2.14-6}$$

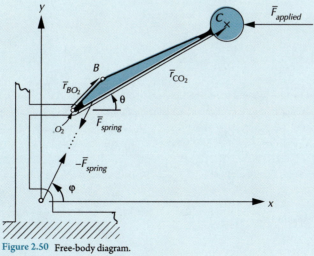

Figure 2.50 Free-body diagram.

EXAMPLE 2.13 Continued

where k is the spring constant. The horizontal and vertical components of the spring force applied to point B are

$$F_{x,spring} = F_{spring}\cos\varphi \ ; \quad F_{y,spring} = F_{spring}\sin\varphi \tag{2.14-7}$$

where

$$\varphi = \tan 2^{-1}(x_B, y_B) \tag{2.14-8}$$

The moment of the spring force about the base pivot is (see Appendix C, Section C.8)

$$\left(\overline{M}_{O_2}\right)_{spring} = \overline{r}_{BO_2} \times \overline{F}_{spring}$$

$$= \begin{vmatrix} \overline{i} & \overline{j} & \overline{k} \\ r_{BO_2}\cos(\theta+\alpha) & r_{BO_2}\sin(\theta+\alpha) & 0 \\ F_{x,spring} & F_{y,spring} & 0 \end{vmatrix}$$

$$= r_{BO_2}\left[\cos(\theta+\alpha)F_{y,spring} - \sin(\theta+\alpha)F_{x,spring}\right]\overline{k} \tag{2.14-9}$$

The moment caused by the applied force is

$$\left(-M_{O_2}\right)_{applied} = \overline{r}_{CO_2} \times \overline{F}_{applied} = r_{CO_2}\sin\varphi\, F_{applied}\,\overline{k} \tag{2.14-10}$$

For each position we consider the latch in static equilibrium and thus we require

$$\Sigma\left(\overline{M}_{O_2}\right) = \left(\overline{M}_{O_2}\right)_{spring} + \left(\overline{M}_{O_2}\right)_{applied} = \overline{0} \tag{2.14-11}$$

Since all of the moments are in the direction of $\overline{k}$, we can consider Equation (2.14-11) as a scalar equation. Substituting Equations (2.14-9) and (2.14-10) in Equation (2.14-11) and solving for the magnitude of the applied force, we obtain

$$F_{applied} = \frac{r_{BO_2}\left[\cos(\theta+\alpha)F_{y,spring} - \sin(\theta+\alpha)F_{x,spring}\right]}{r_{CO_2}\sin\varphi} \tag{2.14-12}$$

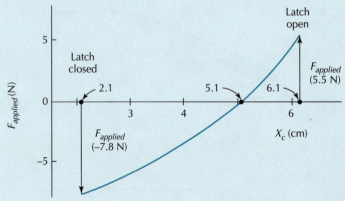

Figure 2.51 Applied force.

(Continued)

EXAMPLE 2.13 Continued

For the given values of the parameters, Figure 2.51 is a plot of the magnitude of the applied force as a function of the horizontal position of point C. A force magnitude of 5.5 N is required to move the pivot arm off of the stop. For values of x_C less than 5.1 cm, the magnitude of the applied force is negative. For these positions, no applied force is needed and the door would close on its own. Negative values correspond to the magnitude of the force that holds the door shut and the magnitude of the force required to pull the door open. A force magnitude of 7.8 N is needed to start moving the door from the closed position.

EXAMPLE 2.14 FORCE ANALYSIS OF A LEVER

Replace the horizontal 300 N force acting on the lever shown in Figure 2.52(a) by an equivalent system consisting of a force at O and a couple.

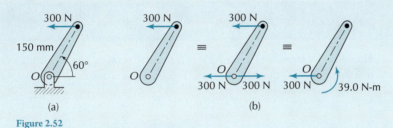

(a)   (b)

Figure 2.52

SOLUTION

We apply two equal and opposite 300 N forces at O and identify the counterclockwise couple.

$$M = Fd = 300\,\text{N} \times 0.15\,\text{m} \times \sin 60° = 39.0\,\text{N-m}$$

Thus, the original force is equivalent to the 300 N force at O and the 39.0 N-m couple as shown in the third of the three equivalent figures shown in Figure 2.52(b).

2.14.2 Dynamic Conditions

Equations (2.12-1) and (2.12-2) can alternatively be expressed as

$$\boxed{\overline{F} + \overline{F}_{IN} = \overline{0}} \tag{2.14-13}$$

and

$$\boxed{M_O + M_{IN} = 0} \tag{2.14-14}$$

where, using Equations (2.13-1), the *inertia force* is defined as

$$\overline{F}_{IN} = -m\overline{a}_G = -\frac{m^*}{g_c}\overline{a}_G \tag{2.14-15}$$

and the *inertia moment* or *torque* is

$$M_{IN} = -I_G \ddot{\theta} - m(x_G \ddot{y}_G - \ddot{x}_G y_G)$$

$$= -\frac{I_G^*}{g_c} \ddot{\theta} - \frac{m^*}{g_c}(x_G \ddot{y}_G - \ddot{x}_G y_G) \tag{2.14-16}$$

Equation (2.14-16) may be simplified if the origin of the coordinate system is placed at the center of mass, and therefore

$$x_G = 0 = y_G \tag{2.14-17}$$

Then

$$M_{IN} = -I_G \ddot{\theta} = -\frac{I_G^*}{g_c} \ddot{\theta} \tag{2.14-18}$$

Equations (2.14-13) and (2.14-14) are referred to as *D'Alembert's Principle*. They resemble those employed for static problems.

EXAMPLE 2.15 DYNAMIC FORCE ANALYSIS OF A DISC ON AN INCLINE

The solid disc shown in Figure 2.53(a) is released from rest on an inclined surface. Assuming that the disc rolls without slipping, and that gravity acts downward, use D'Alembert's Principle to determine its angular acceleration.

$$m^* = 2.0 \text{ kg}; \quad R = 0.3 \text{ m}; \quad \beta = 20°$$

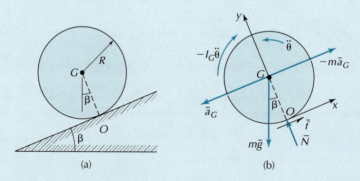

(a)   (b)

Figure 2.53

SOLUTION

Employing Table 2.2, it is recognized from the units of quantities provided for this problem that

$$g_c = 1; \quad m = m^*; \quad I_G = I_G^*$$

The polar mass moment of inertia of the disc (Appendix C, Section C.9) about its center of mass is

$$I_G^* = \tfrac{1}{2} m^* R^2 = \tfrac{1}{2} \times 2.0 \text{ kg} \times 0.3^2 \text{ m}^2 = 0.090 \text{ kg-m}^2$$

(Continued)

EXAMPLE 2.15 Continued

A free-body diagram of the cylinder shown in Figure 2.53(b) includes a frictional force, a normal force between the cylinder and incline, an inertia force, and an inertia moment. A Cartesian coordinate system is placed with its origin at the point of contact between the cylinder and the inclined surface. This placement eliminates the need to account for normal and frictional forces when summing moments. Also

$$x_G = 0; \ y_G = R$$

Also, by inspection for this problem

$$|\bar{a}_G| = |\ddot{x}_G| = R\ddot{\theta}$$

and

$$M_O = mgR\sin\beta \tag{2.14-19}$$

From Equation (2.14-16) we have

$$M_{IN} = -I_G\ddot{\theta} - m(x_G\ddot{y}_G - \ddot{x}_G y_G)$$

$$= -I_G\ddot{\theta} - m(0 - R\ddot{\theta}R) = -(I_G + mR^2)\ddot{\theta} \tag{2.14-20}$$

Combining Equations (2.14-14), (2.14-19), and (2.14-20) yields

$$mgR\sin\beta - (I_G + mR^2)\ddot{\theta} = 0$$

from which

$$\ddot{\theta} = \frac{mgR\sin\beta}{I_G + mR^2} = \frac{2.0\,\text{kg} \times 9.81\,\text{m/sec}^2 \times 0.3\,\text{m} \times \sin 20°}{0.090\,\text{kg-m}^2 + 2.0\,\text{kg} \times 0.3^2\,\text{m}^2} = 7.46\,\frac{\text{rad}}{\text{sec}^2}\,\text{CCW}$$

2.15 MECHANICAL ADVANTAGE

From time to time it is required to take an available force or torque and either increase or decrease its magnitude for application to an adjoining mechanical system. For example, Figure 2.54(a) shows a lever with an applied hand force at one end and in contact with another mechanical system at its other end. Using Figure 2.54(b), we sum moments about the base pivot to maintain static equilibrium and obtain

$$F_{out}b - F_{in}a = 0 \tag{2.15-1}$$

and therefore

$$F_{out} = \frac{a}{b}F_{in} \tag{2.15-2}$$

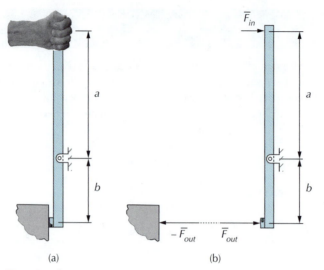

(a)   (b)

Figure 2.54 Lever.

The *mechanical advantage* is defined as the ratio of the magnitude of the output force exerted by a mechanism to the magnitude of the input force applied to it. For the lever, the mechanical advantage is

$$M.A. = \frac{F_{out}}{F_{in}} = \frac{a}{b} \tag{2.15-3}$$

The determination of mechanical advantage involves the analysis of one configuration of a mechanism that is considered to be in static equilibrium. The mechanical advantage was easily determined for the example involving the lever. For more complicated mechanisms, a straightforward means to determine the mechanical advantage may be completed using instantaneous centers (Section 2.10) [5]. In the analysis presented below, we restrict ourselves to mechanisms having a single input, that is a mobility of one. Also, the mechanisms will have a force applied to only one output. We will consider mechanisms in which there is no possibility of losses due to friction and will neglect the effects of gravity and inertia. It can then be assumed that *if* the mechanism were to move from the configuration being analyzed, then the *rate of work* or *power* being supplied from the input force would equal the rate of work being transmitted through the output force.

The lever illustrated in Figure 2.54 has a total of two links, including the base link, and based on Equation (2.10-4), it also includes one instantaneous center that is located at the base pivot. For the procedure presented below, we will consider mechanisms having a minimum of four links and six instantaneous centers. A total of four groups of mechanisms will be considered where the input and output links are either rotating or translating, and an expression of the mechanical advantage will be obtained for each. Expressions for the mechanical advantage will be determined by considering the instantaneous centers of the of the following four links of a mechanism: the base link, 1, the input link, *in*, the output link, *out*, and another link, *k*, that is arbitrary for the mechanism except that it must be distinct from the three other links.

Rotating–Rotating Mechanisms

For this group of mechanisms, both the input and output variables are rotational. A four-bar mechanism is a typical example. Figure 2.55(a) shows the geometric relationship of the six

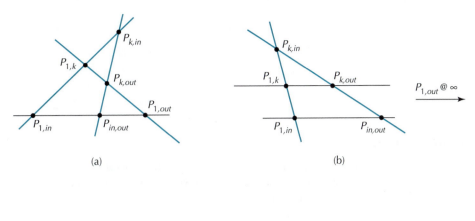

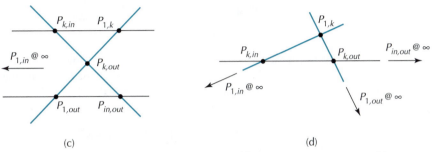

Figure 2.55 Instantaneous centers for four links of a mechanism. (a) Rotating–rotating mechanisms. (b) Rotating–translating mechanisms. (c) Translating–rotating mechanisms. (d) Translating–translating mechanisms.

instantaneous centers associated with the links 1, *in*, *out*, and *k*. Using an expression of power for a system in which both the input and output are rotating (Equation (C.10-7)), we obtain

$$\left(\frac{dW}{dt}\right)_{in} = T_{in}\dot{\theta}_{in} = T_{out}\dot{\theta}_{out} = \left(\frac{dW}{dt}\right)_{out} \tag{2.15-4}$$

where T_{in} and T_{out} are the moments applied by the forces to the input and output links about their base pivots, and $\dot{\theta}_{in}$ and $\dot{\theta}_{out}$ are the rotational speeds of the input and output links, respectively. Using Equation (2.15-4), the mechanical advantage may be expressed as

$$M.A. = \frac{F_{out}}{F_{in}} = \frac{\left(\dfrac{T_{out}}{r_{out}}\right)}{\left(\dfrac{T_{in}}{r_{in}}\right)} = \frac{r_{in}}{r_{out}}\frac{\dot{\theta}_{in}}{\dot{\theta}_{out}} \tag{2.15-5}$$

where r_{in} and r_{out} are the lengths of the input and output links, respectively.

The magnitude of the velocity of the instantaneous center between the input and output links may be written as

$$v_{P_{in,out}} = r_{P_{1,in}P_{in,out}}\dot{\theta}_{in} = r_{P_{1,out}P_{in,out}}\dot{\theta}_{out} \tag{2.15-6}$$

or

$$\frac{\dot{\theta}_{in}}{\dot{\theta}_{out}} = \frac{r_{P_{1,out}P_{in,out}}}{r_{P_{1,in}P_{in,out}}} \tag{2.15-7}$$

Combining Equations (2.15-5) and (2.15-7), we obtain

$$M.A. = \frac{F_{out}}{F_{in}} = \frac{r_{in}}{r_{out}} \frac{{}^r P_{1,out} P_{in,out}}{{}^r P_{1,in} P_{1,out}} \qquad (2.15\text{-}8)$$

Rotating–Translating Mechanisms

For this group of mechanisms, the input variable is rotational and the output variable is translational. A slider crank mechanism is a typical example if the crank is the input and the slider is the output. We may state

$$\left(\frac{dW}{dt}\right)_{in} = T_{in}\dot{\theta}_{in} = F_{in}r_{in}\dot{\theta}_{in} = F_{out}r_{out}\dot{\theta}_{out} = F_{out}v_{out} = \left(\frac{dW}{dt}\right)_{out} \qquad (2.15\text{-}9)$$

where v_{out} is the magnitude of the output velocity. As shown in Figure 2.55(b), instantaneous center $P_{1,out}$ is located at infinity. The magnitude of the velocity of the output link equals the magnitude of the velocity of instantaneous center $P_{in,out}$, and therefore

$$v_{out} = r_{P_{1,in}P_{in,out}} \dot{\theta}_{in} \qquad (2.15\text{-}10)$$

Combining Equations (2.15-9) and (2.15-10) gives

$$F_{in}r_{in}\dot{\theta}_{in} = F_{out}r_{P_{1,in}P_{in,out}} \dot{\theta}_{in} \qquad (2.15\text{-}11)$$

The expression for the mechanical advantage for this group of mechanisms is then

$$M.A. = \frac{F_{out}}{F_{in}} = \frac{r_{in}}{r_{P_{1,in}P_{in,out}}} \qquad (2.15\text{-}12)$$

Translating–Rotating Mechanisms

For this group of mechanisms, the input variable is translational and the output variable is rotational. A slider crank mechanism is a typical example if the slider is the input and the crank is the output. Using a similar analysis as above and employing the arrangement of the instantaneous centers given in Figure 2.55(c) leads to the following expression for the mechanical advantage

$$M.A. = \frac{F_{out}}{F_{in}} = \frac{r_{P_{1,out}P_{in,out}}}{r_{out}} \qquad (2.15\text{-}13)$$

Translating–Translating Mechanisms

For this group of mechanisms, both the input and output variables are translational. Three of the six instantaneous centers of the four links are at infinity as illustrated in Figure 2.55(d). Using a similar procedure as above, we find

$$M.A. = \frac{r_{P_{1,k}P_{k,in}}}{r_{P_{1,k}P_{k,out}}} \qquad (2.15\text{-}14)$$

EXAMPLE 2.16 MECHANICAL ADVANTAGE OF A SLIDER CRANK MECHANISM

For the mechanism shown in Figure 2.56, determine the magnitude of the force $\bar{Q}$ on link 2 required to provide a crushing force of magnitude 20 N from link 4 for the position shown.

$$c = 5.0 \text{ cm}; \qquad d = 15.0 \text{ cm}$$

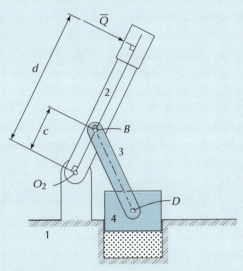

Figure 2.56

SOLUTION

The instantaneous centers are shown in Figure 2.57. This mechanism is in the group of rotating-translating. From Figure 2.57 we have

$$r_{in} = c = 5.0 \text{ cm}; \quad r_{P_{1,2}P_{2,4}} = 4.5 \text{ cm}$$

Using Equation (2.15-12), the mechanical advantage of the mechanism is

$$M.A. = \frac{F_{out}}{F_{in}} = \frac{r_{in}}{r_{P_{1,in}P_{in,out}}} = \frac{c}{r_{P_{1,2}P_{2,4}}} = \frac{5.0}{4.5} = 1.1$$

and therefore

$$F_{in} = \frac{F_{out}}{M.A.} = \frac{20 \text{ N}}{1.1} = 18 \text{ N}$$

which is the magnitude of the force to be applied perpendicular to link 2 at B (see Figure 2.57). Comparing Figures 2.56 and 2.57, the magnitude of the required force $\bar{Q}$ is determined by recognizing that an equivalent moment about the base pivot O_2 is obtained if

$$Qd = F_{in}c$$

and therefore

$$Q = \frac{c}{d}F_{in} = \frac{5.0}{15.0}18 \text{ N} = 6.0 \text{ N}$$

(Continued)

EXAMPLE 2.15 Continued

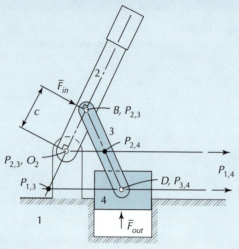

Figure 2.57

An alternative method of force analysis of mechanisms is presented in Chapter 8.

PROBLEMS

Permissible Geometries

P2.1 For the given link lengths of a four-bar mechanism (see Figure 1.9) and the position of link 2, construct the permissible geometries of the mechanism.

$$r_1 = 12.0 \text{ cm}; \quad r_2 = 6.0 \text{ cm}; \quad r_3 = 20.0 \text{ cm}; \quad r_4 = 10.0 \text{ cm}$$

$$\theta_2 = 120°$$

P2.2 For the given link lengths of a five-bar mechanism (see Figure 1.39(b)) and the positions of links 2 and 5, construct the permissible geometries of the mechanism.

$$r_1 = 15.0 \text{ cm}; \quad r_2 = 4.0 \text{ cm}; \quad r_3 = 12.0 \text{ cm}; \quad r_4 = 9.0 \text{ cm};$$

$$r_5 = 6.0 \text{ cm}; \quad \theta_2 = 120°; \quad \theta_5 = 60°$$

P2.3 For the given link lengths of a slider crank mechanism (see Figure 1.7) and the position of link 2, construct the permissible geometries of the mechanism.

$$r_1 = 2.0 \text{ cm}; \quad r_2 = 8.0 \text{ cm}; \quad r_3 = 12.0 \text{ cm}$$

$$\theta_2 = 30°$$

Limit Positions and Zero Speeds

P2.4 For the mechanism shown in Figure P2.4, specify

   (a) the value(s) of θ_2 when the speed of link 4 is zero and determine the corresponding value(s) of the rotational speed of link 3 when this takes place

   (b) the value(s) of θ_2 when the rotational speed of link 3 is zero and determine the corresponding value(s) of the speed of link 4 when this takes place

$$r_{O_2A} = 3.0 \text{ cm}; \quad r_{AB} = 6.0 \text{ cm}; \theta_2 = 100 \text{ rpm CW}$$

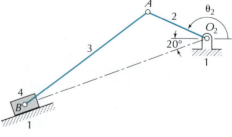

Figure P2.4

P2.5 For the mechanism shown in Figure P2.5, specify

   (a) the value(s) of θ_2 when the linear speed of link 6 is zero

(b) the value(s) of θ_2 when the rotational speed of link 5 is zero and determine the corresponding value(s) of the speed of link 6 when this takes place

$$r_{O_2O_3} = 5.0 \text{ cm}; \quad r_{O_2A_2} = 3.0 \text{ cm}; \quad r_{O_3B} = 10.0 \text{ cm};$$

$$r_{BC} = 4.0 \text{ cm} \quad \dot{\theta}_2 = 10 \text{ rad/sec CCW}$$

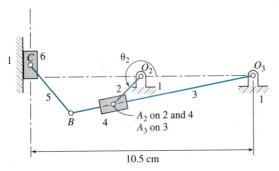

Figure P2.5

P2.6 For the mechanism shown in Figure P2.6, determine

(a) the stroke of point D

(b) the value(s) of θ_2 when the rotational speed of link 3 is zero

(c) the value(s) of θ_2 when the rotational speed of link 5 is zero

(d) the average linear speed of point D to the left while moving between its limit positions

(e) the average linear speed of point D to the right while moving between its limit positions

$$r_{O_2B_2} = 3.0 \text{ cm}; \quad r_{O_2O_3} = 6.0 \text{ cm}; \quad r_{O_3C} = 13.0 \text{ cm}$$

$$\dot{\theta}_2 = 8.0 \text{ rad/sec CW}$$

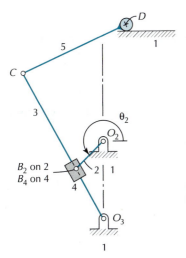

Figure P2.6

Average Speeds, Time Ratio

P2.7 Figure P2.7 shows a crank rocker mechanism. Determine

(a) the average angular speed of the rocker in the clockwise direction

(b) the average angular speed of the rocker in the counterclockwise direction

(c) the time ratio

$$r_1 = 8.0 \text{ cm}; \quad r_2 = 2.0 \text{ cm}; \quad r_3 = 4.0 \text{ cm}; \quad r_4 = 7.0 \text{ cm}$$

$$\dot{\theta}_2 = 40.0 \text{ rad/sec CW(constant)}$$

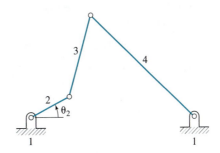

Figure P2.7

P2.8 Figure P2.8 shows a slider crank mechanism. Determine

(a) the average speed of the slider to the right

(b) the average speed of the slider to the left

(c) the time ratio

$$r_1 = 2.0 \text{ cm}; \quad r_2 = 5.0 \text{ cm}; \quad r_3 = 8.0 \text{ cm}$$

$$\dot{\theta}_2 = 30.0 \text{ rad/sec CCW(constant)}$$

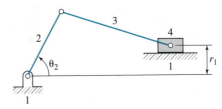

Figure P2.8

P2.9 For the mechanism shown in Figure P2.9, determine

(a) the value(s) of θ_2 when point C is stationary

(b) the stroke of point C

(c) the average linear speed of point C to the left while moving between its limit positions

$$r_{B_2O_2} = 2.5 \text{ cm}; \quad \dot{\theta}_2 = 8.0 \text{ rad/sec CCW(constant)}$$

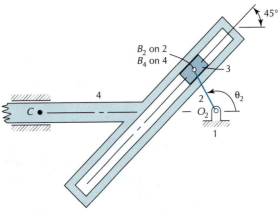

Figure P2.9

Zero Speed, Coriolis Acceleration

P2.10 For the mechanism shown in Figure P2.10, specify

    (a) the value(s) of θ_2 when the rotational speed of link 3 is zero

    (b) the value(s) of θ_2 when the linear speed of point D is zero

    (c) the value(s) of θ_2 when the Coriolis acceleration of B_1 with respect to B_3 is zero

$r_{O_2A} = 3.0$ cm; $r_{O_2B_1} = 5.0$ cm; $r_{AC} = 8.0$ cm; $r_{CD} = 4.0$ cm

$$\dot{\theta}_2 = 10 \text{ rad/sec CCW(constant)}$$

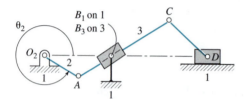

Figure P2.10

P2.11 For the mechanism shown in Figure P2.11

    (a) determine the value(s) of θ_2 when the rotational speed of link 3 is zero

    (b) determine the average rotational speed of link 4 in the clockwise direction between those positions at which the rotational speed of link 3 is zero

    (c) specify all values of the angular position of link 2 in the range $0 \le \theta_2 \le 360°$ when there is no relative Coriolis acceleration between links 3 and 4 at point D_1, and briefly describe your answer

$r_{O_2D_1} = 6.0$ cm; $r_{O_2B} = 2.0$ cm

$$\dot{\theta}_2 = 8.0 \text{ rad/sec CCW (constant)}$$

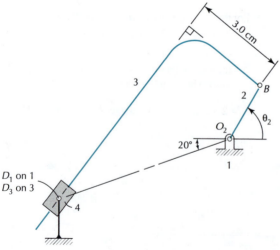

Figure P2.11

P2.12 For the mechanism shown in Figure P2.12, determine all values of θ_2 in the range $0 \le \theta_2 \le 360°$ for which there is no relative Coriolis acceleration between D_1 and the point on the slide, link 3, which instantaneously coincides with D_1.

$r_{O_2B} = 5.0$ cm; $\dot{\theta}_2 = 6.0$ rad/sec CCW (constant)

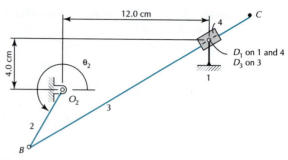

Figure P2.12

Transmission Angle

P2.13 For the given dimensions of a four-bar mechanism (see Figure 1.9), determine the minimum and maximum values of the transmission angle throughout a cycle.

$r_1 = 10.0$ cm; $r_2 = 4.0$ cm; $r_2 = 12.0$ cm; $r_2 = 8.0$ cm

P2.14 For the given dimensions of a slider crank mechanism (see Figure 1.7), determine the minimum and maximum values of the transmission angle throughout a cycle.

$r_1 = 2.0$ cm; $r_2 = 4.0$ cm; $r_3 = 12$ cm

P2.15 For the given dimensions of a four-bar mechanism (see Figure 1.9), generate a plot of the transmission angle throughout a cycle of motion and indicate the locations of the minimum and maximum values.

$r_1 = 10.0$ cm; $r_2 = 2.0$ cm; $r_3 = 12.0$ cm; $r_4 = 8.0$ cm

Instantaneous Centers

P2.16 Locate all instantaneous centers of the mechanism shown in Figure P2.16.

$$\theta_2 = 45°; \quad r_{O_2B} = 2.5 \text{ cm}$$

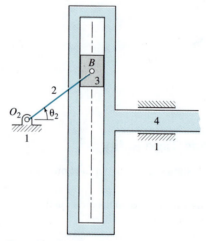

Figure P2.16

P2.17 Locate all instantaneous centers of the mechanism shown in Figure P2.17.

$$r_{O_2B_2} = 2.0 \text{ in}; \quad r_{O_2O_4} = 1.5 \text{ in}; \quad r_{O_4C} = 1.0 \text{ in};$$

$$r_{CD} = 3.0 \text{ in}; \quad \theta_2 = 30°$$

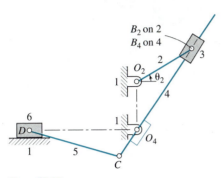

Figure P2.17

P2.18 Locate all instantaneous centers, except $P_{4,5}$, of the mechanism shown in Figure P2.18.

$$r_{O_2O_3} = 6.0 \text{ cm}; \quad r_{O_2B_2} = 4.0 \text{ cm}; \quad r_{O_3C} = 12.0 \text{ cm};$$

$$r_{CD} = 7.0 \text{ cm}; \quad \theta_2 = 30°$$

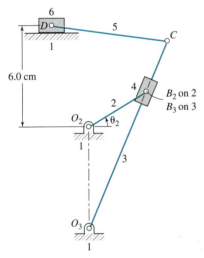

Figure P2.18

P2.19 For the mechanism in the configuration shown in Figure P2.19 determine the positions of all of the instantaneous centers. There is no slippage at the rolling pairs.

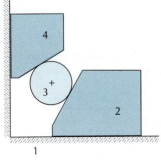

Figure P2.19

P2.20 For the mechanism in the configuration shown in Figure P2.20, determine the locations of all of the instantaneous centers. There is no slippage at the rolling pairs.

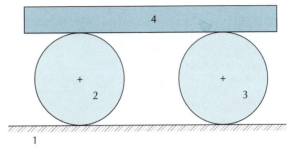

Figure P2.20

P2.21 For the mechanism in the configuration shown in Figure P2.21, determine the locations of all of the instantaneous centers. There is no slippage at the rolling pairs.

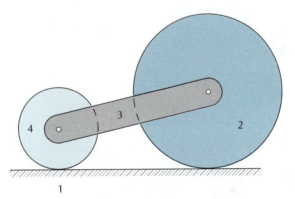

Figure P2.21

P2.22 For the mechanism in the configuration shown in Figure P2.22, determine the locations of all of the instantaneous centers. There is no slippage at the rolling pair.

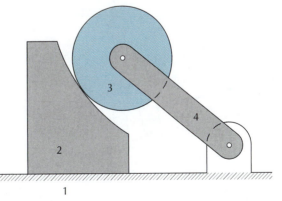

Figure P2.22

P2.23 For the mechanisms in the configurations shown in Figure P2.23, determine the locations of all of the instantaneous centers. There is no slippage at the rolling pairs.

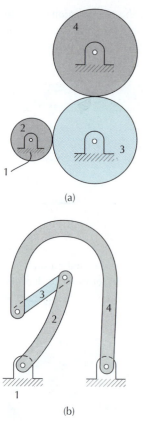

(a)

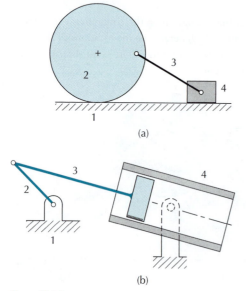

(b)

Figure P2.23

P2.24 For the mechanisms in the configurations shown in Figure P2.24, determine the locations of all of the instantaneous centers. There is no slippage at the rolling pair.

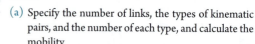

(a)

(b)

Figure P2.24

P2.25 For the mechanisms in the configurations shown in Figure P2.25, determine the locations of all of

the instantaneous centers. There is no slippage at the rolling pair.

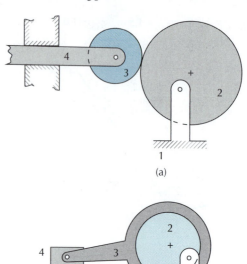

(a)

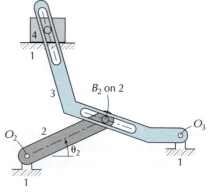

(b)

Figure P2.25

Instantaneous Centers, Mobility

P2.26 For the mechanism in the position shown in Figure P2.26

(a) Specify the number of links, the types of kinematic pairs, and the number of each type, and calculate the mobility

(b) Determine the positions of all instantaneous centers

$$r_{O_2 B_2} = 4.0 \text{ cm}$$

Figure P2.26

P2.27 For the mechanism in the configuration shown in Figure P2.27, four instantaneous centers are also provided.

(a) Specify the types of kinematic pairs, and the number of each type, and calculate the mobility.

(b) Determine the positions of the instantaneous centers that have not been provided.

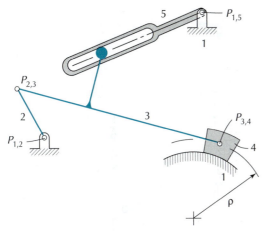

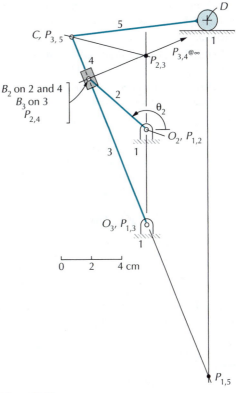

Figure P2.27

P2.28 For the mechanism in the configuration shown in Figure P2.28, six instantaneous centers are also provided.

    (a) Specify the types of kinematic pairs, and the number of each type, and calculate the mobility.

    (b) Determine the positions of the instantaneous centers that have not been provided.

$$r_{O_4C_4} = 1.5 \text{ in}$$

Figure P2.29

P2.30 For the mechanism in the configuration shown in Figure P2.30, three instantaneous centers are also provided.

    (a) Specify the types of kinematic pairs, and the number of each type, and calculate the mobility.

    (b) Determine the positions of the instantaneous centers that have not been provided.

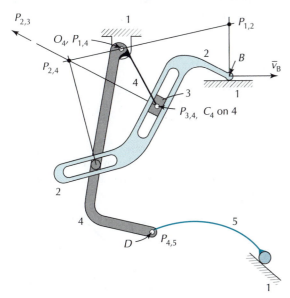

Figure P2.28

P2.29 For the mechanism in the configuration shown in Figure P2.29, seven instantaneous centers are also provided.

    (a) Specify the types of kinematic pairs, and the number of each type, and calculate the mobility.

    (b) Determine the positions of the instantaneous centers that have not been provided.

$$r_{O_2B_2} = 5.0 \text{ cm}$$

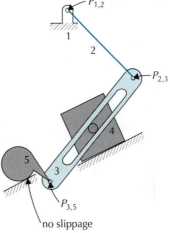

Figure P2.30

P2.31 For the mechanism in the configuration shown in Figure P2.31, six instantaneous centers are also provided.

(a) Specify the types of kinematic pairs, and the number of each type, and calculate the mobility.

(b) Determine the positions of the instantaneous centers that have not been provided.

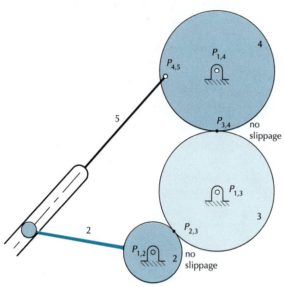

Figure P2.31

Force Analysis

P2.32 For the crane shown in Figure P2.32, calculate the moment about point A exerted by the 120,000 N force supported by the hoisting cable of the crane for the position shown.

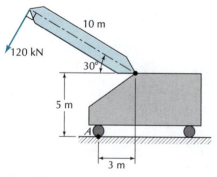

Figure P2.32

P2.33 A construction worker shown in Figure P2.33 uses a 6-kg wheelbarrow to transport a 25-kg bag of cement. What force must he exert on both handles?

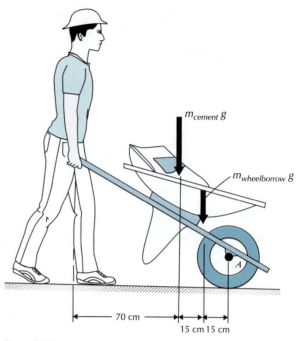

Figure P2.33

P2.34 In raising the massless pole from the position shown in Figure P2.34, the tension T in the cable must supply a moment about O of 72 kN-m. Determine T.

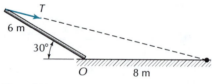

Figure P2.34

P2.35 A truck-mounted crane shown in Figure P2.35 is used to lift a 200-kg safe. (continued on the next page)

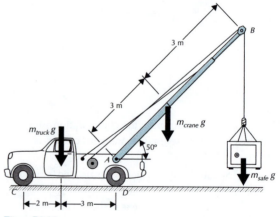

Figure P2.35

Determine the reaction at the

(a) front wheels

(b) rear wheels

$$m_{truck} = 2500 \text{ kg}; \quad m_{crane} = 120 \text{ kg}$$

P2.36 For the link shown in Figure P2.36, replace the couple and force shown by a single force $\bar{F}$ applied at point D. Locate D by determining distance b.

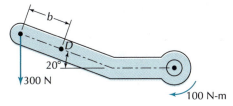

Figure P2.36

P2.37 The lever AB shown in Figure P2.37 is hinged at O and attached to control cable at B. If the lever is subjected at A to a 200 N horizontal force, determine

(a) the tension in the cable

(b) the reaction at O

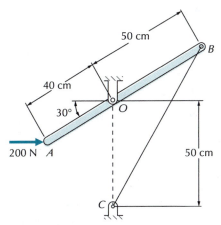

Figure P2.37

P2.38 The lever shown in Figure P2.38 is to be designed to operate with a 200 N force. The force is to provide a counterclockwise couple of 80 N-m at A. Determine dimension a.

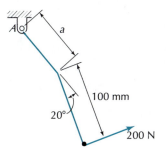

Figure P2.38

P2.39 One end of the rod AB shown in Figure P2.39 rests in the corner A and the other is attached to cable BD. Neglect the effect of gravity. Determine

(a) the reaction at A

(b) the tension in the cable

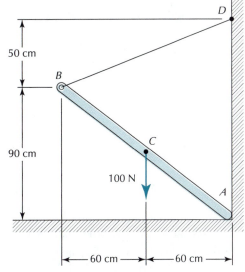

Figure P2.39

P2.40 A box weighing 10 N rests on the floor of an elevator as shown in Figure P2.40. Determine the force exerted by the box on the floor of the elevator if

(a) the elevator starts up with an acceleration of 3 m/sec²

(b) the elevator starts down with an acceleration of 3m/sec²

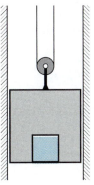

Figure P2.40

P2.41 Rod OB shown in Figure P2.41 is supported by a base pivot at O and rests against a pin at A. Determine the reactions at O and A when a 200 N horizontal force is directed to the right at B. Neglect the effect of gravity.

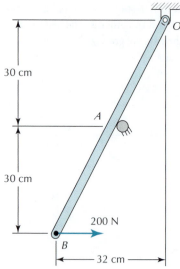

Figure P2.41

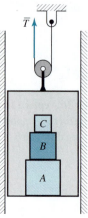

Figure P2.44

P2.42 Determine the constant force required to accelerate an automobile with mass 1000 kg on a level road from rest to 30 m/sec in 5 seconds. Gravity acts vertically downward (Figure P2.42).

Figure P2.42

P2.43 Determine the constant force required to accelerate an automobile with mass 1000 kg on a road with a 10° incline from rest to 30 m/sec in 5 seconds Gravity acts vertically downward (Figure P2.43).

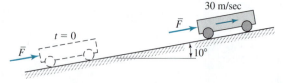

Figure P2.43

P2.44 A freight elevator contains three crates as shown in Figure P2.44. The mass of the cage of the elevator is 1000 kg, and the masses of crates A, B, and C are 200, 100, and 50 kg, respectively. At one instant, the elevator undergoes an upward acceleration of 5 m/sec². At this instant, determine

(a) the tension in the cable

(b) the force exerted by crate B on crate C.

P2.45 The solid disc shown in Figure P2.45 has a mass of 20 kg and a radius of 0.3 m. Block B has a mass of 5 kg. When a torque of $T = 75$ N-m is applied to the disc, determine the acceleration of block B and the tension in the cable. Gravity acts vertically downward.

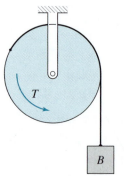

Figure P2.45

Mechanical Advantage

P2.46 Derive Equation (2.15-13), the expression for the mechanical advantage of a translating–rotating mechanism.

P2.47 Derive Equation (2.15-14), the expression for the mechanical advantage of a translating–translating mechanism.

P2.48 For the mechanism shown in Figure P2.48, link 2 is the input, and link 4 is the output. Determine the output force required to maintain static equilibrium for the position shown. Solve the problem by determining the mechanical advantage for the given configuration.

$$r_1 = 13.0 \text{ cm}; \quad r_2 = 6.0 \text{ cm}; \quad r_3 = 5.5 \text{ cm}; \quad r_4 = 10.0 \text{ cm}$$

$$\theta = 70°; \quad F_{in} = 15.0 \text{ N}$$

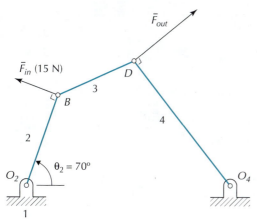

$\bar{F}_{out}$

$\bar{F}_{in}$ (15 N)

D

3

B

4

2

$\theta_2 = 70°$

O_2 O_4

1

Figure P2.48

P2.49 For the mechanism shown in problem P2.49 an external force of 20 N is applied to link 5 to the right. Determine the magnitude of the moment to be applied to link 2 of length 4.0 cm to keep the mechanism is static equilibrium. Solve the problem by determining the mechanical advantage for the given configuration.

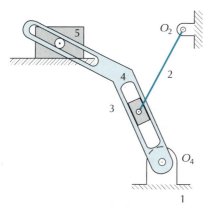

5

O_2

4

2

3

O_4

1

Figure P2.49

P2.50 For the mechanism shown in problem P2.50 an external moment of 40 N-cm counterclockwise is applied to link 2 of length 3.0 cm. Determine the magnitude of the moment to be applied to be applied to link 4 to keep the mechanism is static equilibrium. Solve the problem by determining the mechanical advantage for the given configuration.

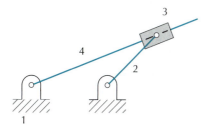

3

4

2

1

Figure P2.50

P2.51 For the mechanism shown in problem P2.51 an external force of 20 N is applied to link 2 to the left. Determine the magnitude of the force to be applied to be applied to link 4 along the line of action of the slide to keep the mechanism is static equilibrium. Solve the problem by determining the mechanical advantage for the given configuration.

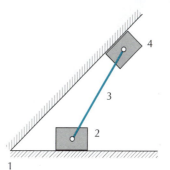

4

3

2

1

Figure P2.51

P2.52 For the mechanism shown in problem P2.26, determine the magnitude of the horizontal force to be applied to be applied to link 4 to keep the mechanism is static equilibrium if a moment of 2.0 N-m counterclockwise is applied to link 2. Solve the problem by determining the mechanical advantage for the given configuration.

$$r_{B_2O_2} = 4.0 \text{ cm}$$

P2.53 The mechanism shown in Figure 2.39 has a total of three links. Derive an expression for the magnitude of the ratio of the torques to be applied to links 2 and 3 about their base pivots to maintain static equilibrium. Use the concept of mechanical advantage and the locations of the instantaneous centers.

P2.54 For the mechanism shown in Figure 2.40, link 6 is the input, and link 2 is the output. Determine the output force required to maintain static equilibrium for the position shown if the input force is 60 N to the right. Solve the problem by determining the mechanical advantage for the given configuration.

CHAPTER 3

GRAPHICAL KINEMATIC ANALYSIS OF PLANAR MECHANISMS

3.1 INTRODUCTION

In this chapter, graphical velocity and acceleration analyses of planar mechanisms are presented. Each analysis is suitable for one configuration of the mechanism and requires that the mechanism be drawn to scale in the position for which the analysis is to be completed. One method for velocity analysis employs the concept of instantaneous centers (Section 2.10). Another method involves construction of vector polygons of velocity and acceleration. All methods presented are based on material presented in Chapter 2.

3.2 VELOCITY ANALYSIS USING INSTANTANEOUS CENTERS

Instantaneous centers of a mechanism may be used to complete a velocity analysis for a single configuration of a planar mechanism. Since the linear velocity at instantaneous center $P_{i,j}$ is identical for links i and j, it can be thought of as a transfer point of motion between links at the instant under consideration.

A velocity analysis of a planar mechanism can be completed by carrying out the following general steps:

1. Draw a diagram of the mechanism to scale, in the configuration under analysis.
2. Obtain the locations of the instantaneous centers, which are required to complete the analysis.
3. Express the velocities at the instantaneous centers based on input motions. For the crank, link 2, of the mechanism shown in Figure 2.37(b), the magnitude of the linear velocity of instantaneous center $P_{2,3}$ may be expressed in terms of the rotational speed of link 2 as

$$|\bar{v}_{P_{2,3}}| = r_{P_{1,2}P_{2,3}}|\dot{\theta}_2| \qquad (3.2\text{-}1a)$$

4. Write the appropriate equations needed in calculating other velocities. When considering the instantaneous center between links i and j, we can state

$$|\bar{v}_{P_{1,i}}| = r_{P_{1,i}P_{i,j}}|\dot{\theta}_i| = r_{P_{1,j}P_{i,j}}|\dot{\theta}_j| \qquad (3.2\text{-}1b)$$

5. Complete the analysis by repeating the process until all required velocities are determined.

For the velocity analysis of a mechanism, it may not be necessary to determine the location of all of the instantaneous centers. This will be demonstrated by examples.

EXAMPLE 3.1 VELOCITY ANALYSIS OF A SLIDER CRANK MECHANISM

For the slider crank mechanism shown in Figure 3.1, determine velocity of the slider as a function of the rotational velocity of link 2 for the position shown. Use the instantaneous centers.

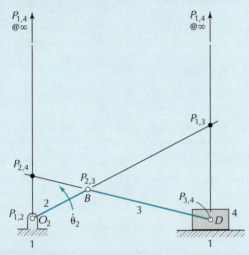

Figure 3.1

SOLUTION

The instantaneous centers for this mechanism were determined in Example 2.6.

The velocity of the slider can be determined by more than one method. For this example, two methods are presented. The first method progresses through the mechanism links and instantaneous centers from the input to the output. The second method determines the final result by considering transfer of motion through one instantaneous center.

Method #1

We consider instantaneous center $P_{2,3}$. It is the transfer point of motion between links 2 and 3. The linear velocity of this point is the same for both links. Considering that link 2 revolves about $P_{1,2}$, and for the instant considered link 3 revolves about $P_{1,3}$,

$$\left|\bar{v}_{P_{2,3}}\right|=\left|\bar{v}_{B}\right|=r_{P_{1,2}P_{2,3}}\left|\dot{\theta}_2\right|=r_{P_{1,3}P_{2,3}}\left|\dot{\theta}_3\right| \tag{3.2-2}$$

which gives

$$\left|\dot{\theta}_3\right|=\frac{r_{P_{1,2}P_{2,3}}}{r_{P_{1,3}P_{2,3}}}\left|\dot{\theta}_2\right| \tag{3.2-3}$$

Instantaneous center $P_{3,4}$, is the motion transfer point between links 3 and 4. Its magnitude of velocity, considering it as a point on link 3, is

$$\left|\bar{v}_{P_{3,4}}\right|=\left|\bar{v}_{D}\right|=r_{P_{1,3}P_{3,4}}\left|\dot{\theta}_3\right| \tag{3.2-4}$$

Combining Equations (3.2-3) and (3.2-4) yields

$$\left|\bar{v}_{D}\right|=r_{P_{1,3}P_{3,4}}\frac{r_{P_{1,2}P_{2,3}}}{r_{P_{1,3}P_{2,3}}}\left|\dot{\theta}_2\right| \tag{3.2-5}$$

The slider, link 4, is undergoing pure translation. Therefore, velocity $\bar{v}_D$ is the same for any point on this link.

In the above analysis, instantaneous center $P_{2,4}$ was not employed.

(Continued)

EXAMPLE 3.1 Continued

Method #2

If we employ instantaneous center $P_{2,4}$, then we consider transfer of motion directly between links 2 and 4. The speed of this instantaneous center expressed as a point on link 2 is

$$|\bar{v}_{P_{2,4}}| = r_{P_{1,2}P_{2,4}} |\dot{\theta}_2|$$

(3.2-6)

Because the slider, link 4, is in pure translation

$$|\bar{v}_{P_{2,4}}| = |\bar{v}_D| = r_{P_{1,2}P_{2,4}} |\dot{\theta}_2|$$

(3.2-7)

In the above analysis, instantaneous centers $P_{1,3}$ and $P_{1,4}$ were not employed. It was also not required to determine the angular velocity of link 3.

EXAMPLE 3.2 VELOCITY ANALYSIS OF AN INVERTED SLIDER CRANK MECHANISM

For the inverted slider crank mechanism shown in Figure 3.2, determine the velocity of point B as a function of the rotational speed of link 2 for the position shown. Use the instantaneous centers.

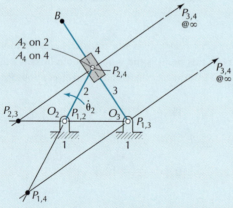

Figure 3.2

SOLUTION

The instantaneous centers for this mechanism were determined in Example 2.7. Two methods for determining the velocity of point B are presented.

Method #1

Instantaneous center $P_{2,4}$ is the motion transfer point between links 2 and 4. Hence

$$|\bar{v}_{P_{2,4}}| = |\bar{v}_{A_2}| = r_{P_{1,2}P_{2,4}} |\dot{\theta}_2| = r_{P_{1,4}P_{2,4}} |\dot{\theta}_4|$$

(3.2-8)

(Continued)

EXAMPLE 3.2 Continued

from which

$$|\dot{\theta}_4| = \frac{r_{P_{1,2}P_{2,4}}}{r_{P_{1,4}P_{2,4}}}|\dot{\theta}_2| \qquad (3.2\text{-}9)$$

By inspection, the rotational speed of link 4 is in the counterclockwise direction. Also, links 3 and 4 are constrained to share the same rotational speed,

$$|\dot{\theta}_3| = |\dot{\theta}_4| \qquad (3.2\text{-}10)$$

The magnitude of the velocity of point B is

$$|\overline{v}_B| = r_{P_{1,3}B}|\dot{\theta}_3| \qquad (3.2\text{-}11)$$

Combining Equations (3.2-9)–(3.2-11), the magnitude of the velocity may be expressed as

$$|\overline{v}_B| = r_{P_{1,3}B}|\dot{\theta}_3| = r_{P_{1,3}B}\frac{r_{P_{1,2}P_{2,4}}}{r_{P_{1,4}P_{2,4}}}|\dot{\theta}_2| \qquad (3.2\text{-}12)$$

By inspection, the velocity is directed downward and to the left.

Method #2

Instantaneous center $P_{2,3}$ is the motion transfer point between links 2 and 3. Hence

$$r_{P_{1,2}P_{2,3}}|\dot{\theta}_2| = r_{P_{1,3}P_{2,3}}|\dot{\theta}_3| \qquad (3.2\text{-}13)$$

Combining Equations (3.2-11) and (3.2-13), the magnitude of the velocity of point B may be expressed as

$$|\overline{v}_B| = r_{P_{1,3}B}|\dot{\theta}_3| = r_{P_{1,3}B}\frac{r_{P_{1,2}P_{2,3}}}{r_{P_{1,3}P_{2,3}}}|\dot{\theta}_2| \qquad (3.2\text{-}14)$$

EXAMPLE 3.3 VELOCITY ANALYSIS OF A MECHANISM INCORPORATING A TWO DEGREE OF FREEDOM PAIR

For the mechanism shown in Figure 3.3, link 2 has a prescribed rotational speed in the clockwise direction. Determine the rotational speed of link 3. Use the instantaneous centers.

SOLUTION

The instantaneous centers for this mechanism were determined in Example 2.8.

The magnitude of the velocity of the instantaneous center $P_{2,3}$ may be expressed as

$$v_{P_{2,3}} = r_{P_{1,2}P_{2,3}}|\dot{\theta}_2| = r_{P_{1,3}P_{2,3}}|\dot{\theta}_3| \qquad (3.2\text{-}15)$$

(Continued)

EXAMPLE 3.3 Continued

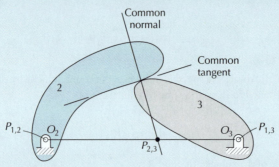

Figure 3.3

from which

$$|\dot{\theta}_3| = \frac{r_{P_{1,2}P_{2,3}}}{r_{P_{1,3}P_{2,3}}} |\dot{\theta}_2| \qquad (3.2\text{-}16)$$

By inspection, the direction of the rotational speed of link 3 is counterclockwise, opposite that of link 2.

EXAMPLE 3.4 VELOCITY ANALYSIS OF A SYSTEM WITH A TWO DEGREE OF FREEDOM PAIR

Link 2 of the mechanism shown in Figure 3.4 has a circular shape with center at C, and executes full rotations about base pivot O_2. Links 2 and 3 remain in direct contact as the mechanism moves. Using the instantaneous centers, determine

(a) the magnitude and direction of rotational speed of link 3 when $\theta_2 = 0$
(b) the magnitude and direction of rotational speed of link 3 when $\theta_2 = 135°$
(c) all values of θ_2 in the range $0 \le \theta_2 \le 360°$ for which the rotational speed of link 3 is zero

$$r_{O_2O_3} = 10.0 \text{ cm} ; \; r_{O_2C} = 2.0 \text{ cm}; \rho = 3.0 \text{ cm}$$
$$\dot{\theta}_2 = 12.0 \text{ rad/sec} \;\; \text{CCW (constant)}$$

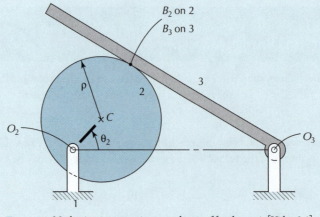

Video 3.4
Mechanism
Incorporating a
Two Degree of
Freedom Pair

Figure 3.4 Mechanism incorporating a two degree of freedom pair [Video 3.4].

(Continued)

EXAMPLE 3.4 Continued

SOLUTION

(a) The mechanism has three instantaneous centers all of which lie on the same straight line. There may be relative sliding and turning between links 2 and 3 and therefore the instantaneous center between these links lies on the common normal that passes through the point of contact. Since link 2 is circular, the common normal always passes through point C, the center of the circle. When $\theta_2 = 0$ (see Figure 3.5(a)),

$$r_{P_{1,2}C} = r_{P_{1,2}P_{2,3}} = 2.0 \text{ cm}$$
$$r_{P_{1,3}P_{2,3}} = r_{P_{1,2}P_{1,3}} - r_{P_{1,2}P_{2,3}} = 8.0 \text{ cm}$$

Using Equation (3.2-1b), the rotational speed of link 3 is

$$|\dot{\theta}_3| = \frac{r_{P_{1,2}P_{2,3}}}{r_{P_{1,3}P_{2,3}}} |\dot{\theta}_2| = \frac{2.0}{8.0} 12.0 = 3.0 \frac{\text{rad}}{\text{sec}}$$

$$\dot{\theta}_3 = 3.0 \frac{\text{rad}}{\text{sec}} \text{CW}$$

(b) When $\theta_2 = 135°$ (see Figure 3.5(b)), we apply the cosine law to triangle $O_2 O_3 C$

$$r^2_{P_{1,3}C} = r^2_{P_{1,2}C} + r^2_{P_{1,2}P_{1,3}} - 2r_{P_{1,2}C} r_{P_{1,3}} \cos\theta_2; \quad \text{Solving: } r_{P_{1,3}C} = 11.5 \text{ cm}$$

and apply the sine law

$$\frac{r_{P_{1,2}C}}{\sin\alpha} = \frac{r_{P_{1,3}C}}{\sin\theta_2}; \quad \text{Solving: } \alpha = 7.1°$$

For triangle $BO_3 C$ we have

$$\sin\beta = \frac{\rho}{r_{P_{1,3}C}}; \quad \text{Solving: } \beta = 15.1°$$

and

$$\cos\beta = \frac{r_{P_{1,3}B}}{r_{P_{1,3}C}}; \quad \text{Solving: } r_{P_{1,3}B} = 11.1 \text{ cm}$$

For triangle $BP_{1,3}P_{2,3}$ we have

$$\cos(\alpha + \beta) = \frac{r_{P_{1,3}B}}{r_{P_{1,3}P_{2,3}}}; \quad \text{Solving: } r_{P_{1,3}P_{2,3}} = 12.0 \text{ cm}$$

From Figure 3.5(b) we obtain

$$r_{P_{1,2}P_{2,3}} = r_{P_{1,3}P_{2,3}} - r_{P_{1,2}P_{2,3}}; \quad \text{Solving: } r_{P_{1,2}P_{2,3}} = 2.0 \text{ cm}$$

Using Equation (3.2-1b), the rotational speed of link 3 is

$$|\dot{\theta}_3| = \frac{r_{P_{1,2}P_{2,3}}}{r_{P_{1,3}P_{2,3}}} |\dot{\theta}_2| = \frac{2.0}{12.0} 12.0 = 2.0 \frac{\text{rad}}{\text{sec}}$$

$$\dot{\theta}_3 = 2.0 \frac{\text{rad}}{\text{sec}} \text{CCW}$$

(Continued)

EXAMPLE 3.4 Continued

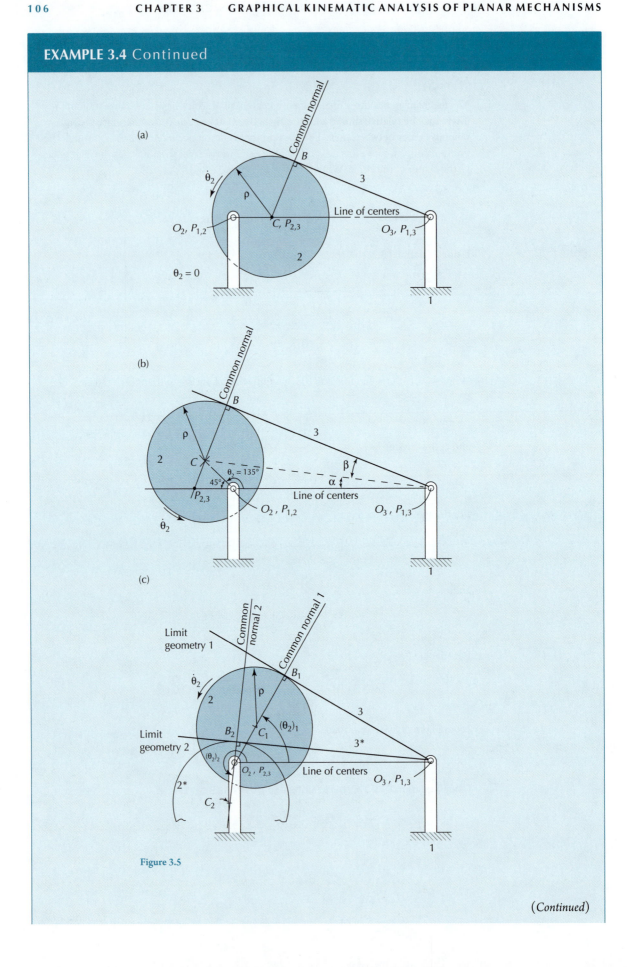

Figure 3.5

(*Continued*)

EXAMPLE 3.4 Continued

(c)   By inspection of Equation (3.2-1b), for the rotational speed of link 3 to be zero we require

$$r_{P_{1,2} P_{2,3}} = 0$$

The two corresponding geometries of the mechanism are shown in Figure 3.5(c). For limit geometry 1, we have

$$r_{P_{2,3} B_1} = \rho + r_{P_{2,3} C_1} = 5.0 \text{ cm}$$

$$(\theta_2)_1 = \cos^{-1}\left(\frac{r_{P_{2,3} B_1}}{r_{P_{2,3} P_{1,3}}}\right) = 60°$$

and for limit geometry 2, we have

$$r_{P_{2,3} B_2} = \rho - r_{P_{2,3} C_2} = 1.0 \text{ cm}$$

$$(\theta_2)_2 = 180° + \cos^{-1}\left(\frac{r_{P_{2,3} B_2}}{r_{P_{2,3} P_{1,3}}}\right) = 264.2°$$

EXAMPLE 3.5 VELOCITY ANALYSIS OF A SIX-LINK MECHANISM

For the mechanism shown in Figure 3.6, determine the velocity of link 6 as a function of the rotational speed of link 2 for the position shown. Employ instantaneous centers.

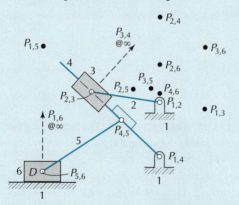

Figure 3.6 Instantaneous centers of a six-link mechanism.

SOLUTION

The instantaneous centers for this mechanism were determined in Example 2.9. The speed of $P_{2,6}$ expressed as a point on input link 2 is

$$v_{P_{2,6}} = r_{P_{1,2} P_{2,6}} |\dot{\theta}_2| \tag{3.2-17}$$

(Continued)

EXAMPLE 3.5: Continued

Because the slider is in pure translation, we obtain

$$v_{P_{2,6}} = v_D = r_{P_{1,2}P_{2,6}} |\dot{\theta}_2| \tag{3.2-18}$$

In the above analysis, it was not necessary to determine the angular velocities of links 3 through 5, inclusive.

EXAMPLE 3.6   VELOCITY ANALYSIS OF A WHEELED SYSTEM

For the mechanism shown in Figure 3.7(a), wheel 2 rotates in the counterclockwise direction at 8.0 rad/sec about the base pivot O_2 while link 3 rotates at 4.0 rad/sec in the clockwise direction about the same base pivot. Wheel 4 is connected to link 3 through a turning pair and to wheel 2 through a rolling pair. For the position shown, locate all instantaneous centers and use them to determine the rotational speed of wheel 4.

$$r_2 = 15.0 \text{ cm}; \quad r_4 = 9.0 \text{ cm}$$

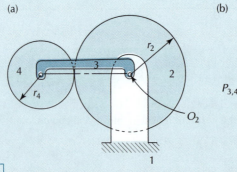
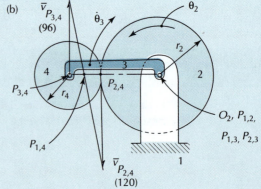

Video 3.7
Wheeled System

Figure 3.7 Wheeled system [Video 3.7].

SOLUTION

The mechanism has a total of six instantaneous centers. The locations of five, $P_{1,2}$, $P_{1,3}$, $P_{2,3}$, $P_{2,4}$, and $P_{3,4}$, are determined by examining the kinematic pairs and are indicated in Figure 3.7(b). For the instantaneous centers $P_{2,4}$ and $P_{3,4}$ we have

$$v_{P_{2,4}} = r_{P_{1,2}P_{2,4}} \dot{\theta}_2 = r_2 \dot{\theta}_2 = 15.0 \times 8.0 = 120 \frac{\text{cm}}{\text{sec}}$$

$$v_{P_{3,4}} = r_{P_{1,3}P_{3,4}} \dot{\theta}_3 = -96.0 \frac{\text{cm}}{\text{sec}}$$

and these results are illustrated in Figure 3.7(b).

The only remaining instantaneous center that still needs to be determined is $P_{1,4}$. Using Kennedy's Theorem, we know that $P_{1,4}$ lies on the line containing $P_{1,2}$ and $P_{2,4}$, as well as on

(Continued)

EXAMPLE 3.6 Continued

the line with $P_{1,3}$ and $P_{3,4}$. However, since these two lines are collinear, there is no unique point of intersection to determine the location of $P_{1,4}$. Instead, we realize that the velocities of the instantaneous centers $P_{2,4}$ and $P_{3,4}$ are parallel and are both associated with link 4. This is an example of the condition illustrated in Figure 2.34(c). There is a linear distribution of velocity on link 4 between the two instantaneous centers and the point of zero velocity on link 4, which is $P_{1,4}$, is shown in Figure 3.7(b). Therefore

$$r_{P_{1,4}P_{2,4}} = \frac{v_{P_{2,4}}}{v_{P_{2,4}} - v_{P_{3,4}}} r_{P_{2,4}P_{3,4}} = \frac{120}{120 - (-96.0)} \, 9.0 \text{ cm} = 5.0 \text{ cm}$$

The rotational speed of link 4 is then

$$|\dot{\theta}_4| = \frac{r_{P_{1,2}P_{2,4}}}{r_{P_{1,4}P_{2,4}}} |\dot{\theta}_2| = \frac{15.0}{5.0} 8.0 = 24.0 \frac{\text{rad}}{\text{sec}}$$

$$\dot{\theta}_4 = 24.0 \frac{\text{rad}}{\text{sec}} \text{CW}$$

Since the center of wheel 4 moves along a circular path relative to the base link, this mechanism is called a *planetary gear train*. Chapter 6 provides a detailed coverage of the many types of planetary gear trains and presents an alternative method of velocity analysis.

3.3 VELOCITY POLYGON ANALYSIS

For a given configuration of a planar mechanism, it is possible to construct an associated velocity polygon. This polygon corresponds to a single configuration of the mechanism and is a graphical representation of all the relative velocities that occur between selected points of interest of the mechanism. Once constructed, it is possible to determine the absolute velocity of any point, the relative velocity between any two points, or the rotational velocity of any link.

To construct a velocity polygon, the following general steps are required:

1. Draw a diagram of the mechanism to scale, in the configuration under analysis.
2. Set up a polar diagram, where every point with zero velocity is at the origin (*pole point*) O_V.
3. Starting from the pole point, add velocity vectors that can be readily calculated. For instance, consider those based on prescribed input motions with respect to the base link.
4. Write the appropriate relative velocity equations to aid in calculating and adding other velocity vectors.
5. Proceed through the mechanism until all points of interest are analyzed.

The velocity polygon can then be employed to determine the absolute and relative velocities. This procedure is illustrated in the following examples.

EXAMPLE 3.7 VELOCITY ANALYSIS OF A FOUR-BAR MECHANISM

For the four-bar mechanism shown in Figure 3.8(a), the link lengths are

$$r_{O_2O_4} = r_1 = 8.0 \text{ cm}; \quad r_{O_2B} = r_2 = 3.5 \text{ cm}$$

$$r_{BD} = r_3 = 10.0 \text{ cm}; \quad r_{O_2D} = r_4 = 9.0 \text{ cm}$$

The input rotational speed is

$$\dot{\theta}_2 = 350 \text{ rpm CCW} = 350 \times \frac{2\pi}{60} \text{rad/sec CCW} = 36.7 \text{ rad/sec CCW}$$

Determine the rotational speeds of links 3 and 4 for the position shown through construction of a velocity polygon.

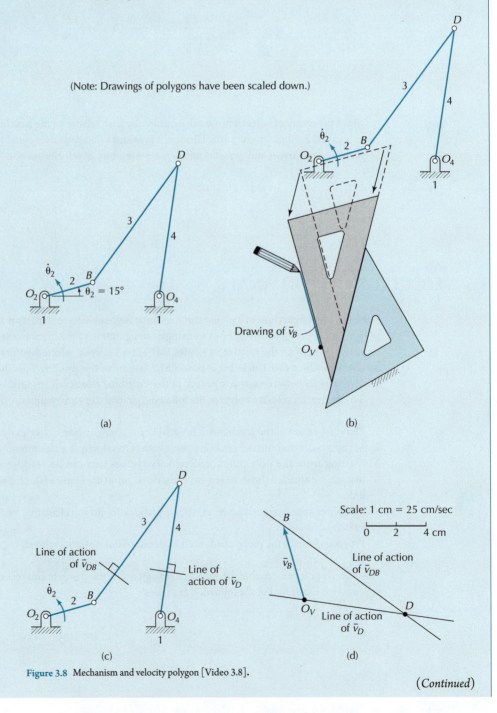

(Note: Drawings of polygons have been scaled down.)

Drawing of $\bar{v}_B$

(a)

(b)

Scale: 1 cm = 25 cm/sec

Line of action
of $\bar{v}_{DB}$

Line of
action of $\bar{v}_D$

Line of action
of $\bar{v}_{DB}$

Line of action
of $\bar{v}_D$

(c)

(d)

Figure 3.8 Mechanism and velocity polygon [Video 3.8].

(Continued)

EXAMPLE 3.7 Continued

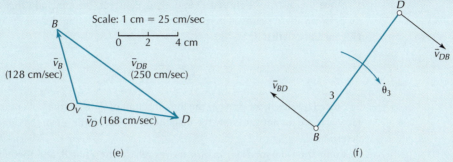

(e)   (f)

Figure 3.8 Mechanism and velocity polygon [Video 3.8] *(continued)*.

SOLUTION

Base pivots are designated as O_2 and O_4. The velocities of these points are zero and will therefore be located at the pole point of the velocity polygon. The two moving turning pairs are designated as B and D.

Construction of the velocity polygon begins with determining the velocity of point B since it is at the distal end of link 2 and moves relative to point O_2. The appropriate relative velocity equation is

$$\bar{v}_B = \bar{v}_{O_2} + \bar{v}_{BO_2} \tag{3.3-1}$$

However, since O_2 has zero velocity, the relative velocity between points B and O_2 also represents the absolute velocity of point B, so the vector is drawn from the pole point. Also, point B is separated from point O_2 by a fixed distance, and therefore from Section 2.5.1 the direction of the relative velocity of point B must be perpendicular to line segment O_2B on the mechanism, that is,

$$\bar{v}_B = \bar{v}_{BO_2} = r_2 \dot{\theta}_2 \bar{i}_{\theta_{BO_2}} \tag{3.3-2}$$

The magnitude of the velocity is

$$v_B = |\bar{v}_B| = |r_2 \dot{\theta}_2| = 3.5 \text{ cm} \times 36.7 \text{ rad/sec} = 128 \text{ cm/sec}$$

The direction of this velocity is determined by inspecting the direction of the rotational speed of link 2. Using a selected scale of

$$1 \text{ cm} = 25 \text{ cm/sec} \tag{3.3-3}$$

vector $\bar{v}_B$ is added to the polygon, and the head of this vector is labeled as B. Figure 3.8(b) illustrates how a pair of drafting triangles is used to draw this velocity vector. One edge of a triangle is placed parallel to line segment O_2B on the mechanism. A second triangle is butted up against another edge of the first triangle as shown. While holding the second triangle stationary, the first triangle is slid into the position whereby an edge touches the pole point. By inspection of Figure 3.8(b), the edge touching the pole point is perpendicular to link 2 and can therefore be used to draw the velocity vector. Triangles can also be easily employed to draw vectors, as required, that are parallel to line segments on the mechanism.

Next, we seek to determine the velocity of point D. We recognize that D moves relative to point O_4, and

$$\bar{v}_D = \bar{v}_{O_4} + \bar{v}_{DO_4} \tag{3.3-4}$$

(Continued)

EXAMPLE 3.7 Continued

Since the distance between D and O_4 is constant, we conclude that the relative velocity between these points is perpendicular to the line joining the points on the mechanism (Figure 3.8(c)), that is,

$$\bar{v}_D = \bar{v}_{DO_4} = r_4\dot{\theta}_4\bar{i}_{\theta_{DO_4}} \qquad (3.3\text{-}5)$$

Since O_4 is fixed, we draw in the line of action of $\bar{v}_D$ from the pole point (Figure 3.8(d)). However, we do not yet know the magnitude of this vector. Also, the direction has yet to be determined.

Point D on the mechanism also moves relative to point B, and therefore we employ the relation

$$\bar{v}_D = \bar{v}_B + \bar{v}_{DB} \qquad (3.3\text{-}6)$$

Since B and D are separated by a fixed distance, the corresponding relative velocity is perpendicular to the corresponding direction on the mechanism (Figure 3.8(c)), that is,

$$\bar{v}_{DB} = r_3\dot{\theta}_3\bar{i}_{\theta_{DB}} \qquad (3.3\text{-}7)$$

Since the velocity of point D in Equation (3.3-6) is expressed with respect to the velocity of point B, the line of action of the relative velocity is added to the polygon starting from B (Figure 3.8(d)). Although we know the line of action, we do not know the magnitude and direction of this vector.

The velocity of point D is unique. In order to satisfy the relative velocity expressions, Equations (3.3-5) and (3.3-7), as well as the lines of action of the velocities, the head of the velocity of point D must lie at the intersection of the two lines of action (Figure 3.8(d)). This is designated as point D on the polygon. The absolute velocity of point D and the relative velocity between points B and D are labeled. The completed velocity polygon is given in Figure 3.8(e). Note also, as illustrated in Figure 3.8(f), that

$$\bar{v}_{DB} = -\bar{v}_{BD} \qquad (3.3\text{-}8)$$

Now that the velocity polygon has been completed, it is possible to determine the magnitudes and directions of angular speeds of links 3 and 4. To do this, we measure speeds from the polygon employing the scale given by Equation (3.3-3). The results are

$$v_{DB} = 250 \text{ cm/sec}; \quad v_D = 168 \text{ cm/sec}$$

Considering the magnitudes of Equation (3.3-7) and rearranging, we obtain

$$\left|\dot{\theta}_3\right| = \frac{v_{DB}}{r_3} = \frac{250}{10.0} = 25.0 \text{ rad/sec}$$

Similarly, from Equation (3.3-5), the magnitude of the rotational speed of link 4 is

$$\left|\dot{\theta}_4\right| = \frac{v_D}{r_4} = \frac{168}{9.0} = 18.7 \text{ rad/sec}$$

The direction of rotational speed of link 3, with points B and D at its ends, may be found by examining the direction of the relative velocity $\bar{v}_{DB}$. This relative velocity is that of D with respect to B. That is, if we were to view point D from B on the mechanism, we would see it moving to the right. This corresponds to the link turning in the clockwise direction. This is illustrated in Figure 3.8(f).

We would obtain the same result if we were to use the relative velocity $\bar{v}_{BD}$. In this case, we see point B moving to the right with respect to point D, once again leading us to the

(*Continued*)

EXAMPLE 3.7 Continued

conclusion that the link is turning in the clockwise direction. A similar analysis for link 4 reveals that it is also turning in the clockwise direction. Therefore, the rotational speeds of links 3 and 4 are

$$\dot{\theta}_3 = 25.0 \text{ rad/sec CW}; \quad \dot{\theta}_4 = 18.7 \text{ rad/sec CW}$$

As an illustration of relative velocity, consider the four-bar mechanism shown in Figure 3.9. The coupler is isolated in Figure 3.9(a). At the ends of the link, the relative velocities are shown. For the configuration shown in Figure 3.9(b), the rotation of the coupler is in the clockwise direction. The angular arrow indicates the direction of rotational speed. Later in the cycle, as for instance shown in Figures 3.9(c) and 3.9(d), rotation of the coupler is in the counterclockwise direction.

[Video 3.8] provides an animated sequence of the construction of the velocity polygon presented in this example.

Video 3.8
Mechanism and Velocity Polygon

Video 3.9
Illustration of Relative Velocities

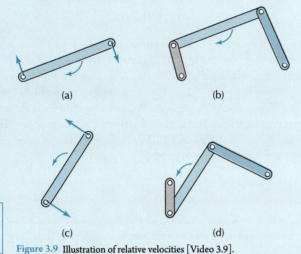

(a)   (b)

(c)   (d)

Figure 3.9 Illustration of relative velocities [Video 3.9].

EXAMPLE 3.8 VELOCITY ANALYSIS OF A FOUR-BAR MECHANISM WITH A COUPLER POINT

Consider the same four-bar kinematic chain as in Example 3.7. Coupler point C is added to link 3 as shown in Figure 3.10. Determine the velocity of the point C.

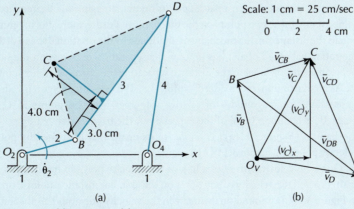

Scale: 1 cm = 25 cm/sec

(a)   (b)

Figure 3.10 (a) Mechanism. (b) Velocity polygon.

(*Continued*)

EXAMPLE 3.8 Continued

SOLUTION

We begin by drawing the portion of the velocity polygon that is the same as that for Example 3.7. Then we consider the velocity of the point C, first with respect to point B. The appropriate relative velocity equation is

$$\overline{v}_C = \overline{v}_B + \overline{v}_{CB} \qquad (3.3\text{-}9)$$

Since points B and C are separated by a fixed distance, the relative velocity between these two points is in a direction perpendicular to the line segment joining these points on the mechanism. Furthermore, since C is moving with respect to B, the line of action of its relative velocity vector is drawn starting from point B on the polygon. However, we do not yet know the magnitude and direction of this vector.

In a similar fashion, the velocity of the point C moves relative to the point D. The appropriate relative velocity equation is

$$\overline{v}_C = \overline{v}_D + \overline{v}_{CD} \qquad (3.3\text{-}10)$$

The relative velocity between C and D on the mechanism is perpendicular to the direction of the line segment joining these points. In this instance, the line of action of this relative velocity is drawn from point D of the polygon.

Since the velocity of the point C is unique and must satisfy both Equations (3.3-9) and (3.3-10), the head of the velocity vector of point C must lie at the intersection of the two lines of action of the relative velocities. This is labeled as point C on the polygon. The relative velocities between points B, C, and D are labeled. The absolute velocity of point C is then found by drawing a vector from the pole point to point C. The result is

$$|\overline{v}_C| = 181 \text{ cm/sec}$$

The direction of this velocity is indicated on the polygon in Figure 3.10(b). The velocity of point C may be broken into its Cartesian components, as shown. The results are

$$(v_C)_x = 87 \text{ cm/sec}; \quad (v_C)_y = 159 \text{ cm/sec}$$

3.4 VELOCITY IMAGE

A useful tool for obtaining the velocity of any point on a link, using a velocity polygon, is referred to as the *velocity image*.

All points within a link are separated by fixed distances as the mechanism moves. Therefore, the relative velocities between a pair of points on the same link will be in a direction perpendicular to a line joining those two points on the mechanism. Furthermore, the magnitude of relative velocity equals the distance on the mechanism multiplied by the rotational speed of the link. However, since the rotational speed of a link is unique, the separation of points on the same link in the velocity polygon is proportional to the distance separating the corresponding points on the mechanism.

As an illustration, reconsider Example 3.8. The velocity polygon is redrawn in Figure 3.11(b). Expressions for magnitudes of relative velocities between points B, C, and D are indicated. Since $\dot{\theta}_3$ is common to all of the expressions, the triangle defined by points B, C, and D on the mechanism (Figure 3.11(a)) is *similar* to its image that is defined by points B, C, and D in the velocity polygon (Figure 3.11 (b)).

Once the velocity image of a link in the velocity polygon has been established, it is a simple matter to find the velocity of any point on the link. For instance, point E on the mechanism shown in Figure 3.11(a) is found on its image in the velocity polygon. The absolute velocity of that point then corresponds to a vector drawn from the pole point to point E on the polygon.

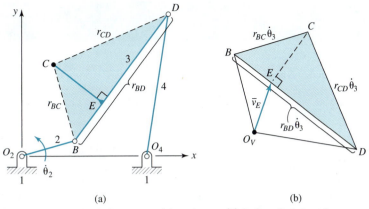

Figure 3.11 Example of velocity image: (a) Mechanism. (b) Outline of velocity polygon.

Velocity images of links 2 and 4 also appear in the velocity polygon. The image of link 2 is between the pole point and point B on the polygon, and the image of link 4 is between the pole point and point D on the polygon.

Further examples of velocity polygon analyses that incorporate velocity images are provided below.

EXAMPLE 3.9 VELOCITY ANALYSIS OF A SLIDER CRANK MECHANISM

For the slider crank mechanism shown in Figure 3.12(a), the link lengths are

$$r_{O_2B} = r_2 = 3.5 \text{ in; } \quad r_{BD} = r_3 = 9.0 \text{ in}$$

The input rotational speed is

$$\dot{\theta}_2 = 500 \text{ rpm CCW} = 52.4 \text{ rad/sec CCW}$$

Determine the rotational speed of link 3 and the velocity of point C for the position shown.

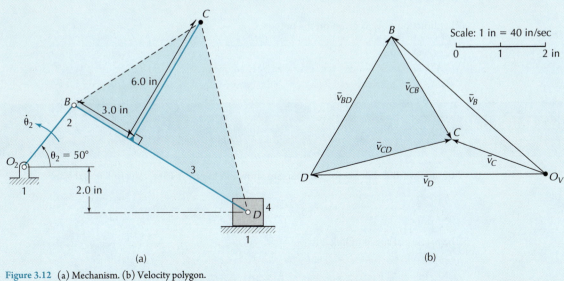

Figure 3.12 (a) Mechanism. (b) Velocity polygon.

(*Continued*)

EXAMPLE 3.9 Continued

SOLUTION

The speed of B is

$$v_B = |r_2\dot{\theta}_2| = 3.5 \text{ in} \times 52.4 \text{ rad/sec} = 183 \text{ in/sec}$$

and the vector is added to the polygon starting from the pole point. Point B is labeled on the polygon in Figure 3.12(b).

The slider, designated as point D, moves with respect to the base link. Therefore, its line of action of velocity is drawn through the pole point. However, at this instant, we do not know the magnitude and direction of this vector.

Point D on the mechanism also moves relative to point B, and therefore we employ the relation

$$\bar{v}_D = \bar{v}_B + \bar{v}_{DB} \tag{3.4-1}$$

The line of action of the relative velocity is added to the polygon, starting from B. The intersection of the two lines of action gives point D on the polygon. Vectors $\bar{v}_D$ and $\bar{v}_{DB}$ are labeled.

Using the same procedure as presented in Example 3.8, point C is determined on the polygon. Alternatively, C may be located by means of the velocity image. Then, $\bar{v}_C$, $\bar{v}_{CB}$, and $\bar{v}_{CD}$ are labeled.

Similar to the analysis performed in Example 3.7, by examining the direction of $\bar{v}_{DB}$, the direction of the rotational speed of link 3 is determined to be in the clockwise direction.

The magnitude of the rotational speed of link 3 is found by measuring

$$v_{DB} = 138 \text{ in/sec}$$

and therefore

$$\dot{\theta}_3 = \frac{v_{DB}}{r_3} = \frac{138}{9.0} = 15.3 \text{ rad/sec CW}$$

Also

$$|\bar{v}_C| = 91 \text{ in/sec}$$

The direction of $\bar{v}_C$ is given on the polygon.

EXAMPLE 3.10 VELOCITY ANALYSIS OF AN INVERTED SLIDER CRANK MECHANISM

For the inverted slider crank mechanism shown in Figure 3.13(a), the link lengths are

$$r_{O_2O_3} = r_1 = 5.0 \text{ in}; \quad r_{O_2B_2} = r_2 = 6.0 \text{ in}$$

The rotational speed of link 2 is

$$\dot{\theta}_2 = 30.0 \text{ rad/sec CW}$$

(Continued)

EXAMPLE 3.10 Continued

Determine the velocity polygon for the mechanism and the rotational speed of link 3 for the given configuration.

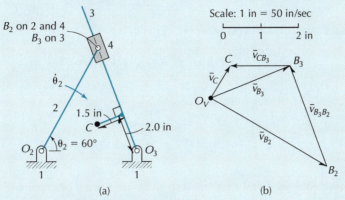

Figure 3.13 (a) Mechanism. (b) Velocity polygon.

SOLUTION

The speed of B_2 is

$$v_{B_2} = |r_2 \dot{\theta}_2| = 6.0 \text{ in} \times 30.0 \text{ rad/sec} = 180 \text{ in/sec}$$

The velocity of B_2 is drawn on the polygon.
Next, we consider the velocity of point B_3, and note that

$$\overline{v}_{B_3} = \overline{v}_{B_3 O_3} = r_{B_3 O_3} \dot{\theta}_3 \overline{i}_{\theta_{B_3 O_3}} \tag{3.4-2}$$

where

$$r_{B_3 O_3} = 5.57 \text{ in}$$

is measured from the scaled diagram.

The line of action of the velocity of B_3 is added to the polygon starting from the pole point. We now examine the motion of B_3 with respect to B_2, employing

$$\overline{v}_{B_3} = \overline{v}_{B_2} + \overline{v}_{B_3 B_2} \tag{3.4-3}$$

In Equation (3.4-3), $\overline{v}_{B_2}$ is known. In addition, from the special case of a slider moving on a straight slide presented in Section 2.5.2, in this instance we have

$$\overline{v}_{B_3 B_2} = \dot{r}_{B_3 B_2} \overline{i}_{r_{B_3 O_3}} \tag{3.4-4}$$

The line of action of this relative velocity, which is along the direction of the slide, is added to the polygon, starting from B_2. The intersection of the two lines of action gives point B_3 on the polygon.

Similar analyses as in previous examples leads to the determination of point C on the polygon. All vectors in the polygon can now be labeled, as shown in Figure 3.13(b).

(Continued)

EXAMPLE 3.10 Continued

Now that the polygon has been completed, the sliding velocity between points B_3 and B_2 has a magnitude

$$v_{B_3B_2} = |\dot{r}_{B_3B_2}| = 140 \text{ in/sec}$$

and the magnitude of the absolute velocity of B_3 is

$$v_{B_3} = v_{B_3O_3} = 113 \text{ in/sec}$$

The rotational speed of link 3 is

$$\dot{\theta}_3 = \frac{v_{B_3O_3}}{r_{B_3O_3}} = \frac{113}{5.57} = 20.3 \text{ rad/sec CW}$$

where the direction of the rotational speed has been found by examining the direction of $\overline{v}_{B_3}$ and the location of B_3 with respect to O_3.

EXAMPLE 3.11 VELOCITY ANALYSIS OF A SIX-LINK MECHANISM

For the mechanism shown in Figure 3.14(a), the link lengths are

$$r_{O_2O_4} = r_1 = 2.0 \text{ in}; \quad r_{BO_2} = r_2 = 1.0 \text{ in}; \quad r_{BD} = r_3 = 3.5 \text{ in}$$

$$r_{BC} = 2.0 \text{ in}; \quad r_{DO_4} = r_4 = 2.0 \text{ in}; \quad r_{E_5C} = 1.5 \text{ in}$$

The input rotational speed is

$$\dot{\theta}_2 = 400 \text{ rpm CCW} = 41.9 \text{ rad/sec CCW}$$

Determine the velocity polygon of the mechanism and the rotational speeds of links 3, 4, and 5 for the given configuration.

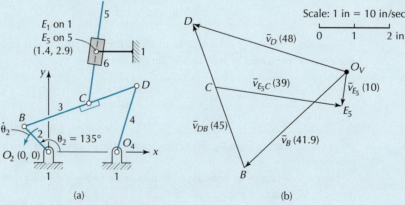

Figure 3.14 (a) Mechanism. (b) Velocity polygon.

(Continued)

EXAMPLE 3.11 Continued

SOLUTION
The following procedure is employed:

- Calculate v_B

$$v_B = |r_2\dot{\theta}_2| = 1.0 \text{ in} \times 41.9 \text{ rad/sec} = 41.9 \text{ in/sec}$$

- Locate B on the polygon.
- Employ

$$\overline{v}_D = \overline{v}_B + \overline{v}_{DB} \qquad (3.4\text{-}5)$$

and draw lines of action of $\overline{v}_D$ and $\overline{v}_{DB}$ to locate D on polygon.

- Locate C using the velocity image.
- Employ

$$\overline{v}_{E_5} = \overline{v}_C + \overline{v}_{E_5 C} \qquad (3.4\text{-}6)$$

and draw the lines of action of $\overline{v}_{E_5}$ and $\overline{v}_{E_5 C}$ to locate E_5 on the polygon.

- Label all vectors. The completed velocity polygon is shown in Figure 3.14(b). Magnitudes of the velocities are indicated in parentheses.
- The rotational speeds are

$$\dot{\theta}_3 = \frac{v_{DB}}{r_3} = \frac{45}{3.5} = 12.9 \text{ rad/sec CCW}$$

$$\dot{\theta}_4 = \frac{v_D}{r_4} = \frac{48}{2.0} = 24.0 \text{ rad/sec CCW}$$

$$\dot{\theta}_5 = \frac{v_{CE_5}}{r_{CE_5}} = \frac{39}{1.5} = 26.0 \text{ rad/sec CW}$$

3.5   ACCELERATION POLYGON ANALYSIS

The acceleration polygon method has many similarities to the velocity polygon method. However, in order to complete an acceleration polygon, results from the velocity polygon analysis must first be obtained. Values of rotational and sliding velocities are needed to determine the normal and Coriolis accelerations.

To construct an acceleration polygon, the following general steps are required:

1. Draw a diagram of the mechanism to scale, in the configuration under analysis.
2. Set up a polar diagram, where every point with zero acceleration is at the origin (pole point) O_A.
3. Starting from the pole point, draw in acceleration vectors that can be readily calculated. For instance, consider those based on prescribed input motions with respect to the base link.

4. Write the appropriate relative acceleration equations to aid in calculating and constructing other acceleration vectors.

5. Proceed through the mechanism until all points of interest are analyzed.

The acceleration polygon can then be employed to determine absolute and relative accelerations. This will be illustrated by the following examples.

EXAMPLE 3.12 ACCELERATION ANALYSIS OF A FOUR-BAR MECHANISM WITH CONSTANT INPUT SPEED

Consider the four-bar mechanism presented in Example 3.7 (refer to Figure 3.8). Pertinent information required from the velocity analysis is illustrated in Figure 3.15(a). The rotational speed of link 2 is constant. Determine the angular accelerations of links 3 and 4 for the position shown by constructing an acceleration polygon.

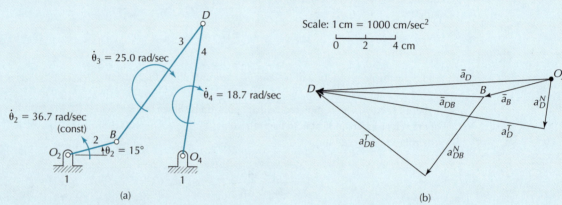

Figure 3.15 (a) Mechanism. (b) Acceleration polygon [Video 3.15].

SOLUTION

For this example, all the kinematic pairs are separated by fixed distances. Therefore, from Section 2.7.1, we need only concern ourselves with normal and tangential components of relative acceleration.

Construction of the acceleration polygon begins by determining the acceleration of point B, which is separated from O_2 by a fixed distance. Using Equation (2.7-4), we obtain

$$\bar{a}_B = \bar{a}_{BO_2} = (-r_2\dot{\theta}_2^2)\bar{i}_{r_{BO_2}} + (r_2\ddot{\theta}_2)\bar{i}_{\theta_{BO_2}}$$

$$= (a_{BO_2}^N)\bar{i}_{r_{BO_2}} + (a_{BO_2}^T)\bar{i}_{\theta_{BO_2}} \qquad (3.5\text{-}1)$$

Furthermore, since

$$\dot{\theta}_2 = \text{constant}; \ddot{\theta}_2 = 0$$

Equation (3.5-1) reduces to

$$\bar{a}_B = (-r_2\dot{\theta}_2^2)\bar{i}_{r_{BO_2}} \qquad (3.5\text{-}2)$$

(Continued)

EXAMPLE 3.12 Continued

The direction of the acceleration of B must be in the negative radial direction of the line segment defined by points O_2 and B. The value is

$$a_B^N = -r_2\dot{\theta}_2^2 = -3.5 \times 36.7^2 = -4710 \text{ cm/sec}^2$$

Using the scale

$$1\text{cm} = 1000 \text{ cm/sec}^2 \tag{3.5-3}$$

the acceleration vector of point B is drawn relative to the pole point.

Next, we seek to determine the acceleration of point D. We recognize that D moves relative to point O_4, so we write the relative acceleration equation

$$\bar{a}_D = \bar{a}_{O_4} + \bar{a}_{DO_4} \tag{3.5-4}$$

Since the distance between D and O_4 is constant, the relative acceleration is

$$\bar{a}_{DO_4} = (-r_4\dot{\theta}_4^2)\bar{i}_{r_{DO_4}} + (r_4\ddot{\theta}_4)\bar{i}_{\theta_{DO_4}}$$

$$= (a_{DO_4}^N)\bar{i}_{r_{DO_4}} + (a_{DO_4}^T)\bar{i}_{\theta_{DO_4}} \tag{3.5-5}$$

Using the value of rotational speed from Example 3.7, the normal acceleration component is

$$a_{DO_4}^N = -r_4\dot{\theta}_4^2 = -9.0 \times 18.7^2 = -3150 \text{ cm/sec}^2$$

From the pole point, we draw the normal component of acceleration, and from the head of the vector for normal acceleration, the line of action of the tangential component. We do not yet know the magnitude of the tangential component. Also, the vector may point in one of two possible directions and still be perpendicular to the line segment defined by point D and O_4. This direction has yet to be determined.

Point D on the mechanism also moves relative to point B. Also, points B and D are separated by a fixed distance. We employ the following relative acceleration equation:

$$\bar{a}_D = \bar{a}_B + \bar{a}_{DB} \tag{3.5-6}$$

where

$$\bar{a}_{DB} = (-r_3\dot{\theta}_3^2)\bar{i}_{r_{DB}} + (r_3\ddot{\theta}_3)\bar{i}_{\theta_{DB}}$$

$$= (a_{DB}^N)\bar{i}_{r_{DB}} + (a_{DB}^T)\bar{i}_{\theta_{DB}} \tag{3.5-7}$$

The normal acceleration component is

$$a_{DB}^N = -r_3\dot{\theta}_3^2 = -10.0 \times 25.0^2 = -6250 \text{ cm/sec}^2$$

The relative acceleration $\bar{a}_{DB}$ in Equation (3.5-6) is with respect to point B. Therefore, the normal acceleration component and the line of action of the tangential acceleration are added to the polygon starting from the point B. For the tangential component, we only know the line of action, not its magnitude and direction.

It must be recognized, however, that the acceleration of point D is unique. In order to satisfy the relative acceleration expressions, Equations (3.5-4) and (3.5-6), as well as the lines of action of the tangential acceleration components, the head of $\bar{a}_D$ must lie at the intersection of

(Continued)

EXAMPLE 3.12 Continued

the two lines of action. This is designated as point D on the polygon. The completed acceleration polygon is given in Figure 3.15(b).

Now that the acceleration polygon has been completed, it is possible to determine the magnitudes and directions of angular accelerations of links 3 and 4. To do this, we measure the tangential components of acceleration from the polygon employing the scale given by Equation (3.5-3). The result is

$$a_{DB}^T = 9080 \text{ cm/sec}^2; \quad a_D^T = 15,300 \text{ cm/sec}^2$$

Considering only the tangential component of Equations (3.5-5) and (3.5-7), expressions for the magnitude of the angular acceleration of links 3 and 4 are

$$|\ddot{\theta}_3| = \frac{a_{DB}^T}{r_3}; \quad |\ddot{\theta}_4| = \frac{a_D^T}{r_4} \tag{3.5-8}$$

Substituting values in Equations (3.5-8) gives

$$|\ddot{\theta}_3| = \frac{a_{DB}^T}{r_3} = \frac{9080}{10.0} = 908 \text{ rad/sec}^2$$

$$|\ddot{\theta}_4| = \frac{a_D^T}{r_4} = \frac{15,300}{9.0} = 1700 \text{ rad/sec}^2 \tag{3.5-9}$$

The direction of the angular acceleration of link 3, with points B and D at its ends, may be found by examining the direction of a_{DB}^T. This relative acceleration is as seen with respect to point B. That is, if we were to view D from B on the mechanism, we would see it accelerating upward and to the left. This corresponds to the link at this instant having an angular acceleration in the counterclockwise direction. This is illustrated in Figure 3.16. Alternatively, we could obtain the same conclusion regarding the direction of the angular acceleration by employing the relative acceleration a_{BD}^T, where

$$a_{BD}^T = -a_{DB}^T \tag{3.5-10}$$

as is illustrated in Figure 3.16.

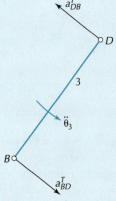

Figure 3.16 Relative components of tangential acceleration between points B and D.

(Continued)

EXAMPLE 3.12 Continued

A similar analysis for link 4 reveals that it has an angular acceleration in the counterclockwise direction at the same instant. Therefore, the angular accelerations of links 3 and 4 are

$$\ddot{\theta}_3 = 908 \ \text{rad/sec}^2 \ \text{CCW}; \quad \ddot{\theta}_4 = 1700 \ \text{rad/sec}^2 \ \text{CCW}$$

Since the direction of angular acceleration for both links (i.e., counterclockwise) is opposite to that of the angular velocity (i.e., clockwise), both links in this position are undergoing an angular *deceleration*—or in other words, slowing down. If directions of angular velocity and angular acceleration for a link were the same, the link would be undergoing an angular *acceleration*—in other words, speeding up.

Video 3.15
Mechanism and
Acceleration
Polygon

[Video 3.15] provides an animated sequence of the construction of the acceleration polygon presented in this example.

EXAMPLE 3.13 ACCELERATION ANALYSIS OF A FOUR-BAR MECHANISM WITH VARIABLE INPUT SPEED

Consider the four-bar mechanism presented in Example 3.12. The configuration of the mechanism and the instantaneous value of the input rotational speed of link 2 are the same as given in Example 3.12. However, link 2 is decelerating at the rate

$$\ddot{\theta}_2 = 600 \ \text{rad/sec}^2 \ \text{CW}$$

Determine the angular accelerations of links 3 and 4 for the position shown by constructing an acceleration polygon.

SOLUTION

Since the configuration of the mechanism and input rotational speed are the same as for Example 3.12, so too are the rotational speeds of links 3 and 4, as well as the normal components of acceleration.

Using Equation (2.7-4), the acceleration of point B is

$$
\begin{aligned}
\bar{a}_B = \bar{a}_{BO_2} &= (-r_2\dot{\theta}_2^2)\bar{i}_{r_{BO_2}} + (r_2\ddot{\theta}_2)\bar{i}_{\theta_{BO_2}} \\
&= (-3.5 \times 36.7^2)\bar{i}_{r_{BO_2}} + [3.5 \times (-600)]\bar{i}_{\theta_{BO_2}} \\
&= (-4710)\bar{i}_{r_{BO_2}} + (-2100)\bar{i}_{\theta_{BO_2}} \\
&= (a_{BO_2}^N)\bar{i}_{r_{BO_2}} + (a_{BO_2}^T)\bar{i}_{\theta_{BO_2}}
\end{aligned}
\tag{3.5-11}
$$

This acceleration is added to the polygon, starting from the pole point. The remainder of the construction of the acceleration polygon is similar to that presented in Example 3.12. From the polygon shown in Figure 3.17(b), we measure

$$a_{DB}^T = 13{,}200 \ \text{cm/sec}^2; \quad a_D^T = 18{,}000 \ \text{cm/sec}^2$$

from which, using Equations (3.5-8), the angular accelerations of links 3 and 4 are

$$\ddot{\theta}_3 = 1320 \ \text{rad/sec}^2 \ \text{CCW}; \quad \ddot{\theta}_4 = 2000 \ \text{rad/sec}^2 \ \text{CCW}$$

(Continued)

EXAMPLE 3.13 Continued

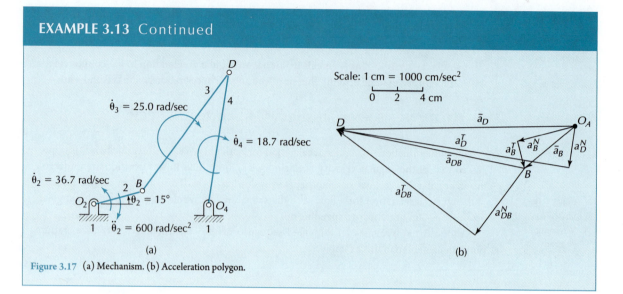

Figure 3.17 (a) Mechanism. (b) Acceleration polygon.

EXAMPLE 3.14 ACCELERATION ANALYSIS OF A FOUR-BAR MECHANISM WITH A COUPLER POINT

Consider the four-bar mechanism presented in Example 3.8 (Figure 3.10), where coupler point C has been added to link 3. The mechanism is redrawn in Figure 3.18(a). The rotational speed of link 2 is constant. Determine the acceleration of the point C.

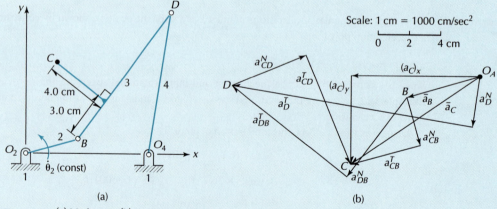

Figure 3.18 (a) Mechanism. (b) Acceleration polygon.

SOLUTION

We begin by drawing the portion of the acceleration polygon that is the same as that given in Example 3.12. Then we consider the acceleration of point C first with respect to point B. The appropriate relative acceleration equation is

$$\bar{a}_C = \bar{a}_B + \bar{a}_{CB} \tag{3.5-12}$$

Since points B and C are separated by a fixed distance, $\bar{a}_{CB}$ has only normal and tangential components. The normal component is in the negative radial direction of the line segment

(Continued)

EXAMPLE 3.14 Continued

connecting the two points on the mechanism. The tangential component is in a direction perpendicular to the line segment. Furthermore, this relative acceleration is with respect to moving point B and will be drawn starting from point B on the polygon.

We start with the normal component, since both its value and direction may be determined. The value of the normal acceleration is

$$a_{CB}^N = -r_{CB}\dot{\theta}_3^2 = -(3.0^2 + 4.0^2)^{1/2} \times 25.0^2 = -3130 \text{ cm/sec}^2$$

which is drawn on the polygon. Also drawn is the line of action of a_{CD}^T. However, at this instant, we do not know its magnitude and direction.

In a similar fashion, the acceleration of point C moves relative to point D. The appropriate relative acceleration equation is

$$\bar{a}_C = \bar{a}_D + \bar{a}_{CD} \tag{3.5-13}$$

where the normal component of $\bar{a}_{CD}$ is

$$a_{CD}^N = -r_{CD}\dot{\theta}_3^2 = -8.06 \times 25.0^2 = -5040 \text{ cm/sec}^2$$

This normal component, along with the line of action of a_{CD}^T, are added to the polygon, starting from point D.

Since the acceleration of point C is unique and must satisfy both of the requirements as given in Equations (3.5-12) and (3.5-13), point C on the polygon must lie at the intersection of the two lines of action of the relative tangential accelerations. The absolute acceleration of point C is then found by drawing a vector from the pole point to point C. The vector drawn gives the direction, and employing the same scale as other vectors in the polygon, the magnitude may be determined. The magnitude is

$$|\bar{a}_C| = 9740 \text{ cm/sec}^2$$

The completed acceleration polygon is shown in Figure 3.18(b). The acceleration of point C maybe broken into its Cartesian components, as shown. The results are

$$(a_C)_x = -8050 \text{ cm/sec}^2; \quad (a_C)_y = -5740 \text{ cm/sec}^2$$

3.6 ACCELERATION IMAGE

A useful tool for obtaining the acceleration of any point on a link from an acceleration polygon is referred to as the *acceleration image*.

All points within a link are separated by fixed distances. Therefore, the relative accelerations between a pair of points on the same link can have only normal and tangential components of relative acceleration. That is, there are no sliding or Coriolis components of relative acceleration between points on the same link. The magnitudes of the normal and tangential components of relative acceleration, as given by Equations (2.6-3) and (2.6-7), are proportional to the distances between the points. In addition, the normal acceleration component is dependent on the rotational speed of the link, whereas the tangential component is proportional to the angular acceleration. However, since there are unique values of rotational speed and angular acceleration for a link, the distance separating points on the same link in a mechanism is proportional to the distance of separation in the acceleration polygon.

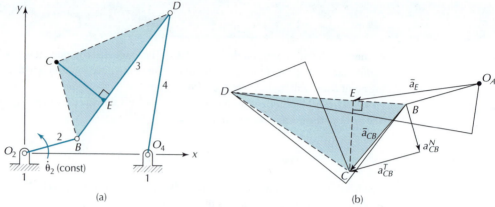

Figure 3.19 Example of acceleration image: (a) Mechanism. (b) Outline of acceleration polygon.

As an illustration, let us reconsider Example 3.14. The acceleration polygon is redrawn in Figure 3.19(b). The relative acceleration between points B an C is indicated, which consists of a normal component and a tangential component. The triangle defined by points B, C, and D on the mechanism (Figure 3.19(a)) is similar to its image that is defined by points B, C, and D as they appear in the acceleration polygon (Figure 3.19(b)).

Once the acceleration image of a link in the acceleration polygon has been established, it is a simple matter to find the acceleration of any point on the link. For instance, point E on the mechanism shown in Figure 3.19(a) is found on its image in the acceleration polygon. The absolute acceleration of E then corresponds to a vector drawn from the pole point to point E on the polygon.

Acceleration images of links 2 and 4 also appear in the acceleration polygon. The image of link 2 is between the pole point and point B on the polygon, and the image of link 4 is between the pole point and point D on the polygon.

Further examples of acceleration polygon analyses that incorporate acceleration images are provided below.

EXAMPLE 3.15 ACCELERATION ANALYSIS OF A SLIDER CRANK MECHANISM

Consider the slider crank mechanism presented in Example 3.9 (refer to Figure 3.12). The mechanism is redrawn in Figure 3.20(a). The rotational speed of link 2 is constant. Determine the acceleration polygon of the mechanism and the angular acceleration of link 3 for the given configuration.

SOLUTION

In addition to normal and tangential components of relative acceleration, there can be a sliding component of acceleration of the slider (link 4) with respect to the slide (link 1). However, since the direction of the slide is fixed, no component of Coriolis acceleration exists.

The normal component of the acceleration of B is

$$a_B^N = -r_2\dot{\theta}_2^2 = -3.5 \times 52.4^2 = -9610 \text{ in/sec}^2$$

and the vector is added to the polygon starting from the pole point. Point B on the polygon is labeled.

(Continued)

EXAMPLE 3.15 Continued

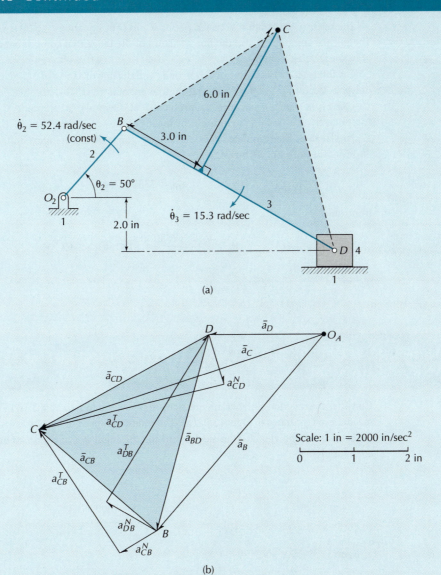

Figure 3.20 (a) Mechanism. (b) Acceleration polygon.

The slider, designated as point D, moves with respect to the base link. Therefore, its line of action of acceleration is drawn with respect to the pole point. However, at this instant, we do not know the magnitude and direction of this vector.

Next, we employ the relation

$$\bar{a}_D = \bar{a}_B + \bar{a}_{DB} \tag{3.6-1}$$

where

$$\bar{a}_{DB} = (-r_3 \dot{\theta}_3^2)\bar{i}_{r_{DB}} + (r_3 \ddot{\theta}_3)\bar{i}_{\theta_{DB}}$$

$$= (a_{DB}^N)\bar{i}_{r_{DB}} + (a_{DB}^T)\bar{i}_{\theta_{DB}} \tag{3.6-2}$$

The normal acceleration component is

$$a_{DB}^N = -r_3 \dot{\theta}_3^2 = -9.0 \times 15.3^2 = -2110 \text{ in/sec}^2$$

(*Continued*)

EXAMPLE 3.15 Continued

The relative acceleration between points B and D in Equation (3.6-2) is with respect to B. The normal component and the line of action of the tangential component are added to the polygon, starting from B.

The intersection of the line of action of $\bar{a}_D$ and the line of action of a_{DB}^T gives point D on the polygon. Vector $\bar{a}_D$ and the tangential component of $\bar{a}_{DB}$ are labeled.

Using the same procedure as presented in Example 3.14, point C is determined. Alternatively, it is possible to locate C by means of the acceleration image. Then, $\bar{a}_C$, $\bar{a}_{CB}$, and $\bar{a}_{CD}$ are labeled. The completed acceleration polygon is shown in Figure 3.20(b).

From the polygon, we measure

$$a_{DB}^T = 7330 \text{ in/sec}^2$$

from which

$$\ddot{\theta}_3 = \frac{a_{DB}^T}{r_3} = \frac{7330}{9.0} = 814 \text{ rad/sec}^2 \text{ CCW}$$

EXAMPLE 3.16 ACCELERATION ANALYSIS OF AN İNVERTED SLIDER CRANK MECHANISM

Consider the inverted slider crank mechanism presented in Example 3.10 (Figure 3.13). The mechanism is redrawn in Figure 3.21(a). The rotational speed of link 2 is constant. Determine the acceleration polygon for the mechanism and the angular acceleration of link 3 in the given configuration.

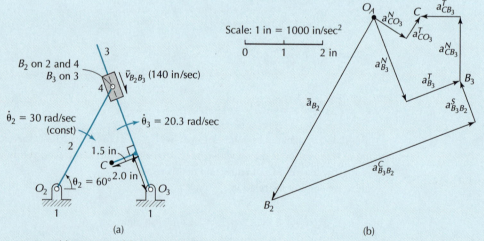

Figure 3.21 (a) Mechanism. (b) Acceleration polygon.

SOLUTION

For this problem, there can be a sliding component of acceleration of the slider with respect to the slide. However, since the slide rotates, there can be also a component of Coriolis acceleration between the slide and slider.

(Continued)

EXAMPLE 3.16   Continued

Construction of the acceleration polygon begins with determining the acceleration of point B_2. Since B_2 and O_2 are separated by fixed distance r_2, we have

$$\bar{a}_{B_2} = \bar{a}_{B_2 O_2} = (-r_2 \dot{\theta}_2^2) \bar{i}_{r_{B_2 O_2}} + (r_2 \ddot{\theta}_2) \bar{i}_{\theta_{B_2 O_2}} \tag{3.6-3}$$

For constant rotational speed, Equation (3.6-3) reduces to

$$\bar{a}_{B_2} = (-r_2 \dot{\theta}_2^2) \bar{i}_{r_{B_2 O_2}} = (-6.0 \times 30^2) \bar{i}_{r_{B_2 O_2}} = -5400 \text{ in/sec}^2 \, \bar{i}_{r_{B_2 O_2}}$$

Acceleration vector $\bar{a}_{B_2}$ is drawn relative to the pole point in the negative radial direction defined by points B_2 and O_2.

Next, we determine the acceleration of point B_3. Recognizing that B_3 moves relative to point O_3, we write the relative acceleration equation

$$\bar{a}_{B_3} = \bar{a}_{O_3} + \bar{a}_{B_3 O_3} \tag{3.6-4}$$

For a fixed distance between B_3 and O_3,

$$\bar{a}_{B_3 O_3} = (-r_{B_3 O_3} \dot{\theta}_3^2) \bar{i}_{r_{B_3 O_3}} + (r_{B_3 O_3} \ddot{\theta}_3) \bar{i}_{\theta_{B_3 O_3}}$$

$$= (a_{B_3 O_3}^N) \bar{i}_{r_{B_3 O_3}} + (a_{B_3 O_3}^T) \bar{i}_{\theta_{B_3 O_3}} \tag{3.6-5}$$

The normal acceleration component is

$$a_{B_3 O_3}^N = -r_{B_3 O_3} \dot{\theta}_3^2 = -5.57 \times 20.3^2 = -2300 \text{ in/sec}^2$$

From the pole point, we draw the normal component of acceleration, and from the head of the vector for normal acceleration, the line of action of the tangential component.

We now examine the relative acceleration between B_3 and B_2. Using the special case of the acceleration of a slider on a straight slide presented in Section 2.7.2, we have in this instance

$$\bar{a}_{B_3} = \bar{a}_{B_2} + \bar{a}_{B_3 B_2} \tag{3.6-6}$$

where

$$\bar{a}_{B_3 B_2} = (\ddot{r}_{B_3 B_2}) \bar{i}_{r_{B_3 O_3}} + (2 \dot{r}_{B_3 B_2} \dot{\theta}_3) \bar{i}_{\theta_{B_3 O_3}}$$

$$= (a_{B_3 B_2}^S) \bar{i}_{r_{B_3 O_3}} + (a_{B_3 B_2}^C) \bar{i}_{\theta_{B_3 O_3}} \tag{3.6-7}$$

The magnitude of the Coriolis acceleration is

$$|a_{B_3 B_2}^C| = |2 \dot{r}_{B_3 B_2} \dot{\theta}_3| = |2 \times 140 \times 20.3| = 5680 \text{ in/sec}^2$$

In Equation (3.6-7), the relative Coriolis and sliding acceleration components are of point B_3 with respect to point B_2. Therefore, these relative acceleration components are added to the polygon starting from point B_2. We can begin by adding the Coriolis acceleration, since we now know its magnitude, and can determine its direction. At this instant, we only know the line of action of the sliding acceleration.

We currently have the acceleration of the slider (point B_2), and will draw a Coriolis acceleration of the slide (point B_3) with respect to the slider. Directions of relative Coriolis acceleration are given in Figure 2.15. Using this figure, with the known combination of relative sliding and rotational speed of the slide, we determine that the direction of the relative Coriolis acceleration is downwards and to the left. However, this figure was set up with the acceleration

(Continued)

EXAMPLE 3.16 Continued

of the slider with respect to the slide. In this instance, we know the acceleration of the slider, and are seeking the acceleration of the slide. Recognizing that

$$a_{B_2B_3}^C = -a_{B_3B_2}^C \tag{3.6-8}$$

we reverse the direction of the acceleration as shown in Figure 2.15. This is illustrated in Figure 3.22.

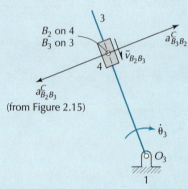

Figure 3.22 Directions of Coriolis acceleration.

The Coriolis component is added to the polygon in Figure 3.21(b). At the head of the vector of the Coriolis acceleration, the line of action of the sliding acceleration is added.

The acceleration of the point B_3 is unique. Therefore, in order to satisfy the relative acceleration expressions, Equations (3.6-4) and (3.6-6), as well as the lines of action of the tangential and sliding acceleration components, the head of the acceleration of point B_3 must lie at the intersection of the two lines of action. This is designated as point B_3 on the polygon.

The acceleration of point C, located on link 3, is now determined by considering the appropriate normal and tangential components with respect to points B_3 and O_3, or by using the acceleration image. The completed acceleration polygon is given in Figure 3.21(b).

From this figure we have

$$a_{B_3O_3}^T = 1490 \text{ in/sec}^2$$

and therefore the angular acceleration of link 3 is

$$\ddot{\theta}_3 = \frac{a_{B_3O_3}^T}{r_{B_3O_3}} = \frac{1490}{5.57} = 268 \text{ rad/sec}^2 \text{ CW}$$

Since this angular acceleration is in the same direction as the rotational speed, link 3 is speeding up.

EXAMPLE 3.17 ACCELERATION ANALYSIS OF A SIX-LINK MECHANISM

Consider the mechanism presented in Example 3.11 (Figure 3.14). The mechanism is redrawn in Figure 3.23(a). The rotational speed of link 2 is constant. Determine the acceleration polygon of the mechanism and the angular acceleration of link 5 for the given configuration.

(Continued)

EXAMPLE 3.17 Continued

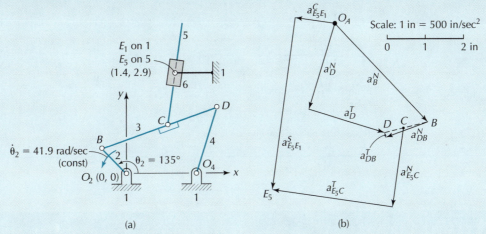

Figure 3.23 (a) Mechanism, (b) Acceleration polygon.

SOLUTION

For this problem, there can be a sliding component of acceleration of the slider (link 6) with respect to the slide (link 5). However, since the slide rotates, a component of Coriolis acceleration can be present.

The following procedure is employed:

(a) Calculate the normal acceleration components:

$$a_{BO_2}^N = -r_{BO_2}\,\dot\theta_2^2 = -1.0 \times 41.9^2 = -1760 \text{ in/sec}^2$$

$$a_{DB}^N = -r_{DB}\,\dot\theta_3^2 = -3.5 \times 12.9^2 = -582 \text{ in/sec}^2$$

$$a_{DO_4}^N = -r_{DO_4}\,\dot\theta_4^2 = -2.0 \times 24.0^2 = -1150 \text{ in/sec}^2$$

$$a_{E_5C}^N = -r_{E_5C}\,\dot\theta_5^2 = -1.5 \times 26.0^2 = -1010 \text{ in/sec}^2$$

(b) Calculate the magnitude of the Coriolis acceleration:

$$a_{E_5E_1}^C = 2\dot r_{E_5E_1}\,\dot\theta_5 = 2 \times 10 \times 26.0 = 520 \text{ in/sec}^2$$

(c) Construct the acceleration polygon as follows:

- Draw a_B^N and locate B.
- Draw a_D^N and a_{DB}^N
- Draw the lines of action of a_D^T and a_{DB}^T and locate D.
- Locate C by using the acceleration image.
- Draw $a_{E_5C}^N$ and $a_{E_5E_1}^C$ (refer to Figure 3.24 for the direction of the Coriolis acceleration).
- Draw the lines of action of $a_{E_5C}^T$ and $a_{E_5E_1}^S$ and locate E_5.
- Label all vectors. The completed acceleration polygon is shown in Figure 3.23(b).

(d) Determine the angular acceleration of link 5.

$$\ddot\theta_5 = \frac{a_{E_5C}^T}{r_{E_5C}} = \frac{1550}{1.5} = 1030 \text{ rad/sec}^2 \text{ CCW}$$

(Continued)

EXAMPLE 3.17 Continued

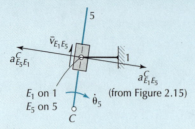

Figure 3.24 Directions of Coriolis acceleration.

EXAMPLE 3.18 VELOCITY AND ACCELERATION ANALYSIS OF A CAM MECHANISM

Consider the cam mechanism shown in Figure 3.25. The rotational speed of link 2 and link lengths are

$$\dot{\theta}_2 = 3.0 \, \text{rad/sec CCW (constant)}$$
$$r_{O_2 B_2} = 1.3 \, \text{cm}; \quad r_{O_4 B_4} = 1.8 \, \text{cm}$$

The radii of the circular cam and roller follower are indicated in the figure.
Determine the velocity and acceleration polygons for the given configuration.

SOLUTION

We consider the center of the roller follower, point B, to be composed of two points, B_2 and B_4. Point B_2 is considered as an imaginary physical extension of link 2. The motion between these two points is equivalent to a slider moving on a curved slide. The radius of curvature is equal to the radius of the cam plus the radius of the roller follower. Between points B_2 and B_4 there can be relative sliding, plus Coriolis and circular components of acceleration.

The calculated values of velocity and acceleration required to complete the polygons are

$$v_{B_2} = r_{B_2 O_2} \dot{\theta}_2 = 1.3 \times 3.0 = 3.9 \, \text{cm/sec}$$
$$\dot{\theta}_4 = \frac{v_{B_4}}{r_{B_4 O_4}} = \frac{2.6}{1.8} = 1.44 \, \text{rad/sec CCW}$$

$$a_{B_2}^N = -r_{B_2 O_2} \dot{\theta}_2^2 = -1.3 \times 3.0^2 = -11.7 \, \text{cm/sec}^2$$
$$a_{B_4}^N = -r_{B_4 O_4} \dot{\theta}_4^2 = -1.8 \times 1.44^2 = -3.7 \, \text{cm/sec}^2$$
$$a_{B_4 B_2}^C = 2 v_{B_4 B_2} \dot{\theta}_2 = 2 \times 2.0 \times 3.0 = 12.0 \, \text{cm/sec}^2$$
$$\frac{v_{B_4 B_2}^2}{\rho} = \frac{v_{B_4 B_2}^2}{r_{CB_2}} = \frac{2.0^2}{(1.0 + 0.25)} = 3.2 \, \text{cm/sec}^2$$

The completed velocity polygon and acceleration polygon are shown in Figures 3.25(b) and 3.25(c), respectively. Magnitudes of the velocities and accelerations are indicated in parentheses.

(Continued)

EXAMPLE 3.18 Continued

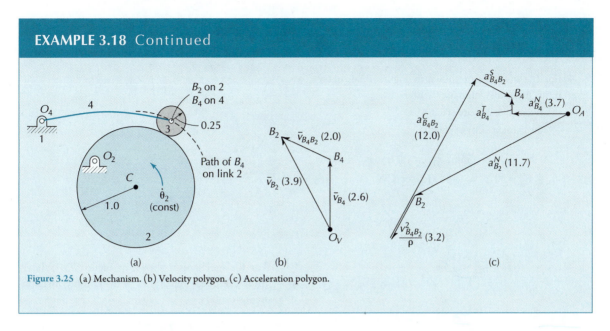

Figure 3.25 (a) Mechanism. (b) Velocity polygon. (c) Acceleration polygon.

PROBLEMS

Instantaneous Centers, Velocity Analysis

P3.1 For the mechanism shown in Figure P3.1, determine the magnitudes of the linear velocities of points B, C, and D using instantaneous centers.

$$r_{O_2O_4} = 4.0 \text{ cm}; \quad r_{O_2B} = 1.0 \text{ cm}; \quad r_{BD} = 3.0 \text{ cm}$$
$$r_{O_4D} = 2.5 \text{ cm}; \quad r_{BC} = 2.0 \text{ cm}; \quad r_{CD} = 1.5 \text{ cm}$$
$$\theta_2 = 45°; \quad \dot{\theta}_2 = 60 \text{ rad/sec CCW}$$

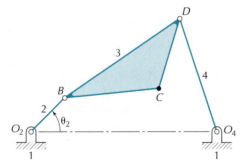

Figure P3.1

P3.2 For the mechanism shown in Figure P3.2, determine the linear velocity of link 3 using instantaneous centers.

$$r_{O_2C} = 1.5 \text{ cm}; \quad \rho = 0.75 \text{ cm}$$
$$\theta_2 = 110°; \quad \dot{\theta}_2 = 600 \text{ rpm CW}$$

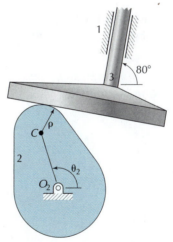

Figure P3.2

P3.3 For the mechanism shown in Figure P3.3, determine the magnitude of the linear velocity of link 4 using instantaneous centers.

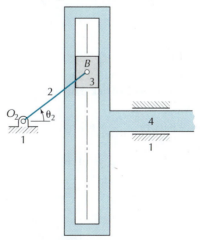

Figure P3.3

$$r_{O_2B} = 2.5 \text{ cm}$$
$$\theta_2 = 45°; \quad \dot{\theta}_2 = 40 \text{ rpm CCW}$$

P3.4 For the mechanism shown in Figure P3.4, determine, using instantaneous centers,

(a) the angular velocity of link 4

(b) the linear velocity of point D

$$r_{O_2B_2} = 2.0 \text{ in}; \quad r_{O_2O_4} = 1.5 \text{ in}; \quad r_{O_4C} = 1.0 \text{ in}; \quad r_{CD} = 3.0 \text{ in}$$
$$\theta_2 = 30°; \quad \dot{\theta}_2 = 36 \text{ rad/sec CW}$$

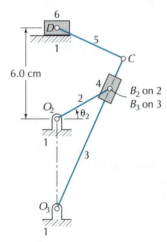

Figure P3.4

P3.5 For the mechanism shown in Figure P3.5, determine the linear velocity of link 6 using instantaneous centers.

$$r_{O_2O_3} = 6.0 \text{ cm}; r_{O_2B_2} = 4.0 \text{ cm}; r_{O_3C} = 11.0 \text{ cm}; r_{CD} = 5.0 \text{ cm}$$
$$\theta_2 = 30°; \dot{\theta}_2 = 20 \text{ rad/sec CCW}$$

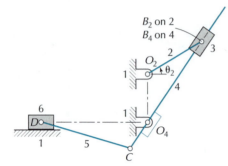

Figure P3.5

P3.6 For the mechanism shown in Figure P3.6, determine the rotational speed of link 6 using instantaneous centers.

$$r_{O_2O_6} = 8.0 \text{ cm}; \quad r_{O_2O_3} = 6.0 \text{ cm};$$
$$r_{O_2B_2} = 3.0 \text{ cm}$$
$$r_{O_3C} = 12.0 \text{ cm}; \quad r_{CD} = 4.0 \text{ cm};$$
$$r_{O_6D} = 6.0 \text{ cm}$$
$$\theta_2 = 300°; \quad \dot{\theta}_2 = 5.0 \text{ rad/sec CCW}$$

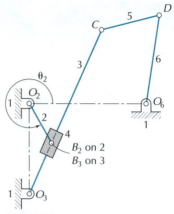

Figure P3.6

P3.7 For the mechanism shown in Figure P3.7, determine the linear velocity of point C using instantaneous centers.

$$r_{O_6A} = r_{O_6C} = 1.0 \text{ in}; \quad r_{AD} = 1.5 \text{ in}; \quad r_{O_2B_2} = 1.0 \text{ in}$$
$$\theta_2 = 315°; \quad \dot{\theta}_2 = 3.0 \text{ rad/sec CW}$$

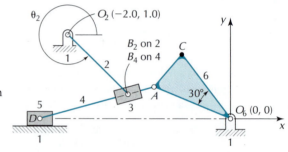

Figure P3.7

P3.8 For the mechanism shown in Figure P3.8, determine the linear velocity of point D using instantaneous centers.

$$r_{O_2O_4} = 1.0 \text{ in}; \quad r_{O_2O_6} = 2.5 \text{ in}; \quad r_{O_2A} = 1.0 \text{ in}$$
$$r_{AB} = 2.5 \text{ in}; \quad r_{O_4B} = 1.5 \text{ in}; \quad r_{O_4C} = 2.0 \text{ in}$$
$$r_{CD} = 1.0 \text{ in}; \quad r_{CE} = 1.5 \text{ in}; \quad r_{O_6E} = 2.0 \text{ in}$$
$$\theta_2 = 45°; \quad \dot{\theta}_2 = 5.0 \text{ rad/sec CCW}$$

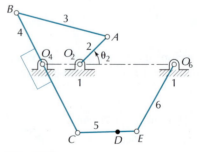

Figure P3.8

P3.9 For the mechanism in the position shown in Figure P3.9, determine, using instantaneous centers,

(a) the rotational speed of link 3

(b) the linear speed of link 4

$$r_{O_2B_2} = 4.0 \text{ cm}; \quad \dot{\theta}_2 = 75 \text{ rpm CW}$$

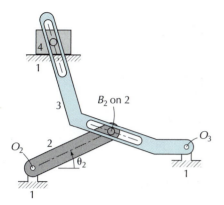

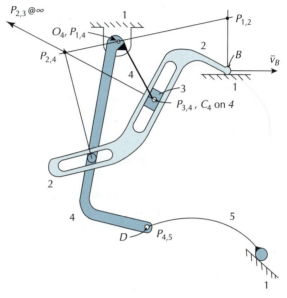

Figure P3.9

P3.10 For the mechanism in the configuration shown in Figure P3.10, seven instantaneous centers are also provided. Using instantaneous centers, determine

(a) the magnitude and direction of rotational speed of link 3

(b) the magnitude and direction of rotational speed of link 5

(c) the velocity of point D

$$r_{O_2B_2} = 5.0 \text{ cm}; \quad \dot{\theta}_2 = 8.0 \text{ rad/sec CW}$$

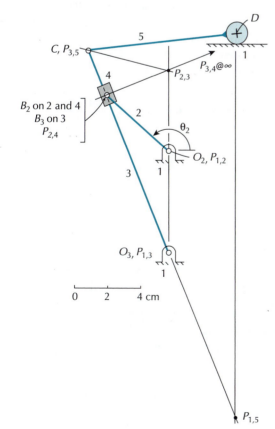

Figure P3.10

P3.11 For the mechanism in the configuration shown in Figure P3.11, six instantaneous centers are also provided. Using only the instantaneous centers which have been provided, determine

(a) the magnitude and direction of rotational speed of link 2

(b) the magnitude and direction of rotational speed of link 4

(c) the velocity of point D

$$r_{O_4C_4} = 1.5 \text{ in}; \quad |\bar{v}_B| = 5.0 \text{ in/sec}$$

Figure P3.11

P3.12 Using the instantaneous centers of the mechanism shown in Figure P3.12, determine

(a) the magnitude and direction of rotational speed of link 3

(b) the Coriolis acceleration of point D_3 with respect to point D_1

(c) the velocity of point C

$$r_{P_{1,2}P_{2,3}} = 3.0 \text{ cm}; \quad \dot{\theta}_2 = 200 \text{ rpm CCW}$$

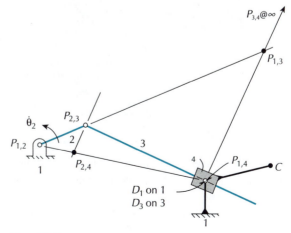

Figure P3.12

P3.13 Using the instantaneous centers of the mechanism shown in Figure P3.13, determine

    (a) the magnitude and direction of rotational speed of link 3

    (b) the velocity of point C

$$r_{P_{2,3}P_{3,4}} = 4.0 \text{ cm}; \quad |\bar{v}_A| = 20 \text{ cm/sec}$$

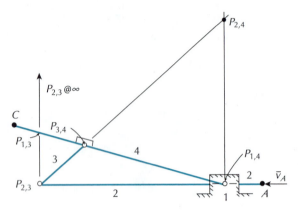

Figure P3.13

P3.14 For the mechanism shown in Figure P3.14, determine using the instantaneous centers

    (a) the magnitude and direction of rotational speed of link 4

    (b) the magnitude and direction of rotational speed of link 5

    (c) the velocity of point C

$$r_{P_{1,2}P_{1,3}} = 6.7 \text{ cm}; \quad \dot{\theta}_2 = 40 \text{ rpm CCW}$$

$$r_{P_{2,3}P_{3,4}}$$

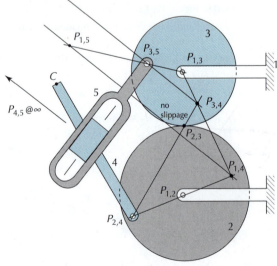

Figure P3.14

P3.15 For the mechanism in the configuration shown in Figure P3.15, using the instantaneous centers, determine

    (a) the magnitude and direction of rotational speed of link 4

    (b) the magnitude and direction of rotational speed of link 5

    (c) the linear velocity of point D

$$r_{P_{1,2}P_{2,3}} = 5.0 \text{ cm}; \quad \dot{\theta}_2 = 8.0 \text{ rad/sec CCW}$$

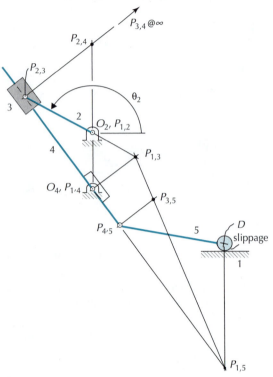

Figure P3.15

P3.16 For the mechanism shown in Figure P3.16, there is no slippage between links 2 and 4. In the configuration shown, determine the locations of all of the instantaneous centers, and use these to find the rotational speed of link 4 when

    (a) $\dot{\theta}_2 = 2.0 \dfrac{\text{rad}}{\text{sec}} \text{ CW}; \quad \dot{\theta}_3 = 5.0 \dfrac{\text{rad}}{\text{sec}} \text{ CW}$

    (b) $\dot{\theta}_2 = 4.0 \dfrac{\text{rad}}{\text{sec}} \text{CW}; \quad \dot{\theta}_3 = 5.0 \dfrac{\text{rad}}{\text{sec}} \text{ CW}$

$$r_2 = 15.0 \text{ cm}; \quad r_4 = 9.0 \text{ cm}$$

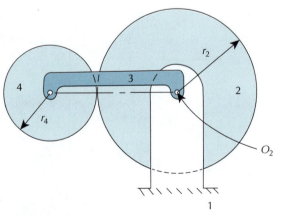

Figure P3.16

P3.17 For the mechanism shown in Figure P3.17, determine the rotational speeds of links 2 and 3 using instantaneous centers.

$$r_{O_2O_4} = 5.75 \text{ in}; \quad r_{O_2B} = 5.75 \text{ in}; \quad \rho_3 = 0.875 \text{ in};$$
$$\rho_4 = 1.75 \text{ in}$$
$$\theta_2 = 171°; \quad \dot{\theta}_4 = 120 \text{ rpm CW}$$

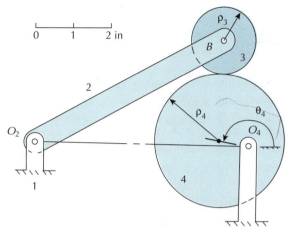

Figure P3.17

P3.18 For the mechanism shown in Figure P3.18, determine the rotational speed of link 3 and the velocity of point C using instantaneous centers.

$$r_{O_2O_3} = 4.5 \text{ in}; \quad r_{O_2B_2} = 2.75 \text{ in}; \quad r_{O_3C} = 8.5 \text{ in}$$
$$\theta_2 = 135°; \quad \dot{\theta}_2 = 120 \text{ rpm CW}$$

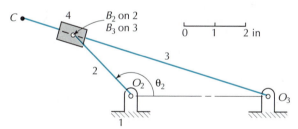

Figure P3.18

P3.19 For the mechanism shown in Figure P3.19, determine the rotational speed of link 3 and the linear velocity of point C using instantaneous centers.

$$r_{O_2B} = 2.5 \text{ in}; \quad r_{BD} = 5.0 \text{ in}; \quad r_{BC} = 3.6 \text{ in}; \quad r_{CD} = 2.6 \text{ in}$$
$$\theta_2 = 40°; \quad \dot{\theta}_2 = 20 \text{ rad/sec CCW}$$

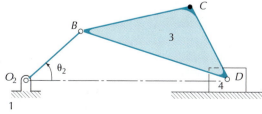

Figure P3.19

P3.20 For the mechanism in the configuration shown in Figure P3.20, determine the rotational speed of link 3 and the linear velocity of point C using instantaneous centers.

$$v_B = 20 \text{ in/sec}$$

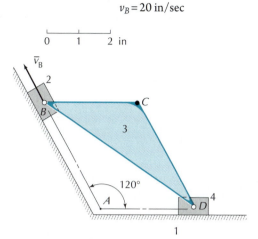

Figure P3.20

Velocity Polygon

P3.21 For the mechanism shown in Figure P3.21, the two links are in contact at point P. For the position shown, determine by drawing a velocity polygon
(a) the absolute velocity of P on 2
(b) the absolute velocity of P on 3
(c) the sliding velocity between links at P
(d) the magnitude of the ratio of rotational speeds of links 2 and 3

$$r_{O_2P_2} = 5.3 \text{ cm}; \quad r_{O_3P_3} = 7.0 \text{ cm}; \quad \dot{\theta}_2 = 10 \text{ rpm CW}$$

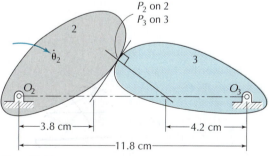

Figure P3.21

P3.22 For the mechanism shown in Figure P3.22, the two links are in contact at point P. For the position shown, determine by drawing a velocity polygon
(a) the absolute velocity of P on 2
(b) the absolute velocity of P on 3
(c) the sliding velocity between links at P
(d) the magnitude of the ratio of rotational speeds of links 2 and 3

$$r_{O_2P_2} = 8.8 \text{ cm}; \quad r_{O_3P_3} = 5.1 \text{ cm}; \quad \dot{\theta}_2 = 20 \text{ rpm CW}$$

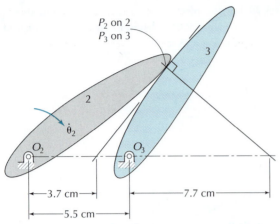

P$_2$ on 2
P$_3$ on 3

←—3.7 cm—→   ←—————7.7 cm—————→

←———5.5 cm———→

Figure P3.22

P3.23 For the mechanism shown in Figure P3.23,

    (a) draw the velocity polygon (employ scale: 1 in = 10 in/sec)

    (b) specify

        (i)   the velocity of point B

        (ii) the velocity of point C

        (iii) the velocity of point D

        (iv) the velocity of point B relative to point D

        (v) the angular velocity of link 3

$$r_{O_2B} = 2.0 \text{ in}; \quad r_{BD} = 3.0 \text{ in}; \quad r_{BC} = 1.0 \text{ in}$$
$$\theta_2 = 135°; \quad \dot{\theta}_2 = 120 \text{ rpm CW}$$

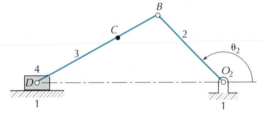

Figure P3.23

P3.24 For the mechanism shown in Figure P3.24,

    (a) draw the velocity polygon (employ scale: 1 in = 10 in/sec)

    (b) specify

        (i)   the velocity of point B

        (ii) the velocity of point C

        (iii) the velocity of point D

        (iv) the angular velocity of link 3

        (v) the angular velocity of link 4

$$r_{O_2B} = 1.5 \text{ in}; \quad r_{BD} = 3.0 \text{ in}; \quad r_{O_2O_4} = 4.0 \text{ in}$$
$$r_{O_4D} = 2.5 \text{ in}; \quad r_{BC} = 3.5 \text{ in}; \quad r_{CD} = 1.0 \text{ in}$$
$$\theta_2 = 60°; \quad \dot{\theta}_2 = 300 \text{ rpm CW}$$

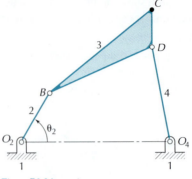

Figure P3.24

P3.25 For the mechanism shown in Figure P3.25,

    (a) draw the velocity polygon (employ scale: 1 cm = 5 cm/sec)

    (b) specify

        (i)   the velocity of point D

        (ii) the velocity of point E

        (iii) the angular velocity of link 3

        (iv) the angular velocity of link 5

$$r_{O_2B} = 4.5 \text{ cm}; \quad r_{BD} = 6.0 \text{ cm}; \quad r_{CD} = 1.5 \text{ cm}; \quad r_{DE} = 3.0 \text{ cm}$$
$$\theta_2 = 135°; \quad \dot{\theta}_2 = 6.0 \text{ rad/sec CW}$$

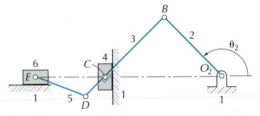

Figure P3.25

P3.26 For the mechanism shown in Figure P3.26,

    (a) draw the velocity polygon (employ scale: 1 cm = 10 cm/sec)

    (b) specify

        (i)   the velocity of point G

        (ii) the Coriolis acceleration of E_5 with respect to E_1

$$r_{O_2B} = 4.0 \text{ cm}; \quad r_{BC} = 4.0 \text{ cm}; \quad r_{BD} = 10.0 \text{ cm}$$
$$r_{CE_5} = 6.0 \text{ cm}; \quad r_{CG} = 9.0 \text{ cm}$$
$$\theta_2 = 30°; \quad \dot{\theta}_2 = 160 \text{ rpm CCW}; \quad \ddot{\theta}_2 = 50 \text{ rad/sec}^2 \text{ CW}$$

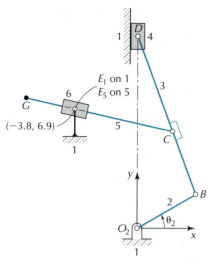

Figure P3.26

P3.27 For the mechanism in the configuration shown in Figure P3.27,

(a) draw the velocity polygon (employ scale: 1 cm = 10 cm/sec)

(b) determine the angular velocity of link 6

$$r_{O_4D} = 8.0 \text{ cm}; \quad |\bar{v}_A| = 60.0 \text{ cm/sec}$$

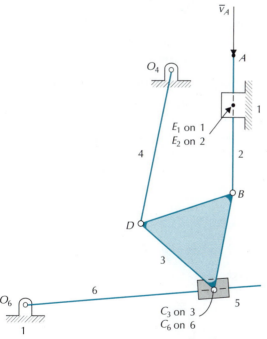

Figure P3.27

P3.28 For the mechanism in the configuration shown in Figure P3.28,

(a) draw the velocity polygon (employ scale: 1 cm = 10 cm/sec)

(b) determine

    (i) the angular velocity of link 3

    (ii) the angular velocity of link 6

    (iii) the Coriolis acceleration of point C_6 with respect to point C_3

$$r_{O_2B} = 4.0 \text{ cm}; \quad \dot{\theta}_2 = 220 \text{ rpm CCW}$$

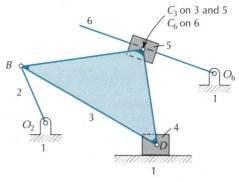

Figure P3.28

P3.29 For the mechanism in the configuration shown in Figure P3.29,

$$r_{O_2B} = 4.0 \text{ cm}; \quad \dot{\theta}_2 = 200 \text{ rpm CCW}$$

(a) draw the velocity polygon (employ scale: 1 cm = 10 cm/sec)

(b) determine

    (i) the angular velocity of link 3

    (ii) the angular velocity of link 6

    (iii) the Coriolis acceleration of point C_6 with respect to point C_3

(c) if the angular velocity of link 2 is changed to a constant rate of 350 rpm CCW for the given configuration, specify the Coriolis acceleration of point C_3 with respect to point C_6

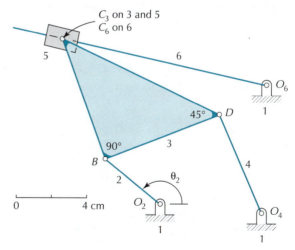

Figure P3.29

P3.30 For the mechanism in the configuration shown in Figure P3.30,

(a) given the input motion

$$r_{O_2A} = 4.0 \text{ cm}; \quad \dot{\theta}_2 = 180 \text{ rpm CW}$$

    (i) draw the velocity polygon (employ scale: 1 cm = 10 cm/sec)

(ii) determine the Coriolis acceleration of D_1 with respect to D_5

(b) given the input motion

$$\bar{v}_B = 30 \text{ cm/sec @ } 0°$$

(The input motion given in part (a) no longer applies.)

(i) determine the magnitude and direction of rotational speed of link 3

(ii) determine the magnitude of the Coriolis acceleration of D_1 with respect to D_5

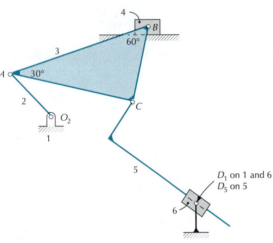

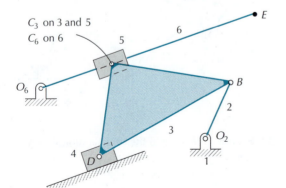

Figure P3.30

P3.31 For the mechanism in the configuration shown in Figure P3.31,

(a) draw the velocity polygon (employing scale: 1 cm = 10 cm/sec)

(b) determine the magnitude and direction of rotational speed of link 3

(c) determine the velocity of point E

(d) determine the Coriolis acceleration of C_6 with respect to C_3

$$r_{O_2B} = 4.0 \text{ cm}; \quad \dot{\theta}_2 = 200 \text{ rpm CW}$$

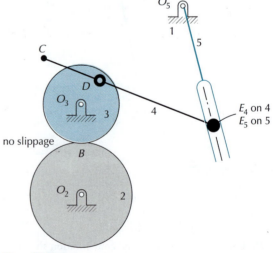

Figure P3.31

P3.32 For the mechanism in the configuration shown in Figure P3.32,

(a) draw the velocity polygon (employ scale: 1 cm = 10 cm/sec)

(b) determine

(i) the angular velocity of link 3

(ii) the angular velocity of link 6

(iii) the velocity of point E

$$r_{O_2B} = 4.0 \text{ cm}; \quad \dot{\theta}_2 = 220 \text{ rpm CW}$$

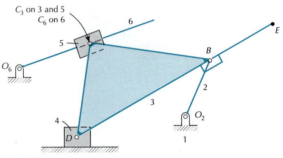

Figure P3.32

P3.33 For the mechanism in the configuration shown in Figure P3.33,

(a) draw the velocity polygon (employ scale: 1 cm = 10 cm/sec)

(b) determine

(i) the angular velocity of link 4

(ii) the linear velocity of point C

$$r_{O_2B} = 2.45 \text{ cm}; \quad \dot{\theta}_2 = 230 \text{ rpm CCW}$$
$$\dot{\theta}_5 = 130 \text{ rpm CCW}$$

Figure P3.33

Velocity and Acceleration Polygons

P3.34 For the mechanism shown in Figure P3.34,

(d) draw the velocity polygon (employ scale: 1 cm = 10 cm/sec)

(e) draw the acceleration polygon (employ scale: 1 cm = 50 cm/sec^2)

(f) specify

(i) the sliding acceleration of point B_2 relative to point B_3

(ii) the angular accelerations of links 3 and 5

$$r_{O_2B_2} = 6.0 \text{ cm}; \quad r_{O_2O_3} = 5.0 \text{ cm}$$
$$r_{O_3C} = 6.0 \text{ cm}; \quad r_{CD} = 9.0 \text{ cm}$$
$$\theta_2 = 10°; \quad \dot{\theta}_2 = 80 \text{ rpm CW (constant)}$$

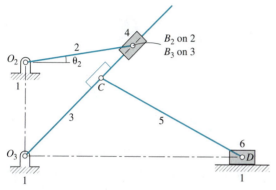

Figure P3.34

P3.35 For the mechanism shown in Figure P3.35,

 (a) draw the velocity polygon (employ scale: 1 cm = 10 cm/sec)

(b) draw the acceleration polygon (employ scale: 1 cm = 100 cm/sec^2)

(c) specify the angular acceleration of link 7

$$r_{O_2A} = 6.0 \text{ cm}; \quad r_{AC} = 8.0 \text{ cm}; \quad r_{CD} = 10.0 \text{ cm}; \quad r_{O_5D} = 4.0 \text{ cm}$$
$$r_{AB_3} = 2.0 \text{ cm}; \quad r_{O_7B_7} = 5.0 \text{ cm}$$
$$\theta_2 = 110°; \quad \dot{\theta}_2 = 75 \text{ rpm CW (constant)}$$
$$\theta_5 = 10°; \quad \dot{\theta}_5 = 100 \text{ rpm CW}; \quad \ddot{\theta}_5 = 75 \text{ rad/sec}^2 \text{ CCW}$$

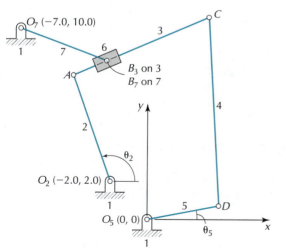

Figure P3.35

P3.36 For the mechanism shown in Figure P3.36,

 (a) draw the velocity polygon (employ scale: 1 cm = 10 cm/sec)

(b) draw the acceleration polygon (employ scale: 1 cm = 100 cm/sec^2)

(c) specify the angular acceleration of link 5

$$r_{O_2A} = 3.0 \text{ cm}; \quad r_{AD_3} = 10.0 \text{ cm}$$
$$r_{AB} = 6.0 \text{ cm}; \quad r_{BC} = 7.0 \text{ cm}$$
$$\theta_2 = 100°; \quad \dot{\theta}_2 = 160 \text{ rpm CCW (constant)}$$

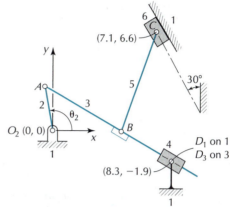

Figure P3.36

P3.37 For the mechanism shown in Figure P3.37,

 (a) draw the velocity polygon (employ scale: 1 cm = 10 cm/sec)

(b) draw the acceleration polygon (employ scale: 1 cm = 100 cm/sec^2)

(c) specify the angular acceleration of link 6

$$r_{O_2A} = 4.0 \text{ cm}; \quad r_{AB} = 9.0 \text{ cm}; \quad r_{O_4B} = 8.0 \text{ cm}$$
$$r_{O_4C_4} = 4.0 \text{ cm}; \quad r_{O_6C_6} = 6.0 \text{ cm}$$
$$\theta_2 = 45°; \quad \dot{\theta}_2 = 140 \text{ rpm CW (constant)}$$

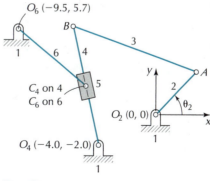

Figure P3.37

P3.38 For the mechanism shown in Figure P3.38,

 (a) draw the velocity polygon (employ scale: 1 cm = 10 cm/sec)

(b) draw the acceleration polygon (employ scale: 1 cm = 100 cm/sec^2)

(c) specify the angular acceleration of link 6

$$r_{O_2A} = 4.0\,\text{cm}; \quad r_{AB} = 9.0\,\text{cm}; \quad r_{O_4B} = 8.0\,\text{cm}$$
$$r_{AC_3} = 6.0\,\text{cm}; \quad r_{O_6C_6} = 6.0\,\text{cm}$$
$$\theta_2 = 140°; \quad \dot\theta_2 = 140\,\text{rpm CCW (constant)}$$

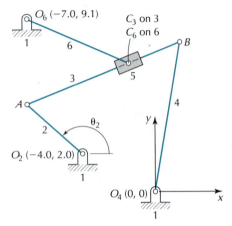

Figure P3.38

P3.39 For the mechanism shown in Figure P3.39,

(a) draw the velocity polygon (employ scale: 1 cm = 10 cm/sec)

(b) draw the acceleration polygon (employ scale: 1 cm = 100 cm/sec²)

(c) specify

   (i)   the acceleration of point B_6

   (ii) the angular accelerations of links 3 and 6

(d) if the angular velocity of link 2 is changed to a constant rate of 80 rpm CCW, specify the new angular acceleration for link 6

$$r_{O_2A} = 5.0\,\text{cm}; \quad r_{AC} = 13.0\,\text{cm}$$
$$r_{AB_3} = 5.0\,\text{cm}; \quad r_{O_6B_6} = 8.0\,\text{cm}$$
$$\theta_2 = 135°; \quad \dot\theta_2 = 130\,\text{rpm CCW (constant)}$$

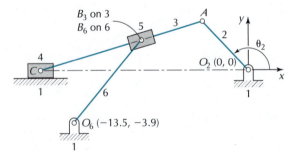

Figure P3.39

P3.40 For the mechanism shown in Figure P3.40,

(a) draw the velocity polygon (employ scale: 1 cm = 10 cm/sec)

(b) draw the acceleration polygon (employ scale: 1 cm = 200 cm/sec²)

(c) specify

   (i)   the angular acceleration of link 4

   (ii) the acceleration of link 6

$$r_{O_2A_2} = 3.0\,\text{cm}; \quad r_{O_4B} = 1.0\,\text{cm}; \quad r_{O_2O_4} = 5.0\,\text{cm}$$
$$r_{BC} = 4.5\,\text{cm}; \quad \theta_2 = 150°$$
$$\dot\theta_2 = 180\,\text{rpm CW (constant)}$$

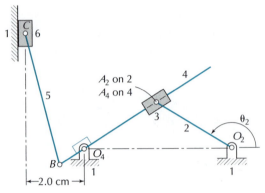

Figure P3.40

P3.41 For the mechanism shown in Figure P3.41,

(a) draw the velocity polygon (employ scale: 1 in = 10 in/sec)

(b) draw the acceleration polygon (employ scale: 1 in = 500 in/sec²)

(c) specify

   (i)   the sliding acceleration between points B_2 and B_4

   (ii) the angular acceleration of link 2

$$r_{O_2B_2} = 4.0\,\text{in}; \quad r_{CB_4} = 0.8\,\text{in}; \quad r_{O_2D} = 4.25\,\text{in}; \quad \rho = 1.5\,\text{in}$$
$$\dot\theta_4 = 100\,\text{rpm CW (constant)}$$

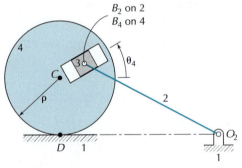

Figure P3.41

P3.42 For the mechanism shown in Figure P3.42,

(a) draw the velocity polygon for the limit position shown (employ scale: 1 cm = 10 cm/sec)

(b) repeat part (a) for the other limit position

(c) draw the acceleration polygon for the limit position shown (employ scale: 1 cm = 200 cm/sec²)

(d) specify the angular accelerations of links 3 and 4

$$r_{O_2O_4} = 8.0\,\text{cm}; \quad r_{O_2B} = 2.5\,\text{cm}$$
$$r_{BD} = 9.5\,\text{cm}; \quad r_{O_4D} = 6.0\,\text{cm}$$
$$\dot\theta_2 = 20\,\text{rad/sec CCW (constant)}$$

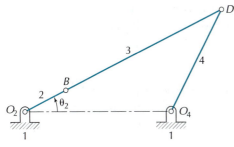

Figure P3.42

P3.43 For the mechanism shown in Figure P3.43,

 (a) draw the velocity polygon (employ scale: 1 cm = 10 cm/sec)

(b) draw the acceleration polygon (employ scale: 1 cm = 100 cm/sec²)

$r_{O_2B} = 3.0$ cm; $r_{BC} = 7.0$ cm; $r_{O_2O_4} = 8.0$ cm
$r_{O_4C} = 4.0$ cm; $r_{O_4D} = 2.0$ cm; $r_{DE} = 8.0$ cm
$\theta_2 = 45°$; $\dot{\theta}_2 = 20$ rad/sec CCW (constant)

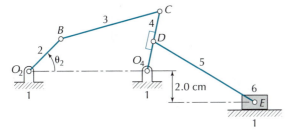

Figure P3.43

P3.44 For the mechanism shown in Figure P3.44,

 (a) draw the velocity polygon (employ scale: 1 in = 5 in/sec)

(b) draw the acceleration polygon (employ scale: 1 in = 50 in/sec²)

$r_{O_2C} = 0.75$ in; $\rho = 1.5$ in
$\theta_2 = 315°$; $\dot{\theta}_2 = 10$ rad/sec CW (constant)

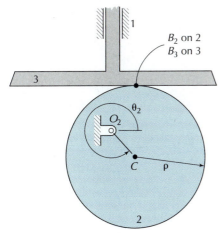

Figure P3.44

P3.45 For the mechanism shown in Figure P3.45,

 (a) draw the velocity polygon (employ scale: 1 cm = 10 cm/sec)

(b) specify

(i) the rotational speeds of links 3 and 4

(ii) the velocities of the midpoints of links 2, 3, and 4 (i.e., points D, E, and F)

(c) draw the acceleration polygon (employ scale: 1 cm = 400 cm/sec²)

(d) specify

(i) the angular accelerations of links 3 and 4

(ii) the accelerations of the midpoints of links 2, 3, and 4

$r_{O_2O_4} = 12.0$ cm; $r_{O_2B} = 4.0$ cm
$r_{BC} = 14.0$ cm; $r_{O_2C} = 8.0$ cm
$\theta_2 = 45°$; $\dot{\theta}_2 = 30$ rad/sec CW (constant)

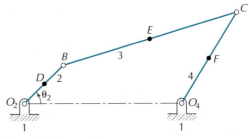

Figure P3.45

P3.46 For the mechanism shown in Figure P3.46,

 (a) draw the velocity polygon (employ scale: 1 cm = 10 cm/sec)

(b) draw the acceleration polygon (employ scale: 1 cm = 100 cm/sec²)

(c) specify the angular acceleration of link 3

$r_{O_2O_3} = 6.0$ cm; $r_{O_2B_2} = 3.0$ cm; $r_{O_3C} = 12.0$ cm
$r_{CD} = 4.5$ cm; $\theta_2 = 30°$
$\dot{\theta}_2 = 20$ rad/sec CCW (constant)

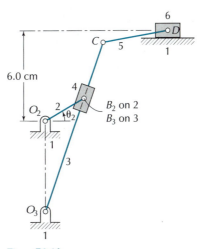

Figure P3.46

P3.47 For the scaled mechanism shown in Figure P3.47,

   (a) draw the velocity polygon (employ scale: 1 cm = 10 cm/sec)

   (b) draw the acceleration polygon (employ scale: 1 cm = 100 cm/sec²)

   (c) specify

      (i) the angular accelerations of links 3 and 5

      (ii) the acceleration of link 6

$$r_{O_2A} = 3.0 \text{ cm}; \quad r_{BA} = 4.0 \text{ cm}$$
$$r_{AC_3} = 8.0 \text{ cm}; \quad r_{BD} = 5.0 \text{ cm}$$
$$\theta_2 = 130°; \quad \dot{\theta}_2 = 180 \text{ rpm CCW (constant)}$$

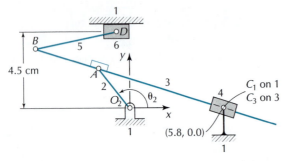

Figure P3.47

P3.48 For the mechanism in the configuration shown in Figure P3.48, determine

   (a) the angular velocity of link 4

   (b) the velocity of point B_2 with respect to point B_4

   (c) the Coriolis acceleration of point B_4 with respect to point B_2

$$r_{O_2O_4} = 8.0 \text{ cm}; \quad r_{O_2B_2} = 8.0\sqrt{2.0} \text{ cm}$$
$$\theta_2 = 45°; \quad \dot{\theta}_2 = 6.0 \text{ rad/sec CW}$$

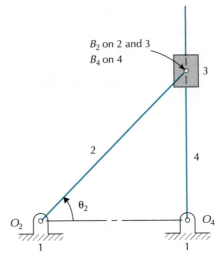

Figure P3.48

CHAPTER 4

ANALYTICAL KINEMATIC ANALYSIS OF PLANAR MECHANISMS

4.1 INTRODUCTION

This chapter presents a method of generating analytical expressions for the positions, velocities, and accelerations for the links in planar mechanisms. These expressions may be programmed on a computer to determine the numerical values of the kinematic quantities. Therefore, when completing analyses at multiple configurations of a mechanism, the results may be obtained more readily than using the graphical techniques presented in Chapter 3.

The method described herein is based on the use of complex numbers. A brief overview of complex numbers is included in Appendix B.

4.2 LOOP CLOSURE EQUATION

A *loop* is defined as a path that traverses a mechanism along links and through kinematic pairs, to arrive back at the starting point. Mechanisms may be classified according to the number of loops required to cover all links and kinematic pairs.

Figure 4.1 illustrates two mechanisms, each having only one loop. In each case, a single loop passes through all links and kinematic pairs of the mechanism. Figure 4.2 illustrates two mechanisms where each requires two loops.

Vectors in a complex plane may be used to represent a mechanism loop. However, a complex vector representation of a loop can be made in various ways. For example, consider the four-bar mechanism illustrated in Figure 4.3(a). The link lengths are designated by r_1, r_2, r_3, and r_4, while the link angles (with respect to the coordinate frame) are θ_2, θ_3, and θ_4. Figures 4.3(b) and 4.3(c) illustrate two different complex vector representations of the mechanism, where R_n ($n = 1, 2, 3, 4$) represent the complex vectors. The representation shown in Figure 4.3(b) has all vectors placed head to tail. The representation shown in Figure 4.3(c) has two vectors, R_1 and R_4, pointing in opposite directions to those shown in Figure 4.3(b). Hence, the corresponding angles (also called the *complex arguments* of the complex numbers or complex vectors), θ_1 and θ_4, are different.

In constructing a complex vector loop, one proceeds by assigning the direction for each vector in the loop. In comparing the two versions presented, one would probably select the second version

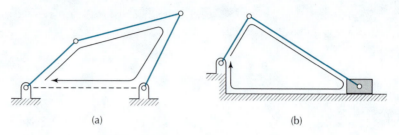

Figure 4.1 Examples of one-loop mechanisms.

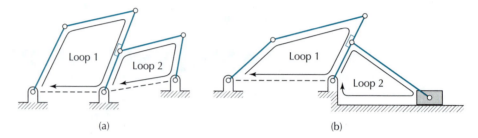

Figure 4.2 Examples of two-loop mechanisms.

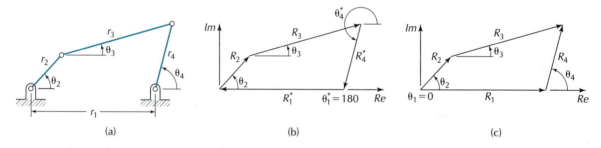

Figure 4.3 Complex vector representation of a four-bar mechanism: (a) Mechanism. (b) Vector representation, version 1. (c) Vector representation, version 2.

as shown in Figure 4.3(c), since the complex arguments correspond to those commonly employed in depicting angular displacements of the links of the mechanism.

It is possible to write a *loop closure equation* for each vector loop to specify the condition that the loop must remain closed as the links of the mechanism move. To generate a loop closure equation, we start at a point between any two vectors in the loop. We then traverse along all of the vectors in the loop to arrive back at the starting point. We have the freedom to proceed in either the clockwise or counterclockwise direction. While traversing through a loop, we note which vectors are encountered and the placement of the head of each vector. The vectors are combined into an equation set equal to zero.

For the vector loop illustrated in Figure 4.3(b), if we start from the origin of the real and imaginary axes and progress through the loop in the clockwise direction, we obtain

$$R_2 + R_3 + R_4^* + R_1^* = 0 \quad \text{or} \quad R_1^* + R_2 + R_3 + R_4^* = 0 \tag{4.2-1}$$

The loop closure equation associated with Figure 4.3(c) is generated as

$$R_2 + R_3 + (-R_4) + (-R_1) = 0 \quad \text{or} \quad R_1 - R_2 - R_3 + R_4 = 0 \tag{4.2-2}$$

For Equation (4.2-2), we placed negative signs in front of R_1 and R_4 because while traversing through the loop we first encountered the heads of these vectors.

Figures 4.3(b) and 4.3(c) are two representations of the same mechanism. Comparing these two representations, vectors R_2 and R_3 are identical, while the relations between the vectors for links 1 and 4 are

$$R_1^* = - R_1; \quad R_4^* = - R_4$$

$$(4.2\text{-}3)$$

For the analysis of a mechanism, it is important to define the *independent variables* and the *dependent variables* that apply to a mechanism. Inputs to a mechanism are considered to be independent variables. Dependent variables are produced as a result of the inputs.

- The number of independent variables is equal to the mobility of the mechanism (see Section 1.5).
- The number of dependent variables is always two, for each mechanism loop.

The dependent variables may be either magnitudes or complex arguments of the vectors. For example, consider the case of the four-bar mechanism shown in Figure 4.3(c). If the independent variable is selected as the angle of link 2, θ_2, then the resulting two dependent variables would be θ_3 and θ_4.

As a second example, consider the complex vector representation of a slider crank mechanism as shown in Figure 4.4(a). The corresponding complex vector representation is shown in Figure 4.4(b). If the independent variable is selected as the angular displacement of link 2, then the dependent variables would be the angular displacement of link 3 and the magnitude of vector R_4. Vector R_1 corresponds to the slider offset, which is fixed, and neither its length nor angle will change as the mechanism operates. Therefore, neither the magnitude nor complex argument of vector R_1 is variable. Vector R_4 is along the slide and its magnitude indicates the position of the slider.

Figure 4.5 shows a five-bar mechanism The mobility of this mechanism is two. If the two independent variables are selected as θ_2 and θ_5, then the two remaining variable quantities are θ_3 and θ_4, which would be the dependent variables.

Figure 4.6 shows a six-bar mechanism. The mobility of this mechanism is one. The single independent variable is selected as θ_2. This mechanism has two loops, and thus there are a total of four dependent variables (two for each loop). Complex vectors are superimposed on the mechanism links in Figure 4.6 to identify these variables more easily. Therefore, the dependent variables would be θ_3, θ_4, θ_7, and θ_8. Also note that

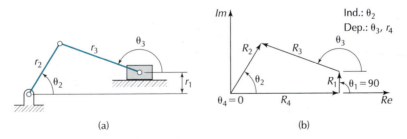

(a)   (b)

Figure 4.4 Slider crank mechanism: (a) Mechanism. (b) Complex vector representation.

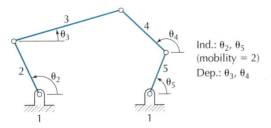

Figure 4.5 Five-bar mechanism.

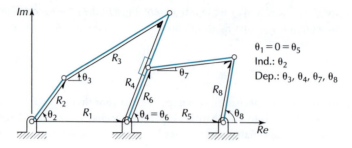

Figure 4.6 Complex vector representation of a six-bar mechanism.

$$\theta_6 = \theta_4$$

and therefore, once the analysis of the first loop has been completed, θ_6 can be considered as a known value in the analysis of the second loop.

Figure 4.7(a) shows a four-bar mechanism with a coupler point. This mechanism may be represented by two vector loops. One vector loop, shown in Figure 4.7(b), traverses the path around the four-bar mechanism. The second loop, shown in Figure 4.7(c), involves the coupler point. For the second loop, we let

$$\theta_8 = 0; \quad \theta_9 = 90°$$

The dependent variables associated with the second vector loop are the magnitudes of vectors R_8 and R_9. Note that the angular displacements of the vectors for the second vector loop can be easily determined if the angles in the first loop are known. The relationships between the angles of these two loops are

$$\theta_5 = \theta_2; \quad \theta_6 = \theta_3; \quad \theta_7 = \theta_3 + 90°$$

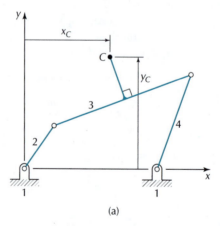

(a)

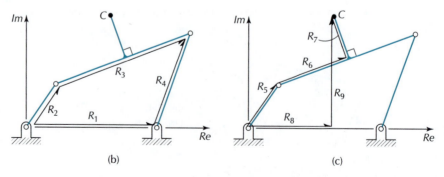

(b)   (c)

Figure 4.7 Complex vector representation of a four-bar mechanism with a coupler point: (a) Mechanism. (b) First vector loop. (c) Second vector loop.

The mechanism used in a front-end loader is illustrated in Figure 4.8(a). This mechanism is more substantial than previous examples and can be modeled by four closed loops. One possible complex vector representation of this mechanism is suggested in the four loop diagrams of Figure 4.8(b). Given this representation, the loop closure equations are

$$\textbf{Loop 1:} \quad R_1 + R_2 - R_3 - R_4 = 0$$

$$\textbf{Loop 2:} \quad R_5 + R_6 - R_7 - R_8 = 0$$

$$\textbf{Loop 3:} \quad R_9 + R_{10} - R_{11} - R_{12} - R_{13} = 0$$

$$\textbf{Loop 4:} \quad R_{14} + R_{15} - R_{16} - R_{17} = 0$$

The mobility of this mechanism is two. In practical terms, this means that two independent inputs are required to control the orientation and position of the bucket, which is represented by vector R_{16}. These two inputs are the length of the lift cylinder and the length of the dump cylinder. In the model, the lift cylinder is represented by vector R_2, and the dump cylinder is represented by vector R_{10}. Therefore, the two independent variables are the magnitudes of vectors R_2 and R_{10}. Since four vector loops are needed to represent this mechanism, eight dependent variables are present.

In the following two sections, analyses are performed for mechanisms that may be represented by one vector loop (Section 4.3), followed by those which require more than one vector loop (Section 4.4).

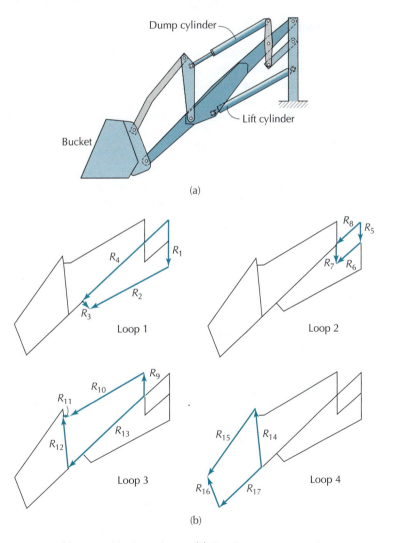

(a)

(b)

Figure 4.8 (a) Front-end-loader mechanism. (b) Complex vector representation.

4.3 COMPLEX VECTOR ANALYSIS OF A PLANAR ONE-LOOP MECHANISM

To perform a complex vector analysis of a planar one-loop mechanism, the following general steps are required:

1. Represent the mechanism links, using complex vectors, to form a polygon.
2. Generate the corresponding loop closure equation from the complex vector representation.
3. Select the independent variable(s), and identify the two dependent variables.
4. Generate two equations from the real and imaginary parts of the loop closure equation and simplify.
5. Solve for the two dependent variables using the equations from step 4.
6. Determine the values of the time derivatives of the dependent variables, using one of the following methods:

   (i) *Method #1*: Use the solved expressions for the dependent variables, obtained in step 5, and differentiate them with respect to time.
   (ii) *Method #2*: Differentiate the real and imaginary components of the loop closure equation, obtained in step 4, and solve for the time derivative quantities.

The above procedure requires the use of trigonometric identities. A listing of several such identities is provided on page 611.

4.3.1 Scotch Yoke Mechanism

Consider the analysis of the Scotch yoke mechanism shown in Figure 4.9(a), where the input motion is the rotation of link 2.

In order to analyze this mechanism, the six-step process outlined above will be applied.

Step 1: A complex vector representation of this mechanism is given in Figure 4.9(b).
Step 2: The corresponding loop closure equation is

$$R_2 - R_3 - R_1 = 0 \qquad (4.3\text{-}1)$$

Step 3: Since the input motion is the rotation of link 2, the independent variable is the angular displacement θ_2, and the dependent variables are the magnitudes of the vectors R_1 and R_3.
Step 4: The two equations can be developed by using the relationship of complex vector R_j, to link values r_j and θ_j, where

$$R_j = r_j(\cos\theta_j + i\sin\theta_j), j = 1, 2, \ldots \qquad (4.3\text{-}2)$$

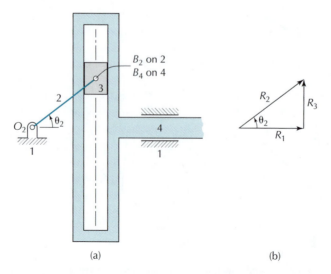

(a)   (b)

Figure 4.9 (a) Scotch yoke mechanism. (b) Complex vector representation.

Applying Equation (4.3-2) to Equation (4.3-1), we obtain

$$r_2\left(\cos\theta_2 + i\sin\theta_2\right) - r_3\left(\cos\theta_3 + i\sin\theta_3\right) - r_1\left(\cos\theta_1 + i\sin\theta_1\right) = 0 \qquad (4.3\text{-}3)$$

Recognizing that $\theta_1 = 0$ and $\theta_3 = 90°$, Equation (4.3-3) may be simplified to

$$r_2\cos\theta_2 - r_1 + i\left(r_2\sin\theta_2 - r_3\right) = 0 \qquad (4.3\text{-}4)$$

The real component of Equation (4.3-4) is

$$\boxed{r_1 = r_2\cos\theta_2} \qquad (4.3\text{-}5)$$

and the imaginary component is

$$\boxed{r_3 = r_2\sin\theta_2} \qquad (4.3\text{-}6)$$

Step 5: The independent variable is the angular displacement θ_2, and the dependent variables are r_1 and r_3. That is, Equations (4.3-5) and (4.3-6) have been arranged to express the dependent variables in terms of the independent variable and a specified link dimension.

Step 6: The time derivative quantities of the dependent variables may now be determined. Since the two dependent variables are expressed explicitly in terms of known quantities, both methods described above are identical.

Taking time derivatives of Equations (4.3-5) and (4.3-6), where we recognize that r_2 is constant, and therefore $\dot{r}_2 = 0$, gives

$$\boxed{\dot{r}_1 = -r_2\dot{\theta}_2\sin\theta_2} \qquad (4.3\text{-}7)$$

and

$$\boxed{\dot{r}_3 = r_2\dot{\theta}_2\cos\theta_2} \qquad (4.3\text{-}8)$$

The second time derivatives are

$$\boxed{\ddot{r}_1 = -r_2\dot{\theta}_2^2\cos\theta_2 - r_2\ddot{\theta}_2\sin\theta} \qquad (4.3\text{-}9)$$

and

$$\boxed{\ddot{r}_3 = -r_2\dot{\theta}_2^2\sin\theta_2 + r_2\ddot{\theta}_2\cos\theta_2} \qquad (4.3\text{-}10)$$

A special case is obtained for Equations (4.3-9) and (4.3-10) if we consider a constant input rotational speed, by setting

$$\ddot{\theta}_2 = 0$$

It is possible to compare expressions for the calculated quantities obtained using the graphical and the analytical techniques. For this example,

$$|\dot{r}_1| = |\bar{v}_{B_4}|; \qquad |\dot{r}_3| = |\bar{v}_{B_4 B_2}|; \qquad |\ddot{r}_1| = |\bar{a}_{B_4}|; \qquad |\ddot{r}_3| = |a_{B_4 B_2}^S|$$

4.3.2 Inverted Slider Crank Mechanism

Consider the analysis of the inverted slider crank mechanism shown in Figure 4.10(a). The input motion is the rotation of link 2.

Step 1: A vector representation of this mechanism is given in Figure 4.10(b).

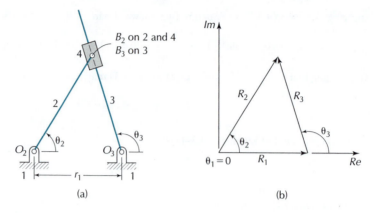

Figure 4.10 (a) Inverted slider crank mechanism. (b) Complex vector representation.

Step 2: The corresponding loop closure equation is

$$R_2 - R_3 - R_1 = 0 \tag{4.3-11}$$

Step 3: Since the input motion is the rotation of link 2, the independent variable is the input angular displacement θ_2, and the dependent variables are r_3 and θ_3.

Step 4: Using Equation (4.3-2), we obtain

$$r_2(\cos\theta_2 + i\sin\theta_2) - r_3(\cos\theta_3 + i\sin\theta_3) - r_1(\cos\theta_1 + i\sin\theta_1) = 0 \tag{4.3-12}$$

Recognizing that $\theta_1 = 0$, Equation (4.3-12) becomes

$$r_2\cos\theta_2 - r_3\cos\theta_3 - r_1 + i(r_2\sin\theta_2 - r_3\sin\theta_3) = 0 \tag{4.3-13}$$

The real component of Equation (4.3-13) is

$$r_2\cos\theta_2 - r_3\cos\theta_3 - r_1 = 0 \tag{4.3-14}$$

and the imaginary component is

$$r_2\sin\theta_2 - r_3\sin\theta_3 = 0 \tag{4.3-15}$$

Step 5: Given the dependent variables r_3 and θ_3, and rearranging Equations (4.3-14) and (4.3-15) to bring these dependent variables onto one side, we obtain

$$r_3\cos\theta_3 = r_2\cos\theta_2 - r_1 \tag{4.3-16}$$

$$r_3\sin\theta_3 = r_2\sin\theta_2 \tag{4.3-17}$$

Squaring and adding Equations (4.3-16) and (4.3-17) and employing the identity

$$\cos^2\theta + \sin^2\theta = 1$$

we obtain

$$r_3^2 = r_1^2 + r_2^2 - 2r_1r_2\cos\theta_2 \tag{4.3-18}$$

and therefore

$$r_3 = \left(r_1^2 + r_2^2 - 2r_1r_2\cos\theta_2\right)^{1/2} \tag{4.3-19}$$

where only the positive value of the square root was retained.

Note that the dependent variable r_3 in Equation (4.3-19) is expressed in terms of the independent variable and the given dimensions of links 1 and 2. It is essential to rearrange the equations given in Equations (4.3-16) and (4.3-17) prior to squaring and adding; otherwise, we would end up with additional cross terms, making it impossible to solve for the dependent variable.

Dividing Equation (4.3-17) by Equation (4.3-16) gives

$$\tan\theta_3 = \left(\frac{r_2 \sin\theta_2}{r_2 \cos\theta_2 - r_1} \right) \tag{4.3-20}$$

or

$$\theta_3^* = \tan^{-1}\left[\frac{\sin\theta_2}{\cos\theta_2 - \left(\dfrac{r_1}{r_2} \right)} \right], \quad -90° < \theta_3^* < 90° \tag{4.3-21}$$

Equation (4.3-21) is expressed in terms of the independent variable θ_2 and the given link dimensions. An asterisk has been added to the left-hand side of Equation (4.3-21) to signify that the equation includes an inverse trigonometric function, and when calculating this expression using a calculator or computer, only the principal values of angles are supplied. The ranges of principal values of inverse trigonometric functions are arcsine, –90° to 90°; arccosine, 0 to 180°; arctangent, –90° to 90°. As a result, the value obtained using Equation (4.3-21) may not correspond to the actual angular displacement of the link. Therefore, when using Equation (4.3-21), it becomes necessary to check the value obtained against a known result. For instance, if we select

$$r_1 = 1.0 \text{ cm}; \quad r_2 = 1.0 \text{ cm}; \quad \theta_2 = 60°$$

the corresponding geometry of the mechanism is shown in Figure 4.11. In this instance, we should obtain

$$\theta_3 = 120°$$

However, from Equation (4.3-21), we have

$$\theta_3^* = -60°$$

Thus, the value of angular displacement obtained by using Equation (4.3-21) is incorrect for the given link lengths and θ_2, calling for an adjustment. Since

$$\tan\theta = \tan(\theta + 180°)$$

we adjust Equation (4.3-21) to obtain

$$\theta_3 = 180° + \theta_3^*$$

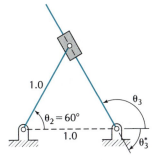

Figure 4.11 Special geometry of an inverted slider crank mechanism.

which will provide the correct value for the angular displacement in the selected geometry. For other configurations of the mechanism shown in Figure 4.11, it can be shown that

$$\theta_3 = \begin{cases} \theta_3^* + 180° & \text{if } 0° < \theta_2 \le 180° \\ \theta_3^* - 180° & \text{if } 180° < \theta_2 \le 360° \end{cases}$$

for the values of θ_3 is to be in the range $-180° < \theta_3 < 180°$. In general, when evaluating an expression for a dependent variable that involves an inverse trigonometric function, it is necessary to check the equation against a known geometry and make adjustments as required.

For this example, checks to determine whether or not an adjustment is needed for the expression of θ_3 can be avoided by employing the inverse trigonometric function $\tan2^{-1}(x, y)$ that keeps track separately of the signs of the numerator and the denominator in the ratio. From Equation (4.3-21) we obtain

$$\theta_3 = \tan2^{-1}\left[\cos\theta_2 - \left(\frac{r_1}{r_2}\right), \sin\theta_2\right]; \quad -180° < \theta_3 < 180° \tag{4.3-22}$$

Step 6: We will now determine the time derivative quantities of the dependent variables. We will employ both methods mentioned earlier in this chapter.

Method #1 (**Use the solved expressions for the dependent variables and differentiate them with respect to time.**)
Taking the time derivative of Equation (4.3-19) gives

$$\dot{r}_3 = \tfrac{1}{2}(r_1^2 + r_2^2 - 2r_1 r_2 \cos\theta_2)^{-1/2}(2r_1 r_2 \sin\theta_2 \dot{\theta}_2) \tag{4.3-23}$$

Substituting Equation (4.3-19) and simplifying, we obtain

$$\dot{r}_3 = \frac{r_1 r_2}{r_3}\sin\theta_2 \dot{\theta}_2 \tag{4.3-24}$$

Equation (4.3-24) is an expression for the time derivative of one of the dependent variables in terms of the independent variable as well as one of the two dependent variables. This is acceptable, since we can solve for the dependent variable first, using Equation (4.3-19), and then determine the time derivative quantity.

Likewise, for the second time derivative, we can show that

$$\ddot{r}_3 = \frac{\dot{r}_3^2}{r_3} + \frac{r_1 r_2}{r_3}(\dot{\theta}_2^2 \cos\theta_2 + \ddot{\theta}_2 \sin\theta_2) \tag{4.3-25}$$

To determine the first time derivative of θ_3, we use Equation (4.3-21) and employ

$$\frac{d}{dt}\tan^{-1}u = \frac{1}{1+u^2}\frac{du}{dt}$$

After simplification we obtain

$$\dot{\theta}_3 = \dot{\theta}_3^* = \dot{\theta}_2\left[\frac{1 - \left(\dfrac{r_1}{r_2}\right)\cos\theta_2}{1 + \left(\dfrac{r_1}{r_2}\right)^2 - 2\left(\dfrac{r_1}{r_2}\right)\cos\theta_2}\right] \tag{4.3-26}$$

Differentiating Equation (4.3-26) with respect to time, we get

$$\dot{\theta}_3 = \dot{\theta}_2^2 \left\{ \frac{\left(\dfrac{r_1}{r_2}\right)\sin\theta_2\left[\left(\dfrac{r_1}{r_2}\right)^2 - 1\right]}{\left[1 + \left(\dfrac{r_1}{r_2}\right)^2 - 2\left(\dfrac{r_1}{r_2}\right)\cos\theta_2\right]^2} \right\} + \ddot{\theta}_2 \left\{ \frac{1 - \left(\dfrac{r_1}{r_2}\right)\cos\theta_2}{\left[1 + \left(\dfrac{r_1}{r_2}\right)^2 - 2\left(\dfrac{r_1}{r_2}\right)\cos\theta_2\right]} \right\} \quad (4.3\text{-}27)$$

For Equations (4.3-26) and (4.3-27), positive values of $\dot{\theta}_3$ and $\ddot{\theta}_3$ correspond to the counterclockwise direction.

A special case of Equations (4.3-25) and (4.3-27) is obtained if we consider a constant input rotational speed by setting $\ddot{\theta}_2 = 0$.

Method #2 (**Differentiate real and imaginary components of the loop closure equation and solve for the time derivative quantities.**)

We return to the real and imaginary components of the loop closure equation, differentiate with respect to time, and solve for the time derivatives of the dependent variables.

We take the time derivative of Equations (4.3-16) and (4.3-17):

$$-r_3 \sin\theta_3 \dot{\theta}_3 + \dot{r}_3 \cos\theta_3 = -r_2 \sin\theta_2 \dot{\theta}_2 \quad (4.3\text{-}28)$$

$$r_3 \cos\theta_3 \dot{\theta}_3 + \dot{r}_3 \sin\theta_3 = r_2 \cos\theta_2 \dot{\theta}_2 \quad (4.3\text{-}29)$$

To solve Equations (4.3-28) and (4.3-29) for $\dot{r}_3$ and to eliminate $\dot{\theta}_3$, we take

$$\text{Equation (4.3-28)} \times \cos\theta_3 + \text{Equation (4.3-29)} \times \sin\theta_3$$

and obtain

$$\dot{r}_3 (\cos^2\theta_3 + \sin^2\theta_3) = r_2 \dot{\theta}_2 (-\sin\theta_2 \cos\theta_3 + \cos\theta_2 \sin\theta_3) \quad (4.3\text{-}30)$$

Employing the identities

$$\cos^2\theta + \sin^2\theta = 1; \quad \sin(\theta_3 - \theta_2) = \sin\theta_3 \cos\theta_2 - \sin\theta_2 \cos\theta_3$$

Equation (4.3-30) becomes

$$\boxed{\dot{r}_3 = r_2 \dot{\theta}_2 \sin(\theta_3 - \theta_2)} \quad (4.3\text{-}31)$$

Similarly,

$$\boxed{\dot{\theta}_3 = \frac{r_2 \dot{\theta}_2}{r_3} \cos(\theta_3 - \theta_2)} \quad (4.3\text{-}32)$$

For the second time derivative quantities, we differentiate Equations (4.3-28) and (4.3-29) with respect to time and solve, to obtain

$$\boxed{\ddot{r}_3 = -r_2 \left[\dot{\theta}_2^2 \cos(\theta_3 - \theta_2) + \ddot{\theta}_2 \sin(\theta_3 - \theta_2) \right] + r_3 \dot{\theta}_3^2} \quad (4.3\text{-}33)$$

$$\boxed{\ddot{\theta}_3 = \frac{r_2}{r_3} \left[\dot{\theta}_2^2 \sin(\theta_3 - \theta_2) + \ddot{\theta}_2 \cos(\theta_3 - \theta_2) \right] - \frac{2\dot{r}_3 \dot{\theta}_3}{r_3}} \quad (4.3\text{-}34)$$

In comparing the two methods for determining time derivative quantities, we see that the first method, although more direct, has the disadvantage of generating complicated derivative quantities.

These two methods appear to provide very different expressions for time derivatives of the dependent variables. This is apparent in comparing Equations (4.3-24) and (4.3-26) with Equations (4.3-31) and (4.3-32). However, the calculated values using these equations will be identical.

It is possible to compare expressions for calculated quantities obtained by use of the graphical and the analytical techniques. For this example, we have

$$\left|\dot{r}_3\right| = \left|\bar{v}_{B_3 B_2}\right|; \quad \left|\ddot{r}_3\right| = \left|a^S_{B_3 B_2}\right|$$

In calculating dependent variables and their time derivatives, the Coriolis acceleration is not directly obtained. However, the Coriolis acceleration may be calculated from the time derivatives of the dependent variables. Using Equation (2.6-8), we have

$$\left|a^C_{B_3 B_2}\right| = \left|2\dot{r}_{B_3 B_2}\dot{\theta}_3\right| = \left|2\dot{r}_3\dot{\theta}_3\right|$$

EXAMPLE 4.1 ANALYSIS OF INVERTED SLIDER CRANK MECHANISM

For the mechanism shown in Figure 4.12(a), use a complex number approach to determine, for the position shown,

(a)   the rotational velocity of link 4
(b)   the Coriolis acceleration of point B_3 with respect to point B_2

$$r_{O_2 O_4} = 3.0 \text{ cm}; \quad r_{O_4 D} = r_4 = 8.0 \text{ cm}; \quad \theta_2 = 60°; \quad \dot{\theta}_2 = 30 \text{ rad/sec CW}$$

SOLUTION

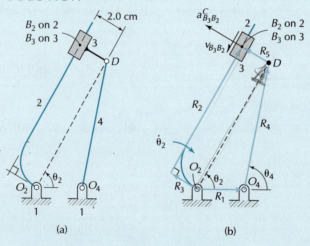

Figure 4.12 Inverted slider crank.

The complex vector representation is shown in Figure 4.12(b). The loop closure equation is

$$R_3 + R_2 - R_5 - R_4 - R_1 = 0 \tag{4.3-35}$$

However, since

$$R_3 = R_5 \tag{4.3-36}$$

Equation (4.3-35) reduces to

$$R_2 - R_4 - R_1 = 0 \tag{4.3-37}$$

The real component of Equation (4.3-37) is

$$r_2 \cos\theta_2 - r_4 \cos\theta_4 - r_1 = 0 \tag{4.3-38}$$

(Continued)

EXAMPLE 4.1 Continued

The imaginary component of Equation (4.3-37) is

$$r_2 \sin\theta_2 - r_4 \sin\theta_4 = 0 \qquad (4.3\text{-}39)$$

The independent variable is θ_2, and the dependent variables are θ_4, and r_2.

Solve for dependent variable θ_4 and eliminate r_2 by taking

$$\text{Equation}(4.3\text{-}38) \times \sin\theta_2 - \text{Equation}(4.3\text{-}39) \times \cos\theta_2$$

which gives

$$r_2 \cos\theta_2 \sin\theta_2 - r_4 \cos\theta_4 \sin\theta_2 - r_1 \sin\theta_2$$
$$- r_2 \sin\theta_2 \cos\theta_2 + r_4 \sin\theta_4 \cos\theta_2 = 0 \qquad (4.3\text{-}40)$$

Simplifying and rearranging Equation (4.3-40) and also solving for θ_4, we obtain

$$\theta_4 = \theta_2 + \sin^{-1}\left(\frac{r_1 \sin\theta_2}{r_4}\right) = 60° + \sin^{-1}\left(\frac{3.0 \sin 60°}{8.0}\right) = 78.9° \qquad (4.3\text{-}41)$$

From Equation (4.3-39) and also solving for r_2, we have

$$r_2 = r_4 \frac{\sin\theta_4}{\sin\theta_2} = 8.0 \text{ cm} \frac{\sin 78.9°}{\sin 60°} = 9.07 \text{cm} \qquad (4.3\text{-}42)$$

The time derivative of Equation (4.3-38) is

$$-r_2 \dot\theta_2 \sin\theta_2 + \dot r_2 \cos\theta_2 + r_4 \dot\theta_4 \sin\theta_4 = 0 \qquad (4.3\text{-}43)$$

The time derivative of Equation (4.3-39) is

$$r_2 \dot\theta_2 \cos\theta_2 + \dot r_2 \sin\theta_2 + r_4 \dot\theta_4 \cos\theta_4 = 0 \qquad (4.3\text{-}44)$$

To solve for the rotational speed of link 4, we take

$$\text{Equation}(4.3\text{-}43) \times \sin\theta_2 - \text{Equation}(4.3\text{-}44) \times \cos\theta_2$$

which gives after simplification

$$\dot\theta_4 = \frac{r_2 \dot\theta_2}{r_4 \cos(\theta_2 - \theta_4)} = \left[\frac{9.07(-30)}{8.0 \cos(60° - 78.9°)}\right] \text{rad/sec} = -35.9 \text{ rad/sec} \qquad (4.3\text{-}45)$$

To solve for the time derivative of the second dependent variable, we take

$$\text{Equation}(4.3\text{-}43) \times \cos\theta_2 + \text{Equation}(4.3\text{-}44) \times \sin\theta_2$$

which gives after simplification

$$\dot r_2 = r_4 \dot\theta_4 \sin(\theta_2 - \theta_4) = [8.0(-35.9)\sin(60° - 78.9°)] \text{ cm/sec}$$
$$= -93.0 \text{ cm/sec} \qquad (4.3\text{-}46)$$

EXAMPLE 4.1 Continued

The solutions to the problem are as follows:

(a) The rotational velocity of link 4 is

$$\dot{\theta}_4 = -35.9 \text{ rad/sec} = 35.9 \text{ rad/sec CW}$$

(b) The magnitude of the Coriolis acceleration of point B_3 with respect to point B_2 is

$$\left| a_{B_3 B_2}^C \right| = 2 \left| \dot{r}_{B_3 B_2} \right| \left| \dot{\theta}_2 \right| = 2 \left| \dot{r}_2 \right| \left| \dot{\theta}_2 \right| = (2 \times 93.0 \times 30.0 \text{ cm/sec}^2) \text{ cm/sec}^2 = 5580 \text{ cm/sec}^2$$

The direction of this relative acceleration is illustrated in Figure 4.12(b).

4.3.3 Four-Bar Mechanism

Consider the analysis of the four-bar mechanism shown in Figure 4.3(a). The input motion is the rotation of link 2.

Step 1: The vector representation of this four-bar mechanism is shown in Figure 4.3(c).
Step 2: The loop closure equation may be written as

$$R_1 + R_4 = R_2 + R_3 \tag{4.3-47}$$

Step 3: Since the input motion is the rotation of link 2, the independent variable is the input angular displacement θ_2. Then the two dependent variables are angular displacements θ_3 and θ_4.
Step 4: Substituting Equation (4.3-2) into Equation (4.3-47) and extracting the real and imaginary components, the real component is

$$r_1 \cos\theta_1 + r_4 \cos\theta_4 = r_2 \cos\theta_2 + r_3 \cos\theta_3 \tag{4.3-48}$$

and the imaginary component is

$$r_1 \sin\theta_1 + r_4 \sin\theta_4 = r_2 \sin\theta_2 + r_3 \sin\theta_3 \tag{4.3-49}$$

Step 5: Given that the dependent variables are θ_3 and θ_4, we now solve Equations (4.3-48) and (4.3-49) for θ_3 and θ_4. Rearranging Equations (4.3-48) and (4.3-49) and also recognizing $\theta_1 = \theta$, we have

$$r_3 \cos\theta_3 = r_1 - r_2 \cos\theta_2 + r_4 \cos\theta_4 \tag{4.3-50}$$

$$r_3 \sin\theta_3 = -r_2 \sin\theta_2 + r_4 \sin\theta_4 \tag{4.3-51}$$

Squaring and adding Equations (4.3-50) and (4.3-51) gives

$$r_3^2 = r_1^2 + r_2^2 + r_4^2 - 2r_2 r_4 \cos\theta_2 \cos\theta_4 - 2r_1 r_2 \cos\theta_2 + 2r_1 r_4 \cos\theta_4 - 2r_2 r_4 \sin\theta_2 \sin\theta_4 \tag{4.3-52}$$

Rearranging gives

$$\cos\theta_4(\cos\theta_2 - h_1) + \sin\theta_2 \sin\theta_4 = -h_3 \cos\theta_2 + h_5 \tag{4.3-53}$$

where

$$\boxed{h_1 = \frac{r_1}{r_2}; \quad h_3 = \frac{r_1}{r_4}; \quad h_5 = \frac{r_1^2 + r_2^2 - r_3^2 + r_4^2}{2r_2 r_4}} \tag{4.3-54}$$

Equation (4.3-53) is in terms of one of the dependent variables, θ_4. However, it appears in the argument of both the sine and cosine functions. To derive an explicit expression for the dependent variable, we employ the identities

$$\sin\theta = \frac{2\tan\left(\dfrac{\theta}{2}\right)}{1+\tan^2\left(\dfrac{\theta}{2}\right)}; \quad \cos\theta = \frac{1-\tan^2\left(\dfrac{\theta}{2}\right)}{1+\tan^2\left(\dfrac{\theta}{2}\right)}$$

in Equation (4.3-53). After simplification, we obtain

$$d\tan^2\left(\frac{\theta_4}{2}\right) + b\tan\left(\frac{\theta_4}{2}\right) + e = 0 \tag{4.3-55}$$

where

$$d = -h_1 + (1-h_3)\cos\theta_2 + h_5 \tag{4.3-56}$$

$$b = -2\sin\theta_2 \tag{4.3-57}$$

$$e = h_1 - (1+h_3)\cos\theta_2 + h_5 \tag{4.3-58}$$

Equation (4.3-55) is a quadratic equation of $\tan(\theta_4/2)$. Solving for θ_4 gives

$$\theta_4 = 2\tan^{-1}\left[\frac{-b \pm \left(b^2 - 4de\right)^{1/2}}{2d}\right] \tag{4.3-59}$$

Similarly, θ_3 may be found by eliminating θ_4 from Equations (4.3-48) and (4.3-49). The result is

$$\theta_3 = 2\tan^{-1}\left[\frac{-b \pm \left(b^2 - 4ac\right)^{1/2}}{2a}\right] \tag{4.3-60}$$

where

$$a = -h_1 + (1+h_2)\cos\theta_2 + h_4 \tag{4.3-61}$$

$$c = h_1 + (1-h_2)\cos\theta_2 + h_4 \tag{4.3-62}$$

$$h_2 = \frac{r_1}{r_3}; \quad h_4 = \frac{-r_1^2 - r_2^2 - r_3^3 + r_4^2}{2r_2 r_3} \tag{4.3-63}$$

Equations (4.3-59) and (4.3-60) indicate that there are two values of θ_3 and θ_4 for a given set of link lengths and value of θ_2. The existence of two possible configurations of the mechanism for a given input angular displacement was also encountered when completing a graphical position analysis of a four-bar mechanism in Section 2.2.2. Figure 4.13 shows two configurations of a mechanism having the same input angular displacement. The angular displacements of links 3 and 4 are

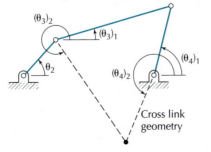

Figure 4.13 Two possible configurations of a four-bar mechanism for a given input angular displacement.

$$(\theta_3)_1 = 2\tan^{-1}\left[\frac{-b - (b^2 - 4ac)^{1/2}}{2a}\right]$$

$$(\theta_4)_1 = 2\tan^{-1}\left[\frac{-b - (b^2 - 4de)^{1/2}}{2d}\right] \tag{4.3-64}$$

and

$$(\theta_3)_2 = 2\tan^{-1}\left[\frac{-b + (b^2 - 4ac)^{1/2}}{2a}\right]$$

$$(\theta_4)_2 = 2\tan^{-1}\left[\frac{-b + (b^2 - 4de)^{1/2}}{2d}\right] \tag{4.3-65}$$

In using Equations (4.3-59) and (4.3-60), one must select either the positive or negative sign. The proper selection is achieved by first calculating both angular displacements, and then selecting the one that matches a known geometry of the mechanism being analyzed. Equations (4.3-59) and (4.3-60) may be applied to all types of four-bar mechanisms.

Step 6: We proceed to determine time derivative quantities of the dependent variables. In this example, we will restrict ourselves to the second method (i.e., differentiating the real and imaginary components, and then solving for the derivative quantities). Taking the time derivatives of Equations (4.3-50) and (4.3-51) gives the following two linear equations in terms of and $\dot{\theta}_3$ and $\dot{\theta}_4$:

$$-r_2\dot{\theta}_2\sin\theta_2 - r_3\dot{\theta}_3\sin\theta_3 = -r_4\dot{\theta}_4\sin\theta_4 \tag{4.3-66}$$

$$r_2\dot{\theta}_2\cos\theta_2 + r_3\dot{\theta}_3\cos\theta_3 = r_4\dot{\theta}_4\cos\theta_4 \tag{4.3-67}$$

For Equations (4.3-66) and (4.3-67), the values of θ_3 and θ_4 can be considered as known, since they may be found using Equations (4.3-59) and (4.3-60). To solve for $\dot{\theta}_3$, we take

$$\text{Equation (4.3-66)} \times \cos\theta_4 + \text{Equation (4.3-67)} \times \sin\theta_4$$

to obtain

$$r_2\dot{\theta}_2(\sin\theta_2\cos\theta_4 - \sin\theta_4\cos\theta_2) + r_3\dot{\theta}_3(\sin\theta_3\cos\theta_4 - \sin\theta_4\cos\theta_3) = 0 \tag{4.3-68}$$

Employing the identity

$$\sin(\theta_i - \theta_j) = \sin\theta_i\cos\theta_j - \sin\theta_j\cos\theta_i$$

and rearranging, we obtain

$$\dot{\theta}_3 = \frac{r_2\dot{\theta}_2}{r_3}\frac{\sin(\theta_2 - \theta_4)}{\sin(\theta_4 - \theta_3)} \tag{4.3-69}$$

Similarly,

$$\dot{\theta}_4 = \frac{r_2\dot{\theta}_2}{r_4}\frac{\sin(\theta_2 - \theta_3)}{\sin(\theta_4 - \theta_3)} \tag{4.3-70}$$

To obtain expressions for $\ddot{\theta}_3$ and $\ddot{\theta}_4$, we differentiate Equations (4.3-66) and (4.3-67) with respect to time and solve, yielding

$$\ddot{\theta}_3 = \frac{-r_2\dot{\theta}_2^2\cos(\theta_2 - \theta_4) - r_2\ddot{\theta}_2\sin(\theta_2 - \theta_4) - r_3\dot{\theta}_3^2\cos(\theta_3 - \theta_4) + r_4\dot{\theta}_4^2}{r_3\sin(\theta_3 - \theta_4)}$$

(4.3-71)

$$\ddot{\theta}_4 = \frac{-r_2\dot{\theta}_2^2\cos(\theta_2 - \theta_3) - r_2\ddot{\theta}_2\sin(\theta_2 - \theta_3) - r_3\dot{\theta}_3^2 + r_4\dot{\theta}_4^2\cos(\theta_3 - \theta_4)}{r_4\sin(\theta_3 - \theta_4)}$$

(4.3-72)

EXAMPLE 4.2 ANALYSIS OF A FOUR-BAR MECHANISM

Figure 4.14(a) depicts a four-bar mechanism having

$$r_1 = 8.0\,\text{cm}; \quad r_2 = 3.5\,\text{cm}; \quad r_3 = 10.0\,\text{cm}; \quad r_4 = 9.0\,\text{cm}$$

$$\theta_2 = 15°; \quad \dot{\theta}_2 = 36.7\,\text{rad/sec (constant)}, \quad \text{therefore } \ddot{\theta}_2 = 0$$

Determine the angular displacements, angular velocities, and angular accelerations of links 3 and 4.

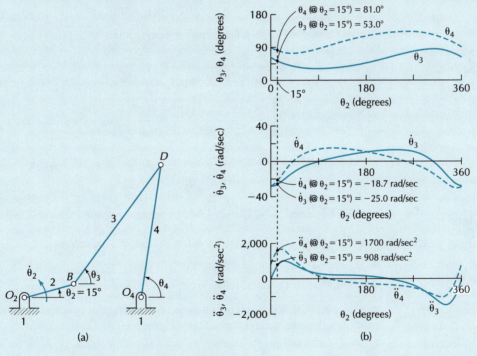

(a)   (b)

Figure 4.14 Angular function plots of a four-bar mechanism.

SOLUTION

Using Equations (4.3-54) and (4.3-63), we first calculate

$$h_1 = \frac{r_1}{r_2} = \frac{8.0}{3.5} = 2.286; \quad h_2 = 0.800; \quad h_4 = -1.361$$

Then, using Equations (4.3-57), (4.3-61), and (4.3-62), we obtain

$$a = -h_1 + (1 + h_2)\cos\theta_2 + h_4$$

$$= -2.286 + (1 + 0.800)\cos 15° - 1.361 = -1.908$$

$$b = -0.518; \quad c = 0.732$$

(*Continued*)

EXAMPLE 4.2 Continued

Using Equations (4.3-64) and (4.3-65), we obtain

$$\left(\theta_3\right)_1 = 2\tan^{-1}\left[\frac{-b - (b^2 - 4ac)^{1/2}}{2a}\right]$$

$$= 2\tan^{-1}\left[\frac{-(-0.518) - \left((-0.518)^2 - 4(-1.908) \times 0.732\right)^{1/2}}{2 \times (-1.908)}\right] = 53.0°$$

$$\left(\theta_3\right)_2 = -75.2°$$

from which we select by viewing the given geometry of the mechanism

$$\theta_3 = (\theta_3)_1 = 53.0°$$

Similarly, the angular displacement of link 4 is

$$\theta_4 = (\theta_4)_1 = 81.0°$$

Using Equation (4.3-69), the angular velocity of link 3 is

$$\dot{\theta}_3 = \frac{r_2\dot{\theta}_2}{r_3}\frac{\sin(\theta_2 - \theta_4)}{\sin(\theta_4 - \theta_3)}$$

$$= \left[\frac{3.5 \times 36.7}{10.0}\frac{\sin(15.0° - 81.0°)}{\sin(81.0 - 53.0°)}\right]\text{rad/sec} = -25.0\text{ rad/sec}$$

Similarly, using Equation (4.3-70), we find

$$\dot{\theta}_4 = -18.7\text{ rad/sec}$$

Using Equations (4.3-71) and (4.3-72), the angular accelerations of links 3 and 4 are $9_3 =$

$$\ddot{\theta}_3 = 908\text{ rad/sec}^2; \quad \ddot{\theta}_4 = 1700\text{ rad/sec}^2$$

A positive value of either rotational speed or angular acceleration is in the counterclockwise direction, whereas a negative value corresponds to the clockwise direction.

Figure 4.14(b) depicts plots of functions of angular displacement, angular velocity, and angular acceleration throughout one cycle of motion.

4.3.4 Slider Crank Mechanism

Consider the slider crank mechanism shown in Figure 4.4(a), for which the input is the rotation of link 2.

Step 1: The complex vector representation is illustrated in Figure 4.4(b).

Step 2: The loop closure equation is

$$R_2 - R_3 - R_1 - R_4 = 0 \tag{4.3-73}$$

Step 3: Since the input motion is the rotation of link 2, the independent variable is θ_2, and the two dependent variables are θ_3 and r_4.

Step 4 and Step 5: Substituting Equation (4.3-2) into (4.3-73), extracting the real and imaginary components, and recognizing that $\theta_1 = 90°$ and $\theta_4 = 0$, we obtain

$$r_2\cos\theta_2 - r_3\cos\theta_3 - r_4 = 0 \tag{4.3-74}$$

$$r_2 \sin\theta_2 - r_3 \sin\theta_3 - r_1 = 0 \qquad (4.3\text{-}75)$$

Equation (4.3-75) contains only one of the dependent variables, θ_3, and therefore it is possible to solve for it explicitly as

$$\theta_3^* = \sin^{-1}\left(\frac{-r_1 + r_2 \sin\theta_2}{r_3}\right), \quad -90° \leq \theta_3^* \leq 90° \qquad (4.3\text{-}76)$$

We consider the special case, where

$$r_1 = 0; \quad r_2 = 1.0; \quad r_3 = 1.0; \quad \theta_2 = 60°$$

In this instance, we should obtain

$$\theta_3 = 120°$$

However, Equation (4.3-76) provides

$$\theta_3^* = 60°$$

The value of the angular displacement as provided by Equation (4.3-76) is thus incorrect, and an adjustment is necessary. Since

$$\sin\theta = \sin(180° - \theta)$$

we let

$$\boxed{\theta_3 = 180° - \theta_3^* = 180° - \sin^{-1}\left(\frac{-r_1 + r_2 \sin\theta_2}{r_3}\right)} \qquad (4.3\text{-}77)$$

Dependent variable θ_3 may now be calculated, and values may be substituted into the following expression, obtained by modifying Equation (4.3-74), to determine r_4:

$$\boxed{r_4 = r_2 \cos\theta_2 - r_3 \cos\theta_3} \qquad (4.3\text{-}78)$$

Note for this example that it was not necessary to combine the real and imaginary components of the loop closure equation to generate equations that explicitly define the dependent variables.

Step 6: Taking the time derivative of Equations (4.3-74) and (4.3-75) gives

$$-r_2 \dot{\theta}_2 \sin\theta_2 + r_3 \dot{\theta}_3 \sin\theta_3 - \dot{r}_4 = 0 \qquad (4.3\text{-}79)$$

$$r_2 \dot{\theta}_2 \cos\theta_2 - r_3 \dot{\theta}_3 \cos\theta_3 = 0 \qquad (4.3\text{-}80)$$

Solving Equations (4.3-79) and (4.3-80) for the first time derivative of the dependent variables gives

$$\boxed{\dot{r}_4 = \frac{r_2 \dot{\theta}_2 \sin(\theta_3 - \theta_2)}{\cos\theta_3}} \qquad (4.3\text{-}81)$$

$$\boxed{\dot{\theta}_3 = \frac{r_2 \dot{\theta}_2 \cos\theta_2}{r_3 \cos\theta_3}} \qquad (4.3\text{-}82)$$

Differentiating Equations (4.3-79) and (4.3-80) with respect to time, along with solving for the second time derivative of the dependent variables, yields

$$\ddot{r}_4 = \frac{-r_2\dot{\theta}_2^2\cos(\theta_3 - \theta_2) + r_2\ddot{\theta}_2\sin(\theta_3 - \theta_2) + r_3\dot{\theta}_3^2}{\cos\theta_3}$$

(4.3-83)

$$\ddot{\theta}_3 = \frac{-r_2\dot{\theta}_2^2\sin\theta_2 + r_2\ddot{\theta}_2\cos\theta_2 + r_3\dot{\theta}_3^2\sin\theta_3}{r_3\cos\theta_3}$$

(4.3-84)

EXAMPLE 4.3 ANALYSIS OF A SLIDER CRANK MECHANISM

Figure 4.15(a) depicts a slider crank mechanism. The link lengths and motion of the crank are

$$r_2 = 3.5\,\text{in};\quad r_3 = 9.0\,\text{in}$$

$$\theta_2 = 50°;\quad \dot{\theta}_2 = 52.4\,\text{rad/sec (constant), and therefore}\quad \ddot{\theta}_2 = 0$$

Determine the angular displacement, angular velocity, and angular acceleration of link 3 and the displacement, velocity, and acceleration of link 4.

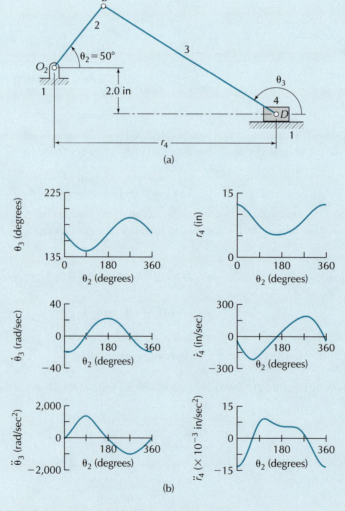

(a)

(b)

Figure 4.15 Function plots of a slider crank mechanism.

(*Continued*)

EXAMPLE 4.3 Continued

SOLUTION

Comparing the mechanism shown in Figure 4.15(a) with that illustrated in Figure 4.4, we let

$$r_1 = -2.0 \text{ in}$$

Using Equation (4.3-77), the angular displacement of link 3 is

$$\theta_3 = 180° - \sin^{-1}\left(\frac{-r_1 + r_2 \sin\theta_2}{r_3}\right)$$

$$= 180° - \sin^{-1}\left(\frac{2.0 + 3.5\sin 50°}{9.0}\right) = 148.7°$$

Using Equation (4.3-78), the displacement of link 4 is

$$r_4 = r_2 \cos\theta_2 - r_3 \cos\theta_3 = (3.5\cos 50° - 9.0\cos 148.7°) \text{ in} = 9.940 \text{ in}$$

Using Equations (4.3-81)–(4.3-84), the velocities and accelerations of links 3 and 4 are

$$\dot\theta_3 = 15.3 \text{ rad/sec}; \quad \dot r_4 = -212 \text{ in/sec}$$

$$\ddot\theta_3 = 814 \text{ rad/sec}^2; \quad \ddot r_4 = -4170 \text{ in/sec}^2$$

Figure 4.15(b) depicts plots of functions of displacement, velocity, and acceleration throughout one cycle of motion.

4.4 COMPLEX VECTOR ANALYSIS OF A PLANAR MECHANISM WITH MULTIPLE LOOPS

For planar mechanisms having multiple loops, it is possible to generate a loop closure equation for each vector loop and solve for the two dependent variables of each loop from the real and imaginary components. In all cases where there are two or more loops, we solve for dependent variables one loop at a time.

The following general steps are required to perform a complex vector analysis of a planar mechanism with multiple loops and having one prescribed input motion:

1. Inspect the mechanism and subdivide it into a number of complex vector loops. Each loop will create a polygon, where each side of the polygon must be represented by its own complex vector.
2. Number each vector in the loops.
3. Begin with the loop associated with the given input motion to the mechanism. Identify the independent variable in that loop. Next, identify the two dependent variables in that same loop.
4. Generate the corresponding loop closure equation for the loop identified in step 3.
5. Generate two equations from the real and imaginary parts of the loop closure equation from step 4 and simplify.
6. Solve for the two dependent variables from the equations from step 5.
7. Proceed to solve adjacent loops, using quantitative information obtained from the previous loop(s), by means of the common angles or lengths of vectors that are shared between loops. This is done by repeating steps 3 through 6, until all the loops and their dependent variables are solved.

8. Determine the values of the time derivatives of the dependent variables, using one of the following methods:

   (i) *Method #1*: Use the solved expressions for the dependent variables, obtained in step 6, and differentiate them with respect to time.

   (ii) *Method #2*: Differentiate the real and imaginary components of the loop closure equation, obtained in step 5, and solve for the time derivative quantities.

This procedure is illustrated by examples.

4.4.1 Inverted Slider Crank Mechanism with a Circle Point

Consider the point C shown in Figure 4.16(a). It is constrained to move on a circular arc about the base pivot of link 3 and is referred to as a *circle point*. For a prescribed angular motion of link 2, we will determine expressions for its Cartesian components of displacement, velocity, and acceleration.

Step 1 and Step 2: For this mechanism, two vector loops are created as shown in Figures 4.16(b) and 4.16(c).

Step 3: Since the input motion is the rotation of link 2, we will begin with the analysis of loop 1, shown in Figure 4.16(b). Here the independent variable is θ_2, and the two dependent variables are θ_3 and r_3.

Step 4: The first loop closure equation is

$$R_2 - R_3 - R_1 = 0 \tag{4.4-1}$$

Step 5: Two equations can now be developed. Recall the relationship of complex vector R_j, to link values r_j and θ_j

$$R_j = r_j(\cos\theta_j + i\sin\theta_j), \quad j = 1, 2, \dots \tag{4.4-2}$$

Applying Equation (4.4-2) to Equation (4.4-1), we obtain

$$r_2(\cos\theta_2 + i\sin\theta_2) - r_3(\cos\theta_3 + i\sin\theta_3)$$
$$- r_1(\cos\theta_1 + i\sin\theta_1) = 0 \tag{4.4-3}$$

Recognizing that $\theta_1 = 0$, Equation (4.4-3) becomes

$$r_2\cos\theta_2 - r_3\cos\theta_3 - r_1 + i(r_2\sin\theta_2 - r_3\sin\theta_3) = 0 \tag{4.4-4}$$

The real component of Equation (4.4-4) is

$$r_2\cos\theta_2 - r_3\cos\theta_3 - r_1 = 0 \tag{4.4-5}$$

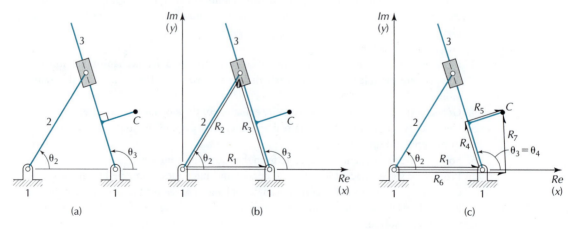

(a)   (b)   (c)

Figure 4.16 Inverted slider crank mechanism with a circle point: (a) Mechanism. (b) First vector loop. (c) Second vector loop.

and the imaginary component is

$$r_2 \sin\theta_2 - r_3 \sin\theta_3 = 0 \tag{4.4-6}$$

Step 6: Given that the dependent variables are r_3 and θ_3, along with rearranging Equations (4.4-5) and (4.4-6) to bring these dependent variables onto one side, we obtain

$$r_3 \cos\theta_3 = r_2 \cos\theta_2 - r_1 \tag{4.4-7}$$

$$r_3 \sin\theta_3 = r_2 \sin\theta_2 \tag{4.4-8}$$

Squaring and adding Equations (4.4-7) and (4.4-8) and employing the identity

$$\cos^2\theta + \sin^2\theta = 1$$

we obtain

$$r_3^2 = r_1^2 + r_2^2 - 2r_1 r_2 \cos\theta_2 \tag{4.4-9}$$

and therefore

$$\boxed{r_3 = (r_1^2 + r_2^2 - 2r_1 r_2 \cos\theta_2)^{1/2}} \tag{4.4-10}$$

where only the positive value of the square root was retained.

Additionally, the expression for θ_3 must be developed. It should be noted that the approach to solve for θ_3 has already been covered in Section 4.3.2. Specifically, θ_3 is solved with Equation (4.3-22), to obtain

$$\boxed{\theta_3 = \tan2^{-1}\left[\cos\theta_2 - \left(\frac{r_1}{r_2}\right), \sin\theta_2\right], \quad -180° < \theta_3 < 180°} \tag{4.4-11}$$

Step 7: Repeat Steps 3 through 6, for the second vector loop. The second loop closure equation is

$$R_1 + R_4 + R_5 - R_7 - R_6 = 0 \tag{4.4-12}$$

The real and imaginary components of this equation are

$$r_1 + r_4 \cos\theta_4 + r_5 \cos(\theta_4 - 90°) - r_6 = 0 \tag{4.4-13}$$

$$r_4 \sin\theta_4 + r_5 \sin(\theta_4 - 90°) - r_7 = 0 \tag{4.4-14}$$

In the second vector loop, r_6 and r_7 are the dependent variables. Also

$$\theta_6 = 0; \quad \theta_7 = 90°$$

Angle θ_4 may be considered as known when analyzing the second vector loop. By inspection we have

$$\theta_4 = \theta_3 \tag{4.4-15}$$

Solving equations for the two dependent variables of the second loop, and simplifying, we obtain

$$r_6 = r_1 + r_4 \cos\theta_3 + r_5 \sin\theta_3 \tag{4.4-16}$$

$$r_7 = r_4 \sin\theta_3 - r_5 \cos\theta_3 \tag{4.4-17}$$

Equations $(4.4\text{-}16)$ and $(4.4\text{-}17)$ also correspond to the Cartesian components that locate the position of point C, that is,

$$x_C = r_6; \quad y_C = r_7 \tag{4.4-18}$$

The position vector of point C is therefore

$$\boxed{\bar{r}_C = x_C \bar{i} + y_C \bar{j}} \tag{4.4-19}$$

Step 8: Differentiating Equations $(4.4\text{-}16)$ and $(4.4\text{-}17)$ with respect to time gives the Cartesian components of velocity as

$$\dot{r}_6 = \dot{x}_C = \dot{\theta}_3(-r_4 \sin\theta_3 + r_5 \cos\theta_3) \tag{4.4-20}$$

$$\dot{r}_7 = \dot{y}_C = \dot{\theta}_3(r_4 \cos\theta_3 + r_5 \sin\theta_3) \tag{4.4-21}$$

The velocity of point C may be expressed as

$$\boxed{\bar{v}_C = \dot{x}_C \bar{i} + \dot{y}_C \bar{j}} \tag{4.4-22}$$

Further, the Cartesian components of acceleration are

$$\ddot{r}_6 = \ddot{x}_C = \dot{\theta}_3^2(-r_4 \cos\theta_3 - r_5 \sin\theta_3) + \ddot{\theta}_3(-r_4 \sin\theta_3 - r_5 \cos\theta_3) \tag{4.4-23}$$

$$\ddot{r}_7 = \ddot{y}_C = \dot{\theta}_3^2(-r_4 \sin\theta_3 + r_5 \cos\theta_3) + \ddot{\theta}_3(r_4 \cos\theta_3 + r_5 \sin\theta_3) \tag{4.4-24}$$

The acceleration of point C is

$$\boxed{\bar{a}_C = \ddot{x}_C \bar{i} + \ddot{y}_C \bar{j}} \tag{4.4-25}$$

The above analysis was completed using the *rectangular form* of a complex number. The other form of a complex number is the *polar form*. (See Appendix B, Section B.4.) Any result may be written in either form. For example, we will determine expressions for the position, velocity, and acceleration of point C in polar form. Referring to Figure 4.17, the position vector of point C may be expressed as the sum of three complex vectors

$$R_C = R_1 + R_4 + R_5 \tag{4.4-26}$$

Substituting the polar form of a complex number for the three vectors on the right-hand side of the equation, we obtain

$$R_C = r_1 e^{i\theta_1} + r_4 e^{i\theta_4} + r_5 e^{i\theta_5} \tag{4.4-27}$$

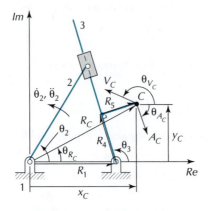

Figure 4.17   Position vector of a circle point in polar form.

Recognizing that

$$\theta_5 = \theta_4 - \frac{\pi}{2} = \theta_3 - \frac{\pi}{2}; \quad \theta_1 = 0 \tag{4.4-28}$$

and therefore Equation (4.4-27) becomes

$$\boxed{R_C = r_1 + r_4 e^{i\theta_3} + r_5 e^{i(\theta_3 - \pi/2)}} \tag{4.4-29}$$

which is the polar form of the position of the point C.

The velocity of point C is

$$V_C = \frac{d}{dt}(R_C) \tag{4.4-30}$$

Substituting Equation (4.4-29) in Equation (4.4-30), we obtain

$$\boxed{\begin{aligned} V_C &= ir_4\dot{\theta}_3 e^{i\theta_3} + ir_5\dot{\theta}_3 e^{i(\dot{\theta}_3 - \pi/2)} \\ &= \dot{\theta}_3 e^{i\theta_3}(r_5 + ir_4) \end{aligned}} \tag{4.4-31}$$

which is the polar form of the velocity of the point C. The acceleration of point C is

$$A_C = \frac{d}{dt}(V_C) \tag{4.4-32}$$

Substituting Equation (4.4-31) in Equation (4.4-32), we get

$$\boxed{\begin{aligned} A_C &= (i\dot{\theta}_3^2 + \ddot{\theta}_3)e^{i\theta_3}(r_5 + ir_4) \\ &= e^{i\theta_3}\left[r_5\ddot{\theta}_3 - r_4\dot{\theta}_3^2 + i(r_5\dot{\theta}_3^2 + r_4\ddot{\theta}_3) \right] \end{aligned}} \tag{4.4-33}$$

which is the polar form of the acceleration of the point C.

Comparing Equation (4.4-31) to Equations (4.4-20)–(4.4-22) and also comparing Equation (4.4-33) to Equations (4.4-23)–(4.4-25), we see that the polar forms of the expressions are far more compact.

We may also express the position vector of point C in polar form as

$$R_C = |R_C|e^{i\theta_{R_C}} \tag{4.4-34}$$

where, using Equation (4.4-18), the magnitude and complex argument of the vector are

$$|R_C| = (x_C^2 + y_C^2)^{1/2}, \quad \theta_{R_C} = \tan2^{-1}(x_C, y_C) \tag{4.4-35}$$

Similarly, polar forms of the velocity and acceleration vectors may be expressed as

$$V_C = |V_C|e^{i\theta_{V_C}} \tag{4.4-36}$$

where

$$|V_C| = (\dot{x}_C^2 + \dot{y}_C^2)^{1/2}, \quad \theta_{V_C} = \tan2^{-1}(\dot{x}_C, \dot{y}_C) \tag{4.4-37}$$

and

$$A_C = |A_C|e^{i\theta_{A_C}} \tag{4.4-38}$$

where

$$|A_C| = (\ddot{x}_C^2 + \ddot{y}_C^2)^{1/2}, \quad \theta_{A_C} = \tan2^{-1}(\ddot{x}_C, \ddot{y}_C) \tag{4.4-39}$$

EXAMPLE 4.4 ANALYSIS OF A TWO-LOOP MECHANISM

For the mechanism shown in Figure 4.18(a), determine the velocity and acceleration of link 6 for the position shown. Use a complex number approach.

$$r_{O_2O_4} = 10.0 \text{ cm}; \quad r_{O_2B} = 4.0 \text{ cm}; \quad r_{BD} = 9.0 \text{ cm}$$

$$r_{O_4D} = 7.0 \text{ cm}; \quad r_{O_4E} = 5.0 \text{ cm}; \quad r_{EF} = 8.0 \text{ cm}$$

$$\theta_2 = 50°; \quad \dot{\theta}_2 = 30 \text{ rad/sec CCW}; \quad \ddot{\theta}_2 = 200 \text{ rad/sec}^2 \text{ CW}$$

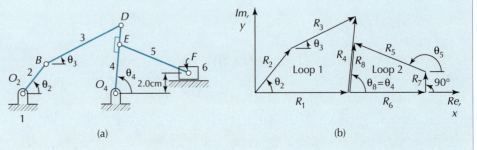

(a) (b)

Figure 4.18 Two-loop mechanism.

SOLUTION

The mechanism is represented by two vector loops in Figure 4.18(b). We start loop 1 because the input motion is associated with one of the vectors of that loop. Loop 1 is a four-bar kinematic chain (Section 4.3.3). Therefore, we employ the equations from that section with

$$r_1 = 10.0 \text{ cm}; \quad r_2 = 4.0 \text{ cm}; \quad r_3 = 9.0 \text{ cm}; \quad r_4 = 7.0 \text{ cm}$$

$$\theta_2 = 50°; \quad \dot{\theta}_2 = 30 \text{ rad/sec}; \quad \ddot{\theta}_2 = -200 \text{ rad/sec}^2$$

to determine

$$\theta_4 = 84.4°; \quad \dot{\theta}_4 = 8.26 \text{ rad/sec}; \quad \ddot{\theta}_4 = 569 \text{ rad/sec}^2$$

We note from the vector representation that vectors 4 and 8 share the same angular motion. The vectors of loop 2 are the same as those for a slider crank mechanism with an offset (Section 4.3.4) and having vector 8 as its input. Therefore, we revise the vector numbers of Section 4.3.4 and employ

$$r_7 = 2.0 \text{ cm}; \quad r_8 = 5.0 \text{ cm}; \quad r_5 = 8.0 \text{ cm}$$

$$\theta_8 = \theta_4 = 84.4°; \quad \dot{\theta}_8 = \dot{\theta}_4 = 8.26 \text{ rad/sec}; \quad \ddot{\theta}_8 = \ddot{\theta}_4 = 569 \text{ rad/sec}^2$$

to obtain

$$\theta_5 = 180° - \sin^{-1}\left(\frac{-r_7 + r_8 \sin\theta_8}{r_5}\right) = 158°$$

$$\dot{r}_6 = \frac{r_8 \dot{\theta}_8 \sin(\theta_5 - \theta_8)}{\cos\theta_5} = -42.7 \text{ cm/sec}$$

$$\dot{\theta}_5 = \frac{r_8 \dot{\theta}_8 \cos\theta_8}{r_5 \cos\theta_5} = -0.543 \text{ rad/sec}$$

(Continued)

EXAMPLE 4.4 Continued

$$\ddot{r}_6 = \frac{-r_8\dot{\theta}_8^2\cos(\theta_5 - \theta_8) + r_8\ddot{\theta}_8\sin(\theta_5 - \theta_8) + r_5\dot{\theta}_5^2}{\cos\theta_5} = -2842 \text{ cm/sec}^2$$

Therefore, the velocity and acceleration of link 6 are

$$\overline{v}_F = 42.7 \text{ cm/sec @180°}; \quad \overline{a}_F = 2842 \text{ cm/sec}^2 \text{ @180°}$$

4.4.2 Four-Bar Mechanism with a Coupler Point

Consider the four-bar mechanism shown in Figure 4.7(a). We will determine the Cartesian components of the displacement, velocity, and acceleration of coupler point C in terms of the angular motion of link 2.

For this mechanism, two vector loops are created as shown in Figures 4.7(b) and 4.7(c). The loop closure equations are

$$R_2 + R_3 - R_4 - R_1 = 0 \tag{4.4-40}$$

and

$$R_5 + R_6 + R_7 - R_9 - R_8 = 0 \tag{4.4-41}$$

The first vector loop is the same as that presented in Section 4.3.3. Results from that analysis may now be employed to determine expressions for the dependent variables of the second vector loop.

For the second vector loop, we have

$$\theta_8 = 0; \quad \theta_9 = 90° \tag{4.4-42}$$

and

$$r_5 = r_2; \quad \theta_5 = \theta_2; \quad \theta_6 = \theta_3; \quad \theta_7 = \theta_3 + 90° \tag{4.4-43}$$

Therefore, θ_5, θ_6, and θ_7 are now known. The dependent variables of the second loop are the magnitudes of vectors R_8 and R_9.

Taking the real and imaginary components of Equation (4.4-41), isolating the dependent variables of the loop onto the left-hand side of the equations, and employing Equations (4.4-42) and (4.4-43), we obtain

$$r_8 = r_2 \cos\theta_2 + r_6 \cos\theta_2 - r_7 \sin\theta_3 \tag{4.4-44}$$

$$r_9 = r_2 \sin\theta_2 + r_6 \sin\theta_3 + r_7 \cos\theta_3 \tag{4.4-45}$$

Equations (4.4-44) and (4.4-45) are also related to the Cartesian components of the position of point C, that is,

$$x_C = r_8; \quad y_C = r_9 \tag{4.4-46}$$

Differentiating Equations (4.4-44) and (4.4-45) with respect to time gives the horizontal and vertical components of velocity as

$$\dot{r}_8 = \dot{x}_C = -r_2\dot{\theta}_2 \sin\theta_2 - \dot{\theta}_3(r_6 \sin\theta_3 + r_7 \cos\theta_3) \tag{4.4-47}$$

$$\dot{r}_9 = \dot{y}_C = r_2 \dot{\theta}_2 \cos\theta_2 + \dot{\theta}_3 (r_6 \cos\theta_3 - r_7 \sin\theta_3) \qquad (4.4\text{-}48)$$

Differentiating Equations (4.4-47) and (4.4-48) with respect to time yields the following expressions for the horizontal and vertical components of acceleration:

$$\ddot{r}_8 = \ddot{x}_C = -r_2 \dot{\theta}_2^2 \cos\theta_2 - r_2 \ddot{\theta}_2 \sin\theta_2$$
$$- \dot{\theta}_3^2 (r_6 \cos\theta_3 - r_7 \sin\theta_3) - \ddot{\theta}_3 (r_6 \sin\theta_3 + r_7 \cos\theta_3) \qquad (4.4\text{-}49)$$

$$\ddot{r}_9 = \ddot{y}_C = -r_2 \dot{\theta}_2^2 \sin\theta_2 + r_2 \ddot{\theta}_2 \cos\theta_2$$
$$- \dot{\theta}_3^2 (r_6 \sin\theta_3 + r_7 \cos\theta_3) + \ddot{\theta}_3 (r_6 \cos\theta_3 - r_7 \sin\theta_3) \qquad (4.4\text{-}50)$$

EXAMPLE 4.5 ANALYSIS OF A FOUR-BAR MECHANISM WITH A COUPLER POINT

Figure 4.19(a) depicts a four-bar mechanism with a coupler point having

$$r_1 = 8.0 \text{ cm}; \quad r_2 = 3.5 \text{ cm}; \quad r_3 = 10.0 \text{ cm}$$
$$r_4 = 9.0 \text{ cm}; \quad r_2 = 3.0 \text{ cm}; \quad r_7 = 4.0 \text{ cm}$$
$$\theta_2 = 15°; \quad \dot{\theta}_2 = 36.7 \text{ rad/sec (constant)}$$

Determine the displacement, velocity, and acceleration of coupler point C with respect to the given coordinate system.

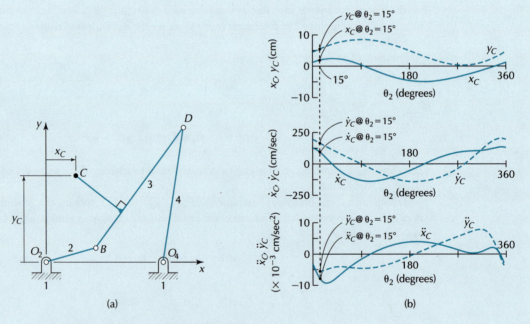

Figure 4.19   Function plots of the coupler point of a four-bar mechanism.

SOLUTION

Some of the required quantities were determined in Example 4.2. Using Equations (4.4-44) and (4.4-46), we obtain

$$x_C = r_2 \cos\theta_2 + r_6 \cos\theta_3 - r_7 \sin\theta_3$$

$$= (3.5 \times \cos 15° + 3.0 \times \cos 53° - 4.0 \times \sin 53°) \text{ cm} = 1.99 \text{ cm}$$

(Continued)

EXAMPLE 4.5 Continued

Similarly,

$$y_C = 5.71 \text{ cm}; \quad \dot{x}_C = 86.6 \text{ cm/sec}; \quad \dot{y}_C = 159 \text{ cm/sec}$$
$$\ddot{x}_C = -8050 \text{ cm/sec}^2; \quad \ddot{y}_C = -5470 \text{ cm/sec}^2$$

From Equations (4.4-19), (4.4-22), and (4.4-25) we have

$$\overline{r}_C = x_C\overline{i} + y_C\overline{j} = (1.99\overline{i} + 5.71\overline{j}) \text{ cm} = 6.05 \text{ cm @ } 70.8°$$

$$\overline{v}_C = (86.6\,\overline{i} + 159\,\overline{j}) \text{ cm/sec} = 181 \text{ cm/sec @ } 64°$$

$$\overline{a}_C = (-8050\,\overline{i} - 5470\,\overline{j}) \text{ cm/sec}^2 = 9730 \text{ cm/sec}^2 \text{ @ } 214°$$

Figure 4.19(b) shows plots of functions of components of displacement, velocity, and acceleration of the coupler point throughout one cycle of motion.

Video 4.20
Parallel-Motion
Mechanism

It is possible to add links to a path-generating four-bar mechanism so that all points on one of the links move with the same shape of path as the coupler point, and therefore the link remains parallel to itself. This is called a *parallel-motion mechanism*. Figure 4.20 is an illustration of such a mechanism. There is a four-bar mechanism, with a coupler point, consisting of links 1 through 4. The additional links are numbered 5 through 8. The additional links may be modeled as two vector loops. Since points $CDFE$ and O_4DFO_8 form parallelograms, links 1, 7, and 5 remain parallel to one another. Since link 1 does not rotate, the orientation of link 5 stays constant as the mechanism moves. Another mechanism having parallel links is given in the following example.

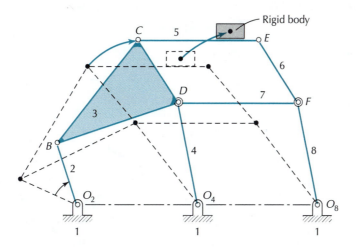

Figure 4.20 Parallel-motion mechanism [Video 4.20].

EXAMPLE 4.6 ANALYSIS OF AN UMBRELLA

Video 4.21
Umbrella

Figure 4.21(a) is an illustration of an umbrella. Figure 4.21(b) shows an umbrella with the membrane cover removed to reveal the mechanisms employed to deploy the membrane. Figure 4.22 is a side view of the one of the several mechanisms employed. The umbrella is brought from a closed to an open position by moving the slider link 4 upward. A schematic of a

(Continued)

EXAMPLE 4.6 Continued

similar mechanism is illustrated in Figure 4.23. The slider can move upward with respect to the sliding shaft to move other links of the umbrella upward and outward during use.

For the mechanism shown in the Figure 4.23, determine the coordinates of point B as a function of the position of A_2 when moving it from point G to point H.

$$r_{DB} = 35.0 \text{ cm}; \quad r_{O_3E} = 16.0 \text{ cm} = r_{A_2E}$$
$$r_{EC} = 13.0 \text{ cm} = r_{DF}; \quad r_{EF} = 3.0 \text{ cm} = r_{CD}$$
$$r_{O_3G} = 30.0 \text{ cm}; \quad r_{O_3H} = 5.0 \text{ cm}$$

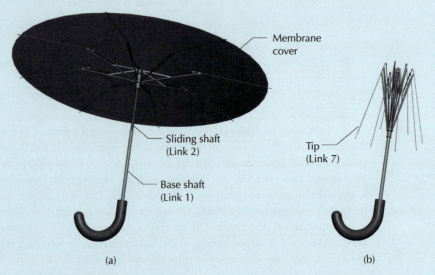

Figure 4.21 Umbrella [Video 4.21].

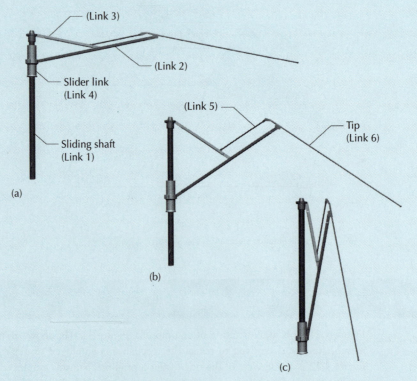

Figure 4.22 Single linkage of umbrella from fully open to fully closed: (a) Fully open. (b) Partially open. (c) Fully closed.

(*Continued*)

EXAMPLE 4.6 Continued

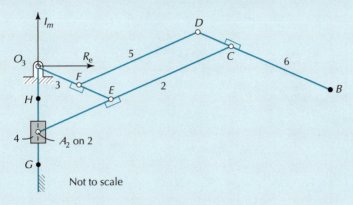

Figure 4.23 Mechanism used in an umbrella.

SOLUTION

Figure 4.24 shows a complex vector representation of the mechanism illustrated in Figure 4.23. Three vector loops are employed. From the given information we conclude

$$r_2 = r_3 = 16.0 \text{ cm};\quad r_4 = r_6 = 3.0 \text{ cm};\quad r_5 = r_7 = r_9 = 13.0 \text{ cm}$$
$$r_8 = 13.0 \text{ cm};\quad r_{10} = 35.0 \text{ cm};\quad 5.0 \text{ cm} \le r_1 \le 30.0 \text{ cm} \tag{4.4-51}$$

The independent variable is r_1. The dependent variables are

$$\textbf{Loop1: } \theta_2, \theta_3$$
$$\textbf{Loop2: } \theta_5, \theta_6$$
$$\textbf{Loop3: } r_{11}, r_{12}$$

Analysis of Loop 1[see Figure 4.24(a)]

The loop closure equation is

$$R_1 + R_3 - R_2 = 0 \tag{4.4-52}$$

Recognizing $\theta_1 = 90°$, the real component is

$$r_3 \cos\theta_3 - r_2 \cos\theta_2 = 0 \tag{4.4-53}$$

Since $r_2 = r_3$, we find

$$\cos\theta_3 = \cos\theta_2 \quad \text{or} \quad |\theta_3| = |\theta_2| \tag{4.4-54}$$

Furthermore, from the arrangement and directions of the vectors in the loop, we conclude

$$\theta_2 = -\theta_3 \tag{4.4-55}$$

The imaginary component of equation (4.4-52) is

$$r_1 + r_3 \sin\theta_3 - r_2 \sin\theta_2 = 0 \tag{4.4-56}$$

Combining Equations (4.4-51), (4.4-55), and (4.4-56), we obtain

$$\theta_2 = \theta_2^* = \sin^{-1}\left(\frac{r_1}{2r_2}\right) \tag{4.4-57}$$

(Continued)

EXAMPLE 4.6 Continued

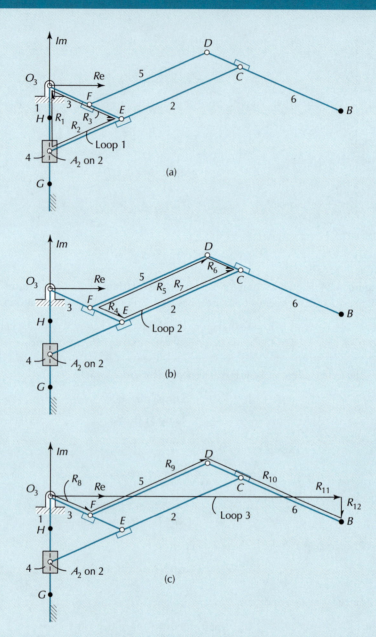

Figure 4.24 Complex vector representation of mechanism used in an umbrella: (a) Loop 1. (b) Loop 2. (c) Loop 3.

Analysis of Loop 2 [see Figure 4.24(b)]

It would be possible to go through a detailed analytical analysis of loop 2. However, since the vectors form a parallelogram, we immediately spot

$$\theta_2 = \theta_6 \quad \text{and} \quad \theta_5 = \theta_7$$

and in relation to the other vector loops,

$$\theta_2 = \theta_5 = \theta_7 = \theta_9; \quad \theta_3 = \theta_4 = \theta_6 = \theta_8 = \theta_{10} \tag{4.4-58}$$

(*Continued*)

EXAMPLE 4.6 Continued

Analysis of Loop 3 [see Figure 4.24(c)]

The loop closure equation is

$$R_{11} + R_{12} - R_{10} - R_9 - R_8 = 0$$

The real and imaginary components provide the horizontal and vertical components of point B, yielding

$$x_B = r_{11} = r_{10}\cos\theta_{10} + r_9\cos\theta_9 + r_8\cos\theta_8$$
$$y_B = -r_{12} = r_{10}\sin\theta_{10} + r_9\sin\theta_9 + r_8\sin\theta_8 \tag{4.4-59}$$

Combining Equations (4.4-51), (4.4-58), and (4.4-59), and simplifying, we have

$$x_B = (r_8 + r_9 + r_{10})\cos\theta_2$$
$$y_B = (-r_8 + r_9 - r_{10})\sin\theta_2 \tag{4.4-60}$$

Using Equations (4.4-51) and (4.4-60), Figure 4.25 provides a plot of the coordinates of point B.

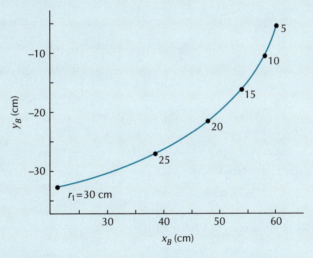

Figure 4.25

EXAMPLE 4.7   ANALYSIS OF A TWO-LOOP MECHANISM

For the mechanism shown in Figure 4.26(a), point A on link 2 is driven upward at a rate of 10 cm/sec. For the configuration shown, use a complex number approach to determine

(a) the angular velocity of link 4

(b) the linear velocity of point E

$$r_{O_4C} = 6.0 \text{ cm}; \quad r_{O_4D} = 8.0 \text{ cm}; \quad r_{BC} = 3.0 \text{ cm}; \quad r_{DE} = 9.0 \text{ cm}$$

(Continued)

EXAMPLE 4.7 Continued

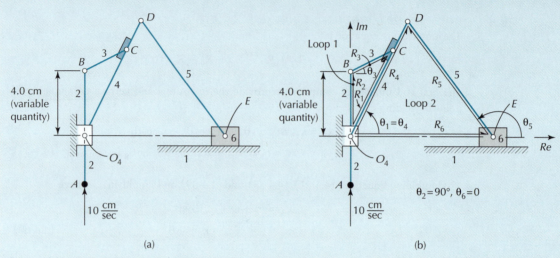

Figure 4.26 Two-loop mechanism.

SOLUTION

The mechanism is represented by two vector loops as illustrated in Figure 4.26(b).

$$r_1 = 6.0 \text{ cm}; \quad r_2 = 4.0 \text{ cm (variable)}; \quad r_3 = 3.0 \text{ cm}$$
$$r_4 = 8.0 \text{ cm}; \quad r_5 = 9.0 \text{ cm}; \quad \dot{r}_2 = 10.0 \text{ cm/sec}$$

(a) Loop 1 (Ind: r_2; Dep: θ_1, θ_3)

Loop Closure Equation: $R_2 + R_3 - R_1 = 0$

Real component: $r_2 \cos\theta_2 + r_3 \cos\theta_3 - r_1 \cos\theta_1 = 0$,   since $\cos\theta_2 = 0$,

$$r_1 \cos\theta_1 = r_3 \cos\theta_3 \tag{4.4-61}$$

Imaginary component: $r_2 \sin\theta_2 + r_3 \sin\theta_3 - r_1 \sin\theta_1 = 0$,   since $\sin\theta_2 = 1$,

$$r_1 \sin\theta_1 = r_2 + r_3 \sin\theta_3 \tag{4.4-62}$$

$$(4.4\text{-}61)^2 + (4.4\text{-}60)^2: \quad r_1^2 = r_2^2 + r_3^2 + 2r_2 r_3 \sin\theta_3$$

$$\theta_3 = \theta_3^* = \sin^{-1}\left(\frac{r_1^2 - r_2^2 - r_3^2}{2r_2 r_3}\right); \quad \text{solve: } \theta_3 = 27.3°$$

$$(4.4\text{-}62): \theta_1 = \theta_1^* = \sin^{-1}\left(\frac{r_2 + r_3 \sin\theta_3}{r_1}\right); \quad \text{solve: } \theta_1 = 63.6°$$

$$\frac{d}{dt}(4.4\text{-}61): \quad -r_1 \dot{\theta}_1 \sin\theta_1 = -r_3 \dot{\theta}_3 \sin\theta_3 \tag{4.4-63}$$

$$\frac{d}{dt}(4.4\text{-}62): \quad r_1 \dot{\theta}_1 \cos\theta_1 = \dot{r}_2 + r_3 \dot{\theta}_3 \cos\theta_3 \tag{4.4-64}$$

(*Continued*)

EXAMPLE 4.7 Continued

$$(4.4\text{-}63) \times \cos\theta_1 + (4.4\text{-}64) \times \sin\theta_1:$$

$$0 = -r_3\dot{\theta}_3\sin\theta_3\cos\theta_1 + \dot{r}_2\sin\theta_1 + r_3\dot{\theta}_3\cos\theta_3\sin\theta_1$$

$$\dot{\theta}_3 = -\frac{\dot{r}_2\sin\theta_1}{r_3\sin(\theta_1 - \theta_3)}; \quad \text{solve: } \dot{\theta}_3 = -5.04\,\frac{\text{rad}}{\text{sec}}$$

and therefore

$$(4.4\text{-}64): \quad \dot{\theta}_1 = \frac{\dot{r}_2 + r_3\dot{\theta}_3\cos\theta_3}{r_1\cos\theta_1}; \quad \text{solve: } \dot{\theta}_1 = -1.29\,\frac{\text{rad}}{\text{sec}}$$

$$\text{since:} \quad \dot{\theta}_4 = \dot{\theta}_1, \quad \text{therefore } \dot{\theta}_4 = 1.29\,\frac{\text{rad}}{\text{sec}}\text{CW}$$

(b) Loop 2 (Ind: θ_4; Dep: θ_5, r_6)

Loop closure equation: $\qquad R_4 - R_5 - R_6 = 0$

Real component: $\qquad r_4\cos\theta_4 - r_5\cos\theta_5 - r_6\cos\theta_6 = 0$, since $\cos\theta_6 = 1$,

$$r_4\cos\theta_4 - r_5\cos\theta_5 - r_6 = 0 \tag{4.4-65}$$

Imaginary component: $\qquad r_4\sin\theta_4 - r_5\sin\theta_5 - r_6\sin\theta_6 = 0$, since $\sin\theta_6 = 0$

$$r_4\sin\theta_4 - r_5\sin\theta_5 = 0 \tag{4.4-66}$$

$$(4.4\text{-}66): \theta_5^* = \sin^{-1}\left(\frac{r_4\sin\theta_4}{r_5}\right); \quad \text{solve: } \theta_5^* = 52.8°$$

$$\text{by inspection,} \quad \theta_5 = 180° - \theta_5^* = 127.2°$$

$$\frac{d}{dt}(4.4\text{-}65): \quad -r_4\dot{\theta}_4\sin\theta_4 + r_5\dot{\theta}_5\sin\theta_5 - \dot{r}_6 = 0$$

$$-r_4\dot{\theta}_4\sin\theta_4 - \dot{r}_6 = -r_5\dot{\theta}_5\sin\theta_5 \tag{4.4-67}$$

$$\frac{d}{dt}(4.4\text{-}66): \quad r_4\dot{\theta}_4\cos\theta_4 - r_5\dot{\theta}_5\cos\theta_5 = 0$$

$$r_4\dot{\theta}_4\cos\theta_4 = r_5\dot{\theta}_5\cos\theta_5 \tag{4.4-68}$$

$$(4.4\text{-}67) \times \cos\theta_5 + (4.4\text{-}68) \times \sin\theta_5:$$

$$-r_4\dot{\theta}_4\sin\theta_4\cos\theta_5 - \dot{r}_6\cos\theta_5 + r_4\dot{\theta}_4\cos\theta_4\sin\theta_5 = 0$$

$$\dot{r}_6 = \frac{r_4\dot{\theta}_4\sin(\theta_5 - \theta_4)}{\cos\theta_5}; \quad \text{solve } \dot{r}_6 = 15.3\,\frac{\text{cm}}{\text{sec}}$$

and therefore

$$\overline{v}_E = 15.3 \text{ cm/sec @ } 0°$$

PROBLEMS

Complex Analysis, Velocity

P4.1 For the mechanism shown in Figure P4.1, determine the rotational speed of link 2 and the sliding velocity of link 3 with respect to link 2 if $\bar{v}_A = 2.0$ in/sec in the direction shown. Use a complex number approach.

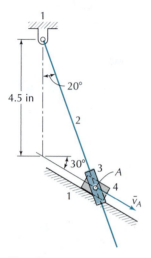

Figure P4.1

P4.2 For the mechanism shown in Figure P4.2, determine the rotational velocity of link 3. Use a complex number approach.

$$r_{O_2B} = 4.0 \text{ cm}; \quad r_{BD} = 8.0 \text{ cm}$$
$$\theta_2 = 225°; \quad \dot{\theta}_2 = 150 \text{ rpm} \quad \text{CW}$$

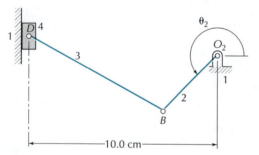

Figure P4.2

P4.3 Figure P4.3 shows an intermittent-motion mechanism known as the Geneva mechanism (Chapter 13, Section 13.10.2). The circular peg P is attached to link 2 and slides in the slots of link 3. Use a complex number approach when $\theta_2 = 160°$ and determine

(a) the rotational velocity of link 3

(b) the sliding velocity of peg P with respect to the slot in link 3

$$r_{O_2P} = 4.0\sqrt{2} \text{ cm}; \quad r_{O_2O_3} = 8.0 \text{ cm}$$
$$\dot{\theta}_2 = 30 \text{ rad/sec} \quad \text{CCW}$$

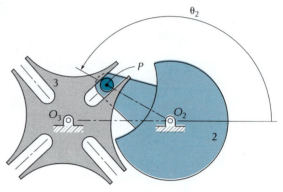

Figure P4.3

P4.4 For the mechanism shown in Figure P4.4, determine

(a) the speed of point C as a function of θ_2

(b) the magnitude of the relative velocity of B_4 with respect to B_2 as a function of θ_2

(c) the magnitude of the sliding acceleration of B_4 with respect to B_2 as a function of θ_2

(d) the maximum magnitude of the sliding velocity of B_4 with respect to B_2 during a complete rotation of link 2, and the corresponding value(s) of θ_2 where this takes place (*Hint*: Use the expression from part (c) to determine value(s) of θ_2 where the sliding acceleration is zero.)

Use a complex number approach.

$$r_{O_2B2} = 1.5 \text{ in}; \quad \dot{\theta}_2 = 75 \text{ rpm} \quad \text{CW (constant)}$$

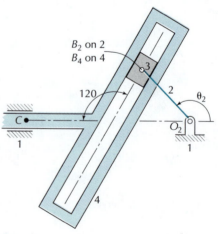

Figure P4.4

P4.5 For the mechanism shown in Figure P4.5, determine

  (a) the angular velocity of link 3

  (b) the velocity of point D_3 with respect to point D_1

Use a complex number approach.

$$r_{O_2B} = 4.0 \text{ cm}; \quad \theta_2 = 135°; \quad \dot{\theta}_2 = 120 \text{ rpm} \quad \text{CW}$$

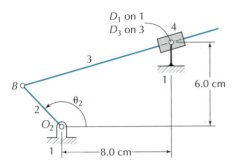

Figure P4.5

P4.6 For the mechanism shown in Figure P4.6, determine

  (a) the speed of point C as a function of θ_2

  (b) the magnitude of the relative velocity of B_4 with respect to B_2 as a function of θ_2

  (c) the magnitude of the sliding acceleration of B_4 with respect to B_2 as a function of θ_2

  (d) the maximum magnitude of the sliding velocity of B_4 with respect to B_2 during a complete rotation of link 2, and the corresponding value(s) of θ_2 when this takes place

  (e) the maximum magnitude of the sliding acceleration of B_4 with respect to B_2 during a complete rotation of link 2, and the corresponding value(s) of θ_2 when this takes place

Use a complex number approach.

$$r_{O_2B_2} = 4.0 \text{ cm}; \quad \dot{\theta}_2 = 200 \text{ rpm} \quad \text{CCW (constant)}$$

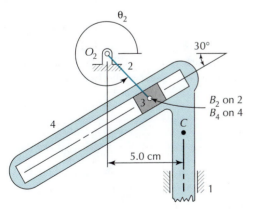

Figure P4.6

Complex Analysis, Velocity and Coriolis Acceleration

P4.7 For the mechanism shown in Figure P4.7, determine

  (a) the angular velocity of link 3

  (b) the velocity of point A_2 with respect to point A_3

  (c) the Coriolis acceleration of point A_2 with respect to point A_3

Use a complex number approach.

$$r_{O_2A_2} = 5.0 \text{ cm}; \quad \theta_2 = 135°; \quad \dot{\theta}_2 = 80 \text{ rpm} \quad \text{CW}$$

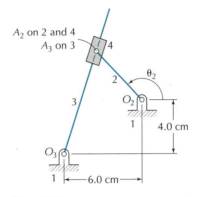

Figure P4.7

P4.8 For the mechanism shown in Figure P4.8, determine

  (a) the rotational velocity of link 3

  (b) the Coriolis acceleration of point B_3 with respect to point B_4

Use a complex number approach.

$$r_{O_2A} = 3.0 \text{ cm}; \quad r_{O_2O_4} = 8.0 \text{ cm}$$
$$\theta_2 = 135°; \quad \dot{\theta}_2 = 30 \text{ rad/sec} \quad \text{CW}$$

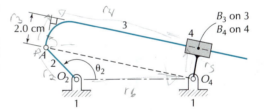

Figure P4.8

Complex Analysis, Velocity and Acceleration

P4.9 For the mechanism shown in Figure P4.9, determine the angular acceleration of link 3. Use a complex number approach.

$$r_{O_2A} = 6.0 \text{ cm}; \quad \theta_2 = 60°$$

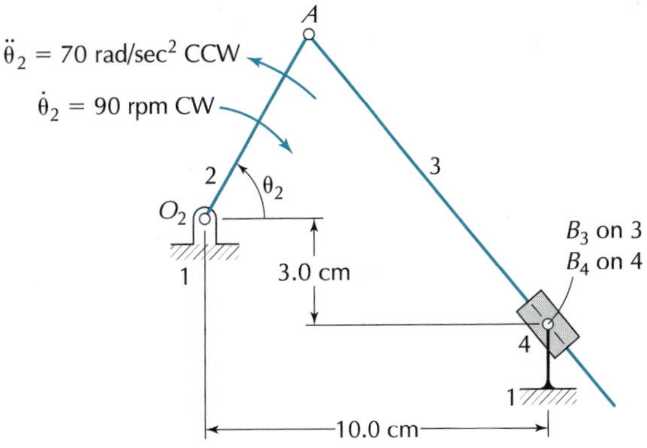

$$\ddot{\theta}_2 = 70 \text{ rad/sec}^2 \text{ CCW}$$
$$\dot{\theta}_2 = 90 \text{ rpm CW}$$

Figure P4.9

P4.10 Figure P4.10 depicts a skeleton diagram representation of an Oldham coupling.(Chapter 13, Section 13.2.3). Determine $r_2, r_3, \dot{r}_2, \dot{r}_3, \ddot{r}_2, \ddot{r}_3$ in terms of $r_1, \theta_2, \dot{\theta}_2, \ddot{\theta}_2$. Use a complex number approach.

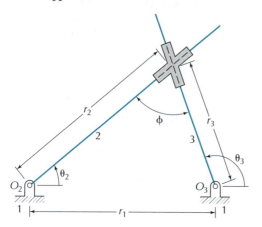

Figure P4.10

P4.11 For the mechanism shown in Figure P4.11, determine

(a) the acceleration of point B

(b) the sliding acceleration of A_2 with respect to A_4

Use a complex number approach.

$$r_{O_2 A_2} = 5.0 \text{ cm}; \quad \theta_2 = 45°$$

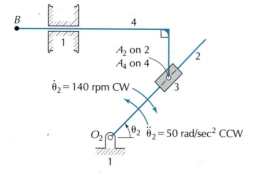

Figure P4.11

P4.12 Determine the rotational speed of link 3 of the mechanism given in Figure P4.12 for the position shown. Use a complex number approach.

$$r_{O_2 A} = 4.0 \text{ cm}; \quad r_{AB} = 10.0 \text{ cm}; \quad \theta_2 = 45°$$

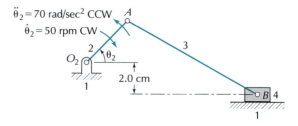

Figure P4.12

P4.13 The nomenclature for this group of problems is given in Figure P4.13, and the dimensions and data are given in Table P4.13. For each, determine the angular displacements, angular velocities, and angular accelerations of links 3 and 4. Use a complex number approach. (**Mathcad program: fourbarkin**)

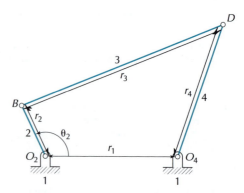

Figure P4.13

P4.14 The nomenclature for this group of problems is given in Figure P4.14, and the dimensions and data are given in Table P4.14. For each, determine the angular displacement, angular

TABLE P4.13

	r_1 (cm)	r_2 (cm)	r_3 (cm)	r_4 (cm)	θ_2 (degrees)	$\dot{\theta}_2$ (rad/sec)	$\ddot{\theta}_2$ (rad/sec^2)
(a)	5	1	4	3	30	30	0
(b)	8	2	5	4	60	45	0
(c)	10	32	20	20	120	−50	40
(d)	5	1	6	4	210	100	−40

TABLE P4.14

	r_1 (cm)	r_2 (cm)	r_3 (cm)	θ_2 (degrees)	$\dot{\theta}_2$ (rad/sec)	$\ddot{\theta}_2$ (rad/sec²)
(a)	0	5	10	30	50	0
(b)	0	3	4	60	45	0
(c)	2	3	6	120	−50	40
(d)	−2	4	10	150	100	−4

velocity, and angular acceleration of link 3, and the displacement, velocity, and acceleration of link 4. Use a complex number approach. (**Mathcad program: slidercrankkin**)

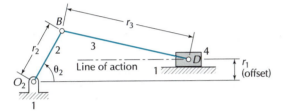

Figure P4.14

Complex Analysis, Multiple Loops

P 4.15 For the mechanism shown in Figure P4.15, determine

   (a) the velocity of point C

   (b) the sliding velocity of point B with respect to the base link

Use a complex number approach.

$$r_{O_2A} = 6.0 \text{ cm}; \quad r_{AB} = 12.0 \text{ cm}; \quad r_{AB} = 9.0 \text{ cm}$$
$$\theta_2 = 315°; \quad \dot{\theta}_2 = 180 \text{ rpm} \quad \text{CW}$$

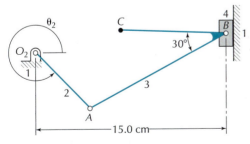

Figure P4.15

P4.16 For the mechanism shown in Figure P4.16, determine

   (a) the angular velocity of link 3

   (b) the velocity of point B_3 with respect to point B_1

   (c) the Coriolis acceleration of point B_3 with respect to point B_1

(d) the equations for the components of velocity of point C in terms of given parameters and quantities determined in parts (a), (b), and (c)

Use a complex number approach.

$$r_{O_2A} = 6.0 \text{ cm}; \quad r_{AB} = 12.0 \text{ cm}; \quad r_{BC} = 9.0 \text{ cm}$$
$$\theta_2 = 315°; \quad \dot{\theta}_2 = 180 \text{ rpm} \quad \text{CW}$$
$$r_{O_2A} = 6.0 \text{ cm}; \quad r_{AB} = 12.0 \text{ cm}$$

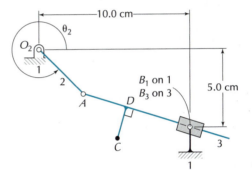

Figure P4.16

P4.17 For the mechanism shown in Figure P4.17, determine the velocity of point C. Use a complex number approach.

$$r_{O_2A} = 1.5 \text{ in}; \quad r_{AB} = 1.5 \text{ in}; \quad r_{BC} = 1.0 \text{ in}; \quad r_{AD} = 4.0 \text{ in}$$
$$\theta_2 = 45°; \quad \dot{\theta}_2 = 150 \text{ rpm} \quad \text{CCW}$$

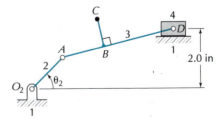

Figure P4.17

P4.18 For the mechanism shown in Figure P4.18, determine

   (a) the angular velocity of link 3

   (b) the velocity of point C

Use a complex number approach.

$$r_{O_2B} = 4.0 \text{ cm}; \quad r_{BC} = 15.0 \text{ cm}$$
$$\theta_2 = 225°; \quad \dot{\theta}_2 = 50 \text{ rad/sec CCW}$$

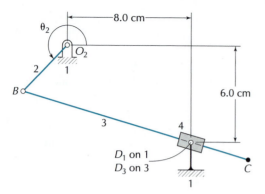

Figure P4.18

P4.19 The nomenclature for this group of problems is given in Figure P4.19, and the dimensions and data are given in Table P4.19. For

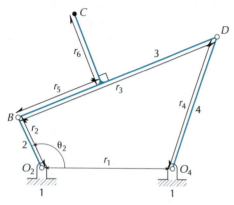

Figure P4.19

each, determine the linear displacement, linear velocity, and linear acceleration of point C. Use a complex number approach. (**Mathcad program: fourbarcpkin**)

P4.20 The nomenclature for this group of problems is given in Figure P4.20, and the dimensions and data are given in Table P4.20. For each, determine the linear displacement, linear velocity, and linear acceleration of point C. Use a complex number approach. (**Mathcad program: slidecrankcpkin**)

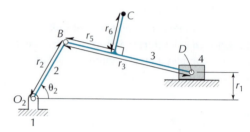

Figure P4.20

P4.21 For the mechanism shown in Figure P4.21, determine

    (a) the angular velocity of link 3

    (b) the velocity of point C

    (c) the angular acceleration of link 3

    (d) the acceleration of point C.

Use a complex number approach.

$$r_{O_2B} = 5.0 \text{ cm}; \quad r_{BC} = 20.0 \text{ cm}$$

$$\theta_2 = 330°; \quad \dot{\theta}_2 = 40.0 \text{ rad/sec CCW (constant)}$$

TABLE P4.19

	r_1 (cm)	r_2 (cm)	r_3 (cm)	r_4 (cm)	r_5 (cm)	r_6 (cm)	θ_2 (degrees)	$\dot{\theta}_2$ (rad/sec)	$\ddot{\theta}_2$ (rad/sec^2)
(a)	1	0.2	0.4	1	0.2	0.5	30	30	0
(b)	5	1	4	3	−2	0	60	45	0
(c)	8	2	10	6	12	−1	120	−50	40
(d)	0.2	1	0.4	0.9	0.2	0.4	210	100	−40

TABLE P4.20

	r_1 (cm)	r_2 (cm)	r_3 (cm)	r_5 (cm)	r_6 (cm)	θ_2 (degrees)	$\dot{\theta}_2$ (rad/sec)	$\ddot{\theta}_2$ (rad/sec^2)
(a)	0	5	10	−4	0	30	30	0
(b)	0	2	6	2	4	60	45	0
(c)	2	4	8	12	0	120	−50	40
(d)	−2	3	7	5	−3	315	100	−40

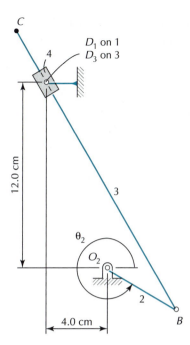

Figure P4.21

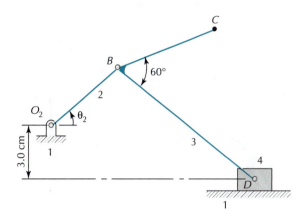

$$r_{O_2B} = 5.0 \text{ cm}; \quad r_{BC} = 6.0 \text{ cm}; \quad r_{BD} = 10.0 \text{ cm}$$
$$\theta_2 = 40°; \quad \dot{\theta}_2 = 15.0 \text{ rad/sec} \quad \text{CW}$$

Figure P4.23

P4.24 For the mechanism shown in Figure P4.24, determine for the position shown
(a) the rotational speed and direction of link 3
(b) the velocity of point C
Use a complex number approach.

$$r_{O_2B} = 2.0 \text{ cm}; \quad r_{BC} = 1.0 \text{ cm}$$
$$\theta_2 = 30°; \quad \dot{\theta}_2 = 200 \text{ rpm} \quad \text{CCW}$$

P4.22 For the mechanism shown in Figure P4.22, determine
(a) the angular velocity of link 3
(b) the velocity of point C
Use a complex number approach.

$$r_{O_2B} = 3.0 \text{ cm}; \quad r_{BC} = 10.0 \text{ cm}; \quad r_{O_2D_1} = 7.0 \text{ cm}$$
$$\theta_2 = 30°; \quad \dot{\theta}_2 = 15.0 \text{ rad/sec} \quad \text{CW}$$

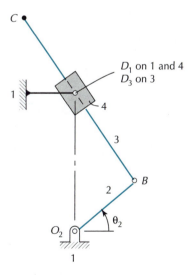

Figure P4.22

P4.23 For the mechanism shown in Figure P4.23, determine
(a) the angular velocity of link 3
(b) the velocity of point C
Use a complex number approach.

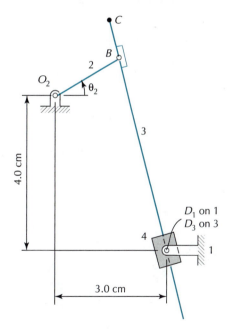

Figure P4.24

P4.25 For the mechanism shown in Figure P4.25, determine
(a) the angular velocity of link 3
(b) the velocity of point C
(c) the angular acceleration of link 3
(d) the acceleration of point C

Use a complex number approach.

$$r_{O_2B} = 5.0 \text{ cm}; \quad r_{BC} = 20.0 \text{ cm}$$
$$\theta_2 = 240°; \quad \dot{\theta}_2 = 40.0 \text{ rad/sec CW};$$
$$\ddot{\theta}_2 = 300 \text{ rad/sec}^2 \text{ CCW}$$

P4.26 For the mechanism shown in Figure P4.26

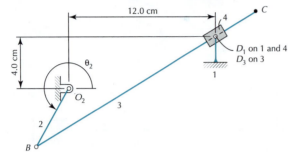

Figure P4.25

$$\theta_2 = 300°; \quad r_{O_2B} = 5.0 \text{ cm}$$

(a) Determine the Coriolis acceleration of D_1 with respect to D_3 if

$$\dot{\theta}_2 = 40.0 \text{ rad/sec CCW (constant)}$$

Use a complex number approach for part (a).

(b) Determine the Coriolis acceleration of D_1 with respect to D_3 if

$$\dot{\theta}_2 = 80.0 \text{ rad/sec CCW (constant)}$$

(c) Determine the Coriolis acceleration of D_1 with respect to D_3 if

$$\dot{\theta}_2 = 80.0 \text{ rad/sec CW (constant)}$$

(*Hint*: For parts (b) and (c), use the result from part (a) as a guide.)

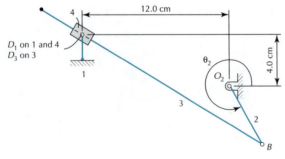

Figure P4.26

P4.27 For the mechanism shown in Figure P4.27, determine the angular position, velocity, and acceleration of link 6. Use a complex number approach.

$$r_{O_2O_4} = 10.0 \text{ cm}; \quad r_{O_2B} = 4.0 \text{ cm}$$
$$r_{BD} = 9.0 \text{ cm}; \quad r_{O_4D} = 7.0 \text{ cm}$$
$$r_{O_4O_6} = 6.0 \text{ cm}; \quad r_{O_4E} = 5.0 \text{ cm}$$
$$r_{EF} = 8.0 \text{ cm}; \quad r_{O_6F} = 5.0 \text{ cm}$$
$$\theta_2 = 50°; \quad \dot{\theta}_2 = 30 \text{ rad/sec CCW (constant)}$$

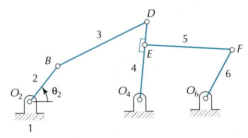

Figure P4.27

P4.28 For the mechanism shown in Figure P4.28, determine the position, velocity, and acceleration of link 6. Use a complex number approach.

$$r_{O_2O_4} = 10.0 \text{ cm}; \quad r_{O_2B} = 4.0 \text{ cm}$$
$$r_{BD} = 9.0 \text{ cm}; \quad r_{O_4D} = 7.0 \text{ cm}$$
$$r_{O_4E} = 5.0 \text{ cm}; \quad r_{EF} = 12.0 \text{ cm}$$
$$\theta_2 = 50°; \quad \dot{\theta}_2 = 30 \text{ rad/sec CCW (constant)}$$

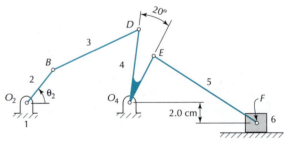

Figure P4.28

CHAPTER 5

GEARS

5.1 INTRODUCTION

This chapter covers both friction and toothed gear systems. The following sections present a mathematical model for analyzing friction gears, descriptions of various types of toothed gears, and common manufacturing methods for gears.

5.2 FRICTION GEARING

A simple means of transmitting rotational motion from one shaft to another is by use of a pair of friction gears, as shown in Figure 5.1(a). Here, transmission relies on the friction force between the cylinders in contact.

A free-body diagram of each cylinder is shown in Figure 5.1(b). Tangent to both cylinders is a force of magnitude

$$F = |\bar{F}| \tag{5.2-1}$$

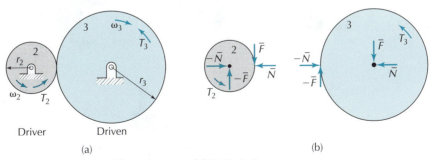

Figure 5.1 Friction gears: (a) Friction gear set, (b) Free-body diagrams.

In the arrangement shown, gear 2 is the *driver* gear, and gear 3 is the *driven* gear. Power is supplied to the system through the shaft connected to gear 2, and power is removed from the system through the shaft attached to gear 3. The torque driving gear 2 is in the same direction as its rotational speed. The magnitude of that torque is

$$|T_2| = Fr_2 \tag{5.2-2}$$

Torque applied to gear 3 is in the opposite direction to its rotation. The magnitude of that torque is

$$|T_3| = Fr_3 \qquad (5.2\text{-}3)$$

Combining Equations (5.2-2) and (5.2-3), we obtain

$$\left|\frac{T_2}{T_3}\right| = \frac{r_2}{r_3} \qquad (5.2\text{-}4)$$

If no slippage occurs, then we have

$$r_2|\omega_2| = r_3|\omega_3| \qquad (5.2\text{-}5)$$

The magnitude of the *speed ratio*, *e*, is

$$|e| = \left|\frac{\omega_3}{\omega_2}\right| = \frac{r_2}{r_3} \qquad (5.2\text{-}6)$$

Combining Equations (5.2-4) and (5.2-6), we get

$$|T_2\omega_2| = |T_3\omega_3| \qquad (5.2\text{-}7)$$

However, multiplication of a torque by a rotational speed yields *power* (Appendix C, Section C.10.3). Therefore, Equation (5.2-7) states that assuming no slippage between the friction gears, no power is lost in the transmission.

Frictional forces tangent to the cylinders are obtained by imposing a normal force of magnitude

$$N = |\overline{N}| \qquad (5.2\text{-}8)$$

Since for the friction model

$$F \leq \mu N \qquad (5.2\text{-}9)$$

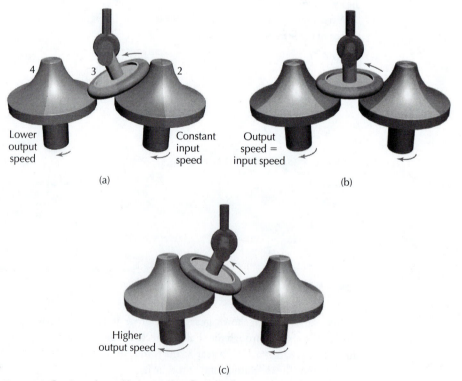

(a)

(b)

(c)

Figure 5.2 Continuously variable traction drive [Video 5.2].

in which the *coefficient of friction*, μ, is generally less than unity, N must therefore be greater than F. These relatively large normal forces must be supported by the bearings, as indicated in Figure 5.1(b), and must be taken into account in a design.

One definite advantage of using friction gearing is the ease of generating a continuous range of speed ratios. Consider the system shown in Figure 5.2. Changes in speed ratio are accomplished by altering the orientation of the idler wheel, link 3, and thereby varying the radii of contact with links 2 and 4. In the illustration, three positions of the idler friction wheel are shown that produce three distinct speed ratios. If the input cone 2 is driven at a constant rate, then the output of cone 4 may have a varying rotational speed.

Figure 5.3 shows an alternative system that can produce a continuous range of speed ratios. This system employs a variation of the open-loop friction drive (Figure 1.17). In this case, a range of speed ratios is obtained by allowing the belt to contact the pulleys at different radii, accomplished by adjusting the positions of the two halves of the pulleys. Two positions are shown in Figures 5.3(a) and 5.3(b), resulting in two distinct speed ratios. Figure 5.4 shows additional illustrations of a similar system.

Video 5.2
Continuously
Variable Traction
Drive

Video 5.4
Continuously
Variable Belt
Drive

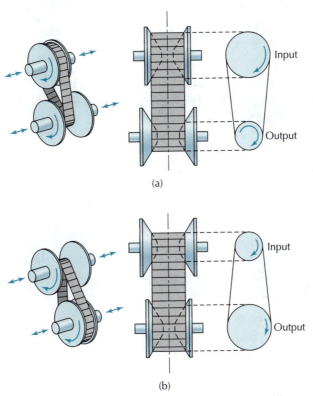

(a)

(b)

Figure 5.3 Continuously variable belt drive: (a) High-speed ratio, (b) Low-speed ratio.

5.3 COMMON TYPES OF TOOTHED GEARS

Video 5.5A
Straight Spur
Gears

Video 5.5B
Helical Spur
Gears

Straight spur gears (Figure 5.5(a)): These gears have straight teeth, parallel to the axis of rotation. They are relatively easy to manufacture.

Helical spur gears (Figure 5.5(b)): The teeth of a helical spur gear are at an oblique angle with respect to the axis of rotation. Compared with straight spur gears, there is a longer line of contact with a meshing gear, resulting in increased strength and durability. Helical spur gears are commonly used in automotive transmissions and machine tools. Figure 13.16 (Chapter 13) shows helical spur gears employed in a power drill.

The *helix angle*, ψ, for helical spur gears is illustrated in Figure 5.6. *Right-hand* and *left-hand* varieties are shown. A pair of meshing helical spur gears with parallel axes of rotation must include a gear with left-hand teeth and another with right-hand teeth. The helix angles must be equal.

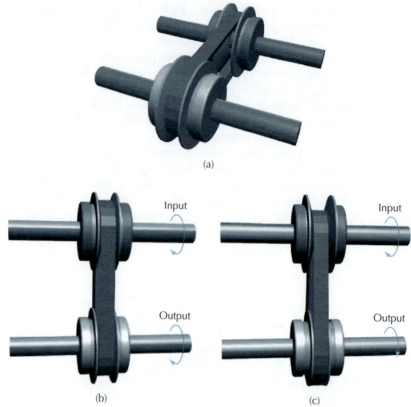

(a)

(b)   (c)

Figure 5.4 (a) Continuously variable belt drive [Video 5.4], (b) Unit speed ratio, (c) Low speed ratio.

(a)   (b)

Figure 5.5 Spur gears: (a) Straight spur gears [Video 5.5A], (b) Helical spur gears [Video 5.5B].

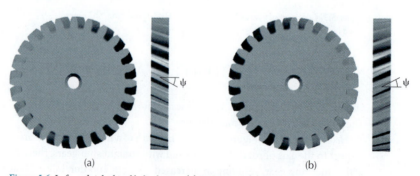

(a)   (b)

Figure 5.6 Left- and right-hand helical gears: (a) Right hand, (b) Left hand.

Video 5.7
Miter Gears

A specific type of a helical spur gear, called a *miter gear*, employs a helix angle of 45°. For miter gears, the centerlines of the axes of rotation can be perpendicular to one another, as shown in Figure 5.7. In this instance, two left-hand gears have been employed.

Figure 5.7 Miter gears [Video 5.7].

(a)   (b)

Figure 5.8 Herringbone gears: (a) Narrow [Video 5.8A], (b) Wide [Video 5.8B].

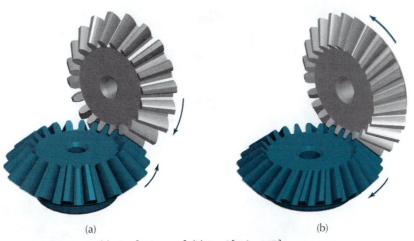

(a)   (b)

Figure 5.9 Bevel gears: (a) Plain [Video 5.9A], (b) Spiral [Video 5.9B].

Because of the orientation of the teeth on a helical spur gear, there is a component of the load from meshing gears directed along their axes of rotation. Devices that incorporate helical spur gears often employ thrust bearings to support these axial loads. Meshing straight spur gears do not have an axial component of the load because here the gear teeth are parallel to the axes of the rotation.

Herringbone gears (Figures 5.8(a) and 5.8(b)): Herringbone gears are equivalent to two helical gears with left- and right-hand teeth. This causes the axial load produced on one side to be counterbalanced by that produced on the other. Herringbone gears are well suited for heavy-load applications such as large turbines and generators.

Plain bevel gears (Figure 5.9(a)): These gears permit transmission of motion between two shafts angled relative to each other.

Video 5.9A
Plain Bevel Gears

Video 5.9B
Spiral Bevel Gears

Video 5.10
Hypoid Gear Set

Video 5.11A
Worm and Wheel Gears, Right-Hand Worm

Video 5.11B
Worm and Wheel Gears, Left-Hand Worm

Video 5.12A
Worm and Wheel, Double-Start Worm, Right-Hand Worm

Spiral bevel gears (Figure 5.9(b)): The teeth of spiral bevel gears are cut obliquely so that the length of the line of contact between meshing gears is longer compared to plain bevel gears, thus providing greater tooth strength and durability.

Hypoid gears (Figure 5.10): A hypoid gear set resembles spiral bevel gears. However, the teeth are cut such that the axes of rotation of the two gears in mesh do not intersect. The smaller input gear is called the *pinion* that meshes with and drives the *ring gear*. A common use of hypoid gears is in the differential assembly (see Chapter 6, Section 6.6) of rear wheel drive vehicles. They allow reduction in height of body styles by lowering the drive shaft to the rear wheels.

Worm and wheel gears (Figures 5.11 and 5.12): A *worm* gear is similar in shape to a screw thread and mates with a gear called the *wheel*. Power is always supplied to the worm. In fact, worm and wheel gear sets are typically self-locking (i.e., motion cannot be transmitted by driving the wheel gear). Worm and wheel gear sets provide high reductions of speed in a compact space. For each worm and wheel gear set shown in Figure 5.11, one rotation of the worm gear will cause the wheel to advance one tooth. Therefore, the magnitude of the speed ratio is

$$|e| = \frac{1}{N_{wheel}} \tag{5.3-1}$$

where N_{wheel} is the number of teeth on the wheel.

The gear set shown in Figure 5.11(a) incorporates a *right-hand worm*, and in Figure 5.11(b) a *left-hand worm* is employed. Although the magnitudes of the speed ratios of both gear sets are the same, for the same input motion of the worms, the direction of the rotation of the wheels is opposite.

Figure 5.12 shows two other versions of worm and wheel sets. Each incorporates a *double-start worm*. Both left- and right-hand worm gears are illustrated. For both cases shown, for each rotation of the worm gear, the wheel advances two teeth. Therefore, the magnitude of the speed ratio is

Figure 5.10 Hypoid gear set [Video 5.10].

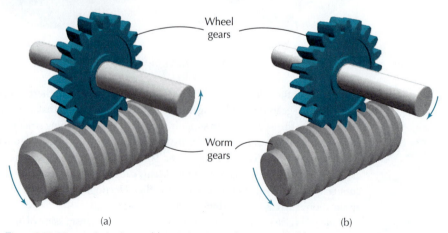

(a)   (b)

Figure 5.11 Worm and wheel gears: (a) Right-hand worm [Video 5.11A], (b) Left-hand worm [Video 5.11B].

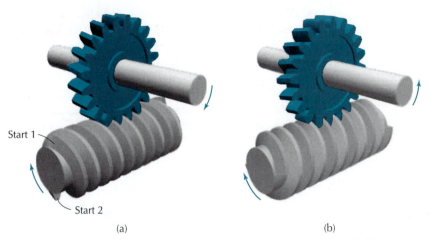

Figure 5.12 Worm and wheel, double-start worm: (a) Right-hand worm [Video 5.12A], (b) Left-hand worm [Video 5.12B].

$$|e| = \frac{2}{N_{wheel}} \qquad (5.3\text{-}2)$$

Worm gears can be designed to have as many as four starts.

5.4 FUNDAMENTAL LAW OF TOOTHED GEARING

For meshing gears, it is usually critical to maintain a constant ratio of speeds. Slight variations in the ratio would lead to unwanted vibrations and noise, caused by fluctuating loads and speeds. This section presents the requirement for the shape of gear teeth necessary to maintain constant speed ratio.

Figure 5.13 shows two members, links 2 and 3, in contact at point Q. The links are pivoted about points O_2 and O_3. The *line of centers* contains O_2 and O_3. Rotational speeds of the links are designated as ω_2 and ω_3. By inspection, ω_2 and ω_3 must be in opposite directions if contact is to be maintained as the links rotate. The point of contact, Q, is actually composed of two points: Q_2 on link 2, and Q_3 on link 3.

When point Q is on the line of centers (Figure 5.13(a)), there is a rolling action between the links. In this case, there is no relative sliding at Q since

$$\overline{v}_{Q_2} = \overline{v}_{Q_3}$$

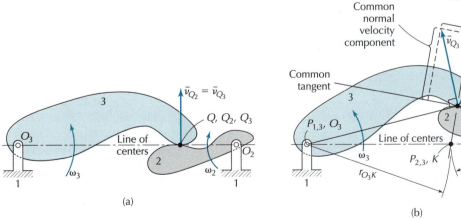

(a)

(b)

Figure 5.13 Two members in contact.

Figure 5.13(b) shows an alternate configuration of the links where point Q is not on the line of centers. In this instance, there is both relative sliding and turning, and point Q is the location of a two degree of freedom pair. The *common tangent* and *common normal* to the surfaces at Q are indicated. The common normal intersects the line of centers at K.

The ratio of rotational speeds between links may be readily found using the concept of instantaneous centers (Section 2.10). As indicated in Figure 5.13(b), two of the instantaneous centers, $P_{1,2}$ and $P_{1,3}$, are at the points O_2 and O_3, respectively. Based on Kennedy's Theorem (Section 2.11), the third instantaneous center coincides with K. For this system we may write

$$\left|\bar{v}_{P_{2,3}}\right| = r_{P_{1,2}P_{2,3}}\left|\omega_2\right| = r_{P_{1,3}P_{2,3}}\left|\omega_3\right| \tag{5.4-1}$$

which is equivalent to

$$r_{O_2K}\left|\omega_2\right| = r_{O_3K}\left|\omega_3\right| \tag{5.4-2}$$

and

$$\left|\frac{\omega_3}{\omega_2}\right| = \frac{r_{O_2K}}{r_{O_3K}} \tag{5.4-3}$$

Equation (5.4-3) indicates that the magnitude of the speed ratio for two links in contact is equal to the ratio of the lengths of the line segments that are along the line of centers. The lengths of the line segments involve the location of K and the axes of rotation of the links.

The above may be applied for the specific case where the links are meshing gears. Also, we may deduce the condition to maintain a constant speed ratio between a pair of meshing gears, known as the *fundamental law of toothed gearing*. The law may be stated as follows: To maintain a constant speed ratio between a pair of gears, the common normal at the point of contact between meshing teeth must always intersect the line of centers at a fixed point.

This fixed point is also called the *pitch point*. Even though the location of the point of contact may change, the fundamental law of toothed gearing can still be satisfied as long as the location of the pitch point remains fixed.

5.5 INVOLUTE TOOTH GEARING

Having established the fundamental law of toothed gearing, it is now possible to search for those shapes of teeth which will provide a constant speed ratio between meshing gears. This section presents gear teeth that have involute form, which is by far the most common shape of gear teeth.

Figure 5.14(a) shows two counter rotating cylinders, also referred to as *base circles*, of radii r_{b2} and r_{b3}. Their rotational speeds are ω_2 and ω_3 in the directions shown. A piece of string is wrapped around both base circles, similar to that of the cross belt friction drive shown in Figure 1.19. A bead is attached to the string between the base circles. In Figure 5.14(a), the bead is located on the line of centers of the base circles, and Figure 5.14(b) shows the bead in an alternate position. As the base circles rotate, the string unwinds from base circle 2 and is taken up by base circle 3. In order for the string to remain taut, the unwinding and take-up speeds must be equal, that is,

$$r_{b2}\left|\omega_2\right| = r_{b3}\left|\omega_3\right| \tag{5.5-1}$$

Video 5.15
Generation of an
Involute Profile

As the base circles rotate, the bead moves on a straight path with respect to the stationary base link. However, motion of the bead relative to a base circle may be observed by holding one base circle stationary, and then either winding or unwinding the string around the base circle, as shown in Figure 5.15(a). Under these conditions, the curve traced out by the bead is called an *involute*. At any position, the direction of the unwound string is normal to the involute.

The model shown in Figure 5.15(b) consists of two base circles. In this instance, the bead traces out two involutes simultaneously, one with respect to each of the base circles. At all positions, the bead corresponds to the point of contact of the involutes, and the orientation of the string between the base circles is normal to both involutes.

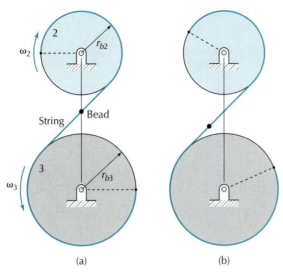

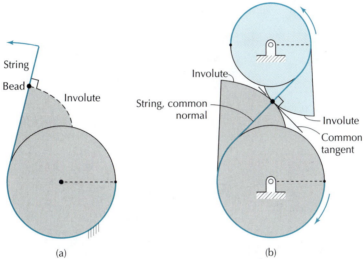

Figure 5.14 Two counterrotating cylinders connected by a string.

Figure 5.15 Generation of an involute profile [Video 5.15]: (a) One-cylinder model (b) Two-cylinder model.

Let us now remove the string from the demonstration model shown in Figure 5.15(b). Instead, as illustrated in Figure 5.16, motion is transmitted by meshing surfaces having involute profiles. These profiles are in contact at point *A*. A common normal to the involute profiles is drawn through the point of contact. It has the same orientation as that which the string had between the base circles. The common normal intersects the line of centers at the pitch point *K*. As the base circles rotate, the location of the pitch point remains fixed, even though the point of contact changes. Since the pitch point does not move, the meshing involute profiles satisfy the fundamental law of toothed gearing: that is, they generate a constant speed ratio.

Involute profiles are incorporated in gear teeth to generate a constant speed ratio between a pair of meshing gears. Each gear employs multiple teeth positioned around its periphery. The shape of both sides of every tooth is an involute. Each tooth, referred to as an *involute gear tooth*, makes use of two involutes. As illustrated in Figure 5.17, the involute on each side of a gear tooth is equivalent to unwrapping a string from the base circle. The beads on string 1 and string 2 trace out involute 1 and involute 2. Figure 5.18 shows a pair of gears with involute gear teeth. All teeth on a gear employ the same involute profile and have uniform height.

For the animation provided through [**Video 5.18**], pairs of teeth in turn come into and out of mesh and maintain a constant speed ratio as the gears rotate. Figure 5.19 shows an external gear meshing with an internal gear. Figure 5.20 shows a rack and pinion gear set.

Video 5.18
Straight Spur
Gears

Video 5.19
External—Internal
Meshing Gears

Video 5.20
Rack and Pinion
Gears

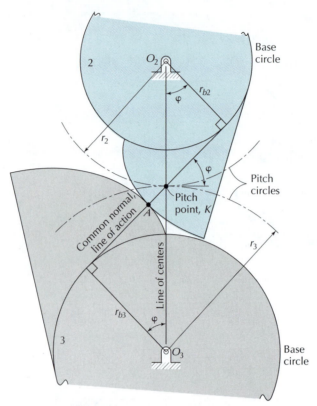

Figure 5.16 Double involute.

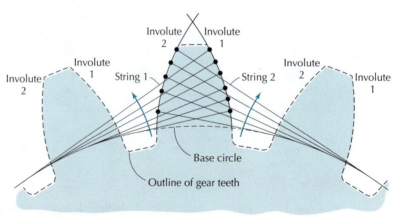

Figure 5.17 Generation of an involute gear.

The common normal at the point of contact for a pair of gears is also called the *line of action*. Gears incorporating involute profiles have a fixed line of action. Assuming no friction, the line of action also represents the orientation of interactive forces between meshing teeth. The angle between the line of action and a direction perpendicular to the line of centers is called the *pressure angle*, φ, as illustrated in Figure 5.16.

Figure 5.16 shows portions of two *pitch circles* associated with the gears. These circles are centered at O_2 and O_3, the same as for the base circles, and both pass through the pitch point. Their radii are r_2 and r_3. Applying Equation (5.4-2) in this instance gives

$$r_2 \left| \omega_2 \right| = r_3 \left| \omega_3 \right| \tag{5.5-2}$$

and therefore the magnitude of the speed ratio is

$$\left| \frac{\omega_3}{\omega_2} \right| = \frac{r_2}{r_3} \tag{5.5-3}$$

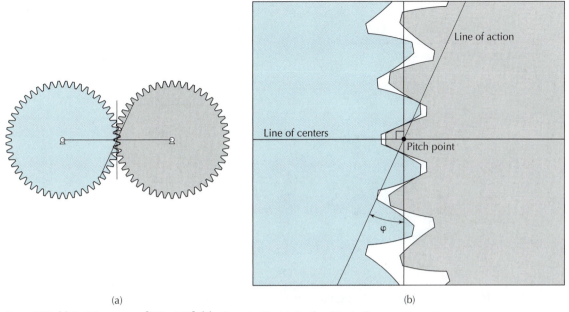

(a)   (b)

Figure 5.18 (a) Straight spur gears [Video 5.18], (b) enlargement in vicinity of meshing teeth.

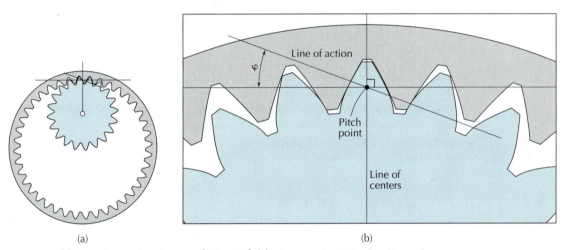

(a)   (b)

Figure 5.19 (a) External–internal meshing gears [Video 5.19], (b) enlargement in vicinity of meshing teeth.

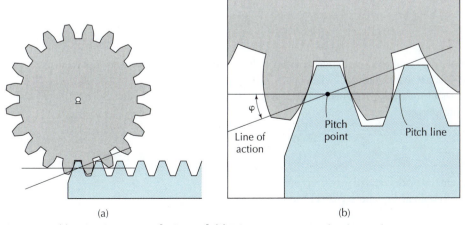

(a)   (b)

Figure 5.20 (a) Rack and pinion gears [Video 5.20], (b) Enlargement in vicinity of meshing teeth.

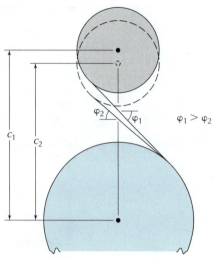

Figure 5.21 Variation of pressure angle.

Equation (5.5-3) is identical to Equations (5.2-6), which was obtained by considering a pair of friction gears of radii r_2 and r_3, for which there was no slippage. Thus, a pair of toothed gears is kinematically equivalent to friction gears with radii equal to the pitch circles. Either side of Equation (5.5-2) represents the magnitude of the *pitch line velocity*, v_p.

Using Figure 5.16, the relation between radii of the base and pitch circles is

$$r_{bj} = r_j \cos\varphi \tag{5.5-4}$$

The shape of an involute for a gear tooth is a function of its base circle diameter and is independent of the *center-to-center distance* between a pair of meshing gears. As an illustration, Figure 5.21 shows two base circles with a string wrapped around both of them. Two center-to-center distances, c_1 and c_2, are shown. For each, the line of action and pressure angle are different. However, the involute profiles generated for each base circle are identical. Thus, as long as contact is kept between gear teeth, even if the center-to-center distance is changed, a constant speed ratio will be maintained.

5.6 SIZING OF INVOLUTE GEAR TEETH

Circular pitch and *diametral pitch* are two means used to designate the size of straight spur gears incorporating an involute profile, using the U.S. Customary system of units.

Circular pitch, p_c, is the curvilinear distance measured along the pitch circle, from a point on one tooth to the corresponding point on an adjacent tooth. It is illustrated in Figure 5.22 and is expressed as

$$p_c = \frac{\pi d}{N} = \frac{2\pi r}{N} \tag{5.6-1}$$

where d is the pitch circle diameter and N is the number of teeth on the gear. The value of circular pitch is stated with units of inches.

Diametral pitch, P, is the number of gear teeth per unit length of pitch circle diameter, and is expressed as

$$P = \frac{N}{d} = \frac{N}{2r} \tag{5.6-2}$$

Diametral pitch is stated with units of in^{-1}.

Combining Equations (5.6-1) and (5.6-2), we obtain

$$Pp_c = \pi \tag{5.6-3}$$

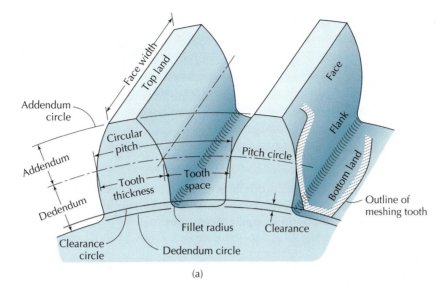

(a)

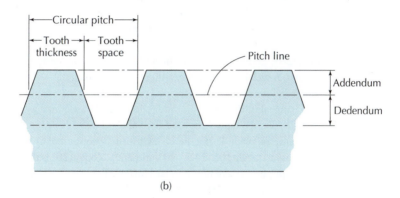

(b)

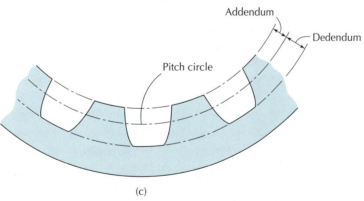

(c)

Figure 5.22 Spur gear terminology: (a) External gear, (b) Rack, (c) Internal gear.

Module is the term used to indicate the size of involute gear teeth in the SI System. It is defined as the pitch circle diameter, in millimeters, divided by the number of teeth on the gear. Module is expressed as

$$m = \frac{d}{N} = \frac{2r}{N} \tag{5.6-4}$$

Two gears in mesh must have the same diametral pitch, circular pitch, and module. This requirement is often simply stated that gears in mesh must have the same pitch. Meshing gears must also have the same pressure angle.

From Equations (5.6-1), (5.6-2), and (5.6-4), we have

$$d = 2r = \frac{Np_c}{\pi} = \frac{N}{P} = mN \tag{5.6-5}$$

Thus, for a given pitch, the number of teeth on a gear is proportional to its pitch circle radius. Using Equation (5.5-3), we conclude

$$\left| \frac{\omega_3}{\omega_2} \right| = \frac{r_2}{r_3} = \frac{N_2}{N_3} \tag{5.6-6}$$

Equation (5.6-6) states that the magnitude of the speed ratio for a pair of gears in mesh equals the inverse of the ratio of the numbers of teeth on the gears. The topic of speed ratio will be considered again in Chapter 6 for a variety of gear trains.

Proportions of a gear tooth are shown in Figure 5.22(a). Included in this figure are

- *addendum*, *a*: the radial distance between the pitch circle and a circle (also called the addendum circle) drawn through the tops of the gear teeth
- *dedendum*, *b*: the radial distance between the pitch circle and a circle (also called the dedendum circle) defining the bottom land between teeth (Note: we require $b > a$).
- *clearance*, $b - a$: the difference between the dedendum and the addendum

Figures 5.22(b) and 5.22(c) show the addendum and dedendum associated with a rack and an internal gear. A rack has a *pitch line*, rather than a pitch circle.

Values of addendum, dedendum, and pressure angle have been standardized in order to facilitate interchangeability of gears. The addendum and dedendum may be expressed in terms of either circular pitch, diametral pitch, or module. Table 5.1 provides a summary of standardized values in terms of diametral pitch. Using the U.S. Customary system of units, there are three standardized values of pressure angle (i.e., 14.5°, 20°, and 25°). Today, 20° is the most common.

Straight spur gears have two standard forms when $\varphi = 20°$: *stub* and *full-depth*. Values are provided for both forms. Full-depth teeth have a larger value of addendum than stub teeth.

Figure 5.23 illustrates teeth on racks with six values of diametral pitch, using the proportions listed in Table 5.1 for full-depth teeth and $\varphi = 20°$.

The proportions in Table 5.1 show that the value of the dedendum is greater than the addendum. The clearance, defined as the difference between the dedendum and the addendum, also corresponds to the distance between the top land of one tooth and the bottom land of a meshing tooth.

TABLE 5.1 STANDARD PROPORTIONS OF GEAR TEETH AS A FUNCTION OF DIAMETRAL PITCH

	Addendum a	Dedendum b	Whole Depth $a + b$	Clearance $b - a$
Full-depth tooth $(\varphi = 14.5°)$	$\dfrac{1.0}{P}$	$\dfrac{1.157}{P}$	$\dfrac{2.157}{P}$	$\dfrac{0.157}{P}$
Stub tooth $(\varphi = 20°)$	$\dfrac{0.8}{P}$	$\dfrac{1.0}{P}$	$\dfrac{1.8}{P}$	$\dfrac{0.2}{P}$
Full-depth tooth $(\varphi = 20°)$	$\dfrac{1.0}{P}$	$\dfrac{1.25}{P}$	$\dfrac{2.25}{P}$	$\dfrac{0.25}{P}$
Full-depth tooth $(\varphi = 25°)$	$\dfrac{1.0}{P}$	$\dfrac{1.25}{P}$	$\dfrac{2.25}{P}$	$\dfrac{0.25}{P}$

For the SI System, the most common form of tooth is full-depth with 20° pressure angle. The corresponding values of the addendum and dedendum expressed in terms of the module are

$$a = 1.0m; \qquad b = 1.25m \qquad (5.6\text{-}7)$$

Typical standard values of the module are

$$m = 1.0, 1.25, 1.5, 2, 2.5, 3, 4, 5, 6, 8, 10, 12, 16, 20, 25, 32, 40, 50 \text{ mm}$$

Figure 5.24 illustrates teeth on racks with three values of module for full-depth teeth and $\varphi = 20°$.

As indicated in Figure 5.22(a), the *tooth thickness* and *tooth space* are measured along the pitch circle. The tooth thickness is nominally one-half of the circular pitch, and thus the tooth thickness and tooth space between two teeth of a gear are equal.

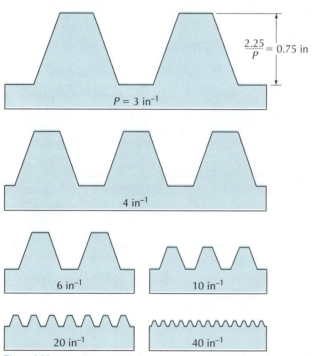

Figure 5.23 Sizes of gear teeth for typical values of diametral pitch (full depth, $\varphi = 20°$).

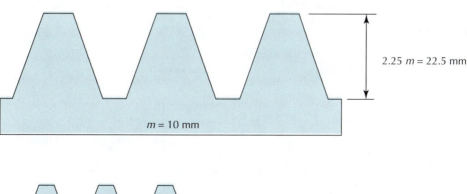

Figure 5.24 Sizes of gear teeth for typical values of module (full depth, $\varphi = 20°$).

In this instance, if the pitch circles of two meshing gears touch at the pitch point, then contact can simultaneously occur on both sides of a gear tooth with the meshing gear. Obviously, the thickness cannot be greater than the tooth space or the gears will not mesh.

EXAMPLE 5.1

A spur gear, having 30 teeth and a diametral pitch of 6 in^{-1}, is rotating at 200 rpm. Determine the circular pitch and the magnitude of the pitch line velocity.

SOLUTION
From Equation (5.6-3), the circular pitch is

$$p_C = \frac{\pi}{P} = \frac{\pi}{6 \text{ in}^{-1}} = 0.524 \text{ in}$$

To find the magnitude of the pitch line velocity, $v_p = r\omega$, we first have to determine the pitch circle diameter of the gear. From Equation (5.6-2)

$$d = \frac{N}{P} = \frac{30}{6 \text{ in}^{-1}} = 5.0 \text{ in}$$

The angular velocity of the gear in terms of radians per second is

$$\omega = 200 \frac{\text{rev}}{\text{min}} \times 2\pi \frac{\text{rad}}{\text{rev}} \times \frac{1 \text{ min}}{60 \text{ sec}} = 20.94 \frac{\text{rad}}{\text{sec}}$$

and finally the magnitude of the pitch line velocity is

$$v_P = r\omega = \frac{d}{2}\omega = 2.5 \text{ in} \times 20.94 \frac{\text{rad}}{\text{sec}} = 52.4 \frac{\text{in}}{\text{sec}}$$

EXAMPLE 5.2

Two spur gears are in mesh. Driven gear 3 has a magnitude of rotational speed that is one-quarter that of driver gear 2. Gear 2 rotates at 500 rpm, has a module of 1.5 mm, and has 24 teeth. Determine

(a) the number of teeth of gear 3
(b) the magnitude of the pitch line velocity

SOLUTION
(a) From Equation (5.6-4)

$$d_2 = mN_2 = 1.5 \times 24 = 36 \text{ mm}; \quad r_2 = 18 \text{ mm}$$

From Equation (5.5-2)

$$r_3 = \frac{r_2 |\omega_2|}{|\omega_3|} = \frac{r_2 \omega_2}{\left(\frac{1}{4}\omega_2\right)} = 4r_2 = 4 \times 18 \text{ mm} = 72 \text{ mm}$$

(Continued)

EXAMPLE 5.2 Continued

$$d_3 = 144 \text{ mm}$$

The number of teeth on gear 3 is then

$$N_3 = \frac{d_3}{m} = \frac{144}{1.5} = 96$$

where, as required, the same value of module has been used for both gears.

(b) The magnitude of the pitch line velocity is

$$v_p = r_2 |\omega_2| = 18 \text{ mm} \times 500 \times \frac{2\pi \text{ rad}}{60 \text{ sec}} = 942 \frac{\text{mm}}{\text{sec}} = r_3 |\omega_3|$$

5.7 BACKLASH AND ANTIBACKLASH GEARS

The limiting condition that permits proper meshing between two gears is when the tooth thicknesses and tooth spaces are equal. For this reason, the tooth thickness of a gear is often manufactured to be slightly less than the tooth space. *Backlash* is the difference between the tooth space and tooth thickness as measured along a pitch circle. Figure 5.25 illustrates the backlash for a pair of meshing gears. Although an insignificant amount of backlash is often incorporated into a design, excessive backlash would cause impacts between teeth, especially if a reversal of rotation occurs. This can give rise to increased noise and tooth loading.

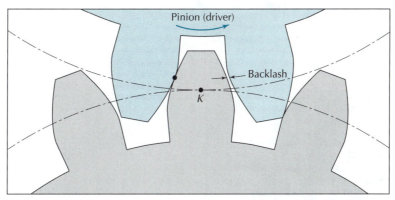

Figure 5.25 Backlash.

An *antibacklash gear*, shown in Figure 5.26(a), is designed to either reduce or eliminate backlash. A close-up of the gear teeth is shown in Figure 5.26(b). The gear is composed of two halves that can move rotationally relative to one another about the same central axis against the action of spring forces. The teeth of the two halves are staggered when the gear is not in mesh. Dashed lines on the top of a tooth on the two halves indicate the stagger. Through meshing, the stagger is either reduced or eliminated. The relative rotation between the two halves is accomplished through forces applied by the meshing teeth.

Figure 5.27 shows a pinion that meshes with two gears. One is a regular gear, and the other is an antibacklash gear. In Figure 5.27(a), the gears are separated. As the animation of [**Video 5.27**] proceeds, the two gears are brought into mesh with the pinion (Figure 5.27(b)). The teeth on the two meshing regular gears do not have simultaneous contact on both sides of a gear tooth. However, for meshing of the pinion and antibacklash gear, contact is maintained on both sides, even when the directions of rotation are reversed.

Video 5.27
Antibacklash Gear

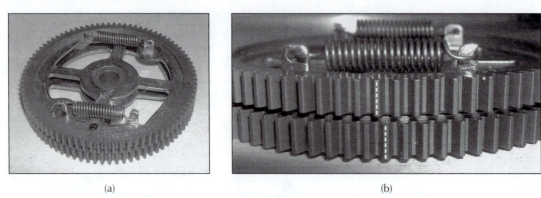

(a)   (b)

Figure 5.26 (a) Antibacklash gear ,(b) Close-up of antibacklash gear.

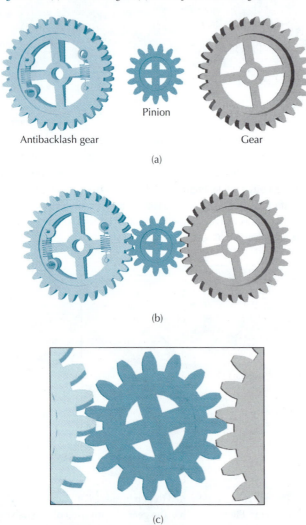

Pinion

Antibacklash gear   Gear

(a)

(b)

(c)

Figure 5.27 Antibacklash gear [Video 5.27]: (a) Separated gears, (b) Gears assembled, (c) Close-up of meshing gears.

5.8 GEOMETRICAL CONSIDERATIONS IN THE DESIGN OF REVERTED GEAR TRAINS

The topic of gear trains will be covered in detail in Chapter 6. For the time being, we will study geometrical considerations that relate to the design of one form of a gear train, called a *reverted gear train*.

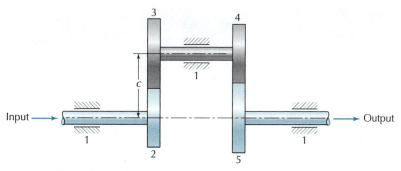

Figure 5.28 Reverted gear train.

Reverted gear trains are characterized by having their input and output axes of rotation collinear. A typical reverted gear train is illustrated in Figure 5.28. Gear 2 is mounted on the input shaft, and gear 5 is mounted on the output shaft. Gears 3 and 4 are rigidly connected through a shaft and have a common axis of rotation. They mesh with gears 2 and 5, respectively. The center-to-center distance, c, between gears 2 and 3 is equal to that between gears 4 and 5, that is,

$$c = \frac{d_2}{2} + \frac{d_3}{2} = \frac{d_4}{2} + \frac{d_5}{2} \tag{5.8-1}$$

where d_2, d_3, d_4, and d_5 denote pitch circle diameters. If the teeth are characterized by their module, then substituting Equation (5.6-4) in Equation (5.8-1) and simplifying,

$$m_2 N_2 + m_3 N_3 = m_4 N_4 + m_5 N_5 \tag{5.8-2}$$

Gears in mesh must have the same pitch. For the gear train shown in Figure 5.28, we require

$$m_2 = m_3 \quad \text{and} \quad m_4 = m_5 \tag{5.8-3}$$

and Equation (5.8-2) becomes

$$m_2 (N_2 + N_3) = m_4 (N_4 + N_5) \tag{5.8-4}$$

Similar expressions to Equations (5.8-2)–(5.8-4) may be generated in cases where the teeth are defined by their circular pitch and diametral pitch.

In the instance when the sizes of all gear teeth are equal (e.g., $m_2 = m_4$), Equation (5.8-4) simplifies to

$$N_2 + N_3 = N_4 + N_5 \tag{5.8-5}$$

EXAMPLE 5.3

Table 5.2 provides information that pertains to the reverted gear train shown in Figure 5.28. All gears have teeth with 20° pressure angle and are full-depth. Determine

(a) the circular pitch of gear 2
(b) the base circle radius of gear 3
(c) the pitch circle diameter of gear 4
(d) the addendum circle diameter of gear 5
(e) the center-to-center distance, c, between gears 2 and 3

(*Continued*)

EXAMPLE 5.2 Continued

> **TABLE 5.2** PARAMETERS OF THE REVERTED GEAR TRAIN SHOWN IN FIGURE 5.28

Gear No.	No. of Teeth	Diametral Pitch (in^{-1})
2	30	6
3	48	—
4	—	8
5	72	—

SOLUTION

We begin by determining the missing information in Table 5.2. Since gears 2 and 3 are in mesh, and gears 4 and 5 are in mesh,

$$P_3 = P_2 = 6 \text{ in}^{-1} \quad \text{and} \quad P_5 = P_4 = 8 \text{ in}^{-1}$$

Also, substituting Equation (5.6-2) in Equation (5.8-1),

$$\frac{N_2}{2P_2} + \frac{N_3}{2P_3} = \frac{N_4}{2P_4} + \frac{N_5}{2P_5}$$

$$\frac{30}{2 \times 6 \text{ in}^{-1}} + \frac{48}{2 \times 6 \text{ in}^{-1}} = \frac{N_4}{2 \times 8 \text{ in}^{-1}} + \frac{72}{2 \times 8 \text{ in}^{-1}}$$

from which

$$N_4 = 32$$

Based on the above, the solutions are as follows:

(a) From Equation (5.6-3) we have

$$p_{c2} = \frac{\pi}{P_2} = \frac{\pi}{6 \text{ in}^{-1}} = 0.524 \text{ in}$$

(b) From Equations (5.5-4) and (5.6-2) we get

$$r_{b3} = r_3 \cos\varphi = \frac{d_3}{2}\cos\varphi = \frac{N_3}{2P_3}\cos\varphi = \frac{48}{2 \times 6 \text{ in}^{-1}}\cos 20° = 3.76 \text{ in}$$

From Equation (5.6-2) we obtain

$$d_4 = \frac{N_4}{P_4} = \frac{32}{8 \text{ in}^{-1}} = 4.00 \text{ in}$$

(d) From Equation (5.6-2) and Table 5.1, the addendum circle diameter of gear 5 is

$$d_5 + 2a_5 = \frac{N_5}{P_5} + 2\frac{1}{P_5} = \frac{72}{8 \text{ in}^{-1}} + \frac{2}{8 \text{ in}^{-1}} = 9.25 \text{ in}$$

(e) From Equations (5.8-1) and (5.6-2) we have

$$c = \frac{d_2}{2} + \frac{d_3}{2} = \frac{N_2}{2P_2} + \frac{N_3}{2P_3} = \frac{30}{2 \times 6 \text{ in}^{-1}} + \frac{48}{2 \times 6 \text{ in}^{-1}} = 6.50 \text{ in}$$

5.9 MANUFACTURING OF GEARS

A variety of methods are used to manufacture gears. They include molding, milling, and generating processes. This section presents some of the main methods used to produce gears.

For a given size of gear teeth (i.e., prescribed value of diametral pitch or circular pitch or module), the shape of an involute depends on the number of teeth on the gear. Figure 5.29 shows shapes of teeth on gears having 6, 20, and 45 teeth. The tooth on the gear having the larger number of teeth has a straighter face. In the limit, as the number of teeth increases to infinity (i.e., a rack), the shape of the side of the gear teeth is a straight line. The dependence of the shape of an involute on the number of teeth on the gear must be taken into account in the manufacturing operation.

5.9.1 Form Milling

Teeth of straight spur gears and racks may be cut using *form milling*. Teeth are produced by repeatedly passing a milling cutter across the face of a gear blank (i.e., workpiece). Such a milling cutter is shown in Figure 5.30. During each cut, the workpiece is held stationary. Figure 5.31 shows a schematic of a form milling operation. The setup for the form milling of a straight spur gear is shown in Figure 5.32. An indexing head rotates the workpiece between cuts by the angle to be subtended by a gear tooth. For a rack, the workpiece is translated between cuts.

This procedure is relatively slow compared to other methods presented below. Also, accuracy of the machining process depends on the correctness of the form cutter. Since the shape of teeth is a function of the number of teeth on a gear, there must be a series of cutters for each tooth pitch and pressure angle. In fact, to be totally accurate, a different cutter would be needed for each number of teeth. However, generally a set of eight cutters is employed for all gears having the same pitch and pressure angle, which leads to approximations of the shape of teeth.

Video 5.32
Milling of a
Straight Spur
Gear

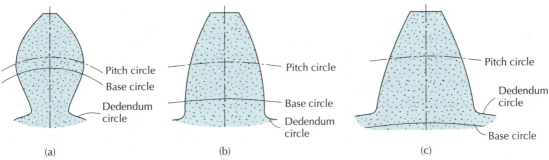

(a)   (b)   (c)

Figure 5.29 Typical shapes of gear teeth: (a) 6 teeth, (b) 20 teeth, (c) 45 teeth.

(a)   (b)

Figure 5.30 (a) Milling cutter, (b) Cross section of cutter tooth.

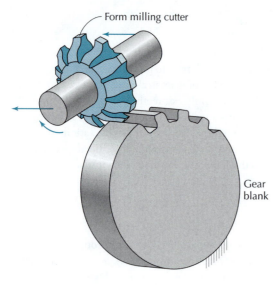

Form milling cutter

Gear
blank

Figure 5.31 Form milling.

Cutter

Indexer

Gear

Figure 5.32 Milling of a straight spur gear [Video 5.32].

5.9.2 Generating Processes

Hobbing

Hobbing is one of the fastest, most versatile methods of making rough and/or finish cuts of gear teeth. A typical cutter, also called a *hob*, is illustrated in Figure 5.33(a). The cross section of a row of cutter teeth of a hob is illustrated in Figure 5.33(b). The *hob pitch*, p_h, is the spacing between the cutter teeth. Figure 5.34 illustrates a hob and the side view of a gear. As the hob rotates, the workpiece (i.e., gear) simultaneously turns in a carefully coordinated manner. For each revolution of the hob, the gear turns about its axis the angle to be subtended by one tooth. Relative motion of the hob and workpiece generates the involute shape. Figure 5.35 shows two series of accumulated images of the positions of the cutter teeth with respect to the workpiece; the first series shows the cutter teeth advancing and generating an involute profile on one side of a gear tooth, and the second series shows the cutter teeth receding while producing an involute profile on an adjacent tooth.

In addition to the relative rotational motion described above, the hob is slowly driven across the face of the workpiece creating involute profiles over the entire face of the gear. Figures 5.36(a) and 5.36(b) respectively show the relative positions of the hob and gear before and after a cutting operation. A corresponding animation which includes both relative rotational motions and translation of the hob across the face is shown through [**Video 5.34**]. In actual operation, prior to the

Video 5.34
Straight Spur
Gear and Hob

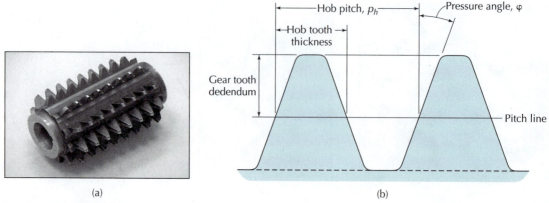

Hob pitch, p_h

Pressure angle, φ

Hob tooth thickness

Gear tooth dedendum

Pitch line

(a)   (b)

Figure 5.33 (a) Gear hob (b) Cross section of hob teeth.

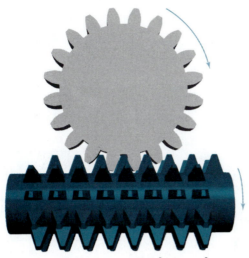

Figure 5.34 Straight spur gear and hob [Video 5.34].

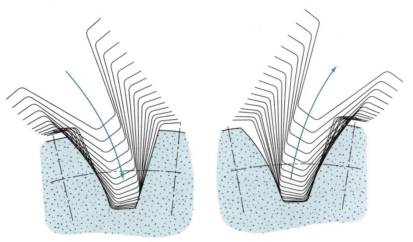

Figure 5.35 Hobbing operation—relative motion between hob and gear blank.

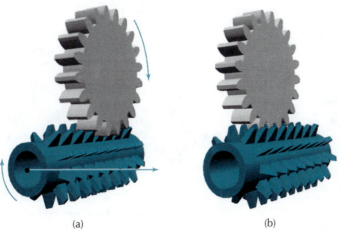

(a)   (b)

Figure 5.36 Straight spur gear and hob [Video 5.34].

Video 5.37
Hobbing of a
Straight Spur
Gear

cut, the gear starts as a circular disc, also called the gear blank, without any teeth. A picture of the hobbing of a straight spur gear is shown in Figure 5.37.

In the hobbing of a gear, spacing between gear teeth is dictated by the hob. For a straight spur gear, the hob will generate a gear with

$$p_c = p_h \qquad (5.9\text{-}1)$$

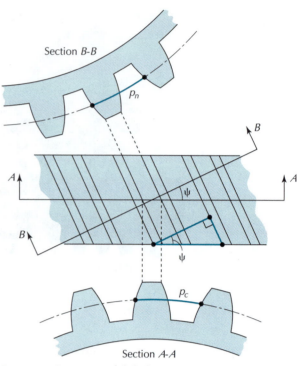

Section B-B

p_n

Figure 5.37 Hobbing of a straight spur gear [Video 5.37].

p_c

Section A-A

Figure 5.38 Cross section of a helical spur gear.

A single hob may be used to produce any size of straight spur gear having a given circular pitch and pressure angle.

A hob can also be used to produce a helical spur gear. In this instance, the hob is rotated by the *helix angle*, ψ, from the position used to cut a straight spur gear (see Figure 5.38). A typical setup for cutting a helical gear is shown in Figure 5.39. A picture of the hobbing of a helical spur gear is shown in Figure 5.40.

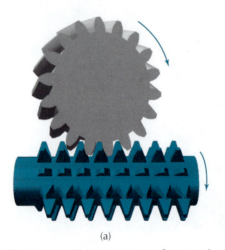

(a)

Figure 5.39 Helical spur gear and hob [Video 5.39].

(b)

Figure 5.40 Hobbing of a helical spur gear [Video 5.40].

For a helical spur gear, the hob generates teeth by having its spacing defined by the *normal pitch*, p_n, as indicated in Figure 5.38. The relationships between the normal pitch, hob pitch, p_h, and circular pitch, p_c, of a helical spur gear are

$$p_n = p_h = p_c \cos \psi$$

(5.9-2)

As an example, the hob pitch of the hob shown in Figure 5.33(a) is 0.375 inches. Therefore, using Equation (5.6-1), the pitch circle diameter of a straight spur gear (i.e., $p_c = p_h$) having $N = 62$ teeth is

$$d = \frac{N p_c}{\pi} = \frac{62 \times 0.375 \text{ in}}{\pi} = 7.400 \text{ in} \tag{5.9-3}$$

If we now wish to employ the same hob to cut a helical spur gear having the same pitch circle diameter. In this instance, spacing between teeth cut by the hob is the normal pitch, that is,

$$p_n = 0.375 \text{ in} \tag{5.9-4}$$

Using Equation (5.9-2), the circular pitch is greater than the normal pitch, and therefore for the same size of gear, we will have fewer teeth. If $N = 60$ on the helical spur gear, and it is to be the same size as the straight spur gear, then the circular pitch is

Video 5.42A
Shaping of a
Straight Spur
Gear, External
Gear

$$p_c = \frac{\pi d}{N} = \frac{\pi \times 7.400 \text{ in}}{60} = 0.3875 \text{ in} \tag{5.9-5}$$

Combining Equations (5.9-2), (5.9-4), and (5.9-5), we obtain

Video 5.42B
Shaping of a
Straight Spur
Gear, Internal
Gear

$$\psi = \cos^{-1}\left(\frac{p_n}{p_c}\right) = \cos^{-1}\left(\frac{0.375}{0.3875}\right) = 14.57° $$

Hobbing is generally limited to producing external gears. However, hobbing may be completed with special spherical hobs to produce internal gears having large diameters.

Shaping

Video 5.43A
Shaping of a
Straight Spur
Gear

Shaping is another type of generating process. A shaping operation is shown in Figure 5.41, where the cutter reciprocates with respect to the workpiece. The cutter and workpiece are indexed between each cutting stroke. Relative motion between the teeth of the cutter and workpiece ensures that the required involute profile is generated.

Video 5.43B
Shaping of an
Internal Gear

Two other illustrations are shown in Figure 5.42 for the cutting of an external and internal gear. The same cutter is used for both. Motions of the animations provided through [**Video 5.42A**] and [**Video 5.42B**] have been exaggerated to illustrate rotation of the gear between strokes of the cutter. Figure 5.43 shows pictures of the cutting of a straight spur gear and an internal gear.

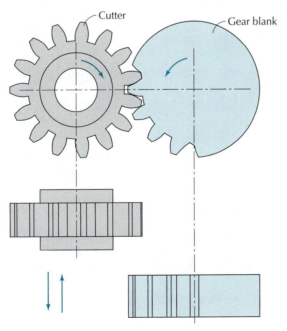

Figure 5.41 Shaping of a straight spur gear.

(a)   (b)

Figure 5.42 Shaping of straight spur gears: (a) External gear [Video 5.42A], (b) Internal gear [Video 5.42B].

(a)   (b)

Figure 5.43 Shaping of gears: (a) Straight spur gear [Video 5.43A], (b) Internal gear [Video 5.43B].

Planing

Video 5.45
Planing of a Plain Bevel Gear

Planing is a cutting operation used to manufacture bevel gears. Figure 5.44 illustrates a schematic for cutting a plain bevel gear. This operation involves two reciprocating cutters that shape both sides of a tooth simultaneously. The two cutters are mounted on a cradle (not shown in the figure). For machining each tooth, the cradle gives the cutters the same relative motion with respect to the workpiece as a meshing gear. Figure 5.45 shows a setup for the cutting of a plain bevel gear.

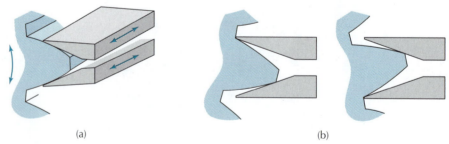

(a)   (b)

Figure 5.44 Schematic of a planing operation.

Rotary Broaching

Video 5.47
Rotary Broaching Operation

Rotary broaching of plain bevel gears has some similarities to form milling of straight spur gears. Figure 5.46(a) shows a setup for a rotary broaching operation. It includes a rotating cutter. A separate picture of the cutter is shown in Figure 5.46(b). There are two distinct portions of cutter teeth on the periphery: a portion where the cutter teeth progressively increase in depth, and a portion in which the cutter teeth have constant depth.

Figure 5.45 Planing of a plain bevel gear [Video 5.45].

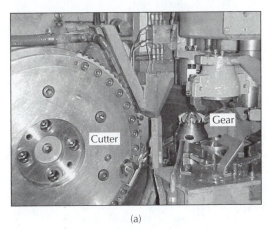

(a)

(b)

Figure 5.46 (a) Rotary broaching of straight bevel gear, (b) Cutter.

Another illustration of rotary broaching is shown in Figure 5.47. Starting from the configuration shown in Figure 5.47(a), the gear blank is swung into position (Figure 5.47(b)). While holding the gear blank stationary, a slot is cut in the gear blank. This is accomplished with the teeth of increasing depth (Figure 5.47(c)). Then, the gear blank is translated vertically upward (Figure 5.47(d)). As the cutter continues to rotate, the gear blank moves downward while at the same time the cutter teeth, with constant depth, cut the slot (Figure 5.47(e)). This relative motion permits the bottom of the slot to be cut in a straight line. A portion of the periphery of the cutting wheel has no teeth.

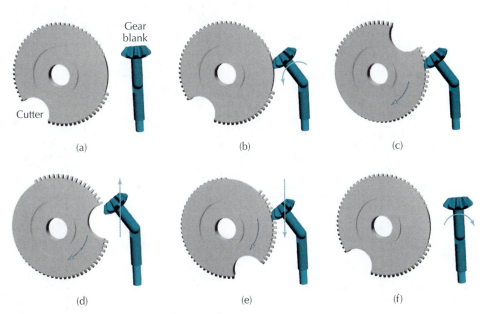

(a)   (b)   (c)

(d)   (e)   (f)

Figure 5.47 Rotary broaching operation [Video 5.47].

This permits the gear blank to be indexed in preparation for cutting of the next slot. While the gear blank is being indexed, the cutter is rotated to the position for the start of another cutting operation. The operation is repeated until all slots are cut and the machined gear is then swung back into its starting position (Figure 5.47(f)).

EXAMPLE 5.4

Table 5.3 provides information that pertains to the reverted gear train shown in Figure 5.28.

TABLE 5.3 PARAMETERS OF THE REVERTED GEAR TRAIN SHOWN IN FIGURE 5.28

Gear No.	No. of Teeth	Type of Gear
2	30	Straight spur
3	48	—
4	29	—
5	47	Helical spur

All gears were manufactured using a hob cutter having 20° pressure angle, 0.375-inch hob pitch, and full depth. Determine

(a) the base circle radius of gear 3
(b) the center-to-center distance between gears 2 and 3
(c) the circular pitch of gear 4
(d) the helix angle of gear 5
(e) the pitch circle diameter of gear 4
(f) the addendum circle diameter of gear 4 (i.e., the diameter of the gear blank)

SOLUTION

By inspection, gear 3 is a straight spur gear, and gear 4 is a helical spur gear.

(a) For gears 2 and 3 we have

$$p_{c2} = p_{c3} = 0.375 \text{ in} = p_h$$

From Equation (5.6-3) we get

$$P_2 = P_3 = \frac{\pi}{p_{c3}} = \frac{\pi}{0.375 \text{ in}} = 8.38 \text{ in}^{-1}$$

From Equations (5.5-4) and (5.6-2) we obtain

$$r_{b3} = r_3 \cos\varphi = \frac{d_3}{2}\cos\varphi = \frac{N}{2P_3}\cos\varphi = \frac{48}{2 \times 8.38 \text{ in}^{-1}}\cos 20° = 2.69 \text{ in}$$

(b) From Equations (5.8-1) and (5.6-2) we have

$$c = \frac{d_2}{2} + \frac{d_3}{2} = \frac{N_2}{2P_2} + \frac{N_3}{2P_3} = \frac{30}{2 \times 8.38 \text{ in}^{-1}} + \frac{48}{2 \times 8.38 \text{ in}^{-1}} = 4.65 \text{ in}$$

(Continued)

EXAMPLE 5.4 Continued

(c) Employing the result from part (b) and then combining Equations (5.8-1) and (5.6-1), we obtain

$$c = 4.65 \text{ in } \frac{d_4}{2} + \frac{d_5}{2} = \frac{N_4 p_{c4}}{2\pi} + \frac{N_5 p_{c5}}{2\pi} = \frac{29 p_{c4}}{2\pi} + \frac{47 p_{c5}}{2\pi}$$

Noting that

$$p_{c4} = p_{c5}$$

we obtain

$$p_{c4} = 0.384 \text{ in}$$

(d) For gears 4 and 5, we recognize that

$$p_{n4} = p_{n5}$$

From Equation (5.9-2) we have

$$p_{n4} = p_{c4} \cos \psi_4 = p_h$$

$$0.375 = 0.384 \cos \psi_4$$

and therefore

$$\psi_4 = 12.7° = \psi_5$$

(e) From Equation (5.6-1) we get

$$d_4 = \frac{N_4 p_{c4}}{\pi} = \frac{29 \times 0.384 \text{ in}}{\pi} = 3.54 \text{ in}$$

(f) For a given hob, the addendum obtained for a helical spur gear is the same as that for a straight spur gear. Therefore

$$a_4 = a_2 = \frac{1}{P_2} = \frac{1}{8.38} \text{ in}$$

Using the results of part (e), the addendum circle diameter of gear 4 is

$$d_4 + 2a_4 = \left(3.54 + \frac{2}{8.38} \right) \text{ in} = 3.78 \text{ in}$$

EXAMPLE 5.5 DESIGN OF A PAIR OF SPUR GEARS

Design a pair of external—external spur gears that will generate a magnitude of the speed ratio of 0.4. The center-to-center distance between the gears is to be 65 mm. The gears are to be cut using a hob with module 1.25 mm.

(Continued)

EXAMPLE 5.5 Continued

SOLUTION

The magnitude of the speed ratio is

$$\left|\frac{\omega_3}{\omega_2}\right| = \frac{N_2}{N_3} = 0.4 \quad \text{or} \quad N_2 = 0.4 N_3 \tag{5.9-6}$$

Combining Equations (5.8-1) and (5.6-4), the center-to-center distance is

$$c = 65\,\text{mm} = \frac{d_2}{2} + \frac{d_3}{2} = \frac{N_2 m_2}{2} + \frac{N_3 m_3}{2}$$

Noting that

$$m_2 = m_3$$

Combining this result with Equation (5.9-6) we obtain

$$2c = 130\,\text{mm} = m_2(N_2 + N_3) = m_2(N_2 + 0.4N_2) = m_2(1.4N_3) \tag{5.9-7}$$

For a given pitch circle diameter and hob, the largest number of teeth are cut on a straight spur gear. In this instance, solving Equations (5.9-6) and (5.9-7) with module of the hob $m_h = m_2 = 1.25\,\text{mm}$

$$N_2 = 29.7; \quad N_3 = 74.3$$

This result is unacceptable because gears must have integer numbers of teeth. We now search for an acceptable solution for a pair of helical spur gears and use the above result as the upper limits of the numbers of teeth. In the following table, we consider various possibilities and employ Equation (5.9-6).

N_3	$N_2 = 0.4\,N_3$
74	29.6
73	29.2
72	28.8
71	28.4
70	28

For the result that has an integer number of teeth for both gears

$$N_2 = 28; \quad N_3 = 70$$

and employing the required center-to-center distance, the circular pitch is determined to be

$$c = 65\,\text{mm} = \frac{N_2 + N_3}{2\pi}p_c = \frac{28 + 70}{2\pi}p_c; \quad p_c = 4.167\,\text{mm}$$

(Continued)

Using Equation (5.6-5), the hob pitch and normal pitch of the gear teeth are

$$p_h = p_n = m_h \pi = 1.25 \, \text{mm} \times \pi = 3.927 \, \text{mm}$$

and therefore, using Equation (5.9-2), the helix angle of both gears is

$$\psi = \cos^{-1}\left(\frac{p_n}{p_c}\right) = \cos^{-1}\left(\frac{3.927}{4.167}\right) = 19.6°$$

This example illustrates that the helix angle may be used as a design variable to achieve a desired center-to-center distance between a pair of meshing gears.

5.10 CONTACT RATIO

In the transmission of rotational motion through two meshing gears, it is essential that at any time one or more pairs of teeth be in contact. Otherwise, there would be instances when there is no smooth transfer of motion.

For a pair of meshing gears turning at a constant rate, *contact ratio* is the average number of pairs of gear teeth in contact over time. Theoretically, the value of contact ratio must be greater than 1.00. However, 1.40 is generally accepted as the practical minimum value.

Figure 5.48 shows a pair of meshing straight spur gears. Gear 2 is the driver gear, and gear 3 is the driven gear. Initial contact, labeled as point *A* in Figure 5.48(a), occurs when the outer tip of the driven gear tooth touches the driver gear. Final contact, labeled as point *B* in Figure 5.48(c), occurs when the outer tip of the driver gear tooth contacts the driven gear. For both of these configurations there is a second pair of teeth in contact. Thus, prior to and after the noted teeth come into and out of mesh, there is another pair of teeth in mesh, which ensures continuous and uniform transmission of motion. Figure 5.48(b) illustrates another configuration where the point of contact is at the pitch point. At this instant, this is the sole point of contact between the gears.

At any instant in time, there is an integer number of pairs of gear teeth in mesh. For the gears shown in Figure 5.48, during a portion of the cycle, one pair of gear teeth is in mesh, whereas at other times there are two pairs. Figure 5.49 shows the function of the number of pairs of gear teeth

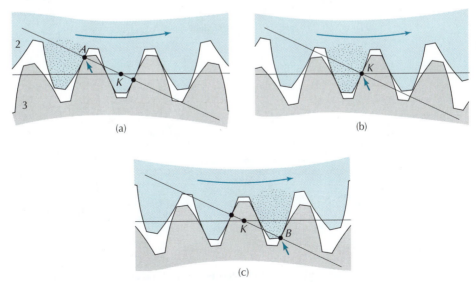

Figure 5.48 Two spur gears in mesh: (a) Initial contact, (b) Contact at pitch point, (c) Final contact.

in mesh over time for the gear set illustrated in Figure 5.48. The time average of the curve, which is the contact ratio, is

$$m_c = \frac{1 \times 0.47\tau + 2 \times 0.53\tau}{\tau} = 1.53$$

Figure 5.50 shows a different pair of gears in mesh. In this instance, for the initial and final points of contact for the noted pair of teeth, there are three points of contact between the gears. For other configurations (not shown), there are two points of contact. This gear set has a higher value of contact ratio than that shown in Figure 5.48.

Figure 5.51 shows an external–internal meshing gear set. The external gear is the driver gear. Configurations for the initial and final points of contact are illustrated.

Figure 5.52 combines the initial and final configurations of an external–external meshing pair of gear teeth in the same drawing. The initial and final points of contact are labeled A and B, respectively. Length r_{AB} is referred to as the *length of action* and is given symbol Z. During meshing action, the point of contact moves along the straight line from A to B. Points S and S' lie at the intersection

Figure 5.49 Illustration of contact ratio.

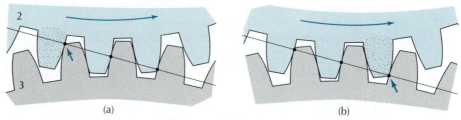

(a)   (b)

Figure 5.50 Two spur gears in mesh: (a) Initial contact, (b) Final contact.

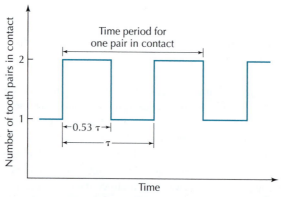

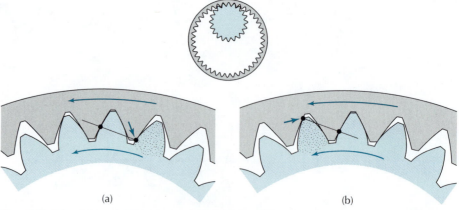

(a)   (b)

Figure 5.51 External–internal meshing gear set: (a) Initial contact, (b) Final contact.

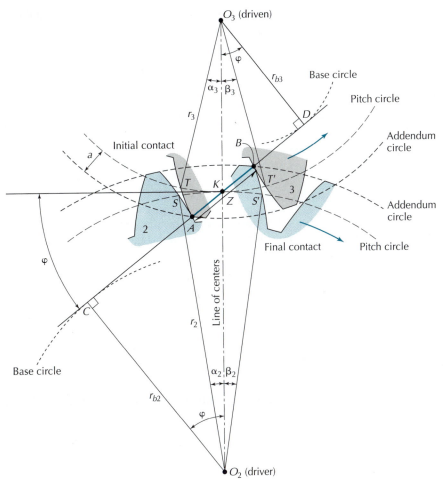

Figure 5.52 Two spur gears in mesh [Video 5.52].

of the pitch circle and side of a tooth of gear 2, when this gear makes initial and final contact with meshing gear 3. Points T and T' have similar definitions pertaining to gear 3. Arcs SS' and TT' are the *arcs of action* and must be equal for the pair of meshing gears. The corresponding angular displacements of the gears between the configurations of initial and final contact may be broken into two parts as shown in Figure 5.52. Quantities α_2 and α_3 are the *angles of approach*, and β_2 and β_3 are the *angles of recess*. The contact ratio may be expressed in terms of the arc of action and the circular pitch as

$$m_c = \frac{\text{arc of action}}{p_c} = \frac{(\alpha_2 + \beta_2)r_2}{p_c} = \frac{(\alpha_3 + \beta_3)r_3}{p_c} \tag{5.10-1}$$

An alternative expression for contact ratio can be found by recognizing that the transmission of rotation between involute gears in mesh is equivalent to a cross belt friction drive. Taking this into consideration and referring to Figure 5.52, the angle turned by either base circle while tooth contact moves from A to B can be obtained by dividing the length of action by the corresponding base circle radius. For either gear, this angle may be expressed as

$$\frac{r_{AB}}{r_b} = \frac{Z}{r_b} \tag{5.10-2}$$

The angle subtended by one gear tooth is

$$\frac{2\pi}{N} \tag{5.10-3}$$

An expression for contact ratio is then found by taking the ratio of the angles given in (5.10-2) and (5.10-3), yielding

$$m_c = \frac{\left(\dfrac{Z}{r_b}\right)}{\left(\dfrac{2\pi}{N}\right)} = \frac{Z}{p_b} \tag{5.10-4}$$

where

$$p_b = \frac{2\pi r_b}{N} \tag{5.10-5}$$

Quantity p_b is called the *base pitch*. The form of the equation for base pitch is the same as for circular pitch, Equation (5.6-1), except that the base circle radius has replaced the pitch circle radius.

The length of action, Z, is found by examining Figure 5.52, where

$$Z = r_{AB} = r_{AK} + r_{KB} \tag{5.10-6}$$

and

$$r_{AK} = r_{AD} - r_{KD} ; \quad r_{KB} = r_{CB} - r_{CK} \tag{5.10-7}$$

To obtain r_{AK}, we determine lengths r_{AD} and r_{KD}. In right triangle O_3DA we have

$$r_{O_3A} = r_3 + a ; \quad r_{O_3D} = r_{b3} = r_3 \cos\varphi \tag{5.10-8}$$

and therefore

$$r_{AD} = [(r_{O_3A})^2 - (r_{O_3D})^2]^{\frac{1}{2}} = [(r_3 + a)^2 - r_3^2 \cos^2\varphi]^{\frac{1}{2}} \tag{5.10-9}$$

Also

$$r_{KD} = r_3 \sin\varphi \tag{5.10-10}$$

Combining Equations (5.10-7)–(5.10-10), we obatain

$$r_{AK} = r_{AD} - r_{KD} = [(r_3 + a)^2 - r_{b3}^2]^{\frac{1}{2}} - r_3 \sin\varphi \tag{5.10-11}$$

Similarly,

$$r_{KB} = [(r_2 + a)^2 - r_{b2}^2]^{\frac{1}{2}} - r_2 \sin\varphi \tag{5.10-12}$$

Combining Equations (5.10-6), (5.10-11), and (5.10-12), we get

$$Z = [(r_2 + a)^2 - r_{b2}^2]^{\frac{1}{2}} + [(r_3 + a)^2 - r_{b3}^2]^{\frac{1}{2}} - (r_2 + r_3)\sin\varphi \tag{5.10-13}$$

For a rack and pinion set, the rack is equivalent to a sector of a gear of infinite radius, and therefore Equation (5.10-13) cannot be used. In this instance, if the pinion is link 2, it can be shown that the length of action is

$$Z = [(r_2 + a)^2 - r_{b2}^2]^{1/2} - r_2 \sin\varphi + \frac{a}{\sin\varphi} \qquad (5.10\text{-}14)$$

For an external–internal pair of meshing gears, for which gear 3 is the internal gear, the length of action is

$$Z = [(r_2 + a)^2 - r_{b2}^2]^{1/2} - [(r_3 - a)^2 - r_{b3}^2]^{1/2} + (r_3 - r_2) \sin\varphi \qquad (5.10\text{-}15)$$

Video 5.52
Two Spur Gears
in Mesh

[Video 5.52] provides an animated sequence of a pair of external-external gears in mesh from the initial contact to the final contact.

EXAMPLE 5.6   CONTACT RATIO FOR MESHING GEARS

Figure 5.53 shows two pitch and addendum circles for two full-depth straight spur gears in mesh. From the information provided, determine the contact ratio.

SOLUTION
The pitch circle radii (Equation (5.6-2)) are

$$r_2 = \frac{d_2}{2} = \frac{N_2}{2P} = \frac{24}{2 \times 5 \ \text{in}^{-1}} = 2.400 \ \text{in}; \quad r_3 = \frac{N_3}{2P} = 3.200 \ \text{in}$$

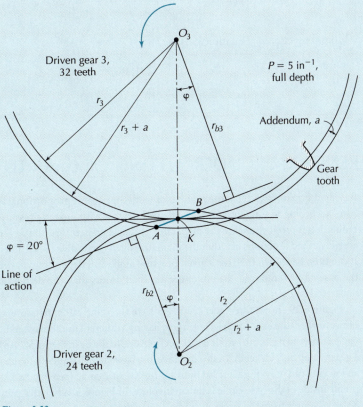

Figure 5.53

(Continued)

EXAMPLE 5.6 Continued

The gear tooth addendum for both gears (Table 5.1) is

$$a = \frac{1.0}{P} = \frac{1}{5 \text{ in}^{-1}} = 0.200 \text{ in}$$

The base circle radii (Equation (5.5-4)) are

$$r_{b2} = r_2 \cos\varphi = 2.4 \cos 20° \text{ in} = 2.255 \text{ in}; \quad r_{b3} = r_3 \cos\varphi = 3.007 \text{ in}$$

The addendum circle radii are

$$r_2 + a = (2.400 + 0.200 \text{ in}) = 2.600 \text{ in}; \quad r_3 + a = 3.400 \text{ in}$$

The length of contact (Equation (5.10-13)) is

$$Z = [(r_2 + a)^2 - r_{b2}^2]^{\frac{1}{2}} + [(r_3 + a)^2 - r_{b3}^2]^{\frac{1}{2}} - (r_2 + r_3)\sin\varphi$$
$$= \{[2.600^2 - 2.255^2]^{\frac{1}{2}} + [3.400^2 - 3.007^2]^{\frac{1}{2}} - (2.400 + 3.200)\sin 20°\}$$
$$= 0.966 \text{ in}$$

The base pitch (Equation (5.10-5)) is

$$p_b = p_{b2} = \frac{2\pi r_{b2}}{N_2} = \frac{2\pi \times 2.255 \text{ in}}{24} = 0.590 \text{ in} = p_{b3}$$

The contact ratio (Equation (5.10-4)) is

$$m_c = \frac{Z}{p_b} = \frac{0.966}{0.590} = 1.64$$

5.11 INTERFERENCE AND UNDERCUTTING OF GEAR TEETH

Based on Equation (5.6-5), the number of teeth on a gear is proportional to its pitch circle radius. Also, according to Equation (5.5-4), an involute gear with a given pressure angle has a base circle radius that is proportional to the pitch circle radius. Furthermore, the dedendum and whole depth are prescribed functions of either the diametral pitch or module (see Table 5.1 or Equation 5.6-7). It can then be shown that, for a given pressure angle and depending on the number of gear teeth, the base circle radius may be either greater than or less than the dedendum circle radius. This is illustrated below.

Figure 5.29 shows three gear teeth. All have the same pitch, pressure angle, and whole depth. However, because each tooth is from a gear having a different number of teeth, the position of the base circle relative to the tooth profile is distinct from one to the next. Figure 5.29(a) illustrates a tooth from a gear having only six teeth. In this instance, the base circle is larger than the dedendum circle. Since an involute can only be generated outward from a base circle (see Figure 5.15), the portion of tooth profile inside of the base circle is not an involute. In Figure 5.29(a), this portion of the tooth profile is drawn simply from the circle as a radial line to the center of the gear, with a fillet at the base of the tooth. Figure 5.29(b) illustrates a tooth on a gear having 20 teeth. Here, there is a reduced portion of the tooth profile inside of the base circle, which is not an involute. Figure 5.29(c) shows a tooth on a gear having 45 teeth. Now, the base circle is smaller than the dedendum circle, and therefore the entire profile of the tooth may be manufactured with an involute profile.

The meshing of a pair of involute gears was shown to be equivalent to a string wound between two base circles (Section 5.5). The bead on the string traces out two involutes, each with respect to its base circle. These involutes match the profiles of the involute gear teeth.

Figure 5.54 shows two base circles and a string (i.e., line of action) wrapped between them. Points E and F are at the intersections of the base circles and the line of action. They represent the limits between which two involutes can be generated. For a corresponding pair of gears, these are the limits within which meshing between two involute profiles can take place. These points are called the *meshing limits*. Outside of these limits, it is impossible to have meshing between two involute profiles. Also shown in Figure 5.54 is the pitch point, K, located on the line of action between the two meshing limits.

A rack is equivalent to a gear of infinite radius. For a rack and pinion set, there is only one meshing limit to be considered. Figure 5.55 indicates the range over which meshing can take place between an involute on the pinion and the rack.

Figure 5.56 illustrates a pinion having eight teeth meshing with a rack. The teeth are full-depth, with a 20° pressure angle (see Table 5.1). The base circle of the pinion is greater than its dedendum circle. If the portion of tooth profile inside of the base circle is a radial line to the center of the pinion, as shown, it would be impossible for the gears to mesh properly. Two regions are depicted in this figure where there is overlapping (i.e., *interference*) of the teeth on the pinion with those on the rack.

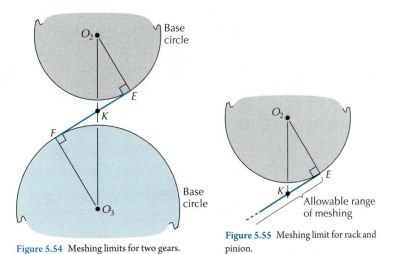

Figure 5.54 Meshing limits for two gears.

Figure 5.55 Meshing limit for rack and pinion.

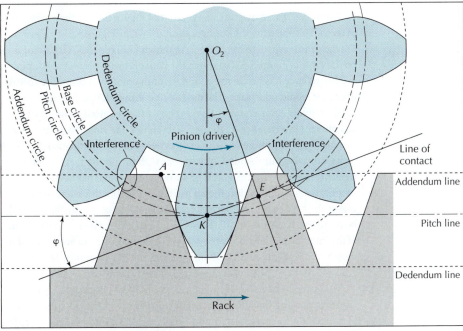

Figure 5.56 Rack and pinion with interference.

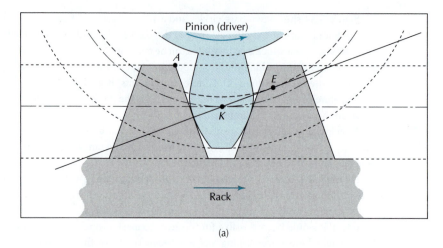

(a)

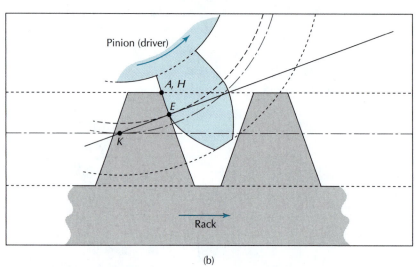

(b)

Figure 5.57 Rack and pinion with interference.

Figure 5.57(a) shows an enlargement of Figure 5.56 in the vicinity of meshing between a pair of teeth. Starting from the configuration shown in Figure 5.57(b), examine the path of point A on the rack with respect to the pinion. As shown in Figure 5.57(b), point A initially contacts point H on the pinion. Figure 5.57(c) shows the path of A with respect to the pinion as a dashed line during meshing. Since it is inside the boundary of the pinion, it therefore reveals the amount of interference that would take place. The dashed line shows the permissible outline of one side of a pinion tooth to avoid interference. *Undercutting* is the required removal of material on the pinion that will prevent interference. Figure 5.57(d) shows one side of a pinion tooth that has been undercut to eliminate its interference with one corner (point A) of the rack tooth. However, point D on the rack would also cause similar interference with the pinion teeth. Therefore, both sides of the pinion teeth must be undercut. Figure 5.58 shows a pinion that has been undercut to prevent interference.

For the undercut pinion shown in Figure 5.58, the thickness of the teeth at the flank is less than that at the base circle. Furthermore, a small portion of the involute outside of the base circle has to be removed. Each gear tooth is essentially a cantilever beam, and motion is transmitted by means of the interactive forces between gear teeth along the line of action. It follows that decreasing thicknesses of the gear teeth near their bases has the detrimental effect of reducing the strength of the gear. Obviously, undercutting in design is to be avoided.

The interference shown in Figure 5.56 may be eliminated by having more teeth on the pinion by increasing the number of teeth, and/or increasing the pitch circle diameter of the pinion. Figure 5.59 shows an alternative means. Here, the pinion is identical to that shown in Figure 5.56. However, the addendum of the rack teeth in Figure 5.59 is smaller than that in Figure 5.56, that

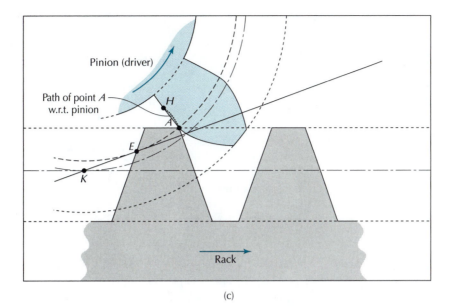

(c)

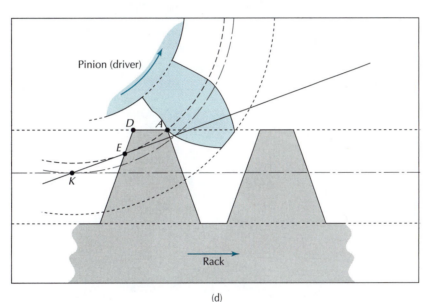

(d)

Figure 5.57 *(Continued)*

is, the rack shown in Figure 5.59 does not possess the standard proportions of teeth as listed in Table 5.1. In Figure 5.59, the addendum of the rack reaches only to the meshing limit, *E*, which is on the base circle. In this instance, points of contact are never inside of the base circle, and it is not necessary to remove a portion of the involute to eliminate interference. If the addendum of the rack were greater than that shown in Figure 5.59, it would be necessary to remove a portion of the involute on the pinion to prevent interference.

When a portion of an involute is cut off in order to prevent interference, expressions for the length of action (Equations (5.10-13)–(5.10-15)) do not apply. Also, reducing the addendum results in a decrease of the length of action as well as the contact ratio.

Figure 5.60 illustrates the same rack and pinion set as shown in Figure 5.59. The outline of a meshing gear is added as a dashed line. The addendum of the gear is identical to that of the rack. The addendum circle of the gear is tangent to the addendum line of the rack at point *X*. Point *J* at the corner of the gear tooth is slightly lower than point *E*. Points of contact between the pinion and the gear never reach the interference limit and always remain in the permissible range for meshing without requiring undercutting. It is therefore concluded that if a pinion can mesh with a

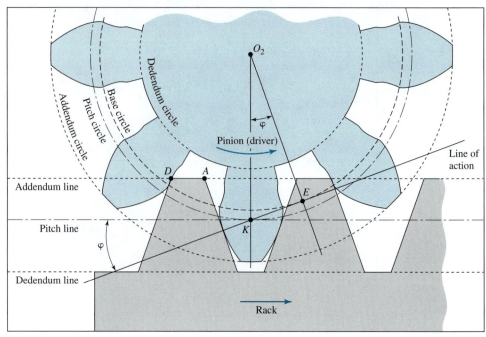

Figure 5.58 Undercut pinion.

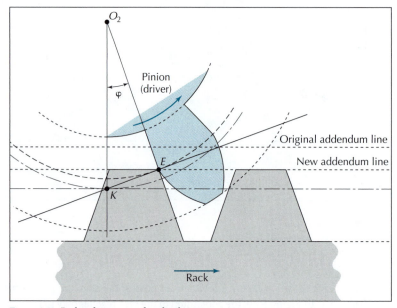

Figure 5.59 Rack and pinion—reduced rack.

rack without interference, it will also properly mesh with another gear having teeth possessing the same specifications as the rack, and containing an identical or larger number of teeth on the pinion.

For a given form of gear tooth, it is possible to determine the minimum number of teeth on a pinion that will mesh with a rack without requiring removal of a portion of the involute profile. The limiting case has the addendum line of the rack passing through the meshing limit. Figure 5.61 includes the essential dimensions of such a rack and pinion set. The pitch point and meshing limit are denoted by K and E, respectively. Therefore

$$\sin \varphi = \frac{r_{KE}}{r} \tag{5.11-1}$$

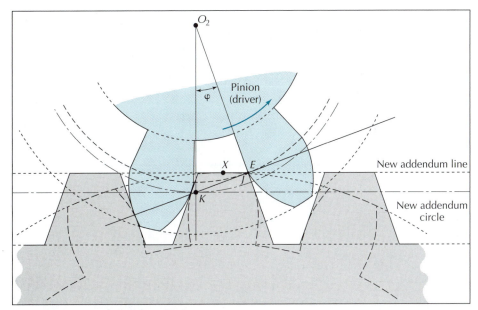

Figure 5.60 Pinion meshing with a rack and gear.

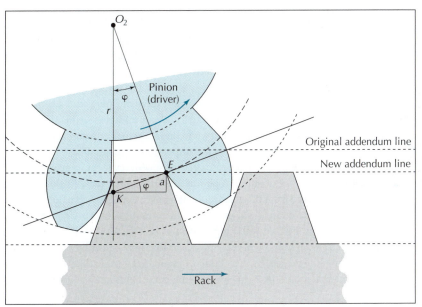

Figure 5.61 Geometry to prevent undercutting.

Also

$$\sin\varphi = \frac{a}{r_{KE}} = \frac{\left(\dfrac{\alpha}{P}\right)}{r_{KE}} \tag{5.11-2}$$

where α is a constant that, when divided by the diametral pitch, P, gives the addendum. Applying this to the standardized tooth dimensions listed in Table 5.1, for full-depth teeth, $\alpha = 1.0$, and for stub teeth, $\alpha = 0.8$.

Multiplying Equations (5.11-1) and (5.11-2) together gives

$$\sin^2\varphi = \frac{\alpha}{rP} \tag{5.11-3}$$

TABLE 5.4 MINIMUM NUMBERS OF TEETH ON A GEAR TO AVOID UNDERCUTTING

	Full-Depth Tooth ($\varphi = 14.5°$)	Stub Tooth ($\varphi = 20°$)	Full-Depth Tooth ($\varphi = 20°$)	Full-Depth Tooth ($\varphi = 25°$)
Meshing with a rack	32	14	18	12
Manufactured using a hob	37	18	22	14

Combining Equations (5.11-3) and (5.6-2) gives

$$\sin^2 \varphi = \frac{2\alpha}{N} \qquad (5.11\text{-}4)$$

or

$$\boxed{N = \frac{2\alpha}{\sin^2 \varphi}} \qquad (5.11\text{-}5)$$

Equation (5.11-5) can be used to calculate the smallest number of teeth on a pinion that will mesh with a rack without requiring the elimination of a portion of the involute. Values are listed in Table 5.4 for common gear tooth systems. Because values in the table were calculated for the pinion meshing with a rack, they can also be conservatively used as minimums for a pinion meshing with a gear of equal or larger size.

Figure 5.62 shows a pinion having five teeth meshing with a larger gear. The base circle of the pinion is illustrated. The pinion teeth have been undercut so that there is no interference. However, the contact ratio is less than unity, and there is actually no contact for the configuration illustrated. As a result, for this pair of gears it is impossible to maintain a constant speed ratio.

The relative motions of hob cutter teeth with respect to a gear blank (Section 5.9.2) has some similarities to the action of a pinion meshing with a rack. In this instance, however, material is being removed from the gear during the relative motions. If a hob is set up to manufacture a gear where the base circle will be greater than the dedendum circle, the required undercutting will be made during the machining process. Also, Equation (5.11-5) may be employed to determine the minimum number of teeth on a gear that may be cut by a hob without removing a portion of the involute profile. By inspection of Figure 5.33, the "addendum" of a hob cutter tooth is equal to the dedendum of a gear tooth being cut. Therefore, α is obtained by examining the values of

Video 5.62
Meshing of Gears Requiring Undercutting

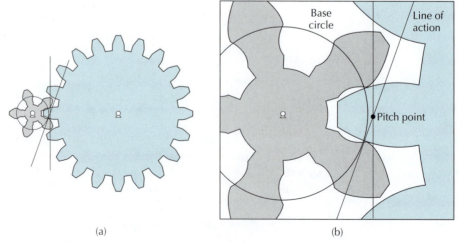

(a)   (b)

Figure 5.62 (a) Meshing of gears requiring undercutting [Video 5.62], (b) Enlargement in vicinity of meshing teeth.

dedendum listed in Table 5.1. Considering full-depth teeth, the values are 1.157 for a 14.5° pressure angle, and 1.25 for a 20° and 25° pressure angle. Calculated minimums are listed in Table 5.4.

5.12 CYCLOIDAL TOOTH GEARING

Gear teeth having the shape of a cycloid, or *cycloidal gear teeth*, also satisfy the fundamental law of toothed gearing. As illustrated in Figure 5.63, a *cycloid* is formed by rolling circles on the inside and outside of a pitch circle, and drawing the path of a point on the periphery of the rolling circles. The curves inside and outside of the pitch circle are called the *hypocycloid* and *epicycloid*, respectively. Sizes of pitch circles for two gears in mesh may differ. However, sizes of the rolling circles for the epicycloid and hypocycloid of one gear must respectively match those generated by the hypocycloid and epicycloid of the meshing gear.

Gears with cycloidal teeth are often employed in positive-displacement pumps. Figure 5.64 shows an oil pump from an automobile engine. A pump of this form is also called a *cycloidal gerotor pump*. Another such pump is shown in Figure 5.65. It consists of two gears with cycloidal teeth, indicated as links 2 and 3. Both gears turn in the same direction. Link 3 is an internal gear, and both gears have a fixed axis of rotation. There are multiple points of contact between the gears. For the configuration shown, we draw the common normals through all points of contact. Since there must be only one constant speed ratio of the gears, then as required, all common normals intersect the line of centers at the same fixed pitch point.

The directions of the common normals at the points of contact are not fixed as the gears rotate. This is distinct from gears having involute gear teeth, for which the direction of the common normal remains fixed.

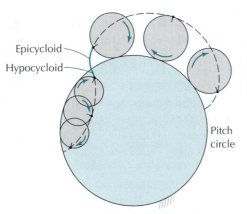

Figure 5.63 Generation of a cycloid [Video 5.63].

Figure 5.64 Cycloidal gerotor pump [Video 5.64].

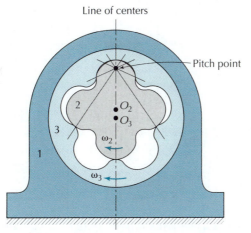

Figure 5.65 Cycloidal gerotor pump.

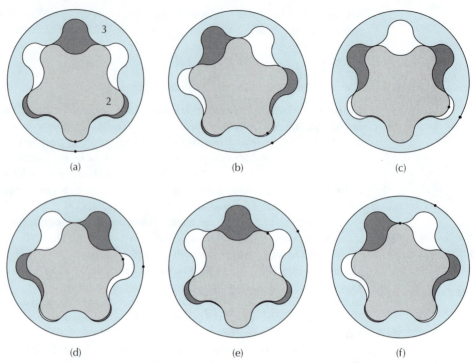

Figure 5.66 Cycloidal gerotor pump [Video 5.66]: (a) $\theta_3 = 0°$, (b) $\theta_3 = 30°$, (c) $\theta_3 = 60°$, (d) $\theta_3 = 90°$, (e) $\theta_3 = 120°$, (f) $\theta_3 = 150°$.

Figure 5.66 shows a series of positions of the gears. The angle indicated for each configuration denotes the rotation of link 3 from the selected starting point. For demonstration purposes, pockets between points of contact are alternately colored gray and white to distinguish one from the next. In actual use, all pockets carry the same fluid. During rotation, each pocket increases in size, drawing fluid in through an inlet port (not shown) that is located on the sides of the gears. Once a pocket reaches its maximum size, the source from the inlet port is cut off. Upon further rotation, the size of the pocket decreases, and fluid is discharged through an outlet port.

Video 5.66
Cycloidal Gerotor
Pump

Another application of cycloidal gear teeth is in a blower employed in superchargers. It consists of two counterrotating lobes as shown in Figure 5.67. The angle shown for each configuration

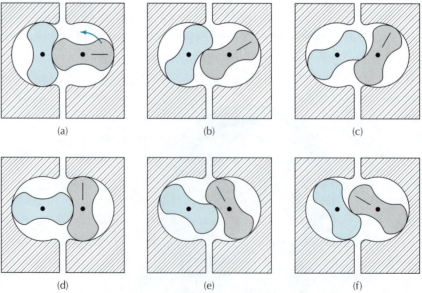

Figure 5.67 Supercharger blower [Video 5.67]: (a) $\theta = 0°$, (b) $\theta = 30°$, (c) $\theta = 60°$, (d) $\theta = 90°$, (e) $\theta = 120°$, (f) $\theta = 150°$.

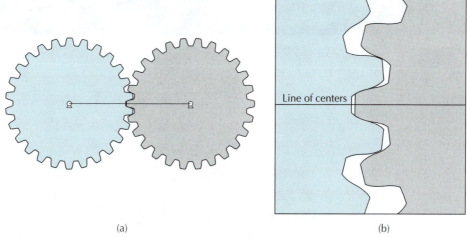

Figure 5.68 (a) Cycloidal spur gears [Video 5.68], (b) Enlargement in vicinity of meshing teeth.

Video 5.67
Supercharger
Blower

indicates the rotation of the right lobe from the selected starting configuration. Each lobe is in fact a gear having two cycloidal teeth. For the starting position illustrated, there cannot be any driving torque between the gears. This is because the point of contact lies on the line of centers, and the common normal at the point of contact coincides with the line of centers. Therefore, in addition to the gears shown, it is necessary to include two other gears of equal size in mesh with their same axes of rotation. The additional gears ensure a continuous torque transmission between the driver and driven shafts.

Video 5.68
Cycloidal Spur
Gears

The above two examples employing cycloidal gears make use of the entire epicycloid and hypocycloid. However, it is also possible to use only a portion of the cycloid, similar to that of a pair of gears using involute profiles (Figure 5.18). Such a pair of gears is shown in Figure 5.68.

EXAMPLE 5.7 ANALYSIS OF A GEROTOR PUMP

Figure 5.69 shows two gears from a cycloidal gerotor pump. Gears 2 and 3 rotate about the base pivots O_2 and O_3, respectively. Gear 3 is an internal gear. The face width of both gears is 2.0 cm.

$$r_2 = 4.6 \text{ cm}; \quad \omega_2 = 300 \text{ rpm CW}$$

(a) Determine the rotational speed of gear 3.
(b) Determine the pitch line velocity.
(c) For the given position of the pump, indicate on the figure all of the instantaneous centers.
(d) For the given position of the pump, indicate on the figure the pressure.angle between the gear teeth at contact point A.
(e) Determine the circular pitch of the teeth on gear 3.
(f) Estimate the volume flow rate through the pump.

SOLUTION
(a) Rotational speed of gear 3

$$\omega_3 = +\frac{N_2}{N_3}\omega_2 = \frac{3}{4}300 \text{ rpm CW} = 225 \text{ rpm CW}$$

(Continued)

EXAMPLE 5.7 Continued

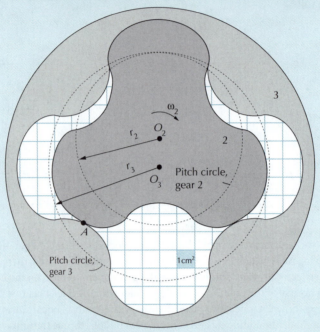

Figure 5.69 Gerotor pump.

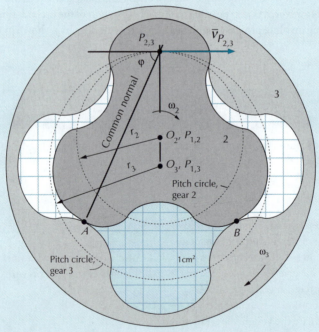

Figure 5.70

(b) Pitch line velocity:

$$v_{P_{2,3}} = \omega_2 r_2 = \left(300 \times \frac{2\pi}{60} \times 4.6 \right) \frac{cm}{sec} = 144 \frac{cm}{sec}; \quad \overline{v}_{P_{2,3}} = 144 \frac{cm}{sec} @ 0°$$

(c) Instantaneous centers: (see Figure 5.70)

(d) Pressure angle at A: (see Figure 5.70)

(e) Circular pitch:

(Continued)

EXAMPLE 5.7 Continued

$$p_{c3} = p_{c2} = \frac{2\pi r_2}{N_2} = \frac{2\pi \times 4.6}{3} = 9.63 \text{ cm}$$

(f)   Volume flow rate through the pump:

For a positive displacement pump the pockets of fluid contained between a pair of adjacent contact points collapse to zero. Therefore, for the calculation of the volume flow rate, we examine the maximum area contained between a pair of contact points (depicted between points A and B in Figure 5.70), then multiply that area by the face width of the gears to determine the maximum volume in the pocket. After that, the maximum volume in the pocket is multiplied by the number of pockets that collapse to zero in a given time period. These calculations are given below.

maximum area in a pocket between A and B (see Figure 5.70) = 32 cm^2
maximum volume in a pocket = 32 cm$^2 \times$ face width = 64 cm^3

$$Q_{pump} = 64 \text{ cm}^3 \times 300 \frac{\text{rev}}{\text{min}} \times 3 \frac{\text{pockets}}{\text{rev}} = 57,600 \frac{\text{cm}^3}{\text{min}} = 57.6 \frac{\text{liters}}{\text{min}}$$

PROBLEMS

Sizing of Gear Teeth, Straight Spur

P5.1 What is the pitch circle diameter of a 50-tooth straight spur gear having a circular pitch of 0.375 in?

P5.2 How many revolutions per minute is a straight spur gear turning at if it has 32 teeth, a circular pitch of 0.75 in, and a magnitude of the pitch line velocity of 15 ft/sec?

P5.3 How many revolutions per minute is a straight spur gear turning at if it has 48 teeth, a module of 3 mm, and a magnitude of the pitch line velocity of 500 mm/sec?

P5.4. A straight spur gear having 40 teeth is rotating at 450 rpm and is to drive another straight spur gear at 600 rpm.

(a) What is the magnitude of the speed ratio?

(b) How many teeth must the second gear have?

P5.5 Two straight spur gears have a diametral pitch of 5 in^{-1}, a magnitude of speed ratio of 0.20, and a center-to-center distance of 12 in. How many teeth do the gears have?

P5.6 Two straight spur gears in mesh have a module of 2.5 mm, a center-to-center distance of 50 mm, and a magnitude of speed ratio of 0.6. How many teeth do the gears have? If the pinion is rotating at 1200 rpm, what is the magnitude of the pitch line velocity?

P5.7 Two straight spur gears in mesh are designated as A and B. Gear A has 40 teeth of 4 in^{-1} diametral pitch, and gear B has a pitch circle diameter of 16 in. The teeth have a pressure angle of 14.5°. Both gears are teeth. Determine

(a) the number of teeth on gear B

(b) the center-to-center distance between the gears

(c) the circular pitch

(d) the outside diameter of gear B

(e) the base circle diameter of gear B

P5.8 Table P5.8 provides information that pertains to the reverted gear train shown in Figure 5.28.

TABLE P5.8 PARAMETERS OF THE REVERTED GEAR TRAIN SHOWN IN FIGURE 5.28

Gear No.	No. of Teeth	Module (mm)
2	24	4
3	—	—
4	32	—
5	62	6

All gears are straight spur with a 20° pressure angle. Determine

(a) the circular pitch of gear 2

(b) the base circle radius of gear 3

(c) the pitch circle diameter of gear 4

(d) the center-to-center distance between gears 2 and 3

P5.9 A straight spur gear *A* has 24 teeth. It meshes with an internal gear *B*. Both gears have a 25° pressure angle, 12-in^{-1} diametral pitch, and full-depth teeth. The center-to-center distance between the gears is 1.50 in. Determine

  (a) the addendum circle diameter of gear *A*

  (b) the number of teeth on gear *B*

  (c) the base circle radius of gear *B*

  (d) the thickness of a gear tooth on gear *B*, measured along its pitch circle

P5.10 Determine the numbers of teeth on a pair of straight spur gears with center-to-center distance of 80 mm, gear teeth having module 2.0 mm and a magnitude of speed ratio 0.6.

P5.11 Determine the numbers of teeth on a pair of straight spur gears with center-to-center distance of 4.0 in, gear teeth having diametral pitch of 10 in^{-1} and a magnitude of speed ratio 0.25.

P5.12 Design a pair of external-internal straight spur gears that will generate a magnitude of speed ratio of 0.4. The center-to-center distance between the gears is to be 42 mm. The external gear is to be cut using a hob with module of 2.0 mm.

P5.13 Each column in Table P5.13 relates to a straight spur gear. Determine the missing information.

TABLE P5. 13

	(a)	(b)	(c)	(d)
Base circle diameter (in)	—	—	—	—
Dedendum circle diameter (in)	—	—	—	—
Pitch circle diameter (in)	—	—	8.0	—
Addendum circle diameter (in)	—	—	—	8.5
Diametral pitch (in^{-1})	2.0	—	4.0	10.0
Circular pitch (in)	—	0.5	—	—
Pressure angle (degrees)	20	20	25	20
Stub/Full-depth (in)	Full	Stub	Full	Full
Number of teeth	24	36	—	—

P5.14 Each column in Table P5.14 relates to a straight spur gear. Determine the missing information.

TABLE P5.14

	(a)	(b)	(c)	(d)
Base circle diameter (mm)	—	—	—	—
Dedendum circle diameter (mm)	—	—	—	40.5
Pitch circle diameter (mm)	—	—	40	—
Addendum circle diameter (mm)	—	100	—	—
Module (mm)	2.0	4.0	0.5	1.0
Pressure angle (degrees)	20	20	20	20
Number of teeth	24	—	—	—

P5.15 Each column in Table P5.15 relates to a pair of meshing external-external straight spur gears (see Figure P5.15). All gear teeth are full-depth. Determine the missing information.

TABLE P5.15

	(a)	(b)	(c)	(d)
Circular pitch (in)	—	$\pi/4$	—	—
Diametral pitch (in^{-1})	2.0	—	8.0	6.0
Pressure angle (degrees)	20	20	20	20
N_2	24	22	—	—
N_3	30	—	—	—
r_2 (in)	—	—	4.0	—
r_3 (in)	—	—	—	—
c (in)	—	7.0	9.0	8.0
$|\omega_2|$ (rad/sec)	60	—	9.0	10
$|\omega_3|$ (rad/sec)	—	40	10	—
v_p (in/sec)	—	—	—	30
Contact ratio	—	—	—	—

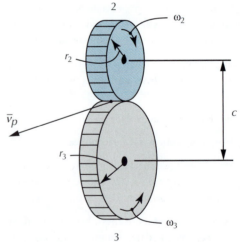

Figure P5.15

P5.16 Each column in Table P5.16 relates to a rack and pinion gear set (see Figure P5.16). The teeth are full-depth and the helix angle is zero. Determine the missing information.

TABLE P5.16

	(a)	(b)	(c)	(d)
Circular pitch (in)	—	—	$\pi/4$	—
Diametral pitch (in^{-1})	4.0	6.0	—	5.0
Pressure angle (degrees)	20	20	25	20
N_2	24	—	—	—
c (in)	—	5.0	—	—
$|\omega_2|$ (rad/sec)	60	—	20	10
v_p (in/sec)	—	60	100	80
Contact ratio	—	—	—	—

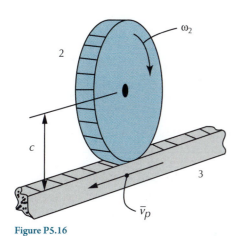

Figure P5.16

Sizing of Gear Teeth, Helical Spur

P5.17 Design a pair of external-external spur gears that will generate a magnitude of speed ratio of 0.4. The center-to-center distance between the gears is to be $8.0/\pi$ in. The gears are to be cut using a hob with hob pitch of 0.25 in.

P5.18 Each column in Table P5.18 relates to a helical spur gear. Determine the missing information.

TABLE P5.18

	(a)	(b)	(c)	(d)
Normal pitch (in)	0.375	0.5	0.25	0.5
Circular pitch (in)	—	0.7	—	—
Pitch circle diameter (in)	—	—	—	8.0
Helix angle (degrees)	30	—	15	—
Number of teeth	24	30	36	40

P5.19 Each column in Table P5.19 relates to a pair of meshing external–external helical spur gears (see Figure P5.19). Determine the missing information.

TABLE P5.19

	(a)	(b)	(c)	(d)
Circular pitch (in)	—	—	—	—
Normal pitch (in)	0.5	0.5	0.25	0.25
Helix angle (degrees)	30	—	—	—
N_2	24	24	24	—
N_3	36	36	—	40
r_2 (in)	—	—	—	3.0
r_3 (in)	—	—	—	—
c (in)	—	5.0	3.0	5.0
$|\omega_2|$ (rad/sec)	60	—	60	—
$|\omega_3|$ (rad/sec)	—	40	40	—
v_p (in/sec)	—	—	—	60

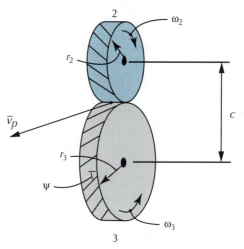

Figure P5.19

P5.20 Each column in Table P5.20 relates to a pair of meshing external–internal helical spur gears (see Figure P5.20). Determine the missing information.

TABLE P5.20

	(a)	(b)	(c)	(d)
Circular pitch (in)	—	—	—	—
Normal pitch (in)	0.5	0.5	0.25	0.50
Helix angle (degrees)	30	—	—	—
N_2	24	24	24	—
N_3	80	80	—	108
r_2 (in)	—	—	—	3.0
r_3 (in)	—	—	—	—
c (in)	—	5.0	1.0	6.0
$|\omega_2|$ (rad/sec)	60	—	80	—
$|\omega_3|$ (rad/sec)	—	40	40	—
v_p (in/sec)	—	—	—	90

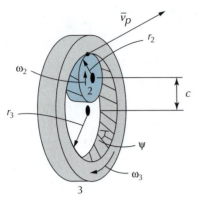

Figure P5.20

Contact Ratio

P5.21 Derive the expression for the length of contact for a rack and pinion gear set (Equation (5.10-14)).

P5.22 Derive the expression for the length of contact for an external-internal gear set (Equation (5.10-15)).

P5.23 A gear with a diametral pitch of 3 in^{-1} and 20° full-depth involute form has 35 teeth and meshes with a 70-tooth gear. The larger gear rotates at 300 rpm CW. Both gears have external teeth. Determine

(a) the pitch circle diameters
(b) the base circle radius of the smaller gear
(c) the circular pitch
(d) the center-to-center distance between the gears
(e) the rotational speed of the smaller gear
(f) the contact ratio

P5.24 A pinion has 32 teeth and has been manufactured using a hob having a 20° pressure angle, 8 in^{-1} diametral pitch, and stub teeth. It meshes with a rack. Determine

(a) the length of action
(b) the contact ratio

P5.25 Determine an expression for the contact ratio between a pair of external gears where gear 2 is stub form and gear 3 is full-depth.

P5.26 Determine an expression for the contact ratio for a rack and pinion set when

(a) the pinion, gear 2, has full-depth teeth, and the rack, gear 3, has stub teeth
(b) the pinion, gear 2, has stub teeth, and the rack, gear 3, has full-depth teeth

P5.27 Determine an expression for the contact ratio of an external–internal pair of meshing gears. for which gear 3 is the internal gear, when

(a) gear 2 has full-depth teeth, and gear 3 has stub teeth
(b) gear 2 has stub teeth, and gear 3 has full-depth teeth

P5.28 Each column in Table P5.28 relates to a pair of meshing external-internal straight spur gears (see Figure P5.28). All gear teeth are full-depth. Determine the missing information.

TABLE P5.28

	(a)	(b)	(c)	(d)
Circular pitch (in)	—	—	$\pi/4$	—
Diametral pitch (in^{-1})	4.0	6.0	—	5.0
Pressure angle (degrees)	20	20	20	25
N_2	24	24	—	—
N_3	80	—	96	—
r_2 (in)	—	—	—	—
r_3 (in)	—	—	—	—
c (in)	—	7.0	8.0	9.0
$|\omega_2|$ (rad/sec)	60	—	—	12
$|\omega_3|$ (rad/sec)	—	40	50	6.0
v_p (in/sec)	—	360	—	—
Contact ratio	—	—	—	—

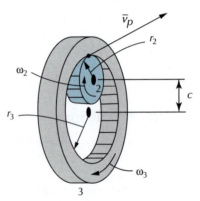

Figure P5.28

Spur and Helical Gears

P5.29 Table P5.29 provides information that pertains to the reverted gear train shown in Figure 5.28.

TABLE P5.29 PARAMETERS OF THE REVERTED GEAR TRAIN SHOWN IN FIGURE 5.28

Gear No.	No. of Teeth	Type of Gear
2	30	—
3	48	Straight spur
4	28	Helical spur
5	46	—

All gears were manufactured using a hob cutter having a 20° pressure angle, 0.375-in hob pitch, and full-depth. Determine

(a) the base circle radius of gear 3

(b) the center-to-center distance between gears 2 and 3

(c) the circular pitch of gear 4

(d) the helix angle of gear 5

(e) the pitch circle diameter of gear 4

(f) the addendum circle diameter of gear 4 (i.e., the diameter of the gear blank)

Cycloidal Gerotor Pump

P5.30 Figure P5.30 shows portions of two gears from a cycloidal gerotor pump. Gears 2 and 3 rotate about points O_2 and O_3, respectively. Gear 2 has five teeth. The face width of both gears is 3.0 cm.

$$r_2 = 6.25 \text{ cm}; \quad \omega_2 = 200 \text{ rpm CW}$$

(a) Determine the rotational speed of gear 3

(b) Specify the mobility of the pump. Explain your answer in terms of the number of links, and the numbers and types of the kinematic pairs.

(c) For the given position of the pump, indicate on the figure all of the instantaneous centers. Determine the velocities of all of the instantaneous centers.

(d) For the given position of the pump, indicate on the figure the pressure angle between the gear teeth at contact point A.

(e) Estimate the volume flow rate through the pump.

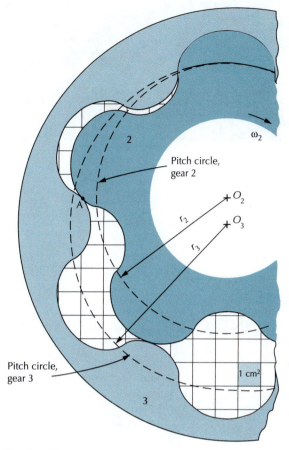

Figure P5.30

CHAPTER 6

GEAR TRAINS

6.1 INTRODUCTION

A combination of gears arranged for the purpose of transmitting torque and rotational motion from an input shaft to an output shaft is called a *gear train*. Gear trains are used to transmit torque and rotary motion within machines, from one location to an alternate location, and/or to change the rotational speed of various machine elements.

The number of gears employed in a gear train can range from two to several dozen. Gears are often keyed to shafts, and they are assembled in a sturdy housing that supports the shafts, using ball or roller bearings. The housing is usually enclosed and provided with ample lubrication. In heavy-duty gear applications, the lubricant is circulated, filtered, and sometimes cooled.

Figure 6.1 shows a gear train employed in a lathe, where an electric motor provides a single input rotational speed. Multiple output turning speeds of the spindle and workpiece may be obtained by changing the arrangement of gears in mesh.

Figure 6.2 illustrates an old-fashioned mechanical clock that incorporates gear trains to perform its functions. One gear train is used to turn the hour and minute hands at the proper rates relative to the input rotation. Another gear train is used to transmit the winding torque into a torsional spring that powers the clock. Energy from the spring is gradually released to drive the gears by using an escapement mechanism (see Chapter 13, Section 13.7). The escapement mechanism also ensures that the gears are turned at a proper rate to keep accurate time. Figure 6.3 illustrates a three-dimensional model of a clock gear train system, showing the major components.

Video 6.3
Mechanical Clock

A gear train has at least one input shaft and one output shaft. However, a gear train may be designed to accommodate multiple inputs, deliver multiple outputs, or both. A gear train is characterized by its power rating, as well as by speed ratio(s) between the input(s) and output(s). The size of a gear train depends on its power rating, which can range from a fraction of a watt, to megawatts.

In this chapter, the main types of gear trains are described and classified. Methods of determining the speeds of the components are also covered.

6.1.1 Speed Ratio

The *speed ratio* is a measure of the relative speed between two rotating components. In a gear train, the speed ratio of gear (i.e., component) j with respect to gear i is defined as

$$e_{j/i} = \frac{\text{rotational speed of component } j}{\text{rotational speed of component } i} \tag{6.1-1}$$

Positive $e_{j/i}$ values indicate that both components j and i turn in the same direction. Negative $e_{j/i}$ values signify an opposite direction of rotation. Equation (6.1-1) may be applied to the overall

Figure 6.1 Gear train in a lathe.

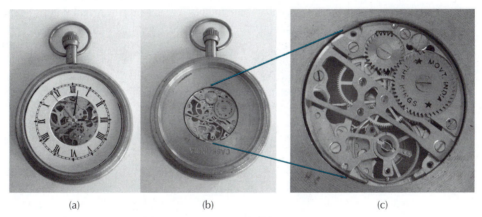

Figure 6.2 (a) Mechanical clock, (b) Back with cover removed, (c) Close up of gear train.

Figure 6.3 Mechanical clock [Video 6.3].

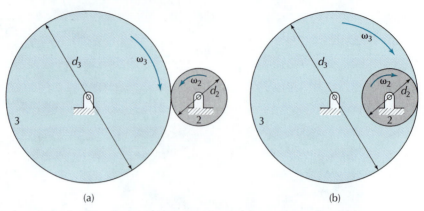

Figure 6.4 Meshing gears represented by their pitch circles: (a) External–external pair, (b) External–internal pair.

speed ratio of the gear train. In that instance, j corresponds to the output component of the gear train, and i corresponds to the input component.

Figure 6.4 illustrates the pitch circles of two gears in mesh. For the external–external meshing pair shown in Figure 6.4(a), the gears rotate in opposite directions, and the speed ratio is negative. For the external–internal pair shown in Figure 6.4(b), the gears rotate in the same direction, and the speed ratio is positive.

The sign convention defined above for speed ratio can only be applied to components that have parallel axes of rotation. For cases where the axes of rotation are not parallel, such as with worm and wheel gears, miter gears, and bevel gears, only the magnitude of the speed ratio is defined by Equation (6.1-1), and the direction of rotation may be determined using a suitably prepared sketch.

From Equation (5.6-6), the magnitude of the speed ratio for a pair of gears in mesh equals the inverse of the ratio of the numbers of teeth on the gears. For the external–external meshing pair shown in Figure 6.4(a), having N_2 and N_3 teeth on gears 2 and 3, respectively, the directions of rotation are opposite. The speed ratio of gear 3 with respect to gear 2 is

$$e_{3/2} = \frac{\omega_3}{\omega_2} = -\frac{N_2}{N_3} \tag{6.1-2}$$

For the external–internal meshing pair shown in Figure 6.4(b), the speed ratio is

$$e_{3/2} = \frac{\omega_3}{\omega_2} = +\frac{N_2}{N_3} \tag{6.1-3}$$

It is usually desired to design a gear train to provide a specific speed ratio. In instances where the magnitude of the specified speed ratio is a fraction involving small whole numbers, for example 2/3, there is a wide selection of the arrangement of gears and numbers of teeth that may be employed. Possible combinations to achieve a 2/3 speed ratio could be 20 and 30 teeth, or 40 and 60 teeth, etc. Since gears must have integer numbers of teeth, some speed ratios are either impossible or impractical to obtain exactly. For instance, a magnitude of speed ratio involving a fraction of two large prime numbers, such as 123/177, would require meshing gears with 123 and 177 teeth to provide the exact speed ratio. Alternative gear trains, however, can be developed if a small deviation of speed ratio is allowed from the desired value. For example, two gears with 36 and 25 teeth provide a magnitude of speed ratio of 0.6944, which is very close to the specified value of 123/177 = 0.6949.

6.1.2 Classification of Gear Trains

Gear trains may be classified into two groups: *ordinary* and *planetary* (or *epicyclic*), (Figure 6.5). In an *ordinary gear train*, axes of all gears are stationary relative to the base link, which is normally the housing (i.e., the base link) of the gear train. In a *planetary gear train*, the axis of at least one gear moves along a circular path relative to the base link. These two groups of gear trains are described further in the following sections.

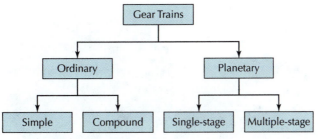

Figure 6.5 Classification of gear trains.

6.2 ORDINARY GEAR TRAINS

There are two types of ordinary gear trains: *simple gear trains* and *compound gear trains*.

6.2.1 Simple Gear Trains

Video 6.6A
Simple Gear
Train

In a simple gear train, two or more gears are in mesh such that they form a consecutive sequence from input to output. Furthermore, there is only one gear mounted on each axis of rotation. Two examples of simple gear trains are shown in Figure 6.6.

The speed ratio for any two gears in mesh of a simple gear train can be obtained using an expression similar to either Equation (6.1-2) or (6.1-3). Thus, for the gear train shown in Figure 6.6(a), the speed ratios of gears 2 and 3, and of gears 3 and 4, are

$$e_{3/2} = \frac{\omega_3}{\omega_2} = -\frac{N_2}{N_3} \quad \text{and} \quad e_{4/3} = \frac{\omega_4}{\omega_3} = -\frac{N_3}{N_4} \tag{6.2-1}$$

respectively. If gear 2 is the input and gear 4 is the output, then by employing Equation (6.2-1) we obtain the speed ratio of the gear train:

$$e_{4/2} = \frac{\omega_4}{\omega_2} = e_{4/3} \times e_{3/2} = \left(-\frac{N_3}{N_4}\right)\left(-\frac{N_2}{N_3}\right) = \frac{N_2}{N_4} \tag{6.2-2}$$

For the gear train shown in Figure 6.6(a), we have

$$N_2 = 31; \quad N_3 = 18; \quad N_4 = 31$$

and using Equation (6.2-2) we get

$$e_{4/2} = +1.0$$

This gear train would have the same magnitude of speed ratio as when gears 2 and 4 are directly in mesh. However, the speed ratio is changed from negative to positive, due to the presence of gear 3, referred to as an *idler gear*.

Even when the number of idler gears in a simple gear train is increased, the magnitude of speed ratio will still depend only on the number of teeth in the gears mounted on the input and output shafts. Whether the speed ratio is positive or negative depends on the number of idler gears. Employing only external gears in a simple gear train, an odd number of idler gears produces a positive speed ratio and with an even number of idler gears, as shown in Figure 6.6(b), the speed ratio is negative, that is

$$e_{5/2} = -\frac{N_2}{N_5} \tag{6.2-3}$$

In this example, gears 3 and 4 are idler gears.

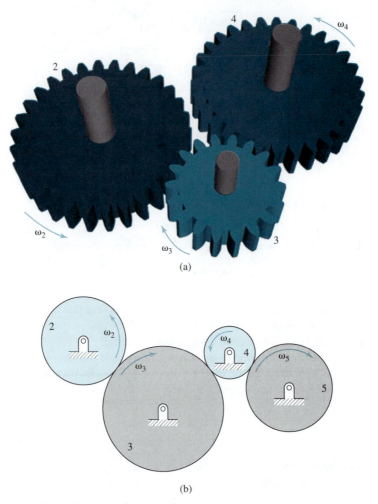

Figure 6.6 Simple gear trains [Video 6.6A].

Idler gears are usually added to a gear train for two purposes. They can be used to change the output direction of rotation and/or to transmit rotary motion from an input shaft to an output shaft that are separated by a specified distance from each other. If the distance between the input and output shafts is large, it may be more economical to use one or more idler gears instead of using large input and output gears to bridge the distance. Figure 6.7 illustrates two alternative simple gear trains that have the same speed ratio and center-to-center distance between the input and output shafts.

6.2.2 Compound Gear Trains

Video 6.8A
Reverted Gear Train

A *compound gear train* has four or more gears of which at least two are attached to the same shaft. Those gears attached to the same shaft are constrained to turn at the same rate and in the same direction. Figure 6.8 shows three examples of compound gear trains. In each example, gears 3 and 4 are attached to the same shaft and both are idler gears. For the gear train shown in Figure 6.8(a), which is also called a *reverted gear train*, the input and output axes of rotation are collinear.

The speed ratio for a compound gear train is found in a similar manner to that used for a simple gear train. Consider the gear train in Figure 6.8(a), where gear 2 is attached to the input shaft, gear 5 to the output shaft, and idler gears 3 and 4 are mounted on the idler shaft and turn at the same speed. Speed ratios for the meshing gears 2 and 3, and gears 4 and 5, are

$$e_{3/2} = \frac{\omega_3}{\omega_2} = -\frac{N_2}{N_3} \quad \text{and} \quad e_{5/4} = \frac{\omega_5}{\omega_4} = -\frac{N_4}{N_5}$$

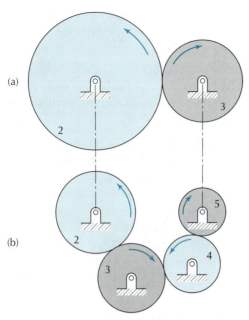

Figure 6.7 Two simple gear trains with the same speed ratio.

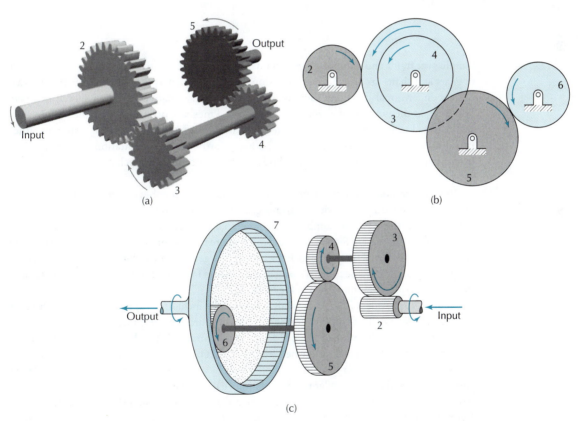

Figure 6.8 Compound gear trains [Video 6.8A].

The speed ratio for the gear train is

$$e_{5/2} = \frac{\omega_5}{\omega_2} = \frac{\omega_5}{\omega_4}\frac{\omega_3}{\omega_2} = \left(-\frac{N_4}{N_5}\right)\left(-\frac{N_2}{N_3}\right) = \frac{N_2 N_4}{N_3 N_5} \tag{6.2-4}$$

In this instance, unlike the case of simple gear trains, the numbers of teeth of the idler gears influence the magnitude of the speed ratio. Using a similar analysis, the speed ratio for the compound gear train in Figure 6.8(b) is

$$e_{6/2} = -\frac{N_2 N_4}{N_3 N_6} \qquad (6.2\text{-}5)$$

Idler gear 5 has no effect on the magnitude of the speed ratio, although it affects the direction of output speed and the distance between the input and output shafts.

The compound gear train illustrated in Figure 6.8(c) incorporates an external–internal meshing pair. The numbers of teeth are

$$N_2 = 9; \quad N_3 = 62; \quad N_4 = 20$$
$$N_5 = 51; \quad N_6 = 15; \quad N_7 = 54$$

The speed ratio is

$$e_{7/2} = \frac{\omega_7}{\omega_2} = \left(+\frac{N_6}{N_7}\right)\left(-\frac{N_4}{N_5}\right)\left(-\frac{N_2}{N_3}\right)$$
$$= \frac{N_2 N_4 N_6}{N_3 N_5 N_7} = \frac{9 \times 20 \times 15}{62 \times 51 \times 54} = 0.0158$$

Video 6.9

Gear Train Used in a Winch

This arrangement of gears is employed in the design of a winch. Figure 6.9 shows a corresponding photograph of the system.

6.2.2.1 Gear Train Diagram Notation

Simple gear trains and compound gear trains are generally easy to illustrate when the number of gears and shafts are few. Three-dimensional diagrams such as that shown in Figure 6.8(a), or two-dimensional diagrams such as Figure 6.8(b), are relatively easy to sketch or draw. However, as the number of elements increases, and meshing occurs on multiple planes, such as illustrated in Figure 6.8(c), it becomes increasingly difficult to sketch the system in either two-dimensional or three-dimensional. To allow for better documentation of complex gear trains, two alternative illustration methods are now introduced, which are: *cross-section diagrams* and *skeleton diagrams*.

These illustration methods can be explained by studying a number of gear and shaft combinations that are typically used within various gear trains. Figure 6.10 provides three-dimensional illustrations of these typical combinations, where all illustrations are labeled (a) through (h). Corresponding to this are Figures 6.11 and 6.12, where illustrations of combinations (a) through (h) are equivalent in all three figures. For example, the gear/shaft combination of Figure 6.10(c) is equivalent to the cross-sectional representation shown in Figure 6.11(c), which is equivalent to the skeleton representation shown in Figure 6.12(c).

Figure 6.10 provides typical gear/shaft combinations including: (a) An input shaft with a gear attached at the end. This combination is a common source of input torque to various types of gear trains. (b) A shaft with two gears, C and D, rigidly attached at each end. This combination is used to transmit rotation and torque between two offset planes of meshing gears. (c) A combination similar to (b), however the combination is held by arm F, where arm F may be connected to another rotating link. (d) Similar to combination (b), however, there is a concentric shaft I that passes through the center. (e) An idler gear that is free spinning on a shaft, connected to arm I. Arm I may be connected to another rotating link. (f) An input shaft I, connected to an internal gear K, also known as a *ring gear* K. (g) A combination similar to (f), however, there is a concentric shaft I that passes through the ring gear K. (h) Two gears that are in mesh and spin freely on their respective shafts. The two shafts are connected to arm I, where arm I may be connected to another rotating link.

6.2.2.2 Compound Gear Train Examples

A compound gear train may be designed to provide multiple speed ratios through one output shaft by transmitting motion through various pairs of gears. This is illustrated in the following example.

Figure 6.9 Gear train used in a winch [Video 6.9].

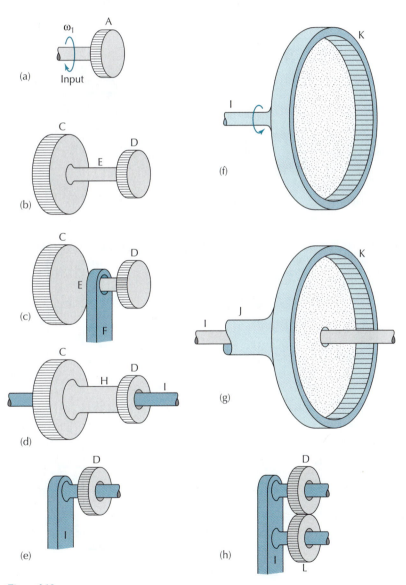

Figure 6.10

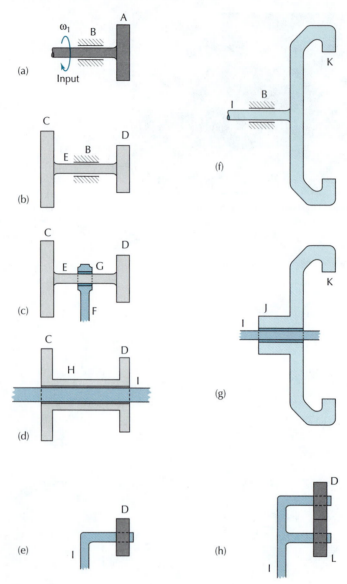

Figure 6.11

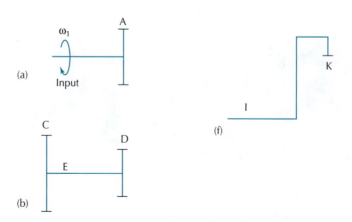

Figure 6.12 Continued

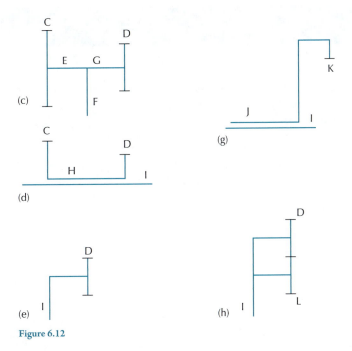

Figure 6.12

EXAMPLE 6.1 SPEED RATIOS OF A MANUAL TRANSMISSION

Figure 6.13 shows a schematic of a manual transmission, which employs a compound gear train. Splines on the input shaft, and gears 1, 3, and 5, that are rigidly connected to one another, ensure the components share the same rotational speed but do permit axial movement of the gears along the *input shaft*. The *shifter* allows pairs of gears to be disengaged and reengaged to obtain three different speed ratios. Determine the speed ratios of the gear train.

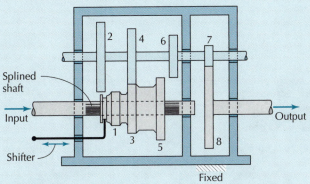

Figure 6.13 Compound gear train.

(*Continued*)

EXAMPLE 6.1 Continued

SOLUTION

Each configuration is equivalent to the reverted compound gear train shown in Figure 6.8(a). Table 6.1 indicates those gears which are in mesh for the three different transmission configurations and the speed ratio for each.

TABLE 6.1 SPEED RATIOS OF THE MANUAL TRANSMISSION GEARBOX SHOWN IN FIGURE 6.13

Output Speed No.	Gears in Mesh	Speed Ratio
1	1, 2, 7, 8	$e_{8/1} = \dfrac{N_1 N_7}{N_2 N_8}$
2	3, 4, 7, 8	$e_{8/3} = \dfrac{N_3 N_7}{N_4 N_8}$
3	5, 6, 7, 8	$e_{8/5} = \dfrac{N_5 N_7}{N_6 N_8}$

A manual automobile transmission incorporates a similar arrangement of gears to that presented in Example 6.1. Figure 6.14 illustrates a typical transmission that has three forward speed ratios and a reverse. Gear 1 is rigidly connected to the *input shaft*. Gears 2 through 5, inclusive, are rigidly connected to the *countershaft*. Gear 6 has its own axis of rotation and is employed to generate the speed ratio for reverse motion. Splines on the *output shaft* and gear 8 ensure both components share the same rotational motion, but do permit axial movement of gear 8 along the output shaft. By employing the *shifter*, gear 8 can be moved from the position shown in Figure 6.14 to mesh with either gear 4 or gear 6. Gear 7, being disengaged, is free to

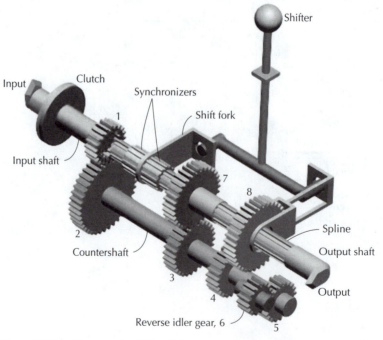

Figure 6.14 Manual transmission [Video 6.14].

rotate relative to the output shaft. When required, one of the *synchronizers* (see Chapter 13, Section 13.3.2) rigidly connects gear 7 to the output shaft.

Figure 6.15 illustrates the configurations for all speed ratios of the manual transmission. While shifting between each speed ratio, the *clutch* (see Chapter 13, Section 13.3.1) is disengaged. For reverse (Figure 6.15(a)), motion is transmitted through gears 1, 2, 5, 6, and 8. Gear 6 is an idler gear and is used solely to change the direction of rotation. For neutral (Figure 6.15(b)), no motion is transmitted to the output shaft. This is because gear 7 is free to rotate with respect to the output shaft, although gears 1, 2, 3, and 7 are in mesh. For the

Video 6.14
Manual
Transmission

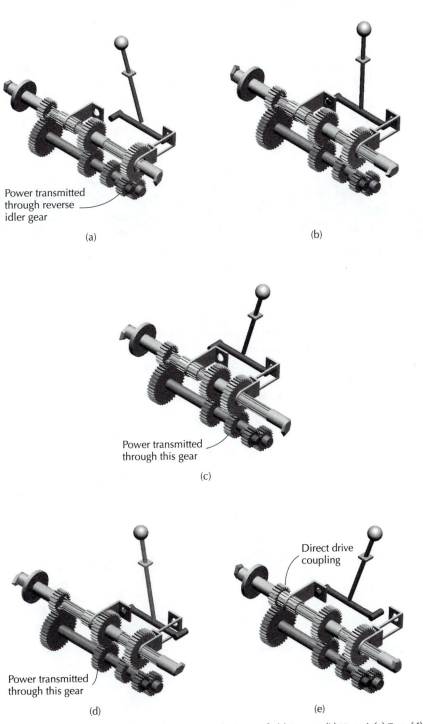

Power transmitted
through reverse
idler gear

(a)

(b)

Power transmitted
through this gear

(c)

Power transmitted
through this gear

(d)

Direct drive
coupling

(e)

Figure 6.15 Configurations of a manual transmission [Video 6.14]: (a) Reverse, (b) Neutral, (c) First, (d) Second, (e) Third.

first forward speed ratio (Figure 6.15(c)), gear 8 has been translated along the splined portion of the output shaft and meshes with gear 4. Motion is transmitted through gears 1, 2, 4, and 8. For the second forward speed ratio (Figure 6.15(d)), gear 8 no longer meshes with any other gear, and motion is transmitted through gears 1, 2, 3, and 7. A synchronizer is used to connect gear 7 to the output shaft. For the third forward speed ratio (Figure 6.15(e)), another synchronizer is employed to provide a direct coupling between the input and output shafts, and the speed ratio is unity. Under these conditions, the countershaft has no influence on the speed ratio.

Through use of a synchronizer, it is not necessary to engage and disengage gears 3 and 7 when shifting into and out of the second forward speed ratio. A synchronizer therefore eliminates the possibility of clashing the teeth of these gears. It also allows the possibility of employing helical spur gears, which are particularly difficult to engage and disengage through relative axial sliding.

The manual transmission shown in Figure 6.14 incorporates a single clutch. The driver can change the speed ratio by pushing the clutch pedal to disengage the clutch, then moving the shift lever to select different gears that transmit power, and finally engaging the clutch. During the shifting process there are distinct intervals when no power is transmitted from the engine to the wheels.

EXAMPLE 6.2 DESIGNING FOR THE NUMBERS OF TEETH OF A MANUAL TRANSMISSION

Video 6.16
Manual
Transmission

Figure 6.16 illustrates a manual transmission. Gears 1 and 2 are always in mesh. Instead of employing sychronizers, gears are moved along the splined shaft and brought into mesh with gears mounted on the countershaft. For the first speed ratio, gears 5 and 8 are in mesh, for the second speed ratio, gears 4 and 7 are in mesh, and for the third speed ratio, gears 3 and 6 are in mesh. A speed ratio of unity is obtained when the input and output shafts are rigidly connected. [Video 6.16] provides an animated sequence of the movements of the gears to obtain the different speed ratios.

Determine suitable tooth numbers to generate the speed ratios 1/4.00, 1/2.50, and 1/1.50. All gears are to have at least 20 teeth. The center-to-center distance, c, between the countershaft and input/output shafts is 75 mm, and all gears are straight spur having a module of 2.5 mm.

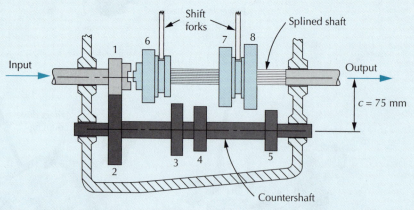

Figure 6.16 Manual transmission [Video 6.16].

(Continued)

EXAMPLE 6.2 Continued

SOLUTION

Expressions for the speed ratios are

$$\text{first: } e_{8/1} = \frac{N_1 N_5}{N_2 N_8}; \quad \text{second: } e_{7/1} = \frac{N_1 N_4}{N_2 N_7}; \quad \text{third: } e_{6/1} = \frac{N_1 N_3}{N_2 N_6}$$

Because all gears are to have the same module, and meshing gears must share a common center-to-center distance, then by employing Equation (5.8-1) for a pair of meshing gears designated as i and j, we obtain

$$c = 75 \, \text{mm} = \frac{d_i}{2} + \frac{d_j}{2} = \frac{m(N_i + N_j)}{2} = \frac{2.5(N_i + N_j)}{2} \, \text{mm} \tag{6.2-6}$$

from which

$$N_i + N_j = 60$$

Therefore

$$N_1 + N_2 = N_3 + N_6 = N_4 + N_7 = N_5 + N_8 = 60 \tag{6.2-7}$$

Subject to the constraint given by Equation (6.2-7), the first desired speed ratio, $e_{8/1}$, can be achieved exactly by selecting

$$N_1 = 20; \quad N_2 = 40; \quad N_5 = 20; \quad N_8 = 40 \tag{6.2-8}$$

It is not possible to obtain exactly the desired values of the second speed ratio, $e_{7/1}$, and third speed ratio, $e_{6/1}$. The numbers of teeth on gears 1 and 2 have already been specified (Equation (6.2-8)), and there are constraints on the numbers of teeth on gears that determine the second and third speed ratios (Equation (6.2-7)). Through a trial and error procedure of testing for various permissible combinations of numbers of teeth, we determine that the closest approximations are obtained using

$$N_3 = 34; \quad N_4 = 27; \quad N_6 = 26; \quad N_7 = 33$$

which yields

$$e_{7/1} = \frac{1}{2.44}; \quad e_{6/1} = \frac{1}{1.53}$$

An alternate set of solutions can be found by selecting

$$N_1 = 21; \quad N_2 = 39$$

The gear tooth numbers that produce values closest to the desired speed ratios, while satisfying the constraint for the minimum number of teeth on a gear, are

$$N_3 = 33; \quad N_4 = 26; \quad N_5 = 20$$

$$N_6 = 27; \quad N_7 = 34; \quad N_8 = 40$$

In this case, the speed ratios are

$$e_{8/1} = \frac{1}{3.71}; \quad e_{7/1} = \frac{1}{2.43}; \quad e_{6/1} = \frac{1}{1.52}$$

(Continued)

EXAMPLE 6.2 Continued

Therefore, we are only able to obtain approximations of the three desired speed ratios. The numbers of gear teeth

$$N_5 = 19; \quad N_8 = 41$$

for the alternate set of solutions would give a better approximation of the first desired speed ratio. However, this solution is not permitted because it violates the constraint of the minimum number of teeth on a gear.

A *dual-clutch transmission* (DCT) can drastically reduce the intervals when no power is transmitted. For a DCT the driver is not required to operate the clutches. Instead, their coordinated engagement and disengagement is controlled automatically by electronics and hydraulics, similar to that used in an automatic transmission. The timing is dictated by the vehicle speed, the engine speed, the load torque, and the position of the accelerator pedal. The DCT is designed to provide smooth acceleration of the vehicle without manual shifting under normal driving conditions. There is usually an option for the driver to override the automatic operation and shift the speeds when desired.

Video 6.17
Dual-Clutch
Transmission

Figure 6.17 shows an illustration of a DCT with three forward speeds and a reverse. Actual DCTs in use typically employ five to seven forward speeds. The design shown incorporates two concentric shafts, where the inner solid shaft has gears for the first and third speeds, while the outer hollow shaft has gears for the second and reverse speeds. Clutch A can connect the input shaft to the inner solid shaft, and clutch B can connect the input shaft to the outer hollow shaft. The synchronizers are used to connect gears to the output shaft, when directed to do so by the control system. DCTs with additional speeds are obtained by adding gears on the concentric shafts that mesh with gears on the output shaft. Gears for the even speeds are added to one of the concentric shafts while the odd speeds are added to the other shaft.

Figure 6.18(a) shows a DCT while in the second speed, with clutch A disengaged and clutch B engaged. One of the synchronizers connects the second gear to the output shaft to complete the path of power flow between the input and output. Prior to shifting to the

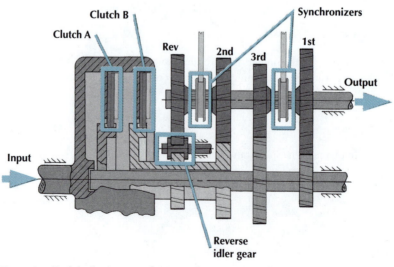

Figure 6.17 Dual-clutch transmission [Video 6.17].

third speed, the other synchronizer connects the third gear to the output shaft as shown in Figure 6.18(b). Even though the third gear was preselected and is now rigidly connected to the output shaft, the DCT is still in the second speed. The actual shift does not take place until both clutch A engages and clutch B disengages (Figure 6.18(c)), which may be completed in minimal time. Thus, continuous power flow is practically maintained. The synchronizer connecting the second gear to the output shaft may subsequently be moved to a neutral location to reduce the drag resistance in the transmission.

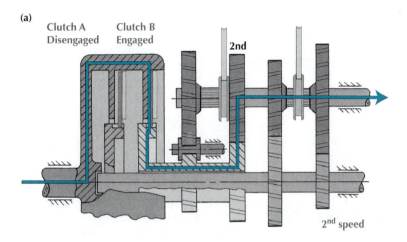

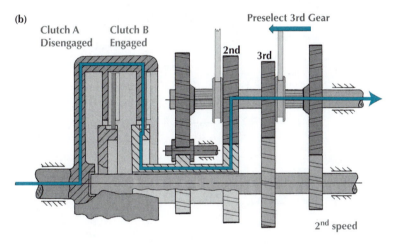

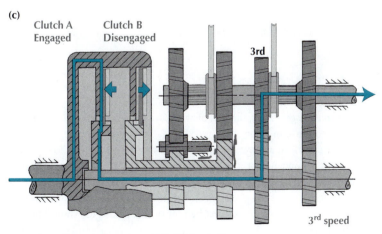

Figure 6.18 Changing of speeds in a DCT [Video 6.17].

6.2.3 Gear Trains with Bevel Gears

For the gear trains presented in Sections 6.2.1 and 6.2.2, all axes of rotation are parallel. However, other ordinary gear trains can be designed by employing meshing gears having nonparallel axes of rotation. Such gear trains can incorporate bevel gears, hypoid gears, and worm and wheel sets.

The magnitude of the speed ratio between any two bevel gears in mesh is inversely proportional to their number of gear teeth. For example, Figure 6.19 shows an ordinary gear train consisting of three bevel gears. Gear 2 is the input, gear 4 is the output, and gear 3 is the idler gear. The magnitudes of the speed ratios between gears 2 and 3, and between gears 3 and 4, are

$$|e_{3/2}| = \left|\frac{\omega_3}{\omega_2}\right| = \frac{N_2}{N_3}; \quad |e_{4/3}| = \left|\frac{\omega_4}{\omega_3}\right| = \frac{N_3}{N_4} \tag{6.2-10}$$

The magnitude of the speed ratio of the gear train is

$$|e_{4/2}| = |e_{4/3}| \times |e_{3/2}| = \left|\frac{\omega_4}{\omega_2}\right| \tag{6.2-11}$$

Combining Equations (6.2-10) and (6.2-11), we obtain

$$e_{4/2} = -\frac{N_3}{N_4}\frac{N_2}{N_3} = -\frac{N_2}{N_4} \tag{6.2-12}$$

A minus sign in Equation (6.2-12) indicates that the input and output turn in opposite directions. The sign convention can be applied in this instance since the input and output shafts of the gear train are parallel to each other.

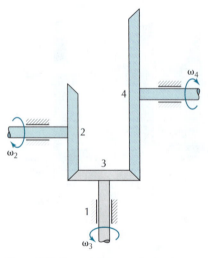

Figure 6.19 Gear train incorporating bevel gears.

6.3 PLANETARY GEAR TRAINS

In a *planetary gear train*, the axis of at least one gear, called a *planet gear*, moves on a circular path relative to the base link.

An example of a planetary gear train is shown in Figure 6.20. It consists of a *sun gear*, having N_1 teeth, a concentric *ring gear*, having N_4 teeth, and one or more planet gears, each with N_3 teeth. Planet gears mesh with the external teeth of the sun gear and with the internal teeth of the ring gear. Shafts of the planet gears are attached to the *crank* or *planet carrier*. The pitch of the sun, planet, and ring gears must be identical for this gear train. Figure 6.20(b) shows the side view of the planetary train illustrated in Figure 6.20(a), and the corresponding skeleton form is shown in Figure 6.20(c).

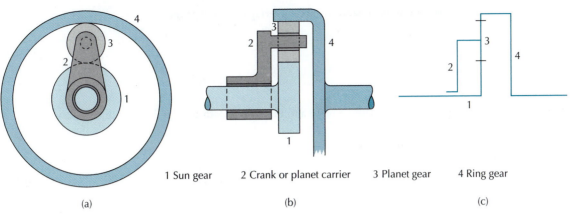

1 Sun gear   2 Crank or planet carrier   3 Planet gear   4 Ring gear

(a)   (b)   (c)

Figure 6.20 (a) Planetary gear train, (b) Side view, (c) Skeleton representation.

A planetary gear train has the following advantages over ordinary gear trains:

- It has compact space requirements, particularly when the input and output axes are collinear.
- Both static and dynamic forces are balanced when multiple planets are equally spaced about the central axis of the gear train.
- High torque capacity is possible, by using multiple planets.
- It can provide a wide range of speed ratios.

6.3.1 Number of Teeth on Gears in Planetary Gear Trains

The numbers of teeth on the gears of a planetary gear train cannot be selected arbitrarily. Consider the planetary gear train shown in Figure 6.20; the following geometrical relationship exists between the pitch circle diameters:

$$d_1 + 2d_3 = d_4 \qquad (6.3\text{-}1)$$

Since in this instance the pitch of all meshing gears must be equal, it follows from Equations (5.6-5) and (6.3-1) that

$$N_1 + 2N_3 = N_4 \qquad (6.3\text{-}2)$$

or

$$N_3 = \frac{N_4 - N_1}{2} \qquad (6.3\text{-}3)$$

6.3.2 Number of Planet Gears

For the planetary gear train shown in Figure 6.20, only one planet gear is needed to transmit motion. However, multiple planet gears are usually employed as in the example shown in Figure 6.21. The use of multiple planets allows more gears to share the transmission of loads through alternate paths simultaneously, which as a result increases the load capacity of the gear train without increasing its overall size. In practice, to improve the dynamic characteristics of the gear train, planet gears are usually equally spaced, so that loads are balanced about the central axis.

The number of equally spaced planet gears, n, that can be placed between a sun and ring gear depends on the space available and the numbers of teeth of the gears. The following two criteria must be satisfied [3]:

1. **Prevention of addendum circle overlap of planet gears:**

$$n < \frac{180°}{\sin^{-1}\left(\dfrac{N_3 + 2\alpha}{N_1 + N_3}\right)} \qquad (6.3\text{-}4)$$

where α takes on values of 0.8 and 1.0 for stub teeth and full-depth teeth, respectively.

2. **Meshing of gear teeth:**

$$\frac{N_1 + N_4}{n} = \text{integer} \qquad (6.3\text{-}5)$$

EXAMPLE 6.3 NUMBER OF EQUALLY SPACED PLANET GEARS

Video 6.21
Planetary Gear
Train

For the planetary gear train shown in Figure 6.21, determine the number of equally spaced planet gears that may be employed. The teeth are to have full-depth. The numbers of teeth are

$$N_1 = 27; \quad N_3 = 18; \quad N_4 = 63$$

Figure 6.21 Planetary gear train [Video 6.21].

SOLUTION

The number of equally spaced planet gears may be found by applying Equations (6.3-4) and (6.3-5). Substituting the numbers of teeth in Equation (6.3-4) and employing $\alpha = 1.0$ for full-depth teeth,

$$n < \frac{180°}{\sin^{-1}\left(\dfrac{N_3 + 2\alpha}{N_1 + N_3}\right)} = \frac{180°}{\sin^{-1}\left(\dfrac{18 + 2.0}{27 + 18}\right)} = 6.82 \qquad (6.3.\text{-}6)$$

Since the number of planet gears must be an integer, according to this constraint, up to six equally spaced planets may be employed. From Equation (6.3-5) we get

$$\frac{N_1 + N_4}{n} = \frac{27 + 63}{n} = \frac{90}{n} = \text{integer} \qquad (6.3\text{-}7)$$

(Continued)

EXAMPLE 6.3 Continued

Substituting the values $n = 1, 2, 3, 4, 5$, and 6 in Equation (6.3-7), we find that the values $n = 1$, 2, 3, 5, and 6 satisfy the equation. Therefore, either 1, 2, 3, 5, or 6 equally spaced planet gears may be employed in this gear train. It is not possible to use four equally placed planet gears.

6.3.3 Classification of Planetary Gear Trains

Lévai [6] identified twelve basic types of *single-stage planetary gear trains* incorporating spur gears, where all axes of rotation are parallel. Their skeleton drawings are shown in Figure 6.22. For example, the planetary gear train shown in Figure 6.20 is Type A. Additional variations of planetary gear trains can be obtained by employing gears having nonparallel axes of rotation. Such examples are presented later in this chapter.

Multiple-stage planetary gear trains employ a combination of single-stage gear trains. Each stage may be one of the types illustrated in Figure 6.22 or may employ other types of gears, such as bevel gears or worm and wheel.

The mobility of each of the planetary gear trains shown in Figure 6.22 equals two. Therefore, the rotational speeds of all components may be determined if two of the rotational speeds are specified. The two input speeds may be equal or unequal, in the same direction or opposite. One of the two input speeds may be zero. For the gear train shown in Figure 6.20, if the crank is not allowed to rotate, all gears have fixed axes of rotation, and the planetary gear train is treated as an ordinary gear train.

6.4 TABULAR ANALYSIS OF PLANETARY GEAR TRAINS

In this section, the *tabular method* is presented to determine speed ratios and speeds of components of planetary gear trains.

A tabular method of a planetary gear train involves the following general steps:

1. Assume all components of the gear train rotate at x rpm about the central axis. This motion constitutes rigid-body rotation of the entire gear train, and there is no meshing movement of the gear teeth.
2. Assume the crank is not permitted to move. Ignore all other motion constraints and choose a component of the gear train other than the crank, such as either a sun or ring gear, and rotate it at y rpm. Determine the rotational speeds of all other components of the gear train. By holding the crank fixed, the planetary gear train becomes an ordinary gear train (Section 6.2).
3. Superimpose the two motions considered in steps 1 and 2 by summing the rotational speeds, that is, the rotational speeds of the components are expressed in terms of x and y.
4. Employ expressions for rotational speeds found in step 3 to determine the output speed of the gear train or speed ratios between the components, as required

Figures 6.23(a) and 6.23(b) portray motions described in steps 1 and 2 as they apply to the planetary gear train illustrated in Figure 6.21.

Table 6.2 summarizes the first three steps of the above procedure for the planetary gear train shown in Figure 6.21. The last row of the table is equivalent to the following equations:

$$\omega_1 = x + y; \quad \omega_2 = x$$

$$\omega_3 = x - \frac{N_1}{N_3} y; \quad \omega_4 = x - \frac{N_1}{N_4} y$$

For the gear train shown in Figure 6.21, if the ring gear is held fixed, then

$$\omega_4 = x - \frac{N_1}{N_4} y = 0 \quad \text{or} \quad y = \frac{N_4}{N_1} x$$

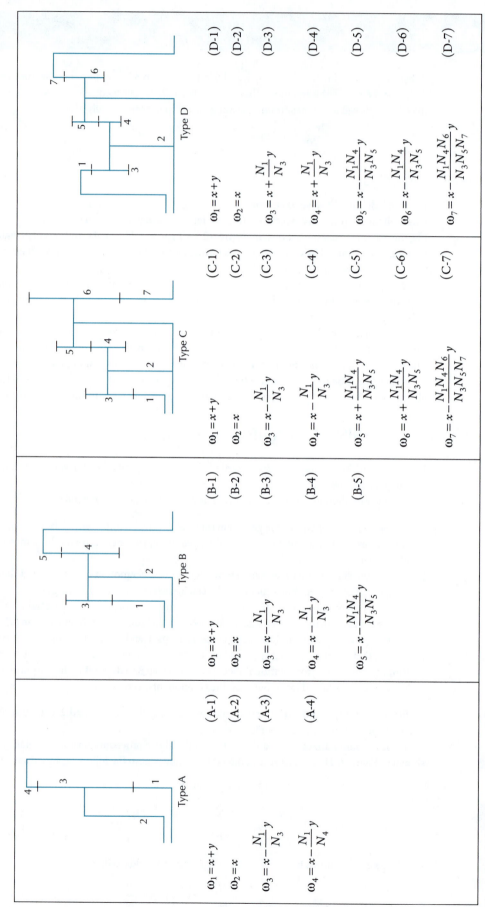

Figure 6.22 Basic planetary gear train types [7].

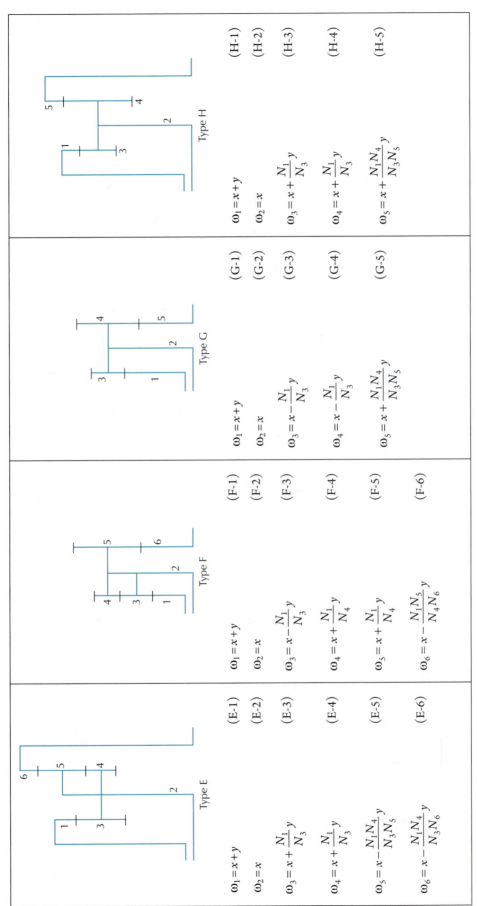

Type E

$$\omega_1 = x + y \quad \text{(E-1)}$$
$$\omega_2 = x \quad \text{(E-2)}$$
$$\omega_3 = x + \frac{N_1}{N_3}y \quad \text{(E-3)}$$
$$\omega_4 = x + \frac{N_1}{N_3}y \quad \text{(E-4)}$$
$$\omega_5 = x - \frac{N_1 N_4}{N_3 N_5}y \quad \text{(E-5)}$$
$$\omega_6 = x - \frac{N_1 N_4}{N_3 N_6}y \quad \text{(E-6)}$$

Type F

$$\omega_1 = x + y \quad \text{(F-1)}$$
$$\omega_2 = x \quad \text{(F-2)}$$
$$\omega_3 = x - \frac{N_1}{N_3}y \quad \text{(F-3)}$$
$$\omega_4 = x + \frac{N_1}{N_4}y \quad \text{(F-4)}$$
$$\omega_5 = x + \frac{N_1}{N_4}y \quad \text{(F-5)}$$
$$\omega_6 = x - \frac{N_1 N_5}{N_4 N_6}y \quad \text{(F-6)}$$

Type G

$$\omega_1 = x + y \quad \text{(G-1)}$$
$$\omega_2 = x \quad \text{(G-2)}$$
$$\omega_3 = x - \frac{N_1}{N_3}y \quad \text{(G-3)}$$
$$\omega_4 = x - \frac{N_1}{N_3}y \quad \text{(G-4)}$$
$$\omega_5 = x + \frac{N_1 N_4}{N_3 N_5}y \quad \text{(G-5)}$$

Type H

$$\omega_1 = x + y \quad \text{(H-1)}$$
$$\omega_2 = x \quad \text{(H-2)}$$
$$\omega_3 = x + \frac{N_1}{N_3}y \quad \text{(H-3)}$$
$$\omega_4 = x + \frac{N_1}{N_3}y \quad \text{(H-4)}$$
$$\omega_5 = x + \frac{N_1 N_4}{N_3 N_5}y \quad \text{(H-5)}$$

Figure 6.22 (*Continued*)

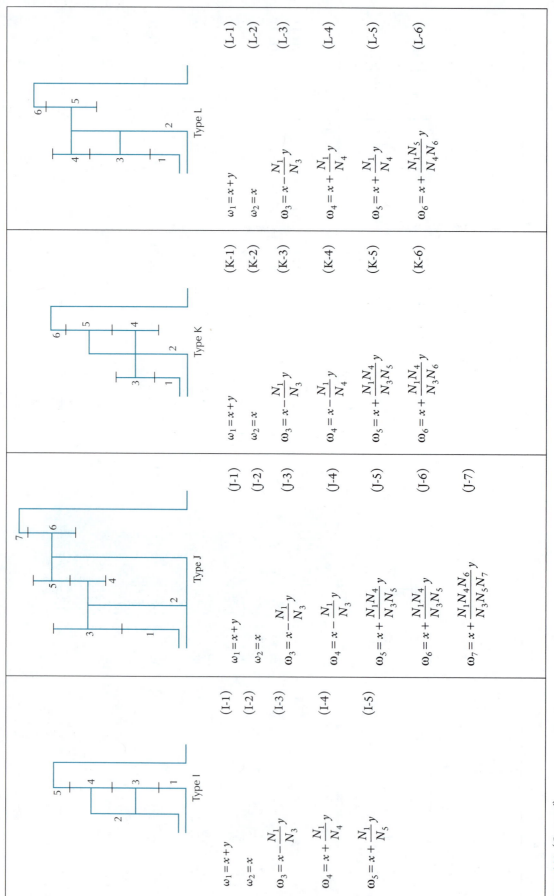

$\omega_1 = x + y$ (I-1)

$\omega_2 = x$ (I-2)

$\omega_3 = x - \dfrac{N_1}{N_3} y$ (I-3)

$\omega_4 = x + \dfrac{N_1}{N_4} y$ (I-4)

$\omega_5 = x + \dfrac{N_1}{N_5} y$ (I-5)

Type I

$\omega_1 = x + y$ (J-1)

$\omega_2 = x$ (J-2)

$\omega_3 = x - \dfrac{N_1}{N_3} y$ (J-3)

$\omega_4 = x - \dfrac{N_1}{N_3} y$ (J-4)

$\omega_5 = x + \dfrac{N_1 N_4}{N_3 N_5} y$ (J-5)

$\omega_6 = x + \dfrac{N_1 N_4}{N_3 N_5} y$ (J-6)

$\omega_7 = x + \dfrac{N_1 N_4 N_6}{N_3 N_5 N_7} y$ (J-7)

Type J

$\omega_1 = x + y$ (K-1)

$\omega_2 = x$ (K-2)

$\omega_3 = x - \dfrac{N_1}{N_3} y$ (K-3)

$\omega_4 = x - \dfrac{N_1}{N_4} y$ (K-4)

$\omega_5 = x + \dfrac{N_1 N_4}{N_3 N_5} y$ (K-5)

$\omega_6 = x + \dfrac{N_1 N_4}{N_3 N_6} y$ (K-6)

Type K

$\omega_1 = x + y$ (L-1)

$\omega_2 = x$ (L-2)

$\omega_3 = x - \dfrac{N_1}{N_3} y$ (L-3)

$\omega_4 = x + \dfrac{N_1}{N_4} y$ (L-4)

$\omega_5 = x + \dfrac{N_1}{N_4} y$ (L-5)

$\omega_6 = x + \dfrac{N_1 N_5}{N_4 N_6} y$ (L-6)

Type L

Figure 6.22 (*Continued*)

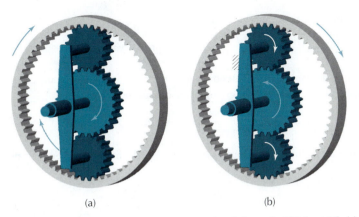

(a)   (b)

Figure 6.23 Example of motions employed using the tabular method [Video 6.21]: (a) All links with same rotational speed, (b) Motion of links with respect to fixed crank.

TABLE 6.2 TABULAR ANALYSIS OF THE PLANETARY GEAR TRAIN SHOWN IN FIGURE 6.21

Component → Operation ↓	Gear 1 ω_1	Crank ω_2	Gear 3 ω_3	Gear 4 ω_4
All Components Turn at x rpm	x	x	x	x
Crank is Fixed, Gear 1 Turns at y rpm	y	0	$-\dfrac{N_1}{N_3}y$	$-\dfrac{N_1}{N_4}y$
Absolute Rotational Speeds	$x+y$	x	$x-\dfrac{N_1}{N_3}y$	$x-\dfrac{N_1}{N_4}y$

Furthermore, if the input motion is applied to the sun gear, and the out motion is obtained from the crank, then the speed ratio of the gear train is

$$e_{2/1}=\frac{\omega_2}{\omega_1}=\frac{x}{x+y}=\frac{x}{x+\dfrac{N_4}{N_1}x}=\frac{N_1}{N_1+N_4} \tag{6.4-1}$$

Employing the following numbers of teeth for the sun and ring gears

$$N_1=27; \quad N_4=63$$

in Equation (6.4-1), we obtain

$$e_{2/1}=\frac{N_1}{N_1+N_4}=\frac{27}{27+63}=\frac{27}{90}=0.30$$

If the input rational speed of the sun gear is $\omega_1=400$ rpm CW , then the rotational speed of the crank is determined by recognizing

$$\omega_2=e_{2/1}\omega_1=0.30(-400)=-120=120 \text{ rpm } \text{ CW}$$

where it was considered that positive rotational speeds are in the counterclockwise direction.

Using the above input motion while holding the ring gear fixed, and employing the prescribed numbers of teeth, we generate the following two linear equation in terms of variables x and y

$$\omega_1=x+y=-400 \text{ rpm } \text{ and}$$

$$\omega_4=x-\frac{N_1}{N_4}y=x-\frac{27}{63}y=0$$

Solving these two equation, yields

$$x = -120 \text{ rpm} \quad : \quad y = -280 \text{ rpm}$$

Having determined the values of x and y, the rotational speed of any component in the gear train may now be evaluated. For example, the output speed is

$$\omega_2 = x = -120 = 120 \text{ rpm} \quad \text{CW}$$

which is the same as previously obtained. Also, the rotational speed of a planet gear ($N_3 = 18$) is

$$\omega_{\text{planet}} = \omega_3 = x - \frac{N_1}{N_3} y$$

$$= (-120) - \frac{27}{18} \times (-280) = 300 = 300 \text{ rpm} \quad \text{CCW}$$

The above calculation gives the absolute value of the rotational speed, that is, with respect to the fixed link, gear 4. The relative rotational speed of a planet gear with respect to the crank is

$$\omega_{\text{planet, relative}} = \omega_3 - \omega_2 = \left(x - \frac{N_1}{N_3} y \right) - x = 420 = 420 \text{ rpm} \quad \text{CCW}$$

A planetary gear train may be used to generate more than one speed ratio by selecting different components to be connected to the input, to be connected to the output, and to be held fixed. For example, Table 6.3 applies to the planetary gear train illustrated in Figure 6.21. The three speed ratios listed are in addition to that determined above. One of the speed ratios is greater than unity (case 1), a second is less than unity but greater than zero (case 2), and a third is less than zero (case 3).

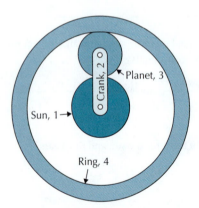

TABLE 6.3 SPEED RATIOS OF A PLANETARY GEAR TRAIN

Operation	Sun, ω_1	Crank, ω_2	Ring, ω_4
all x rpm	x	x	x
Crank fixed, sun y rpm	y	0	$-\dfrac{N_1}{N_4} y$
Absolute rotational speeds	$x+y$	x	$x - \dfrac{N_1}{N_4} y$

(Continued)

TABLE 6.3 Contintued

Case No.	Non-Zero input Component	Fixed Component	Output Component	Speed Ratio
1	Crank, 2	Ring, 4 $x - \dfrac{N_1}{N_4} y = 0; \quad y = \dfrac{N_4}{N_1} x$	Sun, 1	$e_{1/2} = \dfrac{\omega_1}{\omega_2} = \dfrac{x + \dfrac{N_4}{N_1} x}{x} = \dfrac{N_1 + N_4}{N_1}$
2	Ring, 4	Sun, 1 $x + y = 0; \quad y = -x$	Crank, 2	$e_{2/4} = \dfrac{\omega_2}{\omega_4} = \dfrac{x}{\left(x - \dfrac{N_1}{N_4} - (x) \right)} = \dfrac{N_4}{N_1 + N_4}$
3	Sun, 1	Crank, 2 $x = 0$	Ring, 4	$e_{4/1} = \dfrac{\omega_4}{\omega_1} = \dfrac{\left(0 - \dfrac{N_1}{N_4} y \right)}{0 + y} = -\dfrac{N_1}{N_4}$

For all of the cases presented so far, the speeds of the components and the speed ratios between the components were determined for instances that one of the components of the planetary gear train was held fixed. The following example deals with a condition where there are two non-zero input rotational speeds to the planetary gear train.

EXAMPLE 6.4 SPEED RATIO OF A PLANETARY GEAR TRAIN

For the gear train shown in Figure 6.24, gears 6 and 7 are mounted on the input shaft and turn with the same speed ($\omega_6 = \omega_7$). These gears mesh with gears 2 and 5, which are connected to the crank (link 2) and sun gear (gear 1), respectively, and provide the two input motions to the planetary gear train. The output shaft is attached to the ring gear (gear 4). Gears 1 and 5 also turn at the same rate ($\omega_1 = \omega_5$). Determine the speed ratio of the gear train.

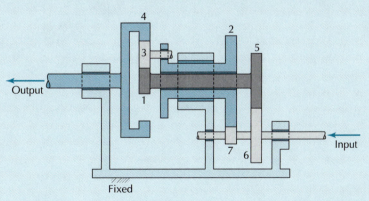

Figure 6.24 Planetary gear train.

SOLUTION

The two input rotational speeds to the planetary gear train are determined by first recognizing that

$$e_{5/6} = \frac{\omega_5}{\omega_6} = -\frac{N_6}{N_5}; \quad e_{2/7} = \frac{\omega_2}{\omega_7} = -\frac{N_7}{N_2}$$

(Continued)

EXAMPLE 6.4 Continued

The rotational speeds of the sun gear and crank are therefore

$$\omega_1 = \omega_5 = \omega_{5/6} \times \omega_6 \tag{6.4-2}$$

$$\omega_2 = \omega_{2/7} \times \omega_7 \tag{6.4-3}$$

Since the arrangement of gears of the planetary gear train in Figures 6.20 and 6.24 are the same, Table 6.2 can also be applied in this example. Equating the input rotational speeds given in Equations (6.4-2) and (6.4-3) with those given in Table 6.2, we obtain

$$\omega_1 = -\frac{N_6}{N_5}\omega_6 = x + y \tag{6.4-4}$$

$$\omega_2 = -\frac{N_7}{N_2}\omega_7 = x \tag{6.4-5}$$

Solving Equations (6.4-4) and (6.4-5) for y and recognizing that

$$\omega_6 = \omega_7 \tag{6.4-6}$$

gives

$$y = -\omega_7\left(\frac{N_6}{N_5} - \frac{N_7}{N_2}\right) \tag{6.4-7}$$

The value of rotational speed of the ring gear (gear 4) is also obtained from Table 6.2:

$$\omega_4 = x - \frac{N_1}{N_4}y \tag{6.4-8}$$

Substituting Equations (6.4-5)–(6.4-7), we get

$$\omega_4 = \omega_7\left[-\frac{N_7}{N_2} + \frac{N_1}{N_4}\left(\frac{N_6}{N_5} - \frac{N_7}{N_2}\right)\right] \tag{6.4-9}$$

And the speed ratio of the gear train employing Equation (6.4-9) is

$$e_{4/7} = \frac{\omega_4}{\omega_7} = -\frac{N_7}{N_2} + \frac{N_1}{N_4}\left(\frac{N_6}{N_5} - \frac{N_7}{N_2}\right) \tag{6.4-10}$$

EXAMPLE 6.5 PLANETARY GEAR TRAIN IN A PENCIL SHARPENER

Video 6.25
Pencil Sharpener

Figure 6.25(a) illustrates a pencil sharpener, and Figure 6.25(b) shows a schematic of the planetary gear train contained within the unit. Ring gear 4 is fixed to the body (base link). The hand crank is rigidly connected to the planet carrier, link 2. The cutter is rigidly connected to planet gear 3, which is in mesh with ring gear 4. Starting from the position shown, determine the position of point A_1 on the planet gear after the hand crank has turned 180° clockwise.

$$N_3 = 11; \quad N_4 = 24$$

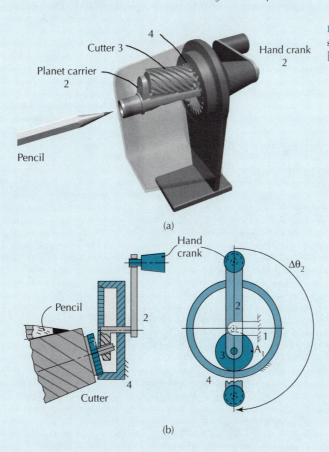

Figure 6.25 Pencil sharpener: (a) Isometric view [Video 6.25], (b) Schematic.

SOLUTION

In this chapter, the tabular method is applied to determine the rotational speeds of the components in planetary gear trains. In this example, a similar procedure is applied to determine the rotations of the links of a gear train.

Table 6.4 is a tabular analysis of the pencil sharpener in terms of the rotations. From this we find that the rotation of planet gear 3, with ring gear 4 fixed, and the rotation of the hand crank, $\Delta\theta_2 = x$,

$$\Delta\theta_3 = \Delta\theta_2\left(1 - \frac{N_4}{N_3}\right)$$

For a hand crank rotation of 180° clockwise (i.e., $\Delta\theta_2 = -180°$), we have

$$\Delta\theta_3 = \Delta\theta_2\left(1 - \frac{N_4}{N_3}\right) = -180°\left(1 - \frac{24}{11}\right) = 212.7°$$

(Continued)

EXAMPLE 6.5 Continued

The positive value indicates the rotation of planet gear 3 was counterclockwise. The geometry of the mechanism after turning the hand crank is illustrated in Figure 6.26, and point A_1 has moved to its new location after the motion and is now labeled as A_2.

All tables presented in this chapter for the tabular analysis of rotational speeds of the components in planetary gear trains may also be used to analyze their rotations. This is accomplished by assigning values of rotations to the variables x and y, rather than rotational speeds, and is permitted because the motions of all links in the gear train take place in the same time interval.

TABLE 6.4 TABULAR ANALYSIS OF A PLANETARY GEAR TRAIN IN A PENCIL SHARPENER

Component → Operation ↓	Hand Crank $\Delta\theta_2$	Gear 3 $\Delta\theta_3$	Gear 4 $\Delta\theta_4$
All Components Given Rotation x degrees	x	x	x
Fix Hand Crank, Gear 4 Given Rotation y degrees	0	$+\dfrac{N_4}{N_3}y$	y
Absolute Rotations	x	$x+\dfrac{N_4}{N_3}y$	$x+y$
Absolute Rotations With Fixed Gear 4 (i.e., $x+y=0$)	x	$x\left(1-\dfrac{N_4}{N_3}\right)$	0

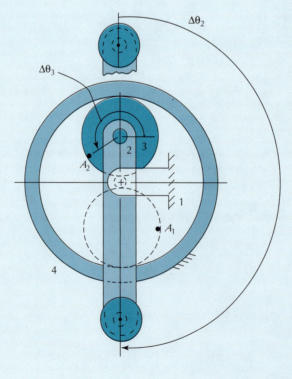

Figure 6.26 Geometry of pencil sharpener after rotation of the crank.

EXAMPLE 6.6 SPEED RATIOS OF AN AUTOMATIC TRANSMISSION

Determine the speed ratios of the automatic transmission illustrated in Figure 6.27. The speed ratios are generated as follows:

- Reverse: Gear 5 is held fixed.
- First speed: Gear 1 is held fixed.
- Second speed: Gears 1 and 6 are locked together.

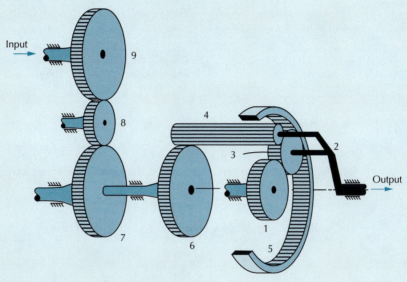

Figure 6.27 Schematic of automatic transmission.

In all cases, the input is connected to gear 9, and the output is connected to crank 2. The numbers of teeth are

$$N_1 = 27; \quad N_3 = 15; \quad N_4 = 9; \quad N_5 = 57;$$

$$N_6 = N_7 = N_9 = 32; \quad N_8 = 16$$

SOLUTION:

TABLE 6.5 TABULAR ANALYSIS OF THE GEAR TRAIN SHOWN IN FIGURE 6.27

Component → Operation ↓	Gear 1 ω_1	Crank ω_2	Gear 5 ω_5	Gear 6 ω_6
All Components Turn at x rpm	x	x	x	x
Crank is Fixed, Gear 5 Turns at y rpm	$-\dfrac{N_5}{N_1}y$	0	y	$+\dfrac{N_5}{N_6}y$
Absolute Rotational Speeds	$x - \dfrac{N_5}{N_1}y$	x	$x + y$	$x + \dfrac{N_5}{N_6}y$

(Continued)

EXAMPLE 6.6 Continued

Table 6.5 was created for this gear train. Using this table, the ratio of speeds between gear 6 and crank 2 is

$$e_{2/6} = \frac{\omega_2}{\omega_6} = \frac{x}{\left(x + \dfrac{N_5}{N_6} y\right)}$$

and therefore the speed ratio of the gear train is

$$e_{2/9} = \frac{\omega_2}{\omega_6} \times \frac{\omega_6}{\omega_9} = \frac{x}{\left(x + \dfrac{N_5}{N_6} y\right)} \times \frac{N_9}{N_7} \tag{6.4.11}$$

where it is recognized that $\omega_6 = \omega_7$.

Reverse
Since for this speed ratio gear 5 is held fixed, then from Table 6.5 we have

$$\omega_5 = x + y = 0; \quad y = -x \tag{6.4-12}$$

Combining Equations (6.4-11) and (6.4-12), the speed ratio is

$$e_{2/9} = \frac{\left(\dfrac{N_9}{N_7} x\right)}{\left(x + \dfrac{N_5}{N_6} y\right)}$$

$$= \frac{\left(\dfrac{N_9}{N_7} x\right)}{x\left(1 - \dfrac{N_5}{N_6}\right)} = \frac{\left(\dfrac{N_9}{N_7}\right)}{\left(1 - \dfrac{N_5}{N_6}\right)} = \frac{1}{\left(1 - \dfrac{57}{32}\right)} = -1.280 \tag{6.4-13}$$

First Speed
Since for this speed ratio gear 1 is held fixed, then from Table 6.5 we have

$$\omega_1 = x - \frac{N_5}{N_6} y = 0; \quad y = \frac{N_1}{N_5} x \tag{6.4-14}$$

Combining Equations (6.4-11) and (6.4-14), the speed ratio is

$$e_{2/9} = \frac{\left(\dfrac{N_9}{N_7} x\right)}{\left(x + \dfrac{N_5}{N_6} y\right)}$$

(Continued)

EXAMPLE 6.6 Continued

$$= \frac{\left(\dfrac{N_9}{N_7}x\right)}{x\left(1+\dfrac{N_5 N_1}{N_6 N_5}\right)} = \frac{\left(\dfrac{N_9}{N_7}\right)}{\left(1+\dfrac{N_1}{N_6}\right)} = \frac{1}{\left(1+\dfrac{27}{32}\right)} = 0.544 \qquad (6.4\text{-}15)$$

Second Speed
Since for this speed ratio gears 1 and 6 are held together, then from Table 6.5 we have

$$\omega_1 = \omega_6$$

$$x - \frac{N_5}{N_1}y = x + \frac{N_5}{N_6}y \qquad (6.4\text{-}16)$$

which can only be satisfied if

$$y = 0 \qquad (6.4\text{-}17)$$

Combining Equations (6.4-11) and (6.4-17), the speed ratio is

$$e_{2/9} = \frac{\left(\dfrac{N_9}{N_7}x\right)}{\left(x + \dfrac{N_5}{N_6}y\right)} = \frac{x}{x} = 1.00 \qquad (6.4\text{-}18)$$

Figure 6.28 is an illustration of this gear train.

Video 6.28
Two-Speed
Automatic
Transmission

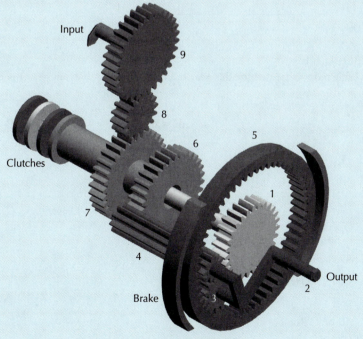

Figure 6.28 Two-speed automatic transmission [Video 6.28].

EXAMPLE 6.7 ANALYSIS OF A WANKEL ENGINE

A Wankel engine, as shown in Figure 6.29 (also described in Chapter 13, Section 13.12.3), is a planetary gear train for which the internal planet gear, having N_3 teeth, is part of the rotor. The sun gear, with N_1 teeth, is fixed. Determine the speed ratio of the crank, link 2, with respect to the planet gear, link 3, if

$$\frac{N_1}{N_3} = \frac{2}{3} \tag{6.4-19}$$

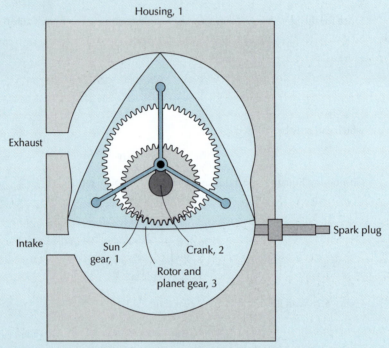

Figure 6.29 Wankel engine.

SOLUTION

Table 6.6 was created for this gear train. Recognizing that sun gear 1 is fixed, its rotational speed, ω_1, is

$$\omega_1 = x + y = 0, \quad y = -x \tag{6.4-20}$$

TABLE 6.6 TABULAR ANALYSIS OF THE WANKEL ENGINE SHOWN IN FIGURE 6.29

Component → Operation ↓	Gear 1 ω_1	Crank ω_2	Gear 3 ω_3
All Components Turn at x rpm	x	x	x
Crank is Fixed, Gear 1 Turns at y rpm	y	0	$\dfrac{N_1}{N_3}y$
Absolute Rotational Speeds	$x + y$	x	

(Continued)

EXAMPLE 6.7 Continued

Obtaining expressions for ω_2 and ω_3 from Table 6.6, and employing Equation (6.4-20), we determine equations for two rotational speeds in terms of one variable, x, as follows:

$$\omega_2 = x; \quad \omega_3 = x + \frac{N_1}{N_3} y = x\left(1 - \frac{N_1}{N_3}\right) \tag{6.4-21}$$

Employing Equations (6.4-19) and (6.4-21), the speed ratio of the crank with respect to the planet gear is

$$e_{2/3} = \frac{\omega_2}{\omega_3} = \frac{x}{x\left(1 - \dfrac{N_1}{N_3}\right)} = \frac{1}{\left(1 - \dfrac{N_1}{N_3}\right)} = \frac{1}{\left(1 - \dfrac{2}{3}\right)} = 3 \tag{6.4-22}$$

that is, for each rotation of the crank, the planet gear (i.e., rotor) makes one-third of a revolution in the same direction.

EXAMPLE 6.8 PLANETARY GEAR TRAIN IN A POWERED SCREWDRIVER

Figure 6.30(a) shows the end of a powered screwdriver that incorporates a planetary gear train. The components of the disassembled gear train are shown in Figure 6.30(b), and Figure 6.30(c) illustrates a skeleton diagram of the gear train. When the motor turns at 1000 rpm CW, determine the rotational speed of the output (i.e., crank 5) if the numbers of gear teeth are

$$N_1 = 6 = N_2; \quad N_3 = 21 = N_6$$

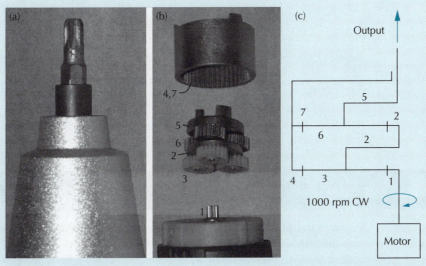

Figure 6.30 (a) Powered screwdriver, (b) Disassembled components, (c) Skeleton diagram of planetary gear train.

(*Continued*)

EXAMPLE 6.8 Continued

SOLUTION

The number of teeth on the ring gears is

$$N_4 = N_1 + 2N_3 = 48 = N_7$$

The planetary gear train has two stages. Gears 1, 3 and 4, and crank 2 make up the first stage. Gears 2, 6 and 7 and crank 5 make up the second stage. Crank 2 of the first stage and gear 2 of the second stage are rigidly connected. Using Table 6.2 and recognizing that for a fixed ring gear 4

$$\omega_4 = x - \frac{N_1}{N_4}y = 0 \quad \text{or} \quad y = \frac{N_4}{N_1}x$$

And therefore,

$$e_{2/1} = \frac{\omega_2}{\omega_1} = \frac{x}{x+y} = \frac{x}{x + \dfrac{N_4}{N_1}x} = \frac{N_1}{N_1 + N_4} = \frac{6}{6+48} = \frac{1}{9}$$

Since the numbers of teeth for the second stage are the same as for the first stage, and the ring gear for the second stage is fixed, we have

$$e_{5/2} = \frac{1}{9}$$

The speed ratio of the gear train is

$$e_{5/1} = e_{5/2} \times e_{2/1} = \frac{1}{81}$$

Therefore, the rotational speed of the output is

$$\omega_5 = \omega_1 \times e_{5/1} = 1000 \times \frac{1}{81} = 12.4 \text{ rpm} \quad \text{CW}$$

Additional examples of planetary gear trains with more than one stage are presented in Section 6.5.

EXAMPLE 6.9 SPEED RATIO OF A PLANETARY GEAR TRAIN

For the gear train shown in Figure 6.31, planet gears 3, 4, and 5 are rigidly connected to a common concentric shaft and turn at the same rotational speed, $\omega_3 = \omega_4 = \omega_5$. Gear 1 is connected to the input, gear 7 is connected to the output, and gear 6 is fixed. The planet carrier, component 2, is connected to neither an input nor an output. Determine the speed ratio between the input and the output.

(Continued)

EXAMPLE 6.9 Continued

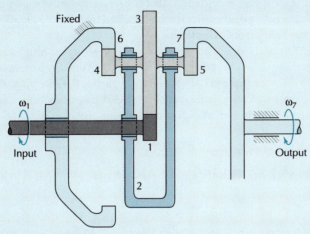

Figure 6.31 Planetary gear train.

SOLUTION

Table 6.7 was created for this gear train. Recognizing that ring gear 6 is fixed, its rotational speed, ω_6, is

$$\omega_6 = x + y = 0; \quad y = -x \tag{6.4-23}$$

Substituting Equation (6.4-23) in the expressions for ω_1 and ω_7 obtained from Table 6.7 allows the rotational speeds to be expressed in terms of one variable, x, as follows:

$$\omega_1 = x - \frac{N_6 N_3}{N_4 N_1} y = x\left(1 + \frac{N_6 N_3}{N_4 N_1}\right) \tag{6.4-24}$$

$$\omega_7 = x + \frac{N_6 N_5}{N_4 N_7} y = x\left(1 - \frac{N_6 N_5}{N_4 N_7}\right) \tag{6.4-25}$$

The speed ratio for this gear train is therefore

$$e_{7/1} = \frac{\omega_7}{\omega_1} = \frac{\left(1 - \dfrac{N_6 N_5}{N_4 N_7}\right)}{\left(1 + \dfrac{N_6 N_3}{N_4 N_1}\right)} \tag{6.4-26}$$

TABLE 6.7 TABULAR ANALYSIS OF THE PLANETARY GEAR TRAIN SHOWN IN FIGURE 6.31

Component → Operation ↓	Gear 1 ω_1	Gear 6 ω_6	Gear 7 ω_7
All Components Turn at x rpm	x	x	x
Crank Is Fixed, Gear 6 Turns at y rpm	$-\dfrac{N_6 N_3}{N_4 N_1} y$	y	$\dfrac{N_6 N_5}{N_4 N_7} y$
Absolute Rotational Speeds	$x - \dfrac{N_6 N_3}{N_4 N_1} y$	$x + y$	$x + \dfrac{N_6 N_5}{N_4 N_7} y$

When setting up the tabular method, it is not always necessary to include all components of the planetary gear train in the table. Unless the rotational speed of a planet gear is required, it need not be included in the table. However, components connected to the output(s) and the input(s) must always be included. This is necessary even when an input is fixed. For Example 6.9, since the planet carrier is not connected to either the inputs or the output, this component has been left out of the table.

Figure 6.22 shows twelve basic types of planetary gear trains. Also listed in this figure are equations for the rotational speeds of the components obtained by performing a tabular analysis on each [7]. These equations can not only be applied to the gear trains shown, but are also useful in the analysis of other more complicated gear trains, which are combinations of the twelve basic types. The analysis of such gear trains is presented in Section 6.5.

6.5 KINEMATIC ANALYSIS OF MULTIPLE-STAGE PLANETARY GEAR TRAINS[†]

It is possible to apply results presented in Figure 6.22 to more complicated systems, made up of combinations of the basic types. Each basic type of gear train that becomes part of the more complicated system will be considered as one *stage* of the gear train. A gear train that comprises a combination of more than one of the twelve basic types is referred to as a *multiple-stage planetary gear train*. Figure 6.32 shows a planetary gear train that is equivalent to a combination of type K and type L gear trains. These basic types of gear trains that make up this system are referred to as stage 1 and stage 2. Table 6.8 compares the numbering schemes used for the planetary gear train shown in Figure 6.32, with the two basic types it comprises (Figure 6.22).

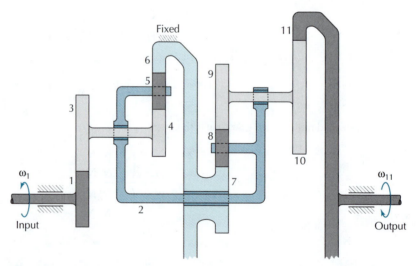

Figure 6.32 Two-stage planetary gear train incorporating the basic planetary gear train types K and L.

[†]The material presented in this section is often not covered in a first course on the mechanics of machines. It does not specifically relate to any to the remaining topics presented in this textbook.

TABLE 6.8 COMPARISON OF THE COMPONENT NUMBERS IN FIGURES 6.32 AND 6.22

Numbering of Components of Two-Stage Planetary Gear Train (Figure 6.32)	Numbering of Components of Individual Stages (Figure 6.22)	
	Stage 1 (Type K)	Stage 2 (Type L)
1	1	—
2	2	2
3	3	—
4	4	—
5	5	—
6	6	—
7	—	1
8	—	3
9	—	4
10	—	5
11	—	6

Determining the speeds of the components of a planetary gear train with k stages requires solving $2k$ equations containing $2k$ unknowns. In a single-stage planetary gear train ($k = 1$), there are two equations with two unknowns, which are normally based on the inputs to the gear train. Solving planetary gear trains for $k > 1$, requires not only that equations be generated based on the inputs, but also that constraint equations be obtained by considering connections of components between stages of the gear train. A total of $2k$ equations need to be generated.

For the two-stage planetary gear train in Figure 6.32, the numbers of teeth are

$$N_1 = 20; \quad N_3 = 25; \quad N_4 = 16; \quad N_5 = 17; \quad N_6 = 95$$
$$N_7 = 20; \quad N_8 = 17; \quad N_9 = 22; \quad N_{10} = 40; \quad N_{11} = 116$$

and the specified input speeds are

$$(\omega_1)_1 = 200 \text{ rpm} \tag{6.5-1}$$

$$(\omega_6)_1 = 0 \tag{6.5-2}$$

where subscript 1 outside of the parentheses indicates the stage number.

Additional constraints are obtained by equating rotational speeds of those components connecting the two stages. The ring gear (component 6) of stage 1 is joined to the sun gear (component 1) of stage 2, giving

$$(\omega_6)_1 = (\omega_1)_2 \tag{6.5-3}$$

and the crank (component 2) of stage 1 is coupled to the crank (component 2) of stage 2, giving

$$(\omega_2)_1 = (\omega_2)_2 \tag{6.5-4}$$

Equations (6.5-1)–(6.5-4) may be expressed in terms of variables x and y using the appropriate equations listed in Figure 6.22, giving

$$x_1 + y_1 = 200 \tag{6.5-5}$$

$$x_1 + \left(\frac{N_1 N_4}{N_3 N_6}\right)_1 y_1 = 0 \tag{6.5-6}$$

$$x_1 + \left(\frac{N_1 N_4}{N_3 N_6}\right)_1 y_1 = x_2 + y_2 \tag{6.5-7}$$

$$x_1 = x_2 \tag{6.5-8}$$

where subscripts of x and y and those outside the parentheses indicate the associated stage number.

Employing Table 6.8, the numbers of the gears may be expressed in accordance with those given in Figure 6.32. Also, substituting the given number of gear teeth, Equations (6.5-6) and (6.5-7) become

$$x_1 + \left(\frac{N_1 N_4}{N_3 N_6}\right)_1 y_1 = x_1 + \left(\frac{N_1 N_4}{N_3 N_6}\right) y_1 = x_1 + \left(\frac{20 \times 16}{25 \times 95}\right) y_1$$

$$= x_1 + 0.1347 \quad y_1 = 0 \tag{6.5-9}$$

$$x_1 + \left(\frac{N_1 N_4}{N_3 N_6}\right) y_1 = x_1 + \left(\frac{20 \times 16}{25 \times 95}\right) y_1$$

$$= x_1 + 0.1347 \, y_1 = x_2 + y_2 \tag{6.5-10}$$

Solving Equations (6.5-5), (6.5-8), (6.5-9), and (6.5-10) gives

$$x_1 = x_2 = -31.14; \quad y_1 = 231.14; \quad y_2 = 31.14$$

The rotational speed of every component in the planetary gear train may now be calculated by substituting values of x_1, x_2, y_1 and y_2, back in the appropriate equations listed in Figure 6.22. The speed ratio of this planetary gear train is

$$e_{11/1} = \frac{\omega_{11}}{\omega_1} = \frac{(\omega_6)_2}{(\omega_1)_1}$$

$$= \frac{x_2 + \left(\frac{N_1 N_5}{N_4 N_6}\right)_2 y_2}{x_1 + y_1} = \frac{x_2 + \left(\frac{N_7 N_{10}}{N_9 N_{11}}\right) y_2}{x_1 + y_1} = -0.107$$

In general, for a planetary gear train with k stages, a set of $2k$ equations can be generated and expressed in the form

$$[A]\{x\} = \{B\} \tag{6.5-11}$$

where

$$\{x\} = [x_1, y_1, x_2, y_2, \dots, x_k, y_k]^T \tag{6.5-12}$$

Equation (6.5-11) is then solved for $\{x\}$.

The equations given in Figure 6.22 are ideally suited for obtaining solutions using a computer. All the types of planetary gear trains shown in Figure 6.22 can be stored in the computer's memory. Upon specification of a planetary gear train type, the computer will work with the appropriate set of equations. As each constraint is entered, matrix $[A]$ and vector $\{B\}$ of Equation (6.5-11) may be constructed.

EXAMPLE 6.10 SPEED RATIO OF A PLANETARY GEAR TRAIN

Determine the speed ratio of the planetary gear train shown in Figure 6.31 by combining the basic types of planetary gear trains shown in Figure 6.22.

SOLUTION

The gear train is equivalent to combining two basic planetary gear train stages, each of type B, shown in Figure 6.22. Components 1, 2, and 3 are shared by both stages. Therefore, they could be considered as linked components between the stages. Table 6.9 compares the numbering schemes presented in Figures 6.22 and 6.31.

TABLE 6.9 COMPARISON OF THE COMPONENT NUMBERS IN FIGURES 6.31 AND 6.22

Numbering of Components of Two-stage Planetary Gear Train (Figure 6.31)	Numbering of Components of Individual Stages (Figure 6.22)	
	Stage 1	Stage 2
	(Type B)	(Type B)
1	1	1
2	2	2
3	3	3
4	4	–
5	–	4
6	5	–
7	–	5

The following equations are generated by considering two specified input rotational speeds, one of which is zero, and constraints imposed by the connected components of the planetary gear train.

$$(\omega_1)_1 = \omega_{\text{input}} = \omega_1 \tag{6.5-13}$$

$$\omega_6 = (\omega_5)_1 = 0 \tag{6.5-14}$$

$$(\omega_1)_1 = (\omega_1)_2 \tag{6.5-15}$$

$$(\omega_2)_1 = (\omega_2)_2 \tag{6.5-16}$$

$$(\omega_3)_1 = (\omega_3)_2 \tag{6.5-17}$$

Expressing Equations (6.5-13)–(6.5-17) in terms of the unknowns given in Figure 6.22, we obtain

$$x_1 + y_1 = \omega_1 \tag{6.5-18}$$

$$x_1 - \left(\frac{N_1 N_4}{N_3 N_5}\right)_1 y_1 = 0 \quad \text{or} \quad x_1 - \frac{N_1 N_4}{N_3 N_6} y_1 = 0 \tag{6.5-19}$$

(Continued)

EXAMPLE 6.10 Continued

$$x_1 + y_1 = x_2 + y_2 \tag{6.5-20}$$

$$x_1 = x_2 \tag{6.5-21}$$

$$x_1 - \frac{N_1}{N_3} y_1 = x_2 - \frac{N_1}{N_3} y_2 \tag{6.5-22}$$

However, Equation (6.5-22) is a linear combination of Equations (6.5-20) and (6.5-21). Therefore, we only need to consider solving Equations (6.5-18)–(6.5-21) with unknowns x_1, x_2, y_1, and y_2, and then Equation (6.5-22) will be automatically satisfied.

Combining Equations (6.5-18)–(6.5-21) in matrix form, we get

$$\begin{bmatrix} 1 & 1 & 1 & 1 \\ 1 & \left(-\dfrac{N_1 N_4}{N_3 N_6}\right) & 0 & 0 \\ 1 & 1 & -1 & -1 \\ 1 & 0 & -1 & 0 \end{bmatrix} \begin{Bmatrix} x_1 \\ y_1 \\ x_2 \\ y_2 \end{Bmatrix} = \begin{Bmatrix} \omega_1 \\ 0 \\ 0 \\ 0 \end{Bmatrix} \tag{6.5-23}$$

Solving Equation (6.5-23) gives

$$x_1 = x_2 = \omega_1 \frac{N_1 N_4}{N_3 N_6 + N_1 N_4} \tag{6.5-24}$$

$$y_1 = y_2 = \omega_1 \frac{N_3 N_6}{N_3 N_6 + N_1 N_4} \tag{6.5-25}$$

The output rotational speed of the gear train is

$$\omega_7 = (\omega_5)_2 = x_2 - \left(\frac{N_1 N_4}{N_3 N_5}\right)_2 y_2 \quad \text{or} \quad \omega_7 = x_2 - \frac{N_1 N_5}{N_3 N_7} y_2 \tag{6.5-26}$$

Substituting Equations (6.5-24) and (6.5-25) in (6.5-26), we obtain

$$\omega_7 = \omega_1 \frac{N_1 N_4}{N_3 N_6 + N_1 N_4} - \frac{N_1 N_5}{N_3 N_7} \omega_1 \frac{N_3 N_6}{N_3 N_6 + N_1 N_4} \tag{6.5-27}$$

The speed ratio of the gear train is obtained from Equation (6.5-27) and yields after simplification

$$e_{7/1} = \frac{\omega_7}{\omega_1} = \frac{\left(1 - \dfrac{N_6 N_5}{N_4 N_7}\right)}{\left(1 + \dfrac{N_6 N_3}{N_4 N_1}\right)} \tag{6.5-28}$$

which is identical to the result obtained in Example 6.9.

EXAMPLE 6.11 KINEMATIC ANALYSIS OF A TWO-STAGE PLANETARY GEAR TRAIN

Video 6.33B
Two-Stage
Planetary Gear
Train

Figure 6.33(a) shows the skeleton representation of a two-stage planetary gear train. The numbers of teeth are

$$N_1 = 24; \quad N_2 = 60; \quad N_3 = 12$$

$$N_4 = 48; \quad N_5 = 12; \quad N_6 = 24$$

Determine the speed ratios of all components with respect to gear 1 when

(a) gear 2 is held fixed
(b) gear 4 is held fixed

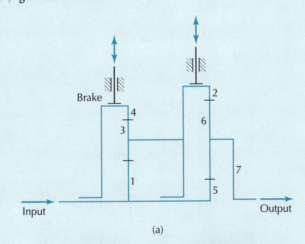

(a)

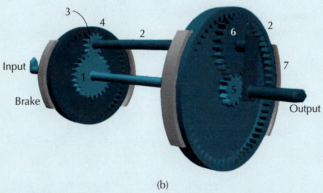

(b)

Figure 6.33 Two-stage planetary gear train: (a) Skeleton representation, (b) Mechanism [Video 6.33B].

SOLUTION

The gear train is equivalent to combining two basic planetary gear train stages, each of type A, shown in Figure 6.22. The sun gears of both stages share the same rotational speed, and the crank of stage 1 is rigidly connected to the ring gear of stage 2.

Table 6.10 compares the numbering schemes presented in Figures 6.22 and 6.33(a).

Using the equations given in Figure 6.22, and the numbering scheme provided in Table 6.10, the rotational speeds of the components may be expressed as

$$\omega_1 = (\omega_1)_1 = x_1 + y_1 \qquad (6.5\text{-}29)$$

(Continued)

EXAMPLE 6.11 Continued

TABLE 6.10 COMPARISON OF THE COMPONENT NUMBERS IN FIGURES 6.33(A) AND 6.22

Numbering of Components of Two-Stage Planetary Gear Train (Figure 6.33(a))	Numbering of Components of Individual Stages (Figure 6.22)	
	Stage 1 (Type A)	Stage 2 (Type A)
1	1	–
2	2	4
3	3	–
4	4	–
5	–	1
6	–	3
7	–	2

$$\omega_2 = (\omega_2)_1 = x_1 \tag{6.5-30}$$

$$\omega_3 = (\omega_3)_1 = x_1 - \left(\frac{N_1}{N_3}\right)_1 y_1 = x_1 - \frac{N_1}{N_3} y_1 \tag{6.5-31}$$

$$\omega_4 = (\omega_4)_1 = x_1 - \left(\frac{N_1}{N_4}\right)_1 y_1 = x_1 - \frac{N_1}{N_4} y_1 \tag{6.5-32}$$

$$\omega_5 = (\omega_1)_2 = x_2 + y_2 \tag{6.5-33}$$

$$\omega_6 = (\omega_3)_2 = x_2 - \left(\frac{N_1}{N_3}\right)_2 y_2 = x_2 - \frac{N_5}{N_6} y_2 \tag{6.5-34}$$

$$\omega_7 = (\omega_2)_2 = x_2 \tag{6.5-35}$$

$$\omega_2 = (\omega_4)_2 = x_2 - \left(\frac{N_1}{N_4}\right)_2 y_2 = x_2 - \frac{N_5}{N_2} y_2 \tag{6.5-36}$$

Speed ratios of components of the gear train relative to gear 1 may be expressed as

$$e_{j/1} = \frac{\omega_j}{\omega_1}, \quad j = 2, \dots, 7 \tag{6.5-37}$$

Considering the case of $j = 7$ and substituting Equations (6.5-29) and (6.5-35) in Equation (6.5-37), we obtain

$$e_{7/1} = \frac{\omega_7}{\omega_1} = \frac{x_2}{x_1 + y_1} \tag{6.5-38}$$

(Continued)

EXAMPLE 6.11 Continued

In addition, the following equations are generated by considering the connections of the components between the stages:

- Connection of the sun gears

$$\omega_1 = (\omega_5) \quad \text{or} \quad (\omega_1)_1 = (\omega_1)_2 \quad \text{or} \quad x_1 + y_1 = x_2 + y_2 \qquad (6.5\text{-}39)$$

- Connection of the crank of stage 1 to the ring gear of stage 2

$$\omega_2 = (\omega_2)_1 = (\omega_4)_2 \quad \text{or} \quad x_1 = x_2 - \frac{N_5}{N_2} y_2 \qquad (6.5\text{-}40)$$

Equations (6.5-29)–(6.5-40) are valid for parts (a) and (b) of this problem. We now proceed to calculate the speed ratios when either gear 2 or gear 4 is held fixed.

(a)   If gear 2 is fixed, then from Equation (6.5-30) we get

$$\omega_2 = x_1 = 0 \qquad (6.5\text{-}41)$$

Solving Equations (6.5-39)–(6.5-41) for y_1 and y_2 in terms of x_2 gives

$$y_1 = \frac{(N_2 + N_5)}{N_5} x_2; \quad y_2 = \frac{N_2}{N_5} x_2 \qquad (6.5\text{-}42)$$

Substituting Equations (6.5-41) and (6.5-42) in Equation (6.5-38) gives

$$\begin{aligned}
e_{7/1} &= \frac{x_2}{x_1 + y_1} \\
&= \frac{x_2}{\left(0 + \dfrac{(N_2 + N_5)}{N_5} x_2\right)} = \frac{N_2}{N_2 + N_5} = \frac{12}{60 + 12} = \frac{1}{6}
\end{aligned} \qquad (6.5\text{-}43)$$

Similar results of analyses for other components of the gear train are given in Table 6.11.

(b)   If gear 4 is fixed, then from Equation (6.5-32) we obtain

$$\omega_4 = x_1 - \frac{N_1}{N_4} y_1 = 0 \qquad (6.5\text{-}44)$$

Solving Equations (6.5-39), (6.5-40), and (6.5-44) for x_2, y_1 and y_2, in terms of x_1 gives

$$x_2 = \left[1 + \frac{N_4 N_5}{N_1(N_2 + N_5)}\right] x_1; \quad y_1 = \frac{N_4}{N_1} x_1 \qquad (6.5\text{-}45)$$

$$y_2 = \frac{N_2 N_4}{N_1(N_2 + N_5)} x_1$$

(Continued)

EXAMPLE 6.11 Continued

Substituting Equations (6.5-45) in Equation (6.5-38) gives

$$e_{7/1} = \frac{x_2}{x_1 + y_1} = \frac{\left[1 + \dfrac{N_4 N_5}{N_1(N_2 + N_5)}\right] x_1}{\left(x_1 + \dfrac{N_4}{N_1} x_1\right)}$$

$$= \frac{N_4 N_5 + N_1(N_2 + N_5)}{(N_1 + N_4)(N_2 + N_5)} = \frac{48 \times 12 + 24(60 + 12)}{(24 + 48)(60 + 12)} = \frac{4}{9} \qquad (6.5\text{-}46)$$

Similar results of analyses for other components of the gear train are given in Table 6.11. Figure 6.33(b) shows another illustration of the same gear train.

Video 6.33B
Two-Stage
Planetary Gear
Train

TABLE 6.11 SPEED RATIOS OF THE PLANETARY GEAR TRAIN WITH RESPECT TO GEAR 1 IN FIGURE 6.33(A)

Speed Ratio	Part (a) (Gear 2 Fixed)	Part (b) (Gear 4 Fixed)
$e_{2/1}$	0	$\dfrac{N_1}{N_1 + N_4} = \dfrac{1}{3}$
$e_{3/1}$	$-\dfrac{N_1}{N_3} = -2$	$-\dfrac{N_1(N_3 - N_4)}{N_3(N_1 + N_4)} = -1$
$e_{4/1}$	$-\dfrac{N_1}{N_4} = -\dfrac{1}{2}$	0
$e_{5/1}$	1	1
$e_{6/1}$	$\dfrac{N_5(N_6 - N_2)}{N_6(N_2 + N_5)} = -\dfrac{1}{4}$	$\dfrac{N_5 N_6(N_1 + N_4) + N_2(N_1 N_6 - N_4 N_5)}{N_6(N_1 + N_4)(N_2 + N_5)} = \dfrac{1}{6}$
$e_{7/1}$	$\dfrac{N_5}{N_2 + N_5} = \dfrac{1}{6}$	$\dfrac{N_4 N_5 + N_1(N_2 + N_5)}{(N_1 + N_4)(N_2 + N_5)} = \dfrac{4}{9}$

6.6 DIFFERENTIALS

Video 6.34
Differential Gear
Train

A *differential* is a mechanism that permits transmission of input power to two separate output components that rotate at unequal speeds. A planetary gear train may be employed to create such a mechanism and can incorporate either bevel gears or spur gears. Consider the gear train illustrated in Figure 6.34. A corresponding sectional top view is given in Figure 6.35. Included in these figures are four bevel gears, numbered 3 through 6, which are the gears of a differential.

Gear trains of the configuration shown in Figure 6.34 are commonly employed in drive trains of vehicles. Gear 1, referred to as the *pinion*, is the input to the gear train. Ring gear 2 is attached to the planet carrier. Gears 1 and 2 may be plain bevel gears as shown. Alternatively, these gears could be either a spiral bevel set or a hypoid gear set, as illustrated in Figures 5.9(b) and 5.10, respectively. Gears 4 and 6 are planet gears because their axes of rotation are not fixed. These planet gears are allowed to rotate with respect to gear 2 and revolve about the axis of rotation of gear 2. Gears 3

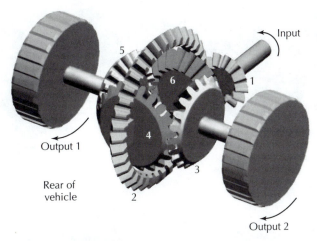

Figure 6.34 Differential gear train [Video 6.34].

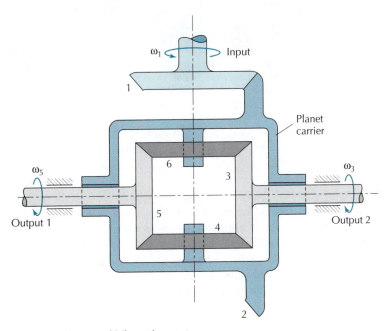

Figure 6.35 Top view of differential gear train.

and 5 are attached to the same shafts as the two drive wheels and share the same axis of rotation as gear 2. Gears 3 and 5 are the same size, and so are gears 4 and 6.

The differential, as part of the gear train, allows the drive wheels of a vehicle to rotate at either the same or unequal speeds. Four possible motions that may be generated are shown in Figure 6.36. In all of these cases, it is considered that there is one input from gear 1 and two outputs to gears 3 and 5. When a vehicle travels straight ahead, as illustrated in Figure 6.36(a), both drive wheels, as well as gears 3 and 5, have the same rotational speed. In this case, gears 2, 3, 4, 5, and 6 all rotate about a common axis as though they formed a rigid body, and there is no meshing movement between these gears. For the three other cases shown in Figure 6.36, there is meshing movement between gears 3, 4, 5, and 6. When a vehicle is turning, as illustrated in Figure 6.36(b), one drive wheel is rotating faster than the other. In the case of a right turn, the drive wheel connected to gear 5 moves along a larger radius of curvature than the drive wheel connected to gear 3. Hence, to avoid slippage of either wheel with the ground, gear 5 must turn at a faster rate than gear 3. The differential permits this difference of the two output speeds. Figure 6.36(c) illustrates the case where one wheel is stopped, while the other wheel continues to spin. Another case is illustrated in Figure 6.36(d), when the input gear 1 is stationary, and gears 3 and 5 rotate in opposite directions. Although this case is allowable, it is unlikely to occur during normal vehicle operation.

Figure 6.36 Motions of a differential gear train [Video 6.34]: (a) Equal outputs (no slip), (b) Cornering or partial slip, (c) Full slip, (d) Wheels rotate in opposite directions (zero input).

As illustrated in Figure 6.36, the rotational speed of gear 2 is always the average of the rotational speeds of gears 3 and 5. This is now verified by employing the tabular method presented in Section 6.4. Completing the first three general steps for a tabular analysis, Table 6.12 was constructed. In this table, gear 1 is listed along with the other gears, even though their axes of rotation are perpendicular to one another. When comparing the speed of gear 1 with those of gears 2, 3 and 5, only the magnitudes should be considered.

From Table 6.12, the rotational speeds of gears 2, 3, and 5 expressed in terms of variables x and y are

$$\omega_2 = \frac{N_1}{N_2}x; \quad \omega_3 = \frac{N_1}{N_2}x - y; \quad \omega_5 = \frac{N_1}{N_2}x + y \tag{6.6-1}$$

TABLE 6.12 TABULAR ANALYSIS OF THE BEVEL DIFFERENTIAL GEAR TRAIN SHOWN IN FIGURE 6.34

Operation ↓ \ Component →	Gear 1 ω_1	Gear 2 ω_2	Gear 3 ω_3	Gear 5 ω_5
Gears 4, and 6 Are Stationary with Respect to Gear 2, Gear 1 Turns at x rpm	x	$\dfrac{N_1}{N_2}x$	$\dfrac{N_1}{N_2}x$	$\dfrac{N_1}{N_2}x$
Gear 1 Is Fixed, Gear 5 Turns at y rpm	0	0	$-y$	y
Absolute Rotational Speeds	x	$\dfrac{N_1}{N_2}x$	$\dfrac{N_1}{N_2}x - y$	$\dfrac{N_1}{N_2}x + y$

The rotational speed of gear 2 is the average of those of gears 3 and 5 because

$$\omega_3 + \omega_5 = \left(\frac{N_1}{N_2}x - y\right) + \left(\frac{N_1}{N_2}x + y\right) = 2\frac{N_1}{N_2}x = 2\omega_2 \quad \text{or} \quad \omega_2 = \frac{\omega_3 + \omega_5}{2} \quad (6.6\text{-}2)$$

When the vehicle is traveling straight ahead (Figure 6.36(a)), the two output rotational speeds are equal. By inspection of Table 6.12, we require $y = 0$, and since $x = \omega_1$, then

$$\omega_3 = \omega_5 = \frac{N_1}{N_2}x = \frac{N_1}{N_2}\omega_1 = \omega_2 \quad (6.6\text{-}3)$$

When the vehicle is in a turn (Figure 6.36(b)), the two drive wheels no longer have the same rotational speed. Here, y takes on a finite value, and $2y$ is the difference of the rotational speeds of the two outputs. If gear 3 stops turning (Figure 6.36(c)), then

$$\omega_3 = 0 = \frac{N_1}{N_2}x - y \quad \text{or} \quad y = \frac{N_1}{N_2}x \quad (6.6\text{-}4)$$

The rotational speed of gear 5 is then

$$\omega_5 = \frac{N_1}{N_2}x + y = 2\frac{N_1}{N_2}x = 2\frac{N_1}{N_2}\omega_1 = 2\omega_2 \quad (6.6\text{-}5)$$

That is, gear 5 rotates at twice the average speed of gear 2.

The condition of having one of the outputs stationary can occur when one drive wheel, say, connected to gear 3, is on dry pavement, while the other, connected to gear 5, is freely spinning on a slippery surface such as ice. In this instance, minimal torque resistance is offered by the spinning wheel, and negligible torque can be transmitted to the other drive wheel even though it is on dry pavement.

The problem of wheel spin on slippery surfaces can be relieved somewhat by employing a *limited-slip differential*. A schematic of such a system is shown in Figure 6.37. The arrangement of gears is the same as a conventional differential (Figure 6.35). However, there are additional friction pads placed between the planet carrier (i.e., link 2 that includes the ring gear) and gears 3 and 5. In this instance, when one of the drive wheels is on ice, a limited amount of torque may still be transmitted to the other wheel. The amount of torque that can be transmitted to one wheel when the other has no traction is dependent on the frictional torque generated through the friction pads.

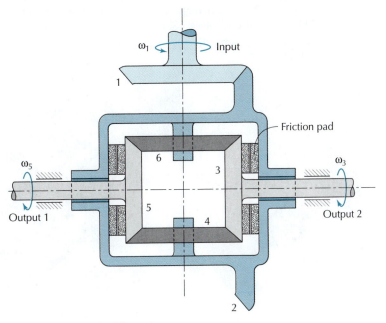

Figure 6.37 Limited-slip differential.

A disadvantage associated with a limited-slip differential is that energy is lost, and wear occurs in the friction pads when relative motion occurs between the output gears and the planet carrier, that is, each time the vehicle changes direction during normal operation.

The arrangement of gears illustrated in Figure 6.34 is employed for rear wheel drive vehicles, where the input connected to gear 1 is perpendicular to the two outputs, gears 3 and 5. Alternatively, Figure 6.38 shows the arrangement commonly used in front wheel drive vehicles, where the engine and transmission (not shown) drive sun gear 1 of the planetary gear train while ring gear 4 is held fixed. This Type A planetary gear train (Figure 6.22) provides one final speed reduction prior to delivering motion to the differential. Component 2 is both the crank of the Type A planetary gear train and the planet carrier of the differential. It rotates at the average of the two output speeds of the differential. Gears 6 and 8 are planet gears of the differential, and gears 5 and 7 are connected to the two output wheels.

A differential may also be created by using a Type I planetary gear train (Figure 6.22) for which

$$N_5 = 2N_1 \tag{6.6-6}$$

Employing Equation (6.6-6) in the equations provided in Figure 6.22,

$$\omega_1 = x + y \tag{6.6-7}$$

$$\omega_2 = x \tag{6.6-8}$$

$$\omega_5 = x + \frac{N_1}{N_5} y = x + \frac{N_1}{2N_1} y = x + \frac{1}{2} y \tag{6.6-9}$$

Therefore, if the input to this gear train is supplied to gear 5, then as required, its value of rotational speed is the average of the two outputs, ω_1 and ω_2.

Figure 6.39 shows such a differential using spur gears. Differentials that incorporate spur gears instead of bevel gears have the advantage of a smaller width of the gear train. For this reason, they are often employed in four wheel drive vehicles, which require a differential for the front wheels, a differential for the rear wheels, and the option to have differential action between the front and rear wheel pairs.

Figure 6.40 shows the skeleton diagram for an all-time four wheel drive gear train. Input from the transmission is supplied to gear 6. One differential, consisting of gears 2, 4, 5, and 6 and crank arm 3, allows a differential action between the front pair and rear pair of wheels. A second differential, consisting of gears 14, 16, 17, and 18 and crank arm 15, allows differential motion between the two front wheels. A third differential (not shown) is employed for the two rear wheels. Gears 11, 12, and 13 are part of a planetary gear train that provides a speed reduction to the front pair of wheels. Gears 7, 8, 9, and 10 and the gears at the rear differential provide the same speed reduction to the rear wheels.

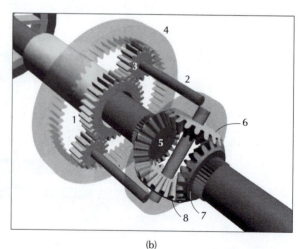

(a)   (b)

Figure 6.38 Front wheel drive gear train [Video 6.38].

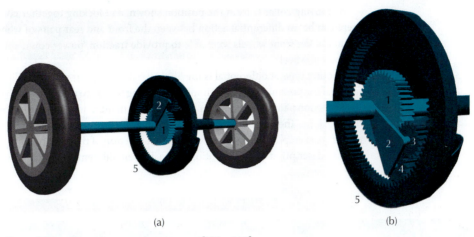

Figure 6.39 Differential incorporating spur gears [Video 6.39].

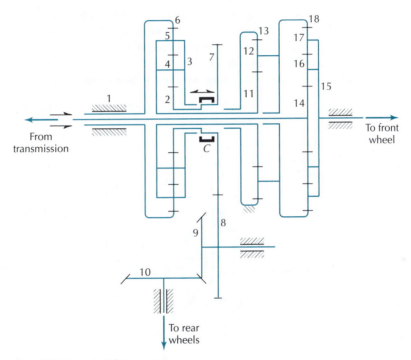

Figure 6.40 Four wheel drive gear train.

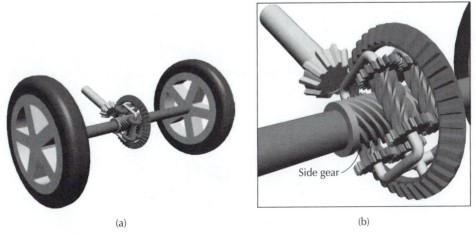

Figure 6.41 Torsen differential [Video 6.41].

By moving collar C from the position shown, and locking together component 3 and gear 7, there can be no differential action between the front and rear pairs of wheels. In this instance, if neither of the front wheels were able to provide traction, power could still be transmitted to the rear pair of wheels.

Another type of differential is the *Torsen*^{TM} *differential* as shown in Figure 6.41. With this differential it is possible to provide improved driving torque to each of the driving wheels without the use of friction pads. Two or more planet gears are in mesh with the central helical gears, also called the *side gears*. The planet gears are interconnected by means of straight spur gears. The number of planet gears employed in a specific design is a function of the required torque capacity. A comprehensive description of the operation of the Torsen differential was prepared by Chocholek [8].

EXAMPLE 6.12 ANALYSIS OF A DIFFERENTIAL GEAR TRAIN

A vehicle is turning along a left-hand curve at 60 km/hr, with a radius of curvature of $\rho = 30$ m. Figure 6.42 provides additional parameters for the example including tire radius r_t, and tire-to-tire distance d. There is no slippage of the tires on the road surface. Determine the relative rotational speed of the inboard tire (i.e., gear 5) with respect to the planet carrier (i.e., gear 2) and the magnitude of the rotational speed of gear 1 The numbers of gear teeth are

$$N_1 = 17; \quad N_2 = 67; \quad N_3 = 24 = N_5; \quad N_4 = 18 = N_6$$

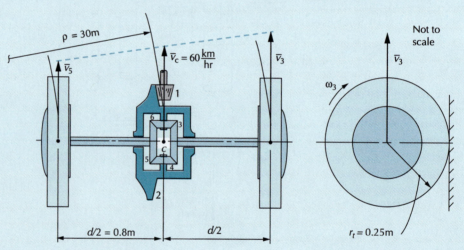

Figure 6.42

SOLUTION

$$v_c = 60 \text{km/hr} = 16.7 \text{m/sec}$$

Since the speed is proportional to the radius of curvature of the path of the tires, we obtain

$$v_3 = \frac{\left(\rho + \dfrac{d}{2}\right)}{\rho} v_C = \frac{30.0 + 0.8}{30.0} 16.67 = 17.17 \text{ m/sec}$$

$$v_5 = \frac{\left(\rho - \dfrac{d}{2}\right)}{\rho} v_C = 16.17 \text{ m/sec}$$

(*Continued*)

EXAMPLE 6.12 Continued

Kinematics of the rolling wheels gives

$$\omega_3 = \frac{v_3}{r_t} = \frac{17.17}{0.25} = 68.67 \text{ rad/sec}; \quad \omega_5 = \frac{v_5}{r_t} = 64.67 \text{ rad/sec}$$

Using Table 6.12, we get

$$\omega_3 = \frac{N_1}{N_2}x - y = 68.67 \text{ rad/sec}; \quad \omega_5 = \frac{N_1}{N_2}x + y = 64.67 \text{ rad/sec}$$

Substituting the given values of the numbers of teeth, substituting the calculated values of the rotational speeds, and solving, we obtain

$$x = 262.7 \text{ rad/sec}; \quad y = -2.0 \text{ rad/sec}$$

The relative rotational speed of the inboard tire with respect to the planet carrier is

$$\omega_{5/2} = \omega_5 - \omega_2 = 64.67 \text{ rad/sec} - \frac{N_1}{N_2}x = -2.0 \text{ rad/sec}$$

The negative sign of the result indicates that the relative rotational speed is in a direction opposite to the direction of the rotational speed of gear 2. The magnitude if the rotational speed of gear 1 is $|\omega_1| = |x| = 262.7$ rad/sec. Note that in the above analysis the numbers of teeth of the differential gears were not required.

EXAMPLE 6.13 SPEED RATIO OF A GEAR TRAIN INCORPORATING BEVEL GEARS

In the planetary gear train shown in Figure 6.43, the input speed is $\omega_2 = 100$ rpm CCW. Determine the magnitude and direction of the output speed ω_{10}. The numbers of teeth are

$$N_2 = 40; N_4 = 30; N_5 = 25; N_6 = 120$$
$$N_7 = 50; N_8 = 20; N_9 = 70; N_{10} = 20$$

SOLUTION

The input shaft drives the two gears of the planetary gear train. Their rotational speeds are

$$\omega_5 = \omega_4 = \frac{N_2}{N_4}\omega_2 = \frac{40}{30} \times 100 = 133 = 133 \text{ rpm} \quad \text{CCW}$$

$$\omega_7 = \omega_8 = -\frac{N_9}{N_8}\omega_2 = -\frac{70}{20} \times 100 = -350 = 350 \text{ rpm} \quad \text{CW}$$

where it has been assumed that positive rotational speeds are in the counterclockwise direction.

Employing the first three steps for the tabular method, Table 6.13 was constructed (note that if the carrier is fixed, gears 5 and 7 turn in opposite directions).

(Continued)

EXAMPLE 6.13 Continued

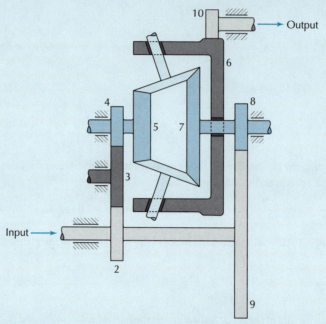

Figure 6.43 Planetary gear train incorporating bevel gears.

TABLE 6.13 TABULAR ANALYSIS OF THE BEVEL DIFFERENTIAL GEAR TRAIN SHOWN IN FIGURE 6.43

Component → Operation ↓	Gear 5 ω_5	Gear 7 ω_7	Gear 6 ω_6
Gears 5, 6, and 7 Turn at x rpm	x	x	x
Gear 6 Is Fixed, Gear 5 Turns at y rpm	y	$-\dfrac{N_5}{N_7}y$	0
Absolute Rotational Speeds	$x + y$	$x - \dfrac{N_5}{N_7}y$	x

Substituting values from Table 6.13, we obtain

$$\omega_5 = 133 = x + y$$

$$\omega_7 = -350 = x - \frac{N_5}{N_7}y = -350 = x - \frac{25}{50}y$$

Solving the above two equations for x and y, we get

$$x = -189; \quad y = 322$$

and therefore

$$\omega_6 = x = -189 = 189 \text{ rpm} \quad \text{CW}$$

(Continued)

EXAMPLE 6.13 Continued

The output rotational speed is then

$$\omega_{10} = -\frac{N_6}{N_{10}}\omega_6 = -\frac{120}{20} \times (-189) = 1134 = 1134 \text{ rpm} \quad \text{CCW}$$

6.7 HARMONIC DRIVES

A *harmonic drive* in exploded form is shown in Figure 6.44. This drive incorporates a unique means of power transmission in which the flexibility of one of its components is utilized. Large reductions or increases of rotational speed are possible with such drives. Due to their relatively small, in-line configuration between the driver and driven components, they have been useful in a variety of applications, including drives of the axes in robots.

Figure 6.45 shows another illustration of a harmonic drive. The *wave generator* has a surface in the shape of an ellipse. This elliptical shape is inserted into a flexible component, called the *flex spline*, which is deflected into the same shape. Through these deflections, external teeth on the flex spline mesh with a rigid internal circular gear known as the *circular spline*. Teeth on the flex spline and the circular spline simultaneously mesh at two locations, 180° apart from one another. This arrangement leads to a balanced torque reaction between the input and output about the central axis of the drive. An advantage of this form of meshing, is that a precisely machined unit can be made to operate with essentially zero backlash.

Although teeth on both splines have the same pitch, the circular spline has two more teeth than the flex spline, that is,

$$N_c = N_f + 2 \tag{6.7-1}$$

where N_c and N_f are the numbers of teeth on the circular spline and flex spline, respectively.

The animation provided through [**Video 6.45**] has two distinct motions. For the first motion, the wave generator is the input component, the flex spline is the output component, and the circular spline is held fixed. As the wave generator rotates, the flex spline forms a traveling deflection wave. Figure 6.46 shows two configurations of the harmonic drive. In Figure 6.46(b), the wave generator has rotated 180° clockwise from the position shown in Figure 6.46(a). Notice that the marking on the flex spline has shifted one tooth with respect to the circular spline as a result of the

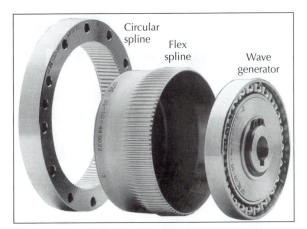

Figure 6.44 Harmonic drive (courtesy of HD Systems, Inc. Hauppauge, NY).

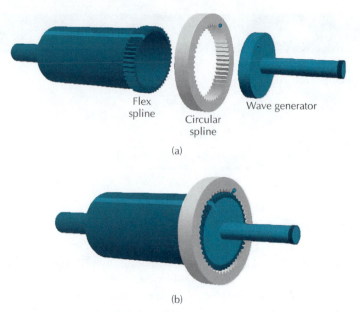

Flex spline

Circular spline

Wave generator

(a)

(b)

Figure 6.45 Harmonic drive [Video 6.45]: (a) Exploded view, (b) Assembled unit.

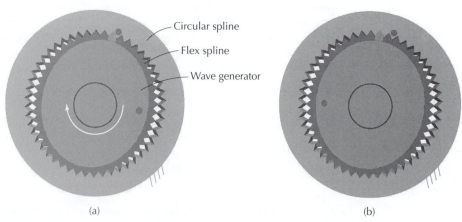

Circular spline

Flex spline

Wave generator

(a)   (b)

Figure 6.46 Harmonic drive [Video 6.45].

rotation of the wave generator. The rotational direction of shift is opposite to that of the traveling deflection wave, that is, for each rotation of the wave generator in the clockwise direction, the flex spline rotates an equivalent of two teeth in the counterclockwise direction. Therefore, the first speed ratio generated is

$$e_{flw} = \frac{\omega_f}{\omega_w} = -\frac{2}{N_f} \tag{6.7-2}$$

where ω_f and ω_w are rotational speeds of the flex spline and wave generator, respectively, and the negative sign indicates that the flex spline and wave generator turn in opposite directions.

For the harmonic drive illustrated in Figure 6.45 we have

$$N_f = 48; \quad N_c = 50$$

and thus for this unit we obtain

$$e_{flw} = -\frac{2}{N_f} = -\frac{1}{24}$$

Video 6.45
Harmonic Drive

TABLE 6.14 TABULAR ANALYSIS OF THE HARMONIC DRIVE SHOWN IN FIGURE 6.45

Component → Operation ↓	Wave Generator ω_w	Flex Spline ω_f	Circular Spline ω_c
All Components Turn at x rpm	x	x	x
Circular Spline Is Fixed, Wave Generator Turns at y	y	$-\dfrac{2}{N_f}y$	0
Absolute Rotational Speeds	$x+y$	$x-\dfrac{2}{N_f}y$	x

Any one of the three components of a harmonic drive can be selected for the input, the output, or the fixed component. Once the selection has been made, the corresponding speed ratio may be determined using a tabular method, similar to that employed for planetary gear trains. Table 6.14 provides such a table. Included in the table is a row for which all three components have the same rotational speed of x rpm. In the subsequent row, the circular spline has zero rotational speed, the wave generator is given y rpm, and the corresponding value of rotational speed of the flex spline is provided. Absolute values of rotational speeds that are a superposition of the two motions are also listed.

For the second portion of the animation provided through [**Video 6.45**], the flex spline is prevented from rotating, and hence, from Table 6.14,

$$\omega_f = x - \frac{2}{N_f} y = 0$$

Therefore

$$x = \frac{2}{N_f} y \quad \text{or} \quad y = \frac{N_f}{2} x \tag{6.7-3}$$

In addition, if the wave generator is the input component, and the circular spline is the output component, then the second speed ratio generated is

$$e_{c/w} = \frac{\omega_c}{\omega_w} = \frac{x}{x+y} \tag{6.7-4}$$

Combining Equations (6.7-4), (6.7-3), and (6.7-1), we obtain

$$e_{c/w} = \frac{x}{x+y} = \frac{x}{\left(x + \dfrac{N_f}{2} x\right)} = \frac{2}{2+N_f} = \frac{2}{N_c} \tag{6.7-5}$$

For the harmonic drive shown in Figure 6.45 we have

$$e_{c/w} = \frac{2}{N_c} = \frac{2}{50} = \frac{1}{25}$$

Here, the circular spline and wave generator rotate in the same direction.

Harmonic drives may be combined in multiple stages, similar to planetary gear trains, to provide even smaller magnitudes of speed ratios.

EXAMPLE 6.14 ANALYSIS OF A TWO-STAGE HARMONIC DRIVE

Figure 6.47 shows the cross section of a two-stage harmonic drive. The numbers of teeth are

$$N_{f_1} = 104; \quad N_{c_1} = 106; \quad N_{f_2} = 108; \quad N_{c_2} = 110$$

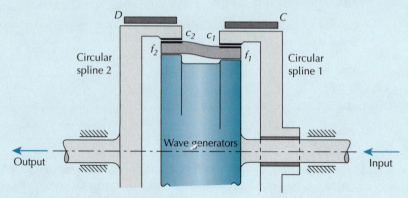

Figure 6.47 Two-stage harmonic drive.

Determine the speed ratio of

(a) circular spline 1 with respect to the wave generator if circular spline 2 is held fixed by brake D
(b) circular spline 2 with respect to the wave generator if circular spline 1 is held fixed by brake C

SOLUTION

The drive is equivalent to combining two harmonic drives. The wave generators of both stages share the same rotational speed, and the flex splines of both stages are rigidly connected. We employ Table 6.14 to generate the following equations:

- Connection of wave generators:

$$x_1 + y_1 = x_2 + y_2 \tag{6.7-6}$$

- Connection of flex splines:

$$x_1 - \frac{2}{N_{f_1}} y_1 = x_2 - \frac{2}{N_{f_2}} y_2 \tag{6.7-7}$$

where subscripts 1 and 2 indicate the stage number. Equations (6.7-6) and (6.7-7) are valid for parts (a) and (b) of this problem. We now proceed to calculate the speed ratios when either circular spline 1 or circular spline 2 is held fixed.

(a) If circular spline 2 is fixed, then from Table 6.14 we have

$$\omega_{c_2} = x_2 = 0 \tag{6.7-8}$$

Solving Equations (6.7-6)–(6.7-8) and (6.7-1) for x_1 and y_1 in terms of y_2, we obtain

$$x_1 = \frac{2\left(N_{f_2} - N_{f_1}\right)}{N_{c_1} N_{f_2}} y_2; \quad y_1 = \frac{N_{c_1} N_{f_2}}{N_{c_1} N_{f_2}} y_2 \tag{6.7-9}$$

(Continued)

EXAMPLE 6.14 Continued

The speed ratio may be expressed as

$$e_{c_1/w} = \frac{\omega_{c_1}}{\omega_w} = \frac{x_1}{x_1 + y_1} \tag{6.7-10}$$

Substituting Equations (6.7-9) in Equation (6.7-10) and simplifying, we obtain

$$e_{c_1/w} = \frac{2\left(N_{f_2} - N_{f_1}\right)}{N_{c_1} N_{f_2}} = \frac{2(108-104)}{106 \times 108} = \frac{1}{1431} \tag{6.7-11}$$

(b) If circular spline 1 is fixed, then from Table 6.14 we have

$$\omega_{c_1} = x_1 = 0 \tag{6.7-12}$$

Solving Equations (6.7-6), (6.7-7), (6.7-12), and (6.7-1) for x_2 and y_1 in terms of y_2, we obtain

$$x_2 = \frac{2\left(N_{f_1} - N_{f_2}\right)}{N_{c_1} N_{f_2}} y_2; \quad y_1 = \frac{N_{f_1} N_{c_2}}{N_{c_1} N_{f_2}} y_2 \tag{6.7-13}$$

The speed ratio may be expressed as

$$e_{c_2/w} = \frac{\omega_{c_2}}{\omega_w} = \frac{x_2}{x_1 + y_1} \tag{6.7-14}$$

Substituting Equations (6.7-12) and (6.7-13) in Equation (6.7-14) and simplifying, we get

$$e_{c_2/w} = \frac{2\left(N_{f_1} - N_{f_2}\right)}{N_{f_1} N_{c_2}} = \frac{2(104-108)}{104 \times 110} = -\frac{1}{1430} \tag{6.7-15}$$

6.8 REACTION TORQUES AND RELATIONS IN GEARBOXES

In a gearbox, torque is transmitted from the input to the output shaft, and if there is a difference in rotational speed between input and output, there will be a related change in torque. This means that the input and output torques are unequal, and to maintain static equilibrium, a reaction torque must be applied to the gearbox housing. Figure 6.48 shows the torques acting on a typical gearbox. The *input torque* and *output torque* are designated as T_i and T_o, respectively, and the input and output rotational speeds are ω_i and ω_o. The input and output powers (see Appendix C, Section C.10.3) to and from the gearbox are

$$P_i = T_i \omega_i \tag{6.8-1}$$

$$P_o = -T_o \omega_o \tag{6.8-2}$$

where it has been recognized that the output torque and output rotational speed must have opposite signs.

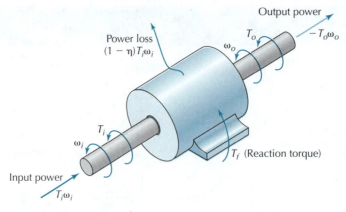

Figure 6.48 Torques on a gearbox.

Due to frictional losses, the power output is less than the power input to the gear train. The *mechanical efficiency* of the gearbox, η, is defined as the ratio of the output power to the input power:

$$\eta = \frac{P_o}{P_i} = \frac{T_o \omega_o}{T_i \omega_i}$$

or

$$T_o \omega_o = -\eta T_i \omega_i \qquad (6.8\text{-}3)$$

In addition, the sum of the moments applied to the gearbox must be zero, that is,

$$T_i + T_o + T_f = 0 \qquad (6.8\text{-}4)$$

where T_f is the *reaction torque* acting on the gearbox, which is required to keep the gearbox housing stationary. The following example illustrates the use of the above equations.

EXAMPLE 6.15 TORQUE ANALYSIS OF A GEAR TRAIN

Determine the torque on the housing of the gearbox in Figure 6.21, with the ring gear fixed, while under the following conditions

$$\text{input power} = P_i = 30{,}000 \frac{\text{N-m}}{\text{sec}} = 30{,}000 \text{ watts} = 30 \text{ kW}$$

$$\omega_i = \omega_1 = 2000 \text{ rpm}; \quad \eta = 0.98$$

SOLUTION
The speed ratio of the gearbox was previously determined (see Equation (6.4-1)) to be

$$e_{o/i} = \frac{\omega_o}{\omega_i} = e_{2/1} = 0.30$$

From Equation (6.8-1) we have

$$T_i = \frac{P_i}{\omega_i} = \frac{30{,}000 \dfrac{\text{N-m}}{\text{sec}}}{\left(2000 \times \dfrac{2\pi \text{ rad}}{60 \text{ sec}}\right)} = 143 \text{ N-m}$$

> ## EXAMPLE 6.15 Continued
>
> From Equation (6.8-3) we have
>
> $$T_o\omega_o = T_o e_{o/i}\omega_i = -\eta T_i\omega_i$$
>
> and therefore
>
> $$T_o = \frac{\eta T_i}{e_{o/i}} = -\frac{0.98\times143}{0.30} = -467\,\text{N-m}$$
>
> Combining the above results with Equation (6.8-4), we obtain
>
> $$T_f = T_4 = -T_o - T_i = 467 - 143 = 324\,\text{N-m}$$
>
> Since T_f is positive, the torque on link 4 is in the same direction as the input rotational speed.

PROBLEMS

Ordinary and Compound Gear Trains

P6.1 For the gear train shown in Figure P6.1, determine

(a) the speed ratio $e_{6/2}$

(b) the output rotational speed, ω_6, if the input rotational speed, ω_2, is 75 rpm CW

$N_2 = 15;\quad N_3 = 25;\quad N_4 = 20;\quad N_5 = 15;\quad N_6 = 15$

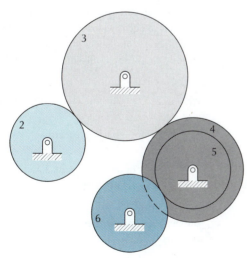

Figure P6.1

P6.2 For a reverted gear train (Figure 5.28), gears 2 and 3 have a module of 5 mm, and gears 4 and 5 have a module of 3 mm. All gears are straight spur. Determine suitable tooth numbers for the gears if the speed ratio is to be

(a) 0.25

(b) approximately 0.42

P6.3 Specify the numbers of teeth for the gears shown in Figure 6.13, having speed ratios of approximately 0.8, 0.5, and 0.3. The smallest gear must have at least 20 teeth, and the module of all teeth is 4 mm.

P6.4 Figure P6.4 (see next page) illustrates a three-speed transmission. Gears 4, 5, and 6 are free to spin about their shaft in the configuration shown, while gears 3, 7, 8, 9, 10, and 11 are keyed to their respective shafts. Gear 10 is an idler gear whose center is not in line with those of gears 6 and 11. C_1 and C_2 are synchroizers that fix one free gear to the shaft each time, so that power is transmitted from the input shaft to the output shaft. The input and output shafts are collinear. Using gears of diametral pitch 12 in⁻¹, and using a center-to-center distance between the input shaft and countershaft of 4 inches, specify the number of teeth of all gears to obtain the following speed ratios: +1, +0.5, +0.2, -0.2 (reverse), subject to the constraint that the minimum number of teeth in any gear is 16.

P6.5 For the gear train of Figure P6.5 (see next page), calculate the speed ratio and determine the speed of rotation of output gear 10, given that gear 3 is driven at 200 rpm in the direction shown.

$N_2 = 20;\quad N_3 = 10;\quad N_4 = 60;\quad N_5 = 72;\quad N_6 = 15$
$N_7 = 22;\quad N_8 = 16;\quad N_9 = 12;\quad N_{10} = 10$

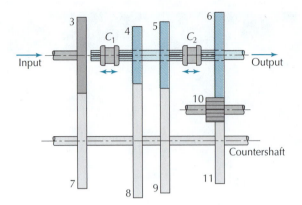

Figure P6.4

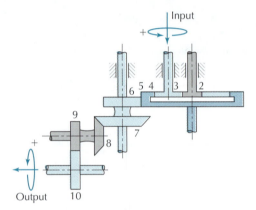

Figure P6.5

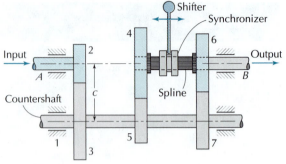

Figure P6.6

TABLE P6.6

Gear No.	No. of Teeth	Type of Gear
2	20	Straight spur
3	40	—
4	—	Helical spur
5	—	—
6	—	—
7	—	—

Ordinary and Compound Gear Trains (requiring material from Chapter 5)

P6.6 Information pertaining to the two-speed transmission shown in Figure P6.6 is given in Table P6.6.

All gears are to be manufactured using a hob cutter having a 20° pressure angle and 0.500-inch hob pitch. All gears are to have full-depth teeth, each with more than 15 teeth and fewer than 45 teeth. Shafts A and B are collinear. The desired speed ratios are +38/156 (synchronizer connects gear 4 to shaft B), and +54/132 (synchronizer connects gear 6 to shaft B). Determine

(a) the base circle radius of gear 2

(b) the center-to-center distance, c, between gears 2 and 3

(c) the contact ratio between gears 2 and 3

(d) the number of teeth on gears 4, 5, 6, and 7 that will generate exactly the desired speed ratios

(e) the circular pitch of gear 4

(f) the helix angle of gear 4

(g) the helix angle of gear 7

(h) the pitch circle diameter of gear 5

(i) addendum circle diameter of gear 5

P6.7 The information in Table P6.7 (see next page) pertains to the two-speed transmission shown in Figure P6.7 (see next page).

All gears are to be manufactured using a hob cutter having a 20° pressure angle and 0.500-inch hob pitch. All gears are to have full-depth teeth, each with more than 15 teeth, and fewer than 45 teeth. Shafts A and B are collinear. The first desired speed ratio is +6/17, and is to occur when gears 4 an 5 are brought into mesh. The second speed ratio is to occur when gears 6 and 7 are brought into mesh. Determine

(a) the base circle radius of gear 5

(b) center-to-center distance, c, between gears 2 and 3

(c) the contact ratio between gears 4 and 5 while in mesh

(d) the numbers of teeth on gears 2 and 3 that will generate exactly the first desired speed ratio

(e) the second speed ratio

(f) the diametral pitch of gear 4

(g) the helix angle of gear 2

(h) pitch circle diameter of gear 2

P6.8 The information in Table 6.8 (see next page) pertains to the two-speed transmission shown in Figure P6.6.

All gears are to be manufactured using a hob cutter having a 20° pressure angle and 0.500-inch hob pitch. All gears are to have full-depth teeth, each having more than 15 teeth but fewer than 45 teeth.

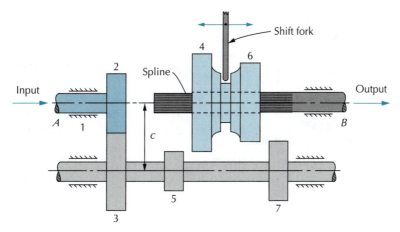

Figure P6.7

TABLE P6.7

Gear No.	No. of Teeth	Type of Gear
2	—	—
3	—	Helical spur
4	40	Straight spur
5	20	—
6	—	—
7	25	Straight spur

Shafts A and B are collinear. The desired speed ratios are $+115/296$ (a synchronizer connects gear 4 to shaft B) and $+155/272$ (a synchronizer connects gear 6 to shaft B). Determine

(a) the base circle radius of gear 2
(b) the center-to-center distance between gears 2 and 3
(c) the contact ratio between gears 2 and 3
(d) the number of teeth on gears 4, 5, 6, and 7 that will generate exactly the desired speed ratios
(e) the circular pitch of gear 4
(f) the helix angle of gear 4
(g) the helix angle of gear 7
(h) the pitch circle diameter of gear 5

TABLE P6.8

Gear No.	No. of Teeth	Type of Gear
2	25	Straight spur
3	40	—
4	—	Helical spur
5	—	—
6	—	—
7	—	—

P6.9 The information in Table P6.9 pertains to the reverted gear train shown in Figure P6.9.

TABLE P6.9

Gear No.	No. of Teeth	Type of Gear
2	20	—
3	40	Straight spur
4	—	—
5	—	—

All gears are to be manufactured using a hob cutter having a 20° pressure angle and 0.500-inch hob pitch. All gears are to have full-depth teeth, each with more than 15 teeth, and fewer than 45 teeth. The desired speed ratio of the gear train is $+6/17$. Determine

(a) the dedendum circle diameter of gear 2
(b) the center-to-center distance, c, between gears 4 and 5
(c) the contact ratio between gears 2 and 3
(d) the numbers of teeth on gears 4 and 5 that will generate exactly the desired speed ratio of the gear train
(e) the helix angle of gear 4
(f) the pitch circle diameter of gear 5

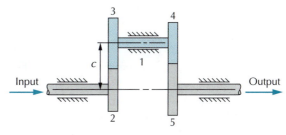

Figure P6.9

P6.10 The information in Table P6.10 pertains to the gear train shown in Figure P6.10:

External gears 1, 3, and 4 are to be manufactured using hob cutters having a 20° pressure angle. All gears are to have full-depth teeth. Shafts A and B are collinear. Determine

(a) the base circle radius of gear 5

(b) the addendum circle diameter of gear 5

(c) the center-to-center distance, c, between gears 3 and 4

(d) the contact ratio between gears 4 and 5

(e) the helix angle of gear 1

(f) the pitch circle diameter of gear 2

P6.11 The information in Table P6.11 pertains to the gear train shown in Figure P6.11:

External gears 1, 3, 5, and 6 are to be manufactured using hob cutters having a 20° pressure angle. Shafts A and B are collinear. Determine

(a) the base circle radius of gear 1

(b) the dedendum circle diameter of gear 4

(c) the center-to-center distance between gears 1 and 5

(d) the contact ratio between gears 3 and 4

(e) the helix angle of gear 6

TABLE P6.11

Gear No.	No. of Teeth	Type of Gear	Hob Pitch (inches)
1	56	—	—
3	20	Straight spur, stub teeth	—
4	—	Straight spur, full-depth teeth	0.25
5	58	Helical spur	0.1875
6	28	—	—

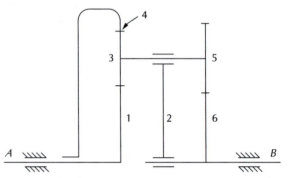

Figure P6.11

TABLE P6.10

Gear No.	No. of Teeth	Type of Gear	Hob Pitch (inches)
1	29	—	0.250
2	159	—	—
3	—	—	—
4	20	Straight spur	0.500
5	100	—	—

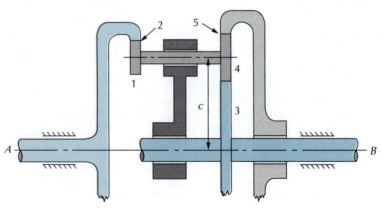

Figure P6.10

P6.12 The information in Table P6.12 pertains to the gear train shown in Figure P6.12:

TABLE P6.12

Gear No.	No. of Teeth	Type of Gear	Hob Pitch (inches)
1	60	—	—
2	—	Straight spur	0.500
3	100	—	—
4	29	—	0.250
5	159	—	—

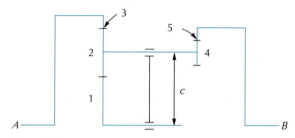

Figure P6.12

External gears 1, 2, and 4 are to be manufactured using hob cutters having a 20° pressure angle. All gears are to have full-depth teeth. Shafts A and B are collinear. Determine

(a) the base circle radius of gear 3
(b) the addendum circle diameter of gear 3
(c) the center-to-center distance, c, between gears 1 and 2
(d) the contact ratio between gears 2 and 3
(e) the helix angle of gear 4
(f) the pitch circle radius of gear 5

P6.13 The information in Table P6.13 pertains to the gear train shown in Figure P6.13:

All gears are straight spur with a 20° pressure angle. Gear 3 is an internal gear. Shafts A and B are collinear. Determine

(a) the numbers of teeth on gears 2 and 5
(b) the speed ratio of the gear train

TABLE P6.13

Gear No.	No. of Teeth	Diametral Pitch, P (in^{-1})	Addendum	Dedendum
2	—	6	$1.0/P$	$1.25/P$
3	120	—	$0.8/P$	$1.25/P$
4	32	8	—	—
5	—	—	—	—

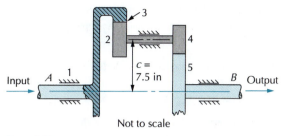

Figure P6.13

(c) the dedendum circle diameter of gear 3
(d) the contact ratio between gears 2 and 3
(e) the arc of action between gears 2 and 3

P6.14 The information in Table P6.14 pertains to the gear train shown in Figure P6.14.

TABLE P6.14

Gear No.	No. of Teeth	Type of Gear
1	18	Helical spur
2	—	—
3	62	—
4	20	—
5	64	Straight spur

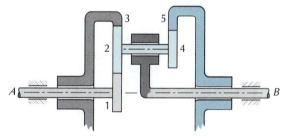

Figure P6.14

External gears 1, 2, and 4 were manufactured using a hob cutter having a 20° pressure angle and 0.500-in hob pitch. Gears 3 and 5 are internal. All gears have full-depth teeth. Shafts A and B are collinear. Determine

(a) the base circle radius of gear 4
(b) the center-to-center distance between gears 1 and 2
(c) the contact ratio between gears 4 and 5
(d) the circular pitch of gear 1
(e) the helix angle of gear 3
(f) the pitch circle diameter of gear 3
(g) the addendum circle diameter of gear 3

Planetary Gear Trains, Spur Gears

P6.15 Using the tabular method, derive the expressions for the component speeds for a type I planetary gear train as given in Figure 6.22.

P6.16 Using the tabular method, derive the expressions for the component speeds in a type J planetary gear train as given in Figure 6.22.

P6.17 Using the tabular method, derive the expressions for the component speeds in a type K planetary gear train as given in Figure 6.22.

P6.18 Determine the expression for the speed ratio of each of the following planetary gear trains (refer to Figure 6.22):
   (a) type D, $\omega_1 = 0$, $\omega_2 = $ input, $\omega_7 = $ output
   (b) type J, $\omega_7 = 0$, $\omega_1 = $ input, $\omega_2 = $ output
   (c) type C, $\omega_7 = 0$, $\omega_1 = $ input, $\omega_2 = $ output
   (d) type L, $\omega_1 = 0$, $\omega_2 = $ input, $\omega_6 = $ output

P6.19 In the gear train of Figure P6.19, shaft A rotates at 200 rpm and shaft B rotates at 300 rpm in the directions indicated. Determine the speed of shaft C and its direction of rotation.

$$N_2 = 35; \quad N_3 = 25; \quad N_4 = 14$$
$$N_5 = 46; \quad N_6 = 20; \quad N_7 = 16$$

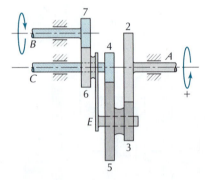

Figure P6.19

P6.20 For a planetary gear train type A (see Figure 6.22), the sun and ring gears have $N_1 = 20$ and $N_4 = 70$ teeth, respectively.
   (a) If the speed of the ring gear is 500 rpm CW, at what speed must the sun gear be driven if the crank is to rotate at
      (i) 75 rpm CW
      (ii) 75 rpm CCW
   (b) Determine the maximum number of equally spaced planet gears that can be employed in this application.

P6.21 Figure P6.21 shows an epicyclic gear train called *Ferguson's paradox*. Gears 2, 3, and 4 are loosely attached to their respective shafts while gear 5 is fixed.
   (a) Find the number of rotations the crank has to turn so that gear 4 rotates five times in the direction shown. How many times does gear 3 rotate, and in which direction?

   (b) Determine the number of turns that gear 2 makes about its own shaft.

$$N_2 = 15; \quad N_3 = 80; \quad N_4 = 81; \quad N_5 = 82$$

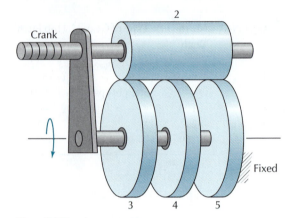

Figure P6.21

P6.22 In Figure P6.22, C and D represent brakes that can be used to stop the rotation of either arm E or gear 4, one at a time. Determine the speed and direction of shaft B when shaft A is rotating at 1000 rpm CW, while
   (a) brake C holds arm E fixed
   (b) brake D holds gear 4 fixed

$$N_2 = 90; N_3 = 32; N_5 = 94 ; N_6 = 28$$

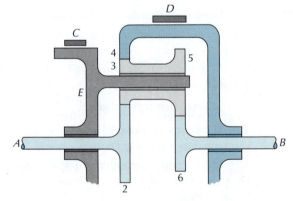

Figure P6.22

P6.23 For the gear train shown in Figure P6.23 (see next page), gear 2 has a module of 2.0 mm with 75 teeth; gear 5 has a module of 4.0 mm with 50 teeth; and gear 4 has 40 teeth. Determine the number of teeth on gear 3 and speed ratio of the gear train.

P6.24 Determine the three speed ratios, $e_{2/1}$, $e_{5/1}$, and $e_{8/1}$ of the planetary gear train shown in Figure P6.24. Employ
   (a) the tabular method, and obtain results with one table

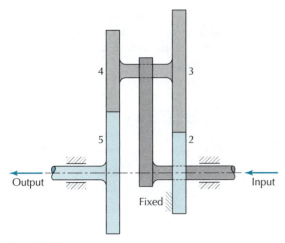

Figure P6.23

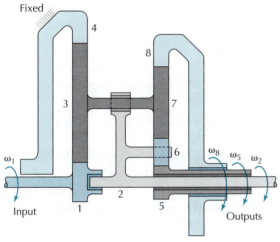

Figure P6.24

(b) the combination of the basic types in Figure 6.22

$N_1 = 15;\quad N_3 = 45;\quad N_4 = 105;\quad N_5 = 13$

$N_6 = 10;\quad N_7 = 27;\quad N_8 = 87$

P6.25 Figure P6.25 illustrates a two-speed transmission. The two output speeds are obtained by fixing either gear 3 or gear 7 by using brakes C and D, respectively. Determine the expressions for the two possible values of the speed ratio, $e_{8/1}$.

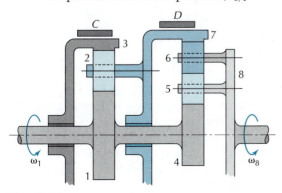

Figure P6.25

P6.26 In the gear train of Figure P6.26, gears 1 and 2 are cut from the same casting and revolve freely about the arms of the planet carrier C. Gears 5 and 6 are also cut from the same casting and are freely attached to shaft A. Gears 3 and 4 are keyed to shaft A. For an output speed of 500 rpm in the direction shown, determine the sense and speed of the input shaft rotation.

$N_1 = 42;\quad N_2 = 44;\quad N_3 = 40;\quad N_4 = 14$

$N_5 = 38;\quad N_6 = 150;\quad N_7 = 50$

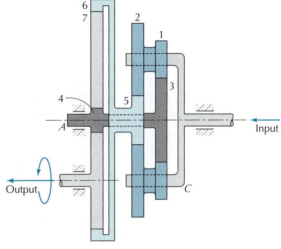

Figure P6.26

P6.27 In the operation of the gear train shown in Figure P6.27, ring gear 1 is fixed. Determine a relation involving N_6 and N_7 to obtain a speed ratio of -0.48

$N_1 = 100;\quad N_2 = 30;\quad N_3 = 10$

$N_4 = 85;\quad N_5 = 88;\quad N_8 = 22$

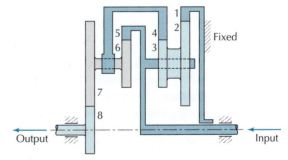

Figure P6.27

P6.28 Determine the output rotational speed, ω_2, for the gear train shown in Figure P6.28 by

(a) using the tabular method

(b) considering the gear train to be a combination of the basic types illustrated in Figure 6.22

$N_1 = 20$; $N_3 = 25$; $N_4 = 70$; $N_5 = 15$; $N_6 = 15$
$N_7 = 35$; $N_8 = 80$; $\omega_1 = 200\,\text{rpm}$ CW

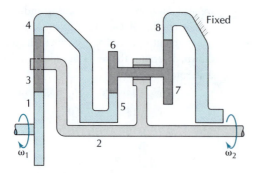

Figure P6.28

P6.29 In the planetary gear train shown on Figure P6.29, internal gear 4 turns at 300 rpm in the direction shown. The diametral pitch of gear 1 is 4 in^{-1}.

(a) Determine the diametral pitch of gear 3 and its pitch diameter.

(b) Determine the speed and direction of rotation of the output shaft.

(c) If the input power is 20 kW, and the mechanical efficiency is 95 percent, determine the external torque on gear 1.

$N_1 = 48$; $N_2 = 24$; $N_3 = 37$; $N_4 = 127$

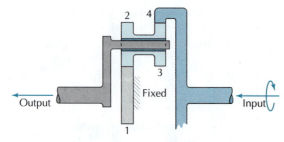

Figure P6.29

P6.30 Figure P6.30 illustrates a planetary gear train. Points A, C, D, and E are used to keep track of the angular positions of links 1, 3, 4, and 5, respectively. Starting from the position shown, gear 5 is rotated 270° counterclockwise (point E moves to point E'), and gear 1 is rotated 90° clockwise (point A moves to point A'). After completion of the motions of gears 1 and 5, do the following:

(a) Specify the angular position of the crank arm 2.

(b) Specify the angular position of planet gear 3 about its own axis.

(c) Specify the angular position of planet gear 4 about its own axis.

(d) Sketch the configuration of the gear train and indicate the positions of points C and D after the motions (indicate as C' and D', respectively).

$N_1 = 54$; $N_3 = 20$; $N_4 = 24$; $N_5 = 108$

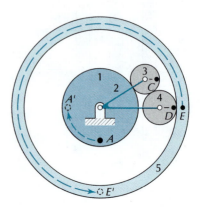

Figure P6.30

P6.31 For the planetary gear train shown in Figure P6.31, all gears have straight spur teeth, full-depth, 20° pressure angle, and module of 3 mm. Determine

(a) the speed and direction of rotation of the output shaft

(b) the relative rotational speed of the planet carrier with respect to gear 7

(c) the maximum number of equally spaced planet gears 7 that may be employed

$N_1 = 35$; $N_2 = 21$; $N_3 = 27$; $N_5 = 56$
$N_6 = 100$; $N_7 = 50$; $N_8 = 180$; $\omega_1 = 500$ rpm CCW

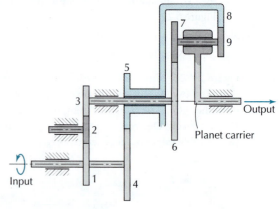

Figure P6.31

P6.32 Figure P6.32 illustrates a toothed planetary gear train. Pitch circles are given for the sun gear (link 1), and the planet gear (link 3). Points A, B, and C are used to keep track of the angular positions of the sun gear, the crank (link 2), and the planet gear, respectively.

(a) Specify the mobility of the gear train. Explain your answer in terms of the number of links, and the numbers and types of the kinematic pairs.

(b) Starting from the position shown, the sun gear is rotated 90 degrees clockwise (point *A* moves to point *A'*), while the crank is held fixed. Then, the crank is rotated 180 degrees counterclockwise (point *B* moves to point *B'*), while the sun gear is held fixed. After the motions of the sun gear and the crank, using the tabular method of planetary gear trains, do the following:

(i) Specify the angular position of the planet gear 3.

(ii) Make a sketch of the configuration of the planetary gear train. Indicate in your sketch the position of point *C* after completion of the motions described (indicate as *C'*).

$$N_1 = 20; \quad N_3 = 60$$

P6.33 In Figure P6.33, *C* and *D* represent brakes, which can be used to stop the rotation of either gear 4 or gear 8, one at a time. All gear teeth have a 20° pressure angle, stub form. Determine

(a) the maximum permissible number of equally spaced planet gears 3

(b) the maximum permissible number of teeth on planet gear 6

(c) the speed and direction of shaft *B* when shaft *A* is rotating at 800 rpm CW, while

(i) brake *C* holds gear 4 fixed

(ii) brake *D* holds gear 8 fixed

$$N_2 = 16; \quad N_4 = 50; \quad N_5 = 30; \quad N_8 = 80$$

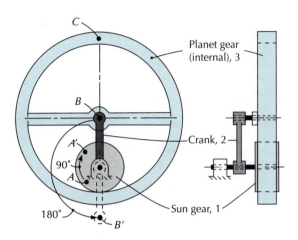

Figure P6.32

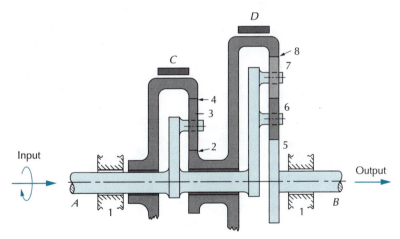

Figure P6.33

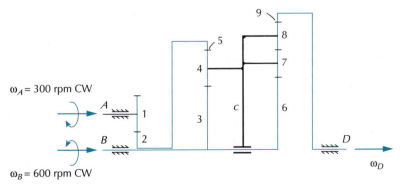

Figure P6.34

P6.34 The skeleton representation of the planetary gear train shown in Figure P6.34 includes collinear shafts B and D, and internal gears 5 and 9. All teeth are stub form, with a 20° pressure angle. Determine

(a) ω_D

(b) the rotational speed of planet gear 4 relative to that of crank arm C

(c) the maximum permissible number of equally spaced planet gears 4

$N_1 = 25;$ $N_2 = 25;$ $N_3 = 40$
$N_5 = 80;$ $N_6 = 50;$ $N_9 = 100$

P6.35 For the gear train in Example 6.5, determine

(a) the rotation required of the hand crank if $\Delta\theta_3 = 90°$ CCW

(b) the rotation required of the planet gear for a point on the cutter originally touching the pencil to come back into contact with the pencil

Planetary Gear Trains Incorporating Bevel Gears

P6.36 If an automobile is making a right-hand turn at 30 km/hr, determine the rotational speed of the planet carrier. The radius of curvature of the curve is 40 m to the center of the automobile, and the automobile tread (the distance between the two drive wheels) is 1.60 m. The outside diameter of the wheels is 850 mm. The differential gear train of the automobile is shown in Figure 6.35.

$N_1 = 30;$ $N_2 = 60;$ $N_3 = N_5 = 30;$ $N_4 = N_6 = 20$

P6.37 The driveshaft of an automobile is turning at 1500 rpm with output 1 on the garage floor and the right (i.e., output 2) jacked up. The bevel gear differential of Figure 6.35 is connected to the wheels. Determine the speed of output 2 and the speed of ring gear 2.

$N_1 = 15;$ $N_2 = 30;$ $N_3 = N_5 = 20;$ $N_4 = N_6 = 18$

P6.38 Refer to the Figure P6.38.

(a) Determine the rotation of the carrier when gear 1 makes 30 rotations clockwise and gear 2 makes 12 rotations counterclockwise. Also, determine the angular displacement of gear 3 about its own axis.

(b) If gear 1 rotates at f rpm and gear 2 at g rpm, determine the rotational speed of the carrier in terms of f and g.

$N_1 = 40;$ $N_2 = 30;$ $N_3 = 24$

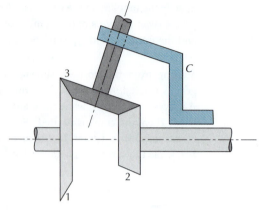

Figure P6.38

P6.39 For the planetary gear train shown in Figure P6.39, shafts B and C are collinear. Gears 4 and 5 are rigidly connected to each other. Gears 5, 6, 7, and 8 are plain bevel. Determine the value of

(a) ω_B, if $\omega_2 = 400$ rpm CW

(b) ω_2, if $\omega_B = 0$

$N_2 = 30;$ $N_3 = 20;$ $N_4 = 40$
$N_5 = 25;$ $N_7 = 55;$ $N_8 = 125$

P6.40 For the planetary gear train shown in Figure P6.40

- shafts A and B are collinear
- gears 2 and 3 are rigidly connected to each other
- gears 4 and 5 are rigidly connected to each other
- gears 3, 4, 5, and 6 are plain bevel

Determine the values of

(a) ω_B, if $\omega_1 = 400$ rpm CW

(b) ω_1, if $\omega_B = 125$ rpm CW

$N_1 = 30;$ $N_2 = 20;$ $N_3 = 50$
$N_4 = 15;$ $N_5 = 40;$ $N_6 = 25$

P6.41 Refer to Figure P6.41. Ring gear 1 is driven at 500 rpm in the direction shown, while gear 4 is held stationary.

(a) Determine the speed ratio of the gear train, and calculate the sense and rotational speed of carrier C.

(b) If 30 kW of power is obtained at the output, and the drive has a mechanical efficiency of 98 percent, calculate the torque required to hold gear 4 fixed

$N_1 = 500;$ $N_2 = 60;$ $N_3 = 60;$ $N_4 = 503$

P6.42 For an input speed $\omega_1 = 600$ rad/sec of the gear train shown in Figure P6.42, calculate the rotational speed of the planet carrier and the output rotational speed, ω_7

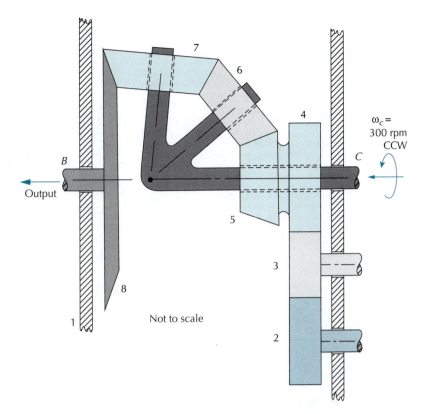

Figure P6.39

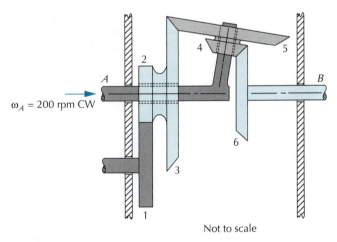

Figure P6.40

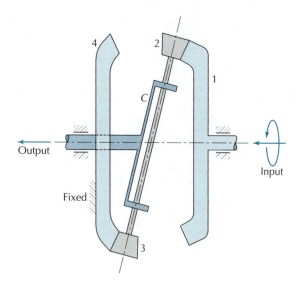

Figure P6.41

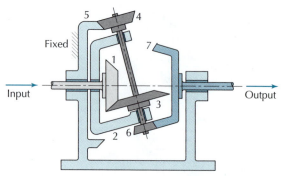

Figure P6.42

$$N_1 = 15; \quad N_3 = 42; \quad N_4 = 33$$
$$N_5 = 150; \quad N_6 = 32; \quad N_7 = 100$$

Planetary Gear Trains with Multiple Stages

P6.43 The three-stage planetary gear train illustrated in Figure P6.43 has the following two input rotational speeds: $\omega_1 = 100$ rpm CW, $\omega_{19} = 6000$ rpm CW. Determine the output rotational speed, ω_2, using the tabular method.

$N_1 = 20;$	$N_3 = 23;$	$N_4 = 21;$	$N_5 = 25$
$N_6 = 115;$	$N_7 = 43;$	$N_8 = 38;$	$N_9 = 17$
$N_{10} = 19;$	$N_{11} = 23;$	$N_{12} = 117;$	$N_{13} = 38$
$N_{14} = 25;$	$N_{15} = 23;$	$N_{16} = 28;$	$N_{17} = 110$
$N_{18} = 120;$	$N_{19} = 10$		

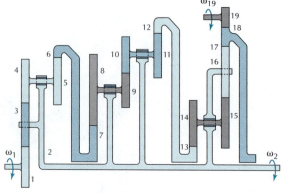

Figure P6.43

P6.44 A two-stage planetary gear train incorporates type A (see Figure 6.22) for the first stage and type H for the second. The ring gear of the first stage is connected to the crank of the second stage, while the crank of the first stage is connected to the ring gear 1 of the second stage. If ring gear 5 of the second stage is fixed, write down, but do not solve, the four governing equations for the motion in terms of $x_1, y_1, x_2,$ and y_2 according to the equations given in Figure 6.22, and determine the condition for the mechanism to function with a nonzero mobility.

P6.45 In the two-stage planetary gear train of Figure P6.45, gear 6 is fixed and crank arms C and D are attached to the output shaft. Gears 3 and 4 form a compound wheel that rotates freely about the output shaft. Determine

(a) the speed and direction of rotation of the output shaft

(b) the relative rotational speed of planet gear 2 with respect to crank arm C

(c) the maximum number of equally spaced planet gears 5 which may be employed

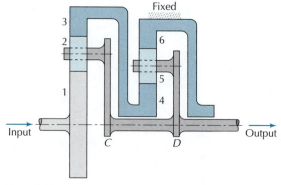

Figure P6.45

$N_1 = 20$; $N_3 = 72$; $N_4 = 24$;
$N_5 = 20$; $\omega_1 = 300$ rpm CW

Harmonic Drive

P6.46 For a harmonic drive similar to that shown in Figure 6.44, determine the rotational speed of the flex spline.

$$N_f = 100; \quad N_c = 102;$$
$$\omega_w = 70 \text{ rpm} \quad \text{CW}; \quad \omega_c = 0$$

Mechanical Efficiency

P6.47 Determine the torque required to hold down the base of the gearbox casing shown on Figure 6.48 while under the following operating conditions:

$\omega_i = 300$ rpm; $T_i = 200$ N-m;

$\eta = 0.95$; $e_{o/i} = -0.60$

P6.48 Refer to Figure P6.48. Determine the speed of rotation of shaft A needed to produce the 300 rpm output rotation of shaft B in the direction shown. If the mechanical efficiency of the gear train is 90 percent, determine the torque T_f that must be applied to the gearbox to convert a 20 kW input to output at shaft B.

$N_2 = 20$; $N_3 = 40$; $N_4 = 12$
$N_6 = 39$; $N_7 = 16$; $N_8 = 60$

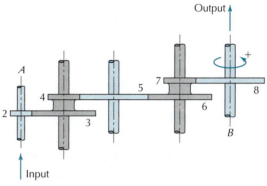

Figure P6.48

CHAPTER 7

CAMS

7.1 INTRODUCTION

In this chapter, the analysis and design of cam mechanisms is presented. Definitions and examples of various planar and spatial cam mechanisms are provided. Graphical and analytical methods are covered for designing planar disc cam mechanisms that determine the shape of the disc cam for a required output motion of the mechanism. The program entitled Cam Design is presented, along with instructions for its use, to help with the design of disc cam mechanisms. Two typical methods for the manufacture of cams are covered.

A *cam* is defined as a link of a mechanism that transmits motion by direct contact with another link, called the *follower*. Generally, the cam provides the input motion to the mechanism, while the follower provides the output motion. The input and output motions may be either translational or rotational. The output motion of the follower can be used to drive other linkages of another machine. Some common types of cam mechanisms are illustrated in Figure 7.1, where the cam is colored blue, and the follower is colored gray. Figures 7.1(a) and 7.1(b) illustrate planar cam mechanisms where all motions are restricted to a single plane, and Figures 7.1(c) and 7.1(d) illustrate spatial cam mechanisms, where motions occurs in 3D space. Cam mechanisms are often able to provide the required output motions with fewer links than by other mechanisms. Cams mechanisms are employed in a vast array of mechanical devices; some examples are presented in Section 1.2.4, and additional applications are described below.

A *ballpoint pen*, as shown in Figure 7.2, incorporates two cam mechanisms that allows the user to move the writing tip from the exposed position to the retracted position, and back again, by repeatedly depressing the pushbutton. While in either position, the writing tip is held firmly in place without external interaction from the user. [Video7.2] provides an animation of the ballpoint pen that demonstrates its operation. Through the translucent wall of the barrel we see the two cams mechanisms. The cams are labeled as cam A and cam B. They are cylindrical in shape and concentric. At first glance, the operation of this device may appear complicated because it consists of spatial mechanisms. Therefore, to better illustrate its operation, a planar version is provided in Figure 7.3. Here, the two cylindrically shaped cam mechanisms are flattened out and laid side by side. The rotations of the spatial cam B shown in Figure 7.2 are now lateral translations of the planar cam B shown in Figure 7.3. As easily seen in the planar representation, cam A is connected to the pushbutton. Follower A is rigidly connected to cam B. Follower B is the barrel of the pen and the base link. The transport arm is connected to the writing tip. The planar system shown is limited in operation because after just a few cycles follower A moves off to one side and all of the required engagements between the components no longer exist. However, in the actual spatial system, because the motion is rotational, the device can operate indefinitely since the same surfaces of the cams and their followers repeatedly come into contact. [Video7.3] provides an animation of the planar representation of the ballpoint pen. When the user depresses the pushbutton, cam

Video 7.2
Ballpoint Pen

Video 7.3
Ballpoint Pen Mechanism—
Planar Representation

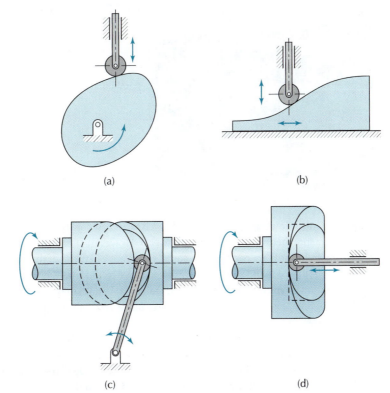

Figure 7.1 Common types of cams: (a) Disc, (b) Wedge, (c) Cylindrical, (d) End or face.

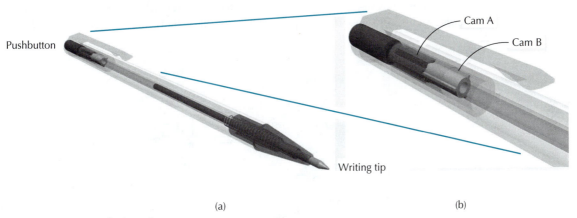

Figure 7.2 Ballpoint pen [Video 7.2].

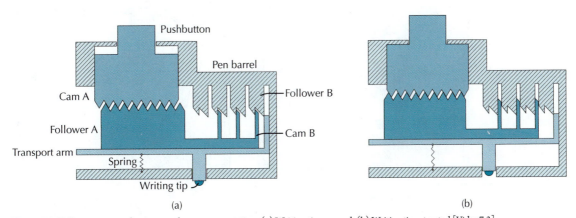

Figure 7.3 Ballpoint pen mechanism — planar representation: (a) Writing tip exposed, (b) Writing tip retracted [Video 7.3].

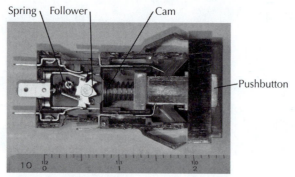

Figure 7.4 Toggle mechanism electric switch.

A engages with follower A. Follower A and cam B are at first constrained to move vertically because of the interaction between cam B and follower B. However, after a certain amount of depression, cam B and follower B are no longer in contact, and now, because of the interaction of cam A and follower A, follower A and cam B are forced to move to the left. Then, when the applied force to the pushbutton is removed, the spring causes follower A and cam B to move laterally and vertically. At the same time, cam A and the transport arm move vertically and the writing tip retracts to the position shown in Figure 7.3(b).

Other common uses of cam mechanisms are in electrical switches. Two applications are described here. First, the *toggle mechanism electric switch* is shown in Figure 7.4. A schematic representation of this switch is illustrated in Figure 7.5(a). The follower may toggle from one side to the other about the base pivot. In one of the positions, the electrical circuit (not shown) is closed (allowing electric flow) and in the other position the circuit is open (preventing electric flow). Starting from the original position shown in Figure 7.5(a) and then depressing the pushbutton causes the cam to contact one of the two recesses on the follower as illustrated in Figure 7.5(b). Depressing the pushbutton further causes the follower to toggle to the other side as illustrated in Figure 7.5(c). The spring keeps

Video 7.5

Toggle Mechanism Electric Switch

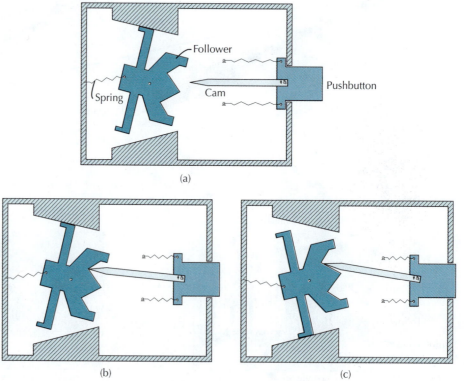

Figure 7.5 Toggle switch [Video 7.5].

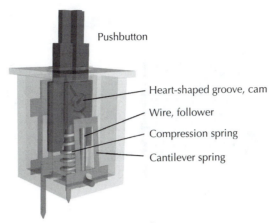

Figure 7.6   Heart switch [Video 7.6].

the follower in either the on position or off position after the pushbutton has retracted. When the pushbutton is depressed again in the next cycle, the cam enters the other recess on the follower and then the switch returns to the original position. Second, the *heart switch* is shown in Figure 7.6. It is operated by depressing the pushbutton against the action of the compression spring. The switch is on when the pushbutton stays in its lower position. The cam is the heart-shaped groove that is recessed in the side surface of the pushbutton. The follower consists of a wire that has a right-angle bend at one end. During operation of the switch, the end of the follower is constrained to move along the track formed by the groove. The cantilever spring ensures that the follower remains in the groove, and its tip remains in contact with the bottom surface of the groove. Inclines and steps on the bottom surface of the groove play a key role in the operation of the switch. When the pushbutton approaches the end of its travel (at either the on position or the off position), the follower drops off a step so that the pushbutton is held in position, without external interaction from the user.

Figure 7.7 illustrates a *door knob assembly* that incorporates a cam mechanism. In this application, the cam is rigidly connected to the door knob, and the follower is rigidly connected to the bolt. When the door knob is turned in either direction, the follower translates in the latch housing. A 3D animation of the door knob assembly in [**Video 7.7**] demonstrates its operation.

Cam-based mechanisms are often used within rotating wheels, to limit or brake motion in certain circumstances. Figure 7.8 shows a *locking caster wheel* that incorporates a cam mechanism. The operation of this mechanism is best understood by viewing the animation of [**Video 7.8**]. A circular side plate is placed beside the caster wheel and is rigidly connected to the cam. The side plate is allowed to move vertically but prevented from rotating with the caster wheel. The follower is on the inner cylindrical surface of the caster wheel and contains eight recesses around the inner surface. When the caster wheel is on a smooth floor, the side plate is pushed up vertically by the floor surface so that the cam and follower do not contact, and the wheel is allowed to turn freely about its center and roll on the floor (Figure 7.8(b)). However, if the side plate is aligned with a straight groove in the floor, the side plate moves vertically downward under the action of a spring. The side plate moves into the groove relative to the caster wheel while the cam enters one of the recesses in the follower (Figure 7.8(c)). This causes the wheel to lock, preventing it

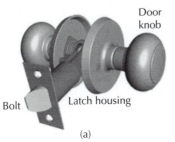

(a)

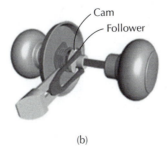

(b)

Figure 7.7   Door knob assembly [Video7.7].

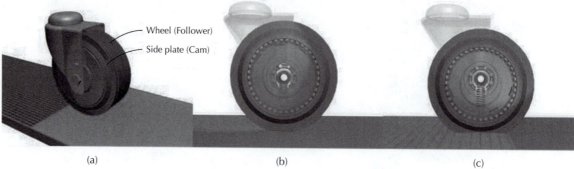

Wheel (Follower)
Side plate (Cam)

(a)   (b)   (c)

Figure 7.8 (a) Caster wheel, (b) Unlocked configuration, (c) Locked configuration [Video 7.8].

from rolling. Locking caster wheels are often employed on shopping carts to fix their location with respect to the surface of a movator.

The above examples illustrate the wide scope of systems that can be classified as cam mechanisms. Although some of the above examples may appear complex at first glance, they employ relatively simple mechanisms when considering the motions they provide. The remainder of this chapter deals primarily with planar disc cam mechanisms, such as shown in Figure 7.1(a), as these are best suited for teaching the basics of the subject. Graphical and analytical methods of determining the shapes of disc cams are presented.

7.2 DISC CAM MECHANISM NOMENCLATURE

Disc cam mechanisms may be classified according to their type of follower. Figure 7.9 shows six examples. Followers are classified according to their shape, location relative to the cam, and whether

Translating   Pivoting

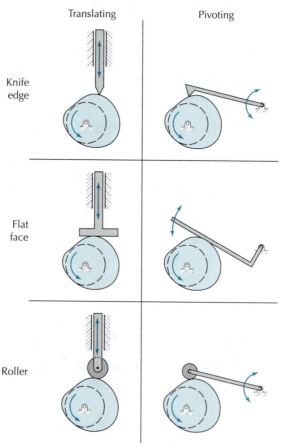

Knife
edge

Flat
face

Roller

Figure 7.9 Disc cam mechanisms [Video 7.9].

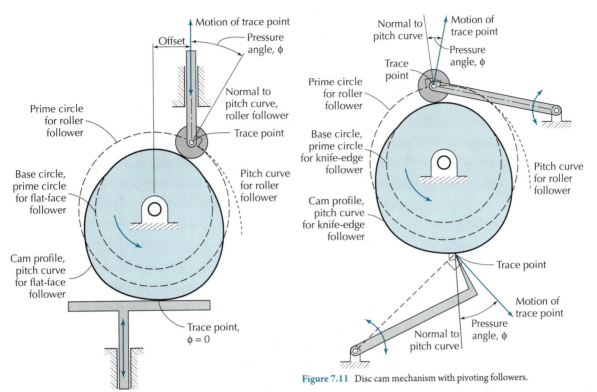

Figure 7.10 Disc cam mechanism with translating followers.

Figure 7.11 Disc cam mechanism with pivoting followers.

the follower motion is translational or rotational. Three of the followers shown in Figure 7.9 execute linear translation, and three execute rotational motion.

Common definitions associated with disc cam mechanisms are

- *Cam profile*: the surface of the cam contacted by the follower.
- *Trace point*: the point of contact of the knife-edge follower, or the center of the roller follower.
- *Pitch curve*: the path of the trace point with respect to the cam.
- *Base circle*: the smallest circle tangent to the cam profile, with its center on the axis of the camshaft.
- *Prime circle*: the smallest circle tangent to the pitch curve, with its center on the axis of the camshaft.
- *Pressure angle*: the angle between the normal to the pitch curve and the direction of motion of the trace point.

This nomenclature for disc cam mechanisms having translating and pivoting followers is illustrated in Figures 7.10 and 7.11. Springs (not shown) between the follower and base link are employed to keep the follower in contact with the cam profile.

7.3 PRESSURE ANGLE

With the exception of a disc cam mechanism with a flat-face translating follower, pressure angle changes as the mechanism moves. Figures 7.10 and 7.11 illustrate pressure angles in mechanisms for the given configurations.

The value of the pressure angle is a critical parameter in design of disc cam mechanisms and should be kept as small as possible. The practical maximum is 30°. Even if the follower and disc cam were a frictionless kinematic pair, there is still a component of unwanted side thrust. This can be illustrated by a free body diagram shown in Figure 7.12. Figure 7.12(a) shows a disc cam and roller follower. Figure 7.12(b) shows the roller with the cam and guide of the follower removed and replaced by the equivalent forces. Larger values of pressure angle cause an increase in the unwanted side thrust on the follower stem. For excessive pressure angles, the mechanism will tend to bind.

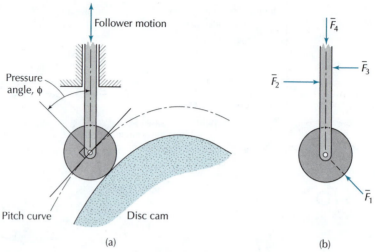

Figure 7.12 Roller follower: (a) Disc cam and follower, (b) Free-body diagram of follower.

Some means by which the pressure angle may be reduced are

- increasing the diameter of the base circle
- increasing the diameter of the roller follower
- changing the offset of the follower (see Figure 7.10)
- changing the motion of the follower

An illustration of the application of the above methods is provided later in this chapter by means of an example.

7.4 THE DISPLACEMENT DIAGRAM

In designing a disc cam, one starts by specifying the required motion of the follower throughout a complete cycle. It may be expressed in the form

$$s = f(\theta); \quad 0 \le \theta \le 2\pi \tag{7.4-1}$$

A *displacement diagram* is a plot of follower displacement versus cam rotation. The motions of the follower may be grouped into the following three categories:

Rise: the follower is moving away from the center of the disc cam.
Dwell: the follower is at rest.
Fall or *return*: the follower is moving toward the center of the disc cam.

Figure 7.13 shows an example of a displacement diagram. It depicts portions of rise, dwell, and return. Rotation of the cam is usually specified in degrees. Displacement of the follower is plotted in terms of the amount of linear movement or rotation, for translating and pivoting cam mechanisms, respectively.

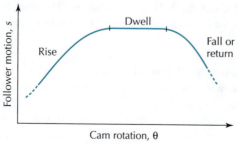

Figure 7.13 Typical displacement function.

EXAMPLE 7.1 DETERMINATION OF A DISPLACEMENT DIAGRAM

A disc cam mechanism is required for an automated screw machine. Starting the position shown in Figure 7.14, the part mounted on the platform (i.e., the follower) is to move 3.0 cm to the right in 0.6 sec, then remain stationary for 1.1 sec to allow a screw to be added to hole A, then move an additional 2.0 cm to the right in 0.4 sec, followed be another interval of 1.1 sec when the platform remains stationary while a screw is added to hole B, and finally return to the starting position in 0.8 sec. The compression spring ensures that contact is maintained between the cam and the follower. Determine

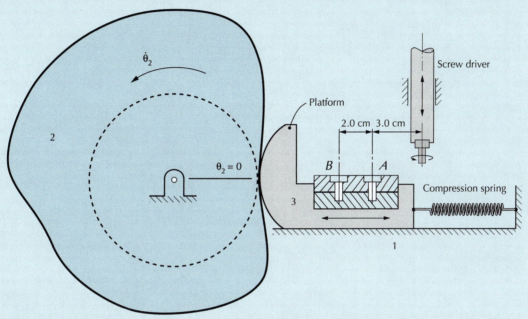

Figure 7.14 Screw machine.

(a)   the required rotational speed of the disc cam
(b)   plot of a suitable displacement diagram.

SOLUTION

The required motions of the platform are summarized in Table 7.1.
(a) Period of time required to carry out a cycle

$$\tau = T_1 + T_2 + T_3 + T_4 + T_5 = 0.6 + 1.1 + 0.4 + 1.1 + 0.8 = 4.0 \text{ sec}$$

TABLE 7.1 REQUIRED PLATFORM MOTIONS

Interval i	Description of Motion of Platform	Lift or Return L_i (cm)	Time T_i (sec)
1	Move platform to right to align hole A with screwdriver	3.0 (lift)	0.6
2	Keep platform stationary	0	1.1
3	Additional motion of platform to right to align hole B with screwdriver	2.0 (lift)	0.4
4	Keep platform stationary	0	1.1
5	Move platform to left, back to starting position	3.0 + 2.0 = 5.0 (return)	0.8

(Continued)

EXAMPLE 7.1 Continued

This period τ corresponds to one rotation of the cam. Therefore, the rotational speed of the cam is

$$\dot{\theta}_2 = \frac{1 \text{ rev}}{\tau} = \frac{1.0 \text{ rev}}{4.0 \text{ sec}} \times 60 \frac{\text{sec}}{\text{min}} = 15 \text{ rpm}$$

(b) The cam rotation for each interval may be calculated using

$$\beta_i = \frac{T_i}{\tau} 360°$$

and therefore

$$\beta_1 = \frac{T_i}{\tau} 360° = \frac{0.6}{4.0} 360° = 54°; \quad \beta_2 = 99°; \quad \beta_3 = 36°; \quad \beta_4 = 99°; \quad \beta_5 = 72°$$

TABLE 7.2 FOLLOWER MOTION FOR EXAMPLE 7.1

Parameter→ Interval↓	Initial Angle for Interval (degrees)	Final Angle for Interval (degrees)	Type of Follower Motion	Amount Follower Motion (cm)
1	0	54	Rise	3.0
2	54	153	Dwell	0
3	153	189	Rise	2.0
4	189	288	Dwell	0
5	288	360	Return	−5.0

Table 7.2 provides a summary of the required motions of the follower and the corresponding rotations of the cam. The corresponding displacement diagram is illustrated in Figure 7.15. Intervals 1, 3, and 5 correspond to the two lifts and one return of the follower. The functions for these intervals are drawn continuous with the start and finish of the dwell intervals 2 and 4. In forthcoming sections, a variety of the forms of the functions for lift and return will be considered and the role that they play on the dynamics of the cam mechanism. Also, both graphical and analytical methods will be presented to determine the profile of the disc cam that give the required motions illustrated in the displacement diagram.

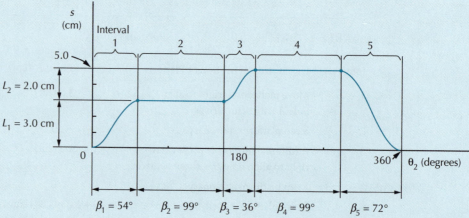

Figure 7.15 Displacement diagram.

7.5 TYPES OF FOLLOWER MOTIONS

A displacement diagram is generally made up of portions of standardized functions. Some common functions adopted for rise and fall portions of the displacement diagram are given below; all of which are of the form

$$s^* = g(\theta^*), \quad 0 \leq \theta^* \leq \beta \tag{7.5-1}$$

where s^* = displacement of the follower in the current portion

$\quad \theta^*$ = rotation of the cam in the current portion

$\quad \beta$ = total rotation in the current portion

Figure 7.16 illustrates application of a standardized function to a displacement diagram. To compare various types of motion, we will consider dwell of the follower just prior to, and just after, its motion. Also, all motions will start at $\theta^* = 0$ and end at $\theta^* = \beta$. The follower lift during the current portion is L, and the constant rotational speed of the cam is $\dot{\theta}$.

In the design of cam mechanisms for high-speed operation, time derivative quantities of the motions of the follower are important. The second time derivative of the displacement is the *acceleration*, which is proportional to the force applied to produce the motion. It is generally desirable to determine extremum values of the peak accelerations, and try to reduce their absolute values. It is also worthwhile to reduce the time rate of change of acceleration, referred to as the *jerk*. Higher values of jerk often result in increased noise levels under operating conditions.

7.5.1 Uniform Motion

Uniform motion is equivalent to *constant velocity*. Figure 7.17 shows a portion of a displacement diagram that incorporates uniform motion. The equations for follower motion during lift and return are given in the same figure.

Expressions for the velocity, acceleration, and jerk during uniform motion are

$$\dot{s}^* = \frac{L}{\beta}\dot{\theta}^* = \text{constant}; \quad \ddot{s}^* = 0; \quad \dddot{s}^* = 0, 0 < \theta^* < \beta \tag{7.5-2}$$

The acceleration and jerk are expressed as zero. However, it must be kept in mind that when $\theta^* = 0$ and $\theta^* = \beta$, acceleration and the jerk are infinite. This is because at the start and end of the motion there is a step change of velocity. In theory, this would correspond to an infinite force and jerk. Because of this undesirable characteristic, uniform motion of the follower without proper blending with the adjoining motions should be avoided in high-speed applications to prevent excessive accelerations.

To improve the dynamic characteristics, constant slope in the displacement profile is sometimes combined with other shapes at its ends to smooth out the sharp corners. As a target, infinite

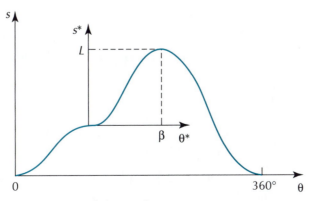

Figure 7.16 Typical displacement function.

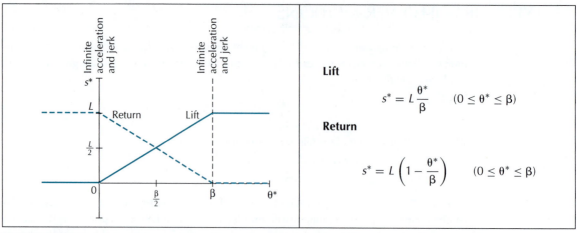

Figure 7.17 Uniform motion.

accelerations may be eliminated. Section 7.5.3 presents a cam motion that incorporates a portion of uniform motion combined with smoothing segments at the beginning and end of the motion.

7.5.2 Parabolic Motion

Parabolic motion is also referred to as *constant acceleration*. Figure 7.18 shows a portion of a displacement diagram that incorporates parabolic motion for the lift and return. At the beginning and end of the motion there is a step change of acceleration and an infinite value of jerk. Equations for follower displacement are given in the same figure.

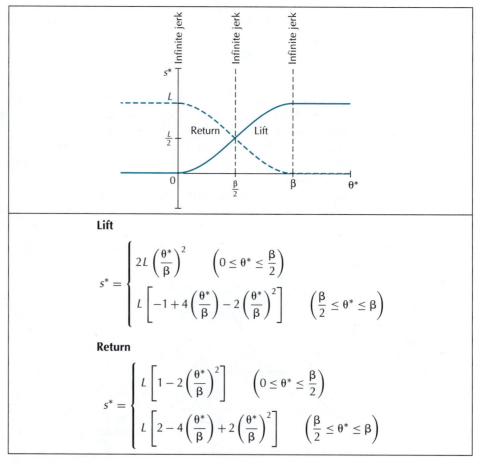

Lift

$$s^* = \begin{cases} 2L\left(\dfrac{\theta^*}{\beta}\right)^2 & \left(0 \le \theta^* \le \dfrac{\beta}{2}\right) \\[2ex] L\left[-1 + 4\left(\dfrac{\theta^*}{\beta}\right) - 2\left(\dfrac{\theta^*}{\beta}\right)^2\right] & \left(\dfrac{\beta}{2} \le \theta^* \le \beta\right) \end{cases}$$

Return

$$s^* = \begin{cases} L\left[1 - 2\left(\dfrac{\theta^*}{\beta}\right)^2\right] & \left(0 \le \theta^* \le \dfrac{\beta}{2}\right) \\[2ex] L\left[2 - 4\left(\dfrac{\theta^*}{\beta}\right) + 2\left(\dfrac{\theta^*}{\beta}\right)^2\right] & \left(\dfrac{\beta}{2} \le \theta^* \le \beta\right) \end{cases}$$

Figure 7.18 Parabolic motion.

The displacement curve shown is made up of two parabolas. For a lift motion, the first half (rotation 0 to $\beta/2$) the follower has a constant positive acceleration, whereas the second half (rotation $\beta/2$ to β) has a constant negative acceleration (i.e., a deceleration). At $\theta = \beta/2$, the displacement and the slope of the displacement curve are matched.

7.5.3 Modified Parabolic Motion

Modified parabolic motion is a combination of uniform motion and parabolic motion. Figure 7.19 shows a portion of a displacement diagram that incorporates modified parabolic motion for the lift and return. At the beginning and end of the rotation of the cam, the follower motion is parabolic. In the central region, the follower motion is uniform. At the transition points C_1 and C_2, between the uniform and parabolic motions, both the displacement and the slope of the displacement curve are matched. Thus, the acceleration remains finite at the transition points. Expressions for follower displacement are included in Figure 7.19.

The expressions for displacement given in Figure 7.19 include the parameter γ. It may take on values

$$1 < \gamma < 2$$

Cases where $y = 1$ and $y = 2$ correspond to uniform motion and parabolic motion respectively

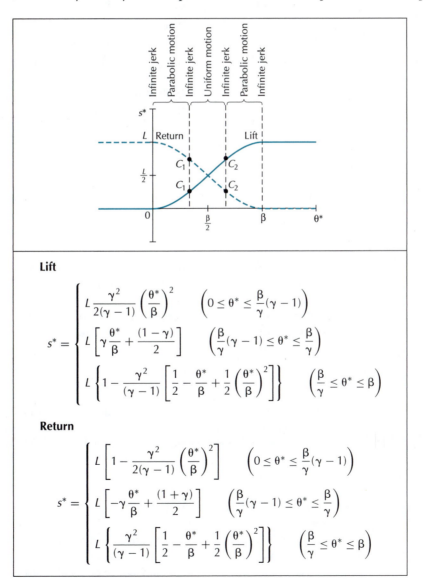

Lift

$$s^* = \begin{cases} L \dfrac{\gamma^2}{2(\gamma - 1)} \left(\dfrac{\theta^*}{\beta} \right)^2 & \left(0 \le \theta^* \le \dfrac{\beta}{\gamma}(\gamma - 1) \right) \\[3mm] L \left[\gamma \dfrac{\theta^*}{\beta} + \dfrac{(1 - \gamma)}{2} \right] & \left(\dfrac{\beta}{\gamma}(\gamma - 1) \le \theta^* \le \dfrac{\beta}{\gamma} \right) \\[3mm] L \left\{ 1 - \dfrac{\gamma^2}{(\gamma - 1)} \left[\dfrac{1}{2} - \dfrac{\theta^*}{\beta} + \dfrac{1}{2} \left(\dfrac{\theta^*}{\beta} \right)^2 \right] \right\} & \left(\dfrac{\beta}{\gamma} \le \theta^* \le \beta \right) \end{cases}$$

Return

$$s^* = \begin{cases} L \left[1 - \dfrac{\gamma^2}{2(\gamma - 1)} \left(\dfrac{\theta^*}{\beta} \right)^2 \right] & \left(0 \le \theta^* \le \dfrac{\beta}{\gamma}(\gamma - 1) \right) \\[3mm] L \left[-\gamma \dfrac{\theta^*}{\beta} + \dfrac{(1 + \gamma)}{2} \right] & \left(\dfrac{\beta}{\gamma}(\gamma - 1) \le \theta^* \le \dfrac{\beta}{\gamma} \right) \\[3mm] L \left\{ \dfrac{\gamma^2}{(\gamma - 1)} \left[\dfrac{1}{2} - \dfrac{\theta^*}{\beta} + \dfrac{1}{2} \left(\dfrac{\theta^*}{\beta} \right)^2 \right] \right\} & \left(\dfrac{\beta}{\gamma} \le \theta^* \le \beta \right) \end{cases}$$

Figure 7.19 Modified parabolic motion.

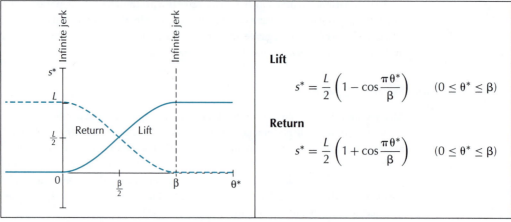

Figure 7.20 Harmonic motion.

Lift

$$s^* = \frac{L}{2}\left(1 - \cos\frac{\pi\theta^*}{\beta}\right) \qquad (0 \le \theta^* \le \beta)$$

Return

$$s^* = \frac{L}{2}\left(1 + \cos\frac{\pi\theta^*}{\beta}\right) \qquad (0 \le \theta^* \le \beta)$$

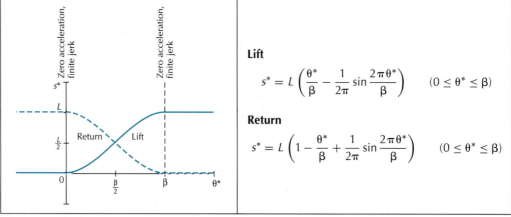

Figure 7.21 Cycloidal motion.

Lift

$$s^* = L\left(\frac{\theta^*}{\beta} - \frac{1}{2\pi}\sin\frac{2\pi\theta^*}{\beta}\right) \qquad (0 \le \theta^* \le \beta)$$

Return

$$s^* = L\left(1 - \frac{\theta^*}{\beta} + \frac{1}{2\pi}\sin\frac{2\pi\theta^*}{\beta}\right) \qquad (0 \le \theta^* \le \beta)$$

7.5.4 Harmonic Motion

Harmonic motion incorporates a portion of a sine wave. A portion of a displacement diagram with harmonic motion is given in Figure 7.20. The equations for lift and return motions in this instance are listed in the same figure.

Acceleration has a finite value at the beginning and end of harmonic motion. Because there is a step change of acceleration, there is infinite jerk at these locations. Other than the start and end points, jerk remains finite.

7.5.5 Cycloidal Motion

Figure 7.21 shows a portion of a displacement diagram that incorporates *cycloidal motion*. The displacement functions for lift and return are also provided. For this motion, the expression for acceleration can be easily shown to be zero at the start and finish. As a result, jerk remains finite throughout the entire motion, including the start and finish.

7.6 COMPARISON OF FOLLOWER MOTIONS

A comparison of the four types of motion presented in Section 7.5 is given in Figure 7.22. The plot for each is nondimensional, that is, results are plotted as a unit lift taking place in one unit of time. For the displacements, only the plot of uniform motion can be immediately identified. However,

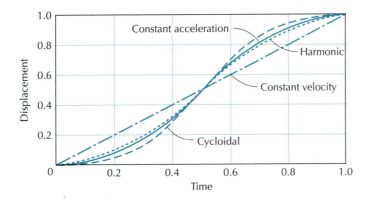

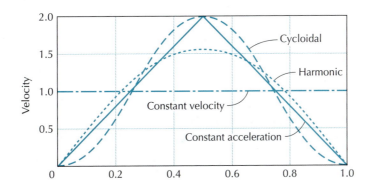

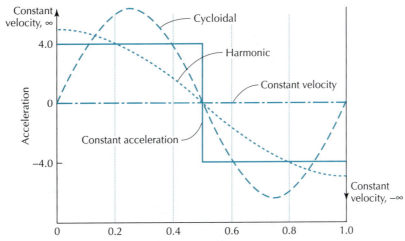

Figure 7.22 Comparison of displacement, velocity, and acceleration for follower motions.

there are significant differences in velocities, accelerations, and jerks. Jerks for the parabolic motion are infinite at the beginning, middle, and end of the motion cycle, and zero everywhere else.

7.7 DETERMINATION OF DISC CAM PROFILE

The basic problem of cam design is to find the cam profile that will produce a desired follower motion. The cam profile depends on the required follower motion, the base circle diameter of the cam, the follower type and its dimensions, and the position of the follower relative to the cam.

Cam profiles may be generated graphically or by evaluating analytical expressions for the particular type of follower under consideration. Both of these techniques will be considered in this section.

7.7.1 Graphical Determination of Disc Cam Profile

To graphically construct a disc cam profile, the following general steps are required:

1. Specify the displacement diagram, base circle diameter, and follower (type and dimensions).
2. Consider the cam as fixed, and move the other links with respect to the cam, that is, invert the mechanism. In order to obtain the same relative motions between the links, the base link and follower are rotated about the cam in the *opposite direction* to cam rotation. The follower is then drawn in numerous positions with respect to the cam, throughout a cycle.
3. Draw the cam profile inside the envelope of the follower positions.

Figure 7.23 illustrates this procedure for six types of followers.

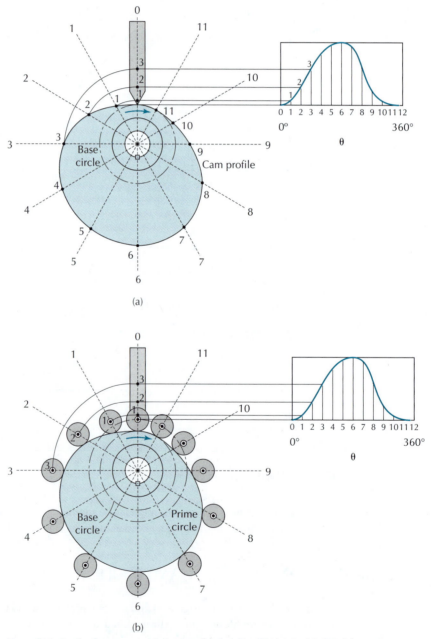

(a)

(b)

Figure 7.23 Graphical construction of a disc cam: (a) Translating knife-edge follower, (b) Translating roller follower, zero offset.

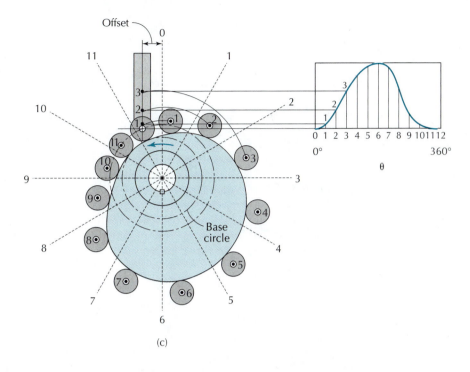

(c)

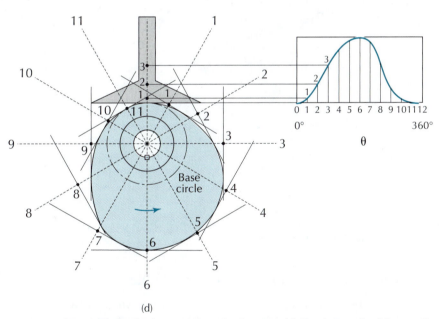

(d)

Figure 7.23 (*Continued*) Graphical construction of a disc cam: (c) Translating roller follower, offset, (d) Translating flat-face follower.

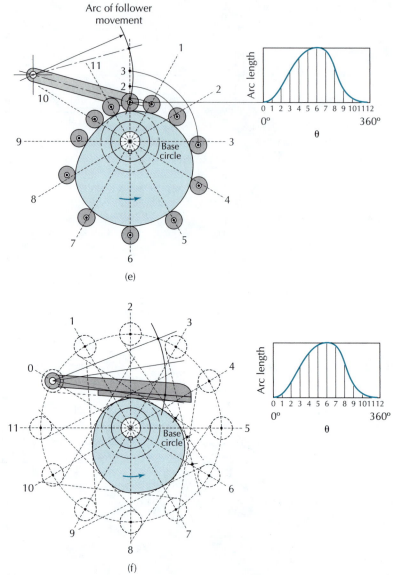

Figure 7.23 (*Continued*) Graphical construction of a disc cam: (c) Pivoting roller follower, (d) Pivoting flat-face follower.

7.7.2 Analytical Determination of Disc Cam Profile

The points on a disc cam may be determined analytically, using the governing mathematical equations presented in reference [9] and that are summarized in Figure 7.24. In these equations, quantities x_c and y_c are the coordinates of a point of contact between the cam and the follower for a given rotation of the cam. Similar to the graphical procedure used in drawing the cam in one position, it is necessary to rotate these calculated points around the center of rotation of the cam, in a direction *opposite* to that of the operational direction of rotation of the cam. The use of these equations is illustrated by the following example.

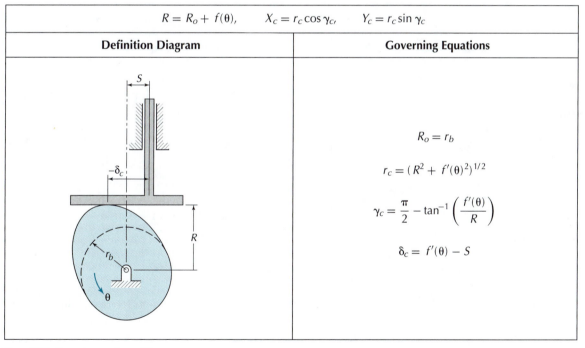

$$R = R_o + f(\theta), \qquad X_c = r_c \cos \gamma_c, \qquad Y_c = r_c \sin \gamma_c$$

Definition Diagram	Governing Equations
	$$R_o = \left(r_b^2 - S^2\right)^{1/2}$$ $$r_c = (S^2 + R^2)^{1/2}$$ $$\gamma_c = \frac{\pi}{2} - \tan^{-1}(S/R)$$ $$\phi = \frac{\pi}{2} - \gamma_c - \tan^{-1}\left(\frac{R f'(\theta)}{r_c^2}\right)$$

$$R = R_o + f(\theta), \qquad X_c = r_c \cos \gamma_c, \qquad Y_c = r_c \sin \gamma_c$$

Definition Diagram	Governing Equations
	$$R_o = r_b$$ $$r_c = (R^2 + f'(\theta)^2)^{1/2}$$ $$\gamma_c = \frac{\pi}{2} - \tan^{-1}\left(\frac{f'(\theta)}{R}\right)$$ $$\delta_c = f'(\theta) - S$$

Figure 7.24 Governing equations of disc cams.

$$R = R_o + f(\theta), \qquad X_c = r_c \cos \gamma_c, \qquad Y_c = r_c \sin \gamma_c$$

Definition Diagram	Governing Equations
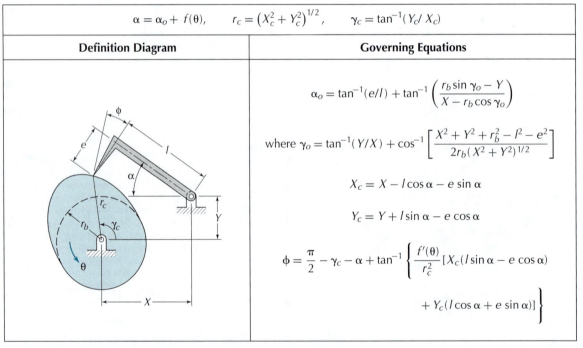	$$R_o = [(r_b + r_r)^2 - S^2]^{1/2}$$ $$r_c = [(R - r_r \cos \phi)^2 + (S + r_r \sin \phi)^2]^{1/2}$$ $$\gamma_c = \frac{\pi}{2} - \tan^{-1}\left(\frac{S + r_r \sin \phi}{R - r_r \cos \phi}\right)$$ where $\phi = \tan^{-1}\left(\dfrac{f'(\theta) - S}{R}\right)$

$$\alpha = \alpha_o + f(\theta), \qquad r_c = \left(X_c^2 + Y_c^2\right)^{1/2}, \qquad \gamma_c = \tan^{-1}(Y_c / X_c)$$

Definition Diagram	Governing Equations
	$$\alpha_o = \tan^{-1}(e/l) + \tan^{-1}\left(\frac{r_b \sin \gamma_o - Y}{X - r_b \cos \gamma_o}\right)$$ where $\gamma_o = \tan^{-1}(Y/X) + \cos^{-1}\left[\dfrac{X^2 + Y^2 + r_b^2 - l^2 - e^2}{2r_b(X^2 + Y^2)^{1/2}}\right]$ $$X_c = X - l\cos\alpha - e\sin\alpha$$ $$Y_c = Y + l\sin\alpha - e\cos\alpha$$ $$\phi = \frac{\pi}{2} - \gamma_c - \alpha + \tan^{-1}\left\{\frac{f'(\theta)}{r_c^2}[X_c(l\sin\alpha - e\cos\alpha)\right.$$ $$\left. + Y_c(l\cos\alpha + e\sin\alpha)]\vphantom{\frac{f'}{r_c^2}}\right\}$$

Figure 7.24 (*Continued*)

$$\alpha = \alpha_o + f(\theta), \qquad r_c = \left(X_c^2 + Y_c^2\right)^{1/2} \qquad \gamma_c = \tan^{-1}(Y_c/X_c)$$

Definition Diagram	Governing Equations
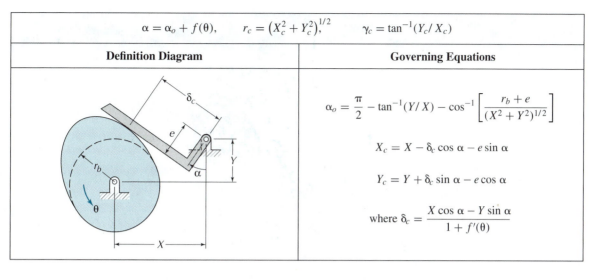	$$\alpha_o = \frac{\pi}{2} - \tan^{-1}(Y/X) - \cos^{-1}\left[\frac{r_b + e}{(X^2 + Y^2)^{1/2}}\right]$$ $$X_c = X - \delta_c \cos\alpha - e\sin\alpha$$ $$Y_c = Y + \delta_c \sin\alpha - e\cos\alpha$$ where $\delta_c = \dfrac{X\cos\alpha - Y\sin\alpha}{1 + f'(\theta)}$

$$\alpha = \alpha_o + f(\theta), \qquad r_c = \left(X_c^2 + Y_c^2\right)^{1/2} \qquad \gamma_c = \tan^{-1}(Y_c/X_c)$$

Definition Diagram	Governing Equations
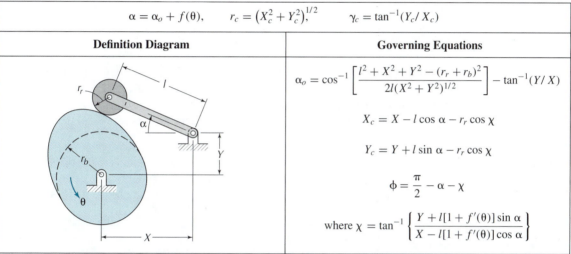	$$\alpha_o = \cos^{-1}\left[\frac{l^2 + X^2 + Y^2 - (r_r + r_b)^2}{2l(X^2 + Y^2)^{1/2}}\right] - \tan^{-1}(Y/X)$$ $$X_c = X - l\cos\alpha - r_r\cos\chi$$ $$Y_c = Y + l\sin\alpha - r_r\cos\chi$$ $$\phi = \frac{\pi}{2} - \alpha - \chi$$ where $\chi = \tan^{-1}\left\{\dfrac{Y + l[1 + f'(\theta)]\sin\alpha}{X - l[1 + f'(\theta)]\cos\alpha}\right\}$

Figure 7.24 (*Continued*)

EXAMPLE 7.2 DETERMINATION OF A DISC CAM PROFILE

Consider the following input parameters of a disc cam to be manufactured:

- translating roller follower
- base circle diameter db = 3.0 cm
- roller circle diameter dr = 1.0 cm
- offset S = 0.6 cm

$$r_b = \frac{d_b}{2} = \frac{3.0}{2} = 1.5 \text{ cm}; \quad r_r = \frac{d_r}{2} = \frac{1.0}{2} = 0.5 \text{ cm}$$

A layout of these dimensions is provided in Figure 7.25.
The displacement parameters are given in Table 7.3, and the corresponding displacement diagram is illustrated in Figure 7.26. Determine

(a) the cam profile
(b) the pressure angle as a function of cam rotation

(*Continued*)

EXAMPLE 7.2 Continued

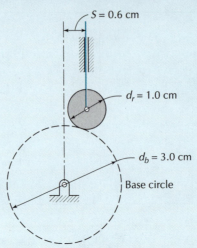

Figure 7.25   Prescribed dimensions of a disc cam.

TABLE 7.3 FOLLOWER MOTION

Interval↓	Parameter → Initial Angle for Interval (degrees)	Final Angle for Interval (degrees)	Type of Motion	Rise (+) or Return (−) (cm)
1	0	100	Harmonic	2.5
2	100	160	Dwell	0
3	160	300	Cycloidal	−1.5
4	300	360	Parabolic	−1.0

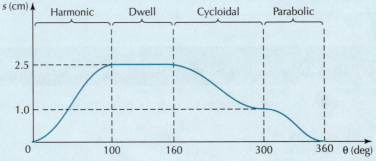

Figure 7.26   Displacement diagram.

SOLUTION

(a) We apply the equations as listed in Figure 7.24. A typical calculated value of the pressure angle is made for $\theta = 30°$. At this rotation, the follower is undergoing lift with harmonic motion, with a lift of 2.5 cm in the interval

$$0 \leq \theta \leq 100°$$

Therefore

$$L = 2.5 \text{ cm}; \quad \beta = 100° - 0° = 100°$$

Using the expressions from Figure 7.24, and recognizing in this instance that

$$\theta = \theta^*$$

(*Continued*)

EXAMPLE 7.2 Continued

yields

$$f(\theta) = s^* = \frac{L}{2}\left(1 - \cos\frac{\pi\theta}{\beta}\right); \quad f(30°) = 0.515 \text{ cm}$$

$$f'(\theta) = \frac{df(\theta)}{d\theta} = \frac{\pi L}{2\beta}\sin\frac{\pi\theta}{\beta}; \quad f'(30°) = 1.82 \text{ cm}$$

$$R_o = \left[(r_b + r_r)^2 - S^2\right]^{1/2} = 1.908 \text{ cm}$$

$$R(\theta) = R_o + f(\theta); \quad R(30°) = 2.423 \text{ cm}$$

$$\varphi(\theta) = \tan^{-1}\left(\frac{f'(\theta) - S}{R(\theta)}\right); \quad \varphi(30°) = 26.7°$$

$$\gamma_c(\theta) = \frac{\pi}{2}\tan^{-1}\left[\frac{S + r_r\sin(\varphi(\theta))}{R(\theta) - r_r\cos(\varphi(\theta))}\right]; \quad \gamma_c(30°) = 67.3°$$

$$r_c(\theta) = \left\{[R(\theta) - r_r\cos(\varphi(\theta))]^2 + [S + r_r\sin(\varphi(\theta))]^2\right\}^{1/2}; \quad r_c(30°) = 2.142 \text{ cm}$$

$$x_c(\theta) = r_c\cos(\gamma_c(\theta)); \quad x_c(30°) = 0.825 \text{ cm}$$

$$y_c(\theta) = r_c\sin(\gamma_c(\theta)); \quad y_c(30°) = 1.98 \text{ cm}$$

The calculated coordinates are rotated about the center of rotation of the cam in an opposite direction to that of the rotation of the cam to locate a point on the cam profile. This is illustrated in Figure 7.27(a). Figure 7.27(b) shows the disc cam mechanism.

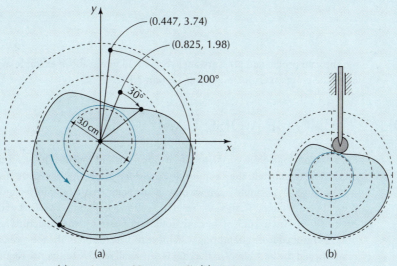

Figure 7.27 (a) Development of disc cam profile (b) Mechanism.

(Continued)

EXAMPLE 7.2 Continued

As a second point on the profile, consider the cam rotation of $\theta = 200°$. At this position, the follower is undergoing return with cycloidal motion. This interval of motion starts at $160°$, and ends at $300°$. The return during the interval is 1.5 cm. Therefore

$$L = 1.5 \text{ cm}; \quad \beta = 300° - 160° = 140°$$
$$\theta^* = \theta - 160° = 200° - 160° = 40°$$
$$f(200°) = f(300°) + s^*(40°) = 2.30 \text{ cm}$$

Employing the equations from Figures 7.21 and 7.24, in this instance

$$x_c(200°) = 0.447 \text{ cm}; \quad y_c(200°) = 3.74 \text{ cm}; \quad \phi = (200°) = -17.8°$$

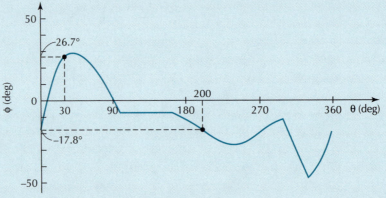

Figure 7.28 Pressure angle as a function of cam rotation.

The point is added to Figure 7.27.

(b) The pressure angle was included in calculating part (a) of this problem. Figure 7.28 shows a plot of pressure angle as a function of cam rotation.

The cam profile and pressure angle presented in the above example may also be obtained using the program Cam Design (see Section 7.10).

7.8 UNDERCUTTING OF A DISC CAM PROFILE

On occasion, for a given geometry of a mechanism and specified follower displacement, it may be impossible to generate the cam profile. Consider employing the same displacement diagram as was employed in Example 7.2. If we now incorporate a flat-face follower and employ the procedure presented in Section 7.7, then the result is as shown in Figure 7.29. Part of the cam profile doubles back on itself. This portion of the cam profile is impossible to manufacture, and is said to be *undercut*. This has resulted from attempting to achieve too great a lift of the follower, within inadequate cam rotation, and the base circle of the cam is too small relative to the rise required.

Undercutting may be eliminated by employing one or more of the following methods:

- increasing the diameter of the base circle
- incorporating a different type of follower
- modifying the motion of the follower

This is illustrated by an example later in this chapter.

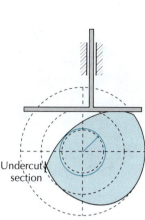

Figure 7.29 Undercut disc cam.

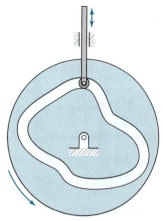

Figure 7.30 Positive-motion disc cam mechanism [Video7.30].

7.9 POSITIVE-MOTION CAMS

Positive-motion cam mechanisms are those where the cam exerts affirmative control of motion of the follower during a complete cycle. The cylindrical cam mechanism shown in Figure 7.1 is also a positive-motion cam mechanism. Such cam mechanisms do not require gravity or spring force to ensure contact between the cam and follower.

Figure 7.30 shows another type of positive-motion cam mechanism. It is very similar to a disc cam mechanism with a roller follower. However, in this instance, a groove of constant width is machined in the cam. As the cam rotates, the roller follower is constrained to move in this groove.

Figure 7.31 shows a *constant-breadth cam mechanism*, which is another type of positive-motion cam mechanism. The cam surface is always in contact with the two parallel surfaces of the translating follower. The distance between parallel surfaces, d, equals the base circle diameter plus the maximum follower rise. Constant-breadth cams may be designed similar to disc cams with a translating flat face, provided that the rise during 180° of cam rotation is the mirror image of the return during the remainder of the cycle.

Figure 7.32 shows an example of a constant-breadth cam employed in a movie projector mechanism. Figure 7.33 gives another illustration of the same system. The function of this mechanism is to intermittently advance film, one frame at a time. While a frame is stationary, light is permitted to shine through the film, and momentarily project a stationary image. While the film is moving, the light must be blocked. This mechanism incorporates two positive-motion cam mechanisms. This includes the constant-breadth cam mechanism used to provide up and down motion of the transport

Video 7.30
Positive-Motion Disc Cam Mechanism

Video 7.31
Constant-Breadth Cam Mechanism

Video 7.33
Movie Projector Mechanism

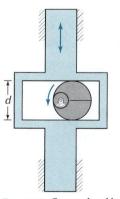

Figure 7.31 Constant-breadth cam mechanism [Video 7.31].

Figure 7.32 Movie projector mechanism.

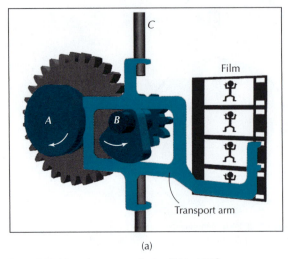
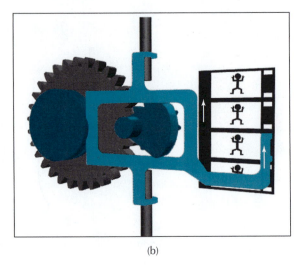

(a)   (b)

Figure 7.33 Movie projector mechanism [Video 7.33].

arm. The constant-breadth cam is located on shaft *B*, as shown in Figure 7.33(a). Shaft *A* is geared to shaft *B*. A cylindrical cam is attached to shaft *A*. One end of the transport arm is in the slot of the cylindrical cam. This cam causes the intermittent rotation of the transport arm about the vertical axis *C*. The rotation causes the other end of the transport arm to enter perforations on the side of the film. While entered into the perforations (Figure 7.33(b)), the arm moves vertically, advancing the film one frame. Every third upward stroke of the transport arm, the end of the arm enters perforations. A *shutter* (shown in animation, but not in Figure 7.33) is attached to shaft *B*.

7.10 PROGRAM CAM DESIGN

The program Cam Design is included on the companion website for this textbook. With this program it is possible to design disc cams having either translating or pivoting followers. The face of the follower may be a knife edge, roller, or flat face. The disc cam profile may be plotted, and an animated motion of the cam mechanism may be displayed. Any of the motions presented in Section 7.5 may be incorporated in the displacement diagram. Values of pressure angle may be displayed. Also, by plotting the profile, we may determine whether or not undercutting is encountered.

As an illustration, consider the following input parameters of a disc cam mechanism to be analyzed:

- translating roller follower
- base circle diameter: 4.0 cm
- roller circle diameter: 1.0 cm
- offset: 0.6 cm
- cam rotational speed: 20 rad/sec CCW

Table 7.4 lists the required motions of the follower. Figure 7.34(a) illustrates the above input parameters, and Figure 7.34(b) shows the displacement diagram.

TABLE 7.4 FOLLOWER MOTION

Interval ↓	Parameter → Initial Angle for Interval (degrees)	Final Angle for Interval (degrees)	Type of Motion	Rise (+) or Return (−) (cm)
1	0	100	Parabolic	0.8
2	100	260	Dwell	0
3	260	360	Harmonic	− 0.8

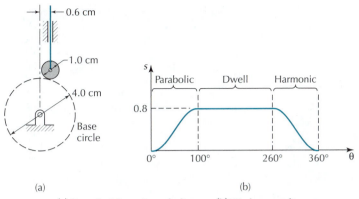

Figure 7.34 (a) Prescribed dimensions of a disc cam, (b) Displacement diagram.

After initiating the program, we obtain the opening screen logo as shown in Figure 7.35. By selecting the **File** menu (Figure 7.36), we have the option to either open an existing file or create a new one. If we select a **New** file, a wizard dialogue box appears. It includes three property pages: **Design Page**, **Motion Page**, and **Display Page**. Each is described separately below.

- **Design Page** (Figure 7.37)

  Figure 7.37(a) shows an illustration of the **Design Page** containing default parameters of a cam mechanism. For this illustration, we enter the following modifications: Replace the knife-edge follower with a roller follower, and replace the values of the base circle diameter, follower offset, and roller diameter. After entering these data, the screen appears as shown in Figure 7.37(b). Then selecting **Next >**, we obtain the **Motion Page**.

- **Motion Page** (Figure 7.38)

  The default **Motion Page** is shown in Figure 7.38(a). Default parameters are replaced by those to be employed for this illustration. This includes the cam rotational speed, along with the number of intervals of motion during a complete cycle (changed from one to three — see Table 7.4). Furthermore, by selecting each interval, we input the cam rotation at the end of the interval, the type of motion, and if needed, the rise or fall of the follower during the interval. For this illustration, we input the parameters listed in Table 7.4. Figures 7.38(b), 7.38(c) and 7.38(d) illustrate the **Motion Page** after the parameters for each interval have been entered. Then, selecting **Next >**, we obtain the **Display Page**.

- **Display Page** (Figure 7.39)

  This page outlines a variety of display modes that may be obtained in either tabular or graphical form. Figure 7.39(a) shows the default **Display Page**, which specifies a displacement diagram. By clicking on **Finish**, the wizard dialogue box is closed, and the screen as shown in Figure 7.39(b) appears. It includes a displacement diagram in nondimensional form.

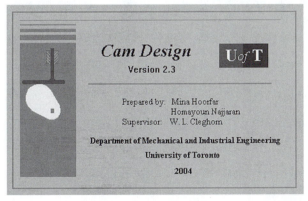

Figure 7.35 Cam Design example—introductory logo.

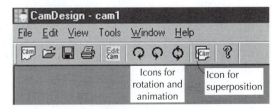

Figure 7.36 Cam Design—menu icons.

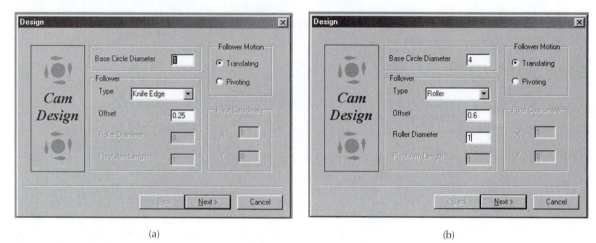

Figure 7.37 Cam Design example—design page.

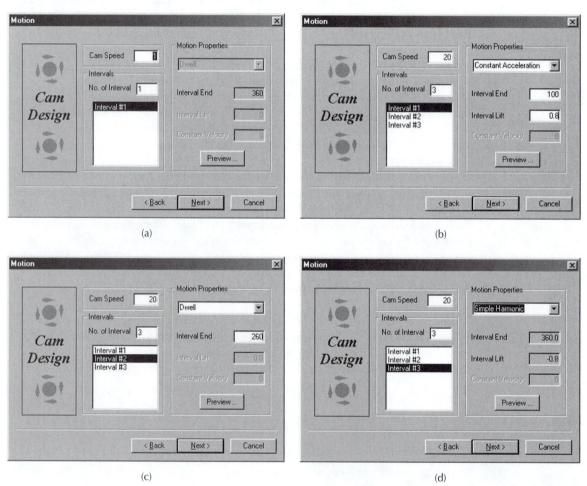

Figure 7.38 Cam Design example—motion page.

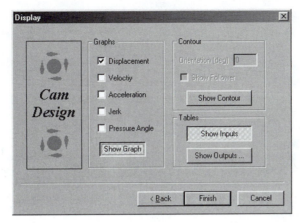

(a)

Absolute Maximum Values

Function	Absolute Maximum	Maximum at (deg)
Follower Displacement	8.000e−001	100.0
Follower Velocity	1.833e+001	50.0
Follower Acceleration	5.184e+002	260.0
Follower Jerk	1.866e+004	310.0
Pressure Angle	2.544e+001	318.0

Output Graphs (normalized with the maximum values)

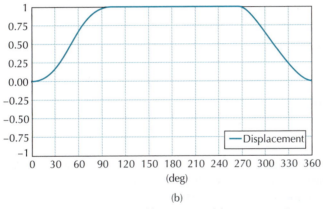

(b)

Figure 7.39 Cam Design example: (a) Display page, (b) Displacement diagram.

Maximum absolute values of displacement, velocity, acceleration, jerk, and pressure angle throughout a cycle are also provided in a table contained within the figure.

By clicking the **Edit Cam** icon (see Figure 7.36), the **Properties Sheet** appears, as shown in Figure 7.40(a). We now have the choice to return to the **Design Page**, the **Motion Page** or the **Display Page**. It is possible to modify the parameters or repeat the displays. We have the option to change the type and dimensions of the cam mechanism through the **Design Page**. Alternatively, by selecting the **Motion Page**, we may alter the follower motions. A further option is to select the **Display Page**, and either review or select outputs. We may go to the **Display Page**, deselect the **Show Graph** button, and select the **Show Contour** button. We also have the option of including the follower with the cam in the illustration by clicking on the appropriate box (Figure 7.40(b)). Then, clicking **Apply** gives the result as shown in Figure 7.41. Once again clicking the **Edit Cam** icon, then the **Display Page** button, deselecting the **Show Contour** button, clicking on the **Show Graph** button, selecting the **Pressure Angle** graph as shown in Figure 7.42(a), and clicking **Apply** gives the result shown in Figure 7.42(b).

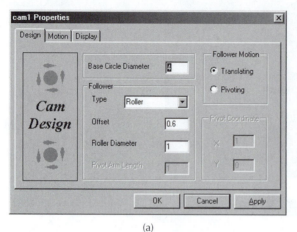

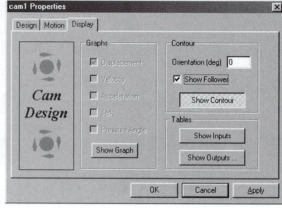

(a)   (b)

Figure 7.40 Cam Design example—properties sheet.

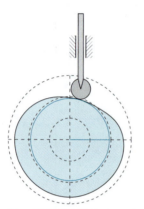

Figure 7.41 Cam Design example—
cam and follower.

An animated motion of the cam mechanism may also be obtained. Animations may be generated with or without the follower. For animations, it is necessary to deselect all other inputs and outputs, except for **Show Contour** (see Figure 7.43(a)). Then, using the icons shown in Figure 7.36, we have the choice to step the motion in increments of 5° of cam rotation in either the clockwise or counterclockwise direction. Alternatively, the motion may be animated. Figures 7.43(b) and 7.43(c) illustrate two positions of the cam mechanism.

Under the **File** menu, we may save the above parameters using a desired file name. The above cam mechanism is saved as **cam1**. Then we proceed to return to the **Motion Page** and change the displacement diagram. The rotation of the cam dividing the second and third intervals is altered from 260° to 300° (Figure 7.44(a)). This modified cam mechanism and displacement diagram are shown in Figures 7.44(b) and 7.44(c), respectively. It is then saved as **cam2**.

Two or more files may be opened simultaneously. Figure 7.45(a) shows the case when both **cam1** and **cam2** are opened at the same time. When more than one file is open, it is possible to superimpose graphs of displacement and the time derivatives of displacement, as well as the cam contours. The icon shown in Figure 7.36 is used to activate the superimpose option. Then, the screen as shown in Figure 7.45(b) appears. We then indicate the desired files to be superimposed by first highlighting them and then clicking the **Add >>** button. If **cam1** and **cam2** are added, we obtain the screen as shown in Figure 7.45(c). If we select to superimpose graphs by clicking **OK**,

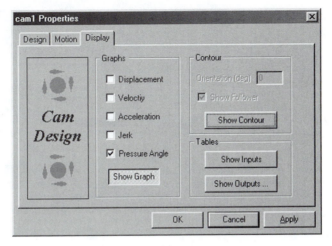

(a)

Function	Absolute Maximum	Maximum at (deg)
Follower Displacement	8.000e–001	100.0
Follower Velocity	1.833e+001	50.0
Follower Acceleration	5.184e+002	260.0
Follower Jerk	1.866e+004	310.0
Pressure Angle	2.544e+001	318.0

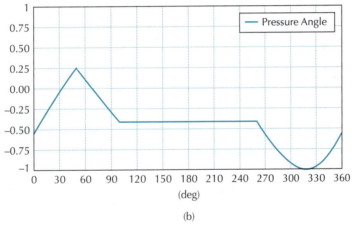

(b)

Figure 7.42 Cam Design example—display of pressure angle.

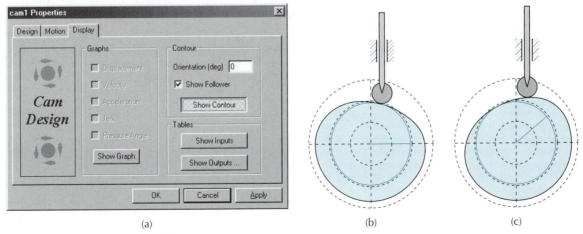

(a)   (b)   (c)

Figure 7.43 Cam Design example—cam animation.

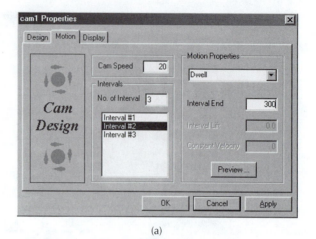

(a)

Function	Absolute Maximum	Maximum at (deg)
Follower Displacement	8.000e−001	100.0
Follower Velocity	2.400e+001	330.0
Follower Acceleration	1.440e+003	300.0
Follower Jerk	8.640e+004	330.0
Pressure Angle	3.288e+001	334.0

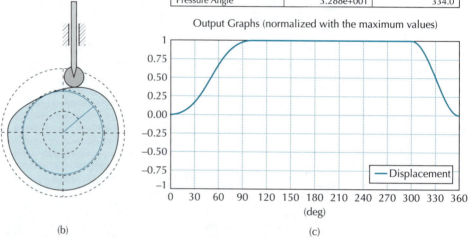

(b)

(c)

Figure 7.44 Cam Design example—modification of displacement diagram.

then we obtain the result shown in Figure 7.45(d). Alternatively, if we select to superimpose the contours, we obtain the result shown in Figure 7.45(e).

This program can be used effectively as a design tool to search for a cam mechanism that has a prescribed motion while subject to a variety of constraints, such as space and maximum pressure angle. It may also be used to check for the existence of undercutting, and to design a cam mechanism where undercutting does not exist.

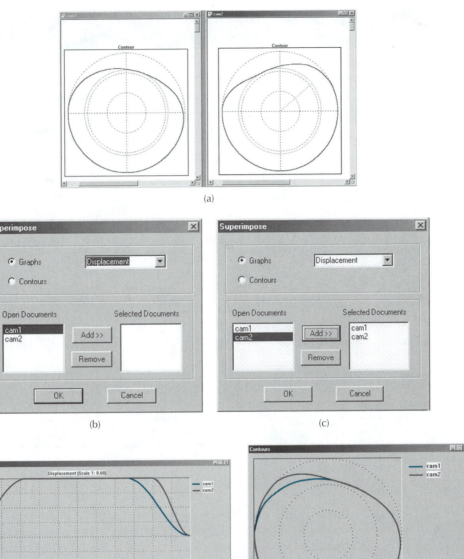

Figure 7.45 Cam Design example—superposition of cam profiles and displacement diagrams.

EXAMPLE 7.3 DESIGN OF A DISC CAM TO REDUCE PRESSURE ANGLE

Consider the displacement parameters given in Table 7.5. The corresponding displacement diagram is illustrated in Figure 7.46.

In addition, we have the following parameters:

- translating roller follower
- base circle diameter: 4.0 cm
- roller circle diameter: 1.0 cm
- offset: 0.0 cm

(*Continued*)

EXAMPLE 7.3 Continued

TABLE 7.5 FOLLOWER MOTION

Parameter → Interval ↓	Initial Angle for Interval (degrees)	Final Angle for Interval (degrees)	Type of Motion	Rise (+) or Return (−) (cm)
1	0	50	Dwell	0
2	50	120	Parabolic	1.5
3	120	160	Dwell	0
4	160	270	Harmonic	−0.7
5	270	360	Cycloidal	−0.8

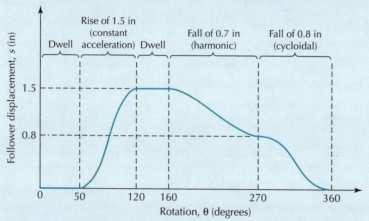

Figure 7.46 Displacement diagram.

Based on the above parameters and motions, Figure 7.47(a) shows the corresponding cam mechanism, and Figure 7.47(b) gives a plot of the pressure angle as a function of cam rotation. The maximum pressure angle during a cycle is 37.1°. Figure 7.47(a) illustrates the configuration of the cam mechanism where the pressure angle is maximum.

Function	Absolute Maximum	Maximum at (deg)
Follower Displacement	1.500e+000	120.0
Follower Velocity	2.456e+000	85.0
Follower Acceleration	4.020e+000	50.0
Follower Jerk	8.149e+000	270.0
Pressure Angle	3.707e+001	85.0

Output Graphs (normalized with the maximum values)

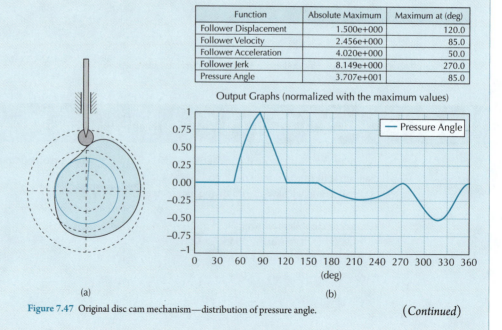

(a)

(b)

Figure 7.47 Original disc cam mechanism—distribution of pressure angle.

(*Continued*)

EXAMPLE 7.3 Continued

Design a cam mechanism for which the pressure angle is always less than 30° through changing

(a) the diameter of base circle
(b) the diameter of roller follower
(c) the offset of follower
(d) the displacement diagram

Employ the program Cam Design.

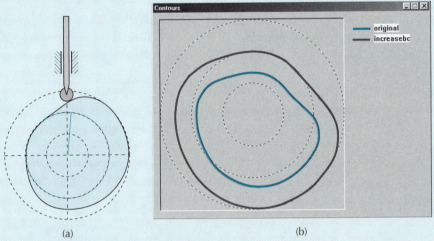

(a)   (b)

Figure 7.48 Disc cam mechanism with increased base circle diameter: (a) Configuration of maximum pressure angle, (b) Superposition of original and modified profiles.

SOLUTION

We consider the disc cam profile shown in Figure 7.47(a) as the *original profile*.

(a) Increasing the diameter of the base circle will always decrease the maximum value of pressure angle. However, this will be accomplished at the expense of creating a larger mechanism. Through trial and error we are able to reduce the maximum pressure angle from 37.1° to 30.0° by increasing the base circle diameter from 4.0 cm to 6.01 cm. Figure 7.48(a) shows the configuration of the mechanism where the pressure angle is maximum. Figure 7.48(b) shows a superposition of the original and modified cam profiles.

(b) We return the diameter of the base circle back to 4.0 cm and now increase the diameter of the roller follower. Through trial and error, by increasing the roller diameter from 1.0 cm to 3.01 cm, the maximum pressure angle has been reduced from 37.1° to 30.0°. Figure 7.49(a) illustrates the disc cam mechanism in the configuration of maximum pressure angle. Figure 7.49(b) shows a superposition of the original and modified cam profiles. The greatest difference between the profiles occurs during the steep rise of the follower.

(c) We return the diameter of the roller follower back to 1.0 cm. Several values of offset are now tried in a search for the minimum value of the maximum pressure angle throughout a cycle. The optimum offset is 0.606 cm. Figure 7.50(a) shows the corresponding cam mechanism. The maximum pressure angle has been reduced from 37.1° to 30.2°. Figure 7.50(b) shows a superposition of the original and modified profiles. In this instance, it was not possible to reduce the maximum pressure angle to 30.0° solely by changing the follower offset. However, a designer has the option to vary more than one parameter in order to achieve a desired result. By increasing the base circle diameter from 4.0 cm

(Continued)

EXAMPLE 7.3 Continued

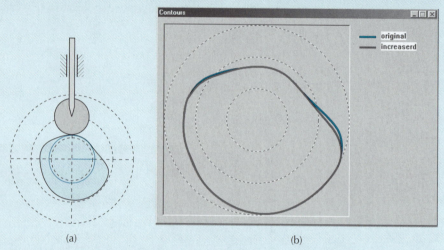

(a)   (b)

Figure 7.49 Disc cam mechanism with increased roller diameter: (a) Configuration of maximum pressure angle, (b) Superposition of original and modified profiles.

to 4.5 cm, with zero offset, the maximum pressure angle reduces to 35.1°. Then, through trial and error, the pressure angle may be reduced to 30.0° by incorporating an offset of 0.45 cm. This result is illustrated in Figure 7.51.

(d) We return the offset back to zero and the base circle diameter to 4.0 cm. For the results presented in Figure 7.47(b), the configuration corresponding to the largest value of pressure angle occurs near the steepest slope of the displacement diagram. Thus, in order to reduce this large pressure angle, we may reduce the slope of the displacement diagram. This may be accomplished by reducing the initial dwell which spans 0° to 50°. Through trial and error, if the initial dwell is reduced to span 0° to 29.0°, then the maximum pressure angle is reduced to 30.0°. Figure 7.52 shows the modified displacement diagram. Figure 7.53(a) shows the cam mechanism in the configuration of maximum pressure angle. Figure 7.53(b) shows a superposition of the new and original profiles.

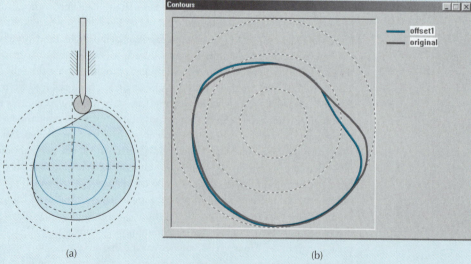

(a)   (b)

Figure 7.50 Disc cam mechanism with offset: (a) Configuration of maximum pressure angle, (b) Superposition of original and modified profiles.

(Continued)

EXAMPLE 7.3 Continued

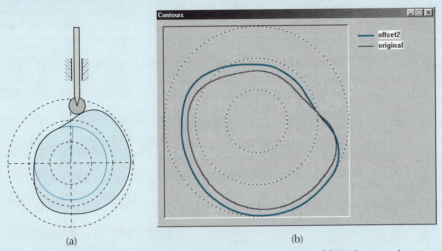

Figure 7.51 Disc cam mechanism with offset and increased base circle diameter: (a) Configuration of maximum pressure angle, (b) Superposition of original and modified profiles.

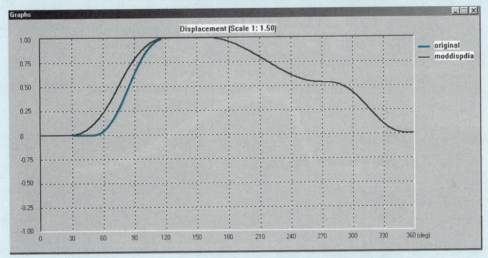

Figure 7.52 Modified displacement diagram.

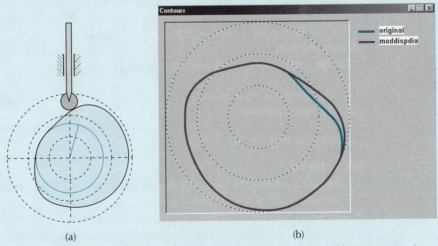

Figure 7.53 Disc cam mechanism incorporating modified displacement diagram: (a) Configuration of maximum pressure angle, (b) Superposition of original and modified profiles.

EXAMPLE 7.4 DESIGN OF A DISC CAM TO ELIMINATE UNDERCUTTING

Consider the displacement parameters given in Table 7.6. The corresponding displacement diagram is illustrated in Figure 7.54. In addition, we have the following parameters:

- translating flat-face follower
- base circle diameter: 15 cm

TABLE 7.6 FOLLOWER MOTION (EXAMPLE 7.4)

Parameter → Interval ↓	Initial Angle for Interval (degrees)	Final Angle for Interval (degrees)	Type of Motion	Rise (+) or Return (−) (cm)
1	0	50	Dwell	0
2	50	100	Harmonic	2.5
3	100	360	Parabolic	− 2.5

Function	Absolute Maximum	Maximum at (deg)
Follower Displacement	2.500e+000	100.0
Follower Velocity	4.500e+000	75.0
Follower Acceleration	1.620e+001	50.0
Follower Jerk	5.832e+001	75.0
Length	4.500e+000	75.0

Output Graphs (normalized with the maximum values)

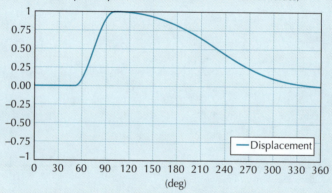

Figure 7.54 Displacement diagram.

The cam mechanism is shown in Figure 7.55(a). There is clearly a section of the profile that has been undercut.

Design a cam for which undercutting is eliminated by changing

(a)   the diameter of the base circle
(b)   the type of follower

Employ the program Cam Design.

SOLUTION

(a) Through increasing the base circle diameter to 50 cm we obtain a disc cam that does not have undercutting. It is illustrated in Figure 7.55(b).

(b) We return the value of the base circle back to 15 cm and now replace the flat-face follower with knife edge follower. The result is shown in Figure 7.55(c). There is no undercutting but

(Continued)

EXAMPLE 7.4   Continued

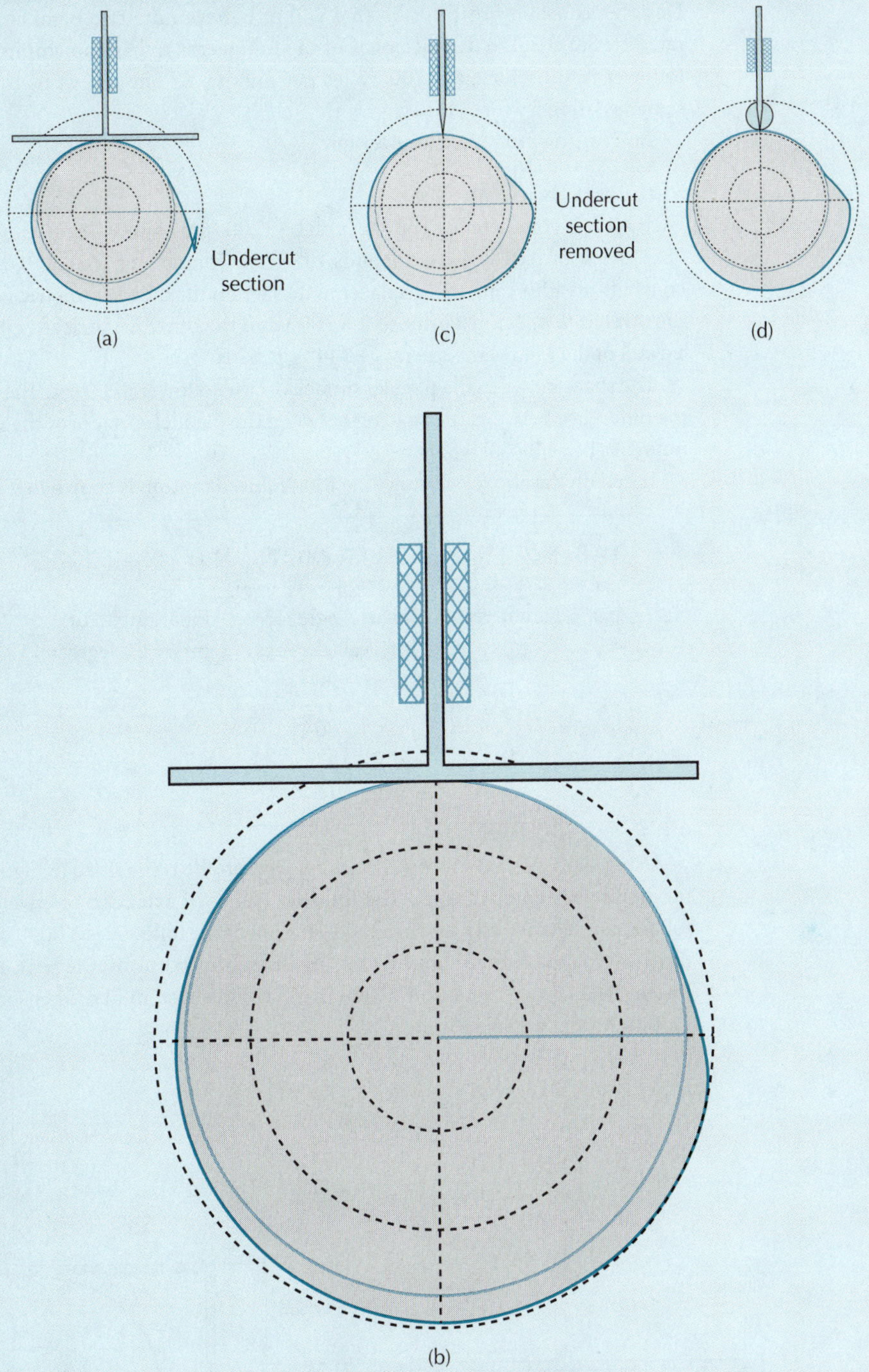

Figure 7.55 (a) Knife edge follower with undercut profile, (b) Undercut section removed, (c) Knife-edge follower, (d) Roller follower.

the maximum pressure angle is 33 degrees, which is above the recommended maximum value. This could be reduced by increasing the base circle diameter. Alternatively, we employ a roller follower with a roller diameter of 3.0 cm as illustrated in Figure 7.55(d). Here the maximum pressure angle is to 28.9 degrees. The disc cam mechanism with the roller follower provides the most compact design without having an excessive pressure angle.

EXAMPLE 7.5 DESIGN OF A CONSTANT-BREADTH CAM

Design a constant-breadth cam that will fit between the two parallel surfaces that are separated by 8.0 cm. The displacement of the follower is to have an amplitude of 3.0 cm and the follower is to oscillate at 1200 cycles per minute. At one end of its travel, the follower is to remain stationary for 0.01 sec.

Employ the program Cam Design.

SOLUTION

Each cycle is completed in $(60 \text{ sec/min})/(1200 \text{ rpm}) = 0.050$ sec/rev.

A dwell of 0.01 sec corresponds to a cam rotation of $(0.010/0.050)360° = 72°$. For a constant-breadth cam, the displacement diagram must be symmetric about $\theta = 180°$. Half of the dwell is to occur just prior to $\theta = 180°$, and the other half just after it. That is, there is to be a dwell of the follower in the range $144° \le \theta \le 216°$.

The space between the parallel surfaces of the follower is 8.0 cm. The base circle diameter is the difference between the spacing between the parallel surfaces of the follower and the maximum lift, i.e., 8.0 - 3.0 = 5.0 cm.

Based on the above, we conclude the required motion is shown in Table 7.7.

TABLE 7.7 FOLLOWER MOTION (EXAMPLE 7.5)

Interval ↓	Parameter→ Initial Angle for Interval (degrees)	Final Angle for Interval (degrees)	Rise (+) or Return (−) (cm)
1	0	144	3.0
2	144	216	0
3	216	360	− 3.0

Using Cam Design, the plot of the disc cam profile is provided in Figure 7.56 when employing simple harmonic motion. The follower has been extended to envelope both sides of the follower simultaneously. The peak acceleration of the follower is 37,000 cm/sec². Alternatively, if we use cycloidal motion during the lift and return portions, the peak acceleration of the follower has a higher value of 47,100 cm/sec². However, unlike when using simple harmonic motion, the jerk remains finite.

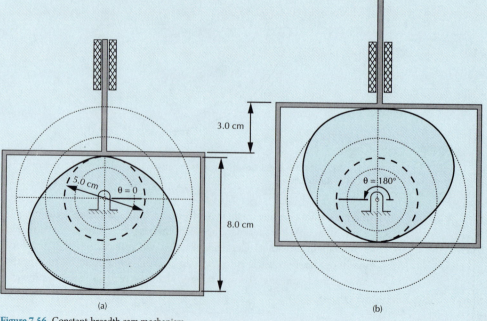

(a)   (b)

Figure 7.56 Constant-breadth cam mechanism.

7.11 MANUFACTURING OF CAMS

Cams are manufactured using a wide range of methods. Depending on the quantity to be produced, speed of operation, load rating, and required accuracy of the motion, cams are made from a variety of materials and may be fabricated by hand, or molded, or machined. This section presents two methods of manufacturing cams; one for low-volume production and the other for mass production.

Figure 7.57 shows a moving-headstock milling machine, used in machining a wide variety of small metal components employed in high precision mechanical devices. This machine is an excellent example of implementing different types of cam mechanisms. Here, the followers transmit motions to the cutting tools. Motions can be readily modified by changing the shapes of the cams. Figure 7.58 shows some of the disc cam mechanisms used in this machine. The photographs depict different types of followers. Figures 7.59 and 7.60 show a cylindrical cam mechanism and a face cam mechanism, respectively.

For low-volume production of disc cams, such as for a moving-headstock milling machine (Figure 7.57), the shape of the cam profile may be transcribed by hand onto a cam blank. The steps are illustrated in Figure 7.61. First, radial lines are drawn from a center on the blank to segregate portions of motions and dwells of the follower (Figure 7.61(a)). Circular arcs are then scribed, using the common center. These arcs correspond to the periods of dwell. The contours for the rises and falls are then added (Figure 7.61(b)). This is followed by completing a rough cut of the profile using a band saw (Figure 7.61(c)). Then a fine cut is taken using a shaping machine (Figure 7.61(d)). Finally, the cam profile is smoothed as necessary by hand.

Cams for internal combustion engines are usually machined from a solid shaft, so all cams on the shaft are rigidly connected and will be driven at the same speed. Figure 1.1 shows a one-cylinder engine with two cams on the camshaft. Most internal combustion engines are designed with either two or four cams for each cylinder.

In a modern facility for the mass production of camshafts, the entire process is highly automated. The process starts with a solid cylindrical shaft, which is accurately cut to length by machining the two end surfaces. Typically four holes are then drilled in one end of the shaft, all

Figure 7.57 Moving-headstock milling machine [Video 7.57].

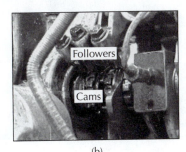

(a)   (b)   (c)

Figure 7.58 Disc cam mechanisms: (a) Single knife-edge follower [Video 7.58A], (b) Multiple knife-edge followers [Video 7.58B], (c) roller follower [Video 7.58C].

Figure 7.59 Cylindrical cam mechanism [Video 7.59].

Figure 7.60 Face cam mechanism [Video 7.60].

Video 7.61A1,
Video 7.61A2,
Video 7.61B,
Video 7.61C,
Video 7.61D
Manufacture of a
Disc Cam

offset from its centerline. Three of the holes are tapped and will be used later for attaching a pulley or timing gear to drive the camshaft. A tapered pin is press-fit in the hole that is not tapped. This pin will be used for controlling the rotations of the shaft in the subsequent manufacturing operations.

Multiple annular slots are cut on the cylindrical surface, spaced along the length of the shaft. These slots will become the gaps between the cams. The cam profiles are then machined. In machining each cam profile, the shaft is given one revolution at a constant rate, taking approximately two seconds. During rotation, a milling cutter machines a cam profile. The milling cutter, rotating on its own axis, is actuated to move radially with respect to the centerline of the camshaft. This radial motion is computer controlled and mimics the motion of the follower in the engine. This process is carried out for all cams on the camshaft. Between the machining of cam profiles, the milling cutter is repositioned by moving it parallel to the axis of the camshaft.

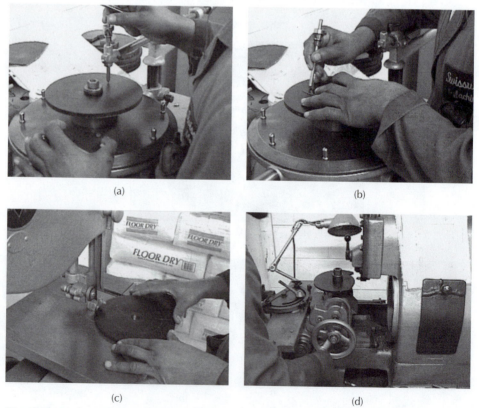

(a)

(b)

(c)

(d)

Figure 7.61 Manufacture of a disc cam: (a) Scribing radial lines [Video 7.61A1], and circular arcs [Video 7.61A2], (b) Scribing rises and falls [Video 7.61B], (c) Rough cut using band saw [Video 7.61C], (d) Fine cut using shaper [Video 7.61D].

The camshafts are automatically transferred to other machines for the finish grinding of the cam profiles and camshaft bearings. There is a separate grinding wheel for each cam profile and bearing. The grinding wheels are also computer controlled, similar to the radial movements used for the milling operation.

At a subsequent station, the cam profiles only are heat treated. The camshafts are later fed through a machine where they are wire brushed to remove burs and scaling caused by the heat treatment process. The camshafts are then transferred to a polishing machine. Here, a polishing compound is applied, and the camshaft is rotated several revolutions in both directions. The polishing is carried out using long bands of polishing cloth that are applied to the working surfaces. In the next machine, the camshafts are washed to remove the polishing compound and subjected to final inspection.

PROBLEMS

Generation of Displacement Diagram

P7.1 A cam mechanism is required for an automated screw machine. The part (i.e., follower) is to move to the right 5.0 cm on 1.0 sec, then remain stationary for 1.25 sec, then move an additional 3.0 cm to the right on 0.75 sec, followed be another dwell of 1.25 sec, and finally return to the original position in 0.75 sec. Determine the required rotational speed of the disc cam and plot a suitable displacement diagram.

P7.2 A disc cam mechanism is required for a machine. A part (i.e., cam follower) is to move upward 5.0 cm in 0.2 sec, then remain stationary for 0.1 sec, then move downward 3.0 cm in 0.3 sec, followed be another dwell of 0.2 sec, and finally return to the original position in 0.2 sec. Determine the required rotational speed of the disc cam and plot a suitable displacement diagram.

P7.3 A disc cam mechanism is required for a machine. A part (i.e., cam follower) is to move to the upward 3.0 cm in 0.5 sec, then remain stationary for 0.6 sec, and finally return to the original position in 0.9 sec. Determine the required rotational speed of the disc cam and plot a suitable displacement diagram.

Generation of Cam Profile, Graphical

P7.4 A disc cam's translating roller follower with zero offset is to rise 3.0 cm with simple harmonic motion in 160° of cam rotation and return with simple harmonic motion in the remaining 200°. If the roller radius is 0.50 cm, and the prime circle radius is 3 cm, construct the displacement diagram, pitch curve, and cam profile for counterclockwise cam rotation.

P7.5 The follower shown in Figure P7.5 is to be pivoted 35° clockwise with harmonic motion during a 120° counterclockwise turn of the cam, then allowed to dwell during a 30° counterclockwise turn, then return to the starting point shown, with uniformly

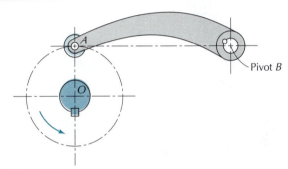

Figure P7.5

accelerated and decelerated motion. Sketch the cam profile surface during the rise motion of the follower, by considering six equally spaced rotations of the cam during the rise.

P7.6 A disc cam with a translating flat-face follower is to have the same motion as in Problem 7.4. The prime circle is to have a radius of 3.0 cm, and the cam is to rotate clockwise. Construct the displacement diagram and the cam profile, offsetting the follower stem by 1.0 cm in the direction that reduces the bending stress in the follower during the rise.

P7.7 Construct the displacement diagram and the cam profile for the disc cam with a pivoting flat-face follower that rises through 10° with cycloidal motion in 150° of counterclockwise cam rotation, then dwells for 30°, returns with cycloidal motion in 120°, and dwells for 60°. Graphically determine the necessary length of the follower face, allowing 5.0 mm clearance at each end. The prime circle is to have a radius of 40 mm. The follower pivot arm is 120 mm to the right. The cam rotation is counterclockwise.

P7.8 The motion of the mechanism shown in Figure P7.8 (on the next page) is to be generated by means of a disc cam, rotating counterclockwise about point O_6. Starting from the configuration shown, link 4 is to move from position A to position B during 120° of cam rotation. Each number on the path of link 4 represents 30° of cam rotation. After link 4 reaches B,

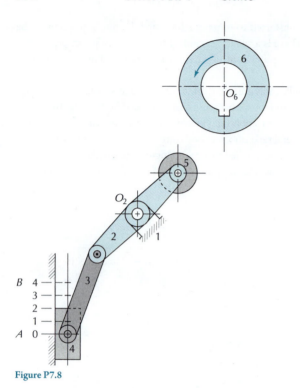

Figure P7.8

it is to remain stationary for 60° of cam rotation, and then return as quickly as possible to position A, where it stays during the remainder of the cycle.

(a) Construct the cam profile.

(b) Determine the pressure angles of the cam follower corresponding to cam rotations of 30°, 60°, and 90°.

P7.9 A disc cam mechanism is to be used for a vibration platform. The follower motion is to have rise of 2.0 cm with constant acceleration in 0.5 sec, dwell for 0.3 sec, and fall with constant acceleration in 0.7 sec. There is a translating roller follower with zero offset.

(a) Determine the rotational speed of the disc cam.

(b) Plot the displacement diagram.

(c) Graphically construct the cam profile.

(d) Determine analytically the maximum follower velocity and acceleration.

$$d_b = 3.0 \text{ cm}; \quad d_r = 1.0 \text{ cm}$$

P7.10 A disc cam mechanism is to provide the motion given in Table P7.10. There is a translating roller follower with zero offset.

(a) Plot the displacement diagram.

(b) Graphically construct the cam profile.

$$d_b = 4.0 \text{ cm}; \quad d_r = 1.0 \text{ cm}$$

P7.11 A disc cam mechanism is to provide the motion given in Table P7.11. There is a translating flat-face follower.

(a) Plot the displacement diagram.

(b) Graphically construct the cam profile.

$$d_b = 4.0 \text{ cm}$$

TABLE P7.10 FOLLOWER MOTION

Parameter → Interval ↓	Initial Angle for Interval (degrees)	Final Angle for Interval (degrees)	Type of Motion	Rise (+) or Return (−) (cm)
1	0	100	Harmonic	1.2
2	100	240	Dwell	0
3	240	360	Parabolic	−1.2

TABLE P7.11 FOLLOWER MOTION

Parameter→ Interval ↓	Initial Angle for Interval (degrees)	Final Angle for Interval (degrees)	Type of Motion	Rise (+) or Return (−) (cm)
1	0	120	Cycloidal	1.2
2	120	200	Dwell	0
3	200	360	Uniform	−1.2

TABLE P7.12 FOLLOWER MOTION

Interval ↓	Parameter → Initial Angle for Interval (degrees)	Final Angle for Interval (degrees)	Type of Motion	Rise (+) or Return (−) (cm)
1	0	80	Harmonic	1.5
2	80	200	Dwell	0
3	200	360	Parabolic	−1.5

TABLE P7.13 FOLLOWER MOTION

Interval ↓	Parameter → Initial Angle for Interval (degrees)	Final Angle for Interval (degrees)	Type of Motion	Rise (+) or Return (−) (cm)
1	0	100	Harmonic	1.4
2	100	220	Dwell	0
3	220	360	Uniform	− 1.4

TABLE P7.14 FOLLOWER MOTION

Interval	Initial angle for interval (degrees)	Final angle for interval (degrees)	Type of Motion	Rise (+) or Return (−) (cm)
1	0	130	Harmonic	3.6
2	130	280	Dwell	0
3	280	360	Harmonic	− 3.6

TABLE P7.15 FOLLOWER MOTION

Interval ↓	Parameter → Initial Angle for Interval (degrees)	Final Angle for Interval (degrees)	Type of Motion	Rise (+) or Return (−) (cm)
1	0	90	Harmonic	1.3
2	90	200	Dwell	0
3	200	360	Uniform	− 1.3

P7.12 A disc cam mechanism is to provide the motion given in Table P7.12. There is a translating roller follower with offset.

(a) Plot the displacement diagram.

(b) Graphically construct the cam profile.

$$d_b = 4.0 \text{ cm}; \quad d_r = 1.0 \text{ cm}; \quad S = 0.5 \text{ cm}$$

Generation of Cam Profile, Cam Design

P7.13 A disc cam mechanism is to provide the motion given in Table P7.13. There is a translating roller follower. The nominal dimensions are provided. Employ the program Cam Design and reduce the peak value of pressure angle to 30° by:

(a) increasing the base circle diameter

(b) increasing the roller diameter

(c) adjusting the offset

$$d_b = 2.0 \text{ cm}; \quad d_r = 1.0 \text{ cm}; \quad S = 0$$

P7.14 A disc cam mechanism is to provide the motion given in Table P7.14. There is a translating flat-face follower. The nominal base circle diameter is provided. For these parameters there is undercutting of the cam profile. Employ the program Cam Design and eliminate the undercutting by increasing the base circle diameter.

$$d_b = 4.0 \text{ cm}$$

P7.15 A disc cam mechanism is to provide the motion given in Table P7.15. There is a translating

knife-edge follower. The nominal dimensions are provided. Employ the program Cam Design and reduce the peak value of pressure angle to 30° by

(a) increasing the base circle diameter

(b) adjusting the offset

$$d_b = 3.0 \text{ cm}; \quad S = 0$$

P7.16 Design a constant-breadth cam that will fit between the two parallel surfaces that are separated by 4.0 cm. The displacement of the follower is to have an amplitude of 1.5 cm and the follower is to oscillate at 1200 cycles per minute. At one end of the travel, the follower is to remain stationary for 0.0050 sec. Employ the program Cam Design.

CHAPTER 8

GRAPHICAL FORCE ANALYSIS OF PLANAR MECHANISMS

8.1 INTRODUCTION

The operation of a mechanism normally results in the transmission of forces between its links. Each link is also subjected to internal loads. It is essential in the design of a mechanism to be aware of the magnitudes and directions of the forces that will be encountered during operation. Knowing the level of the forces can guide the design engineer to appropriately size the cross sections of the links and bearings to withstand the loads. A link that is not designed to be strong enough could break and lead to a failure of the entire system. On the other hand, a link that is obviously overdesigned would result in wastage of material and higher cost.

This chapter presents a method of completing force analyses of mechanisms under static or dynamic conditions. It is possible to determine the driving force or torque required either to hold the mechanism in a stationary configuration or to generate a specified motion. In addition, internal loads in links, and forces across kinematic pairs may be determined.

To make a dynamic analysis, the mechanism will be considered to be in a state of dynamic equilibrium. Thus, similar procedures may be employed for static and dynamic analyses.

If a mechanism is in static or dynamic equilibrium, every link within the mechanism must also be in equilibrium. Sections 8.2 and 8.3 describe two basic cases for which links of a mechanism are in equilibrium. These basic cases are subsequently used in graphical force analyses of mechanical systems.

8.2 TWO-FORCE MEMBER

A link acted on by only two forces is known as a *two-force member*. In order for a link to qualify as a two-force member, no moments can be applied. For instance, Figure 8.1(a) illustrates a four-bar mechanism in which moments M_{O_2} and M_{O_4} are applied from the base link to links 2 and 4, respectively. Therefore, these two links do not qualify as two-force members. However, neglecting gravity, inertia, and friction in the turning pairs at A and B, link 3 can only be subjected to forces applied through the turning pairs. A free body diagram of the link is shown in Figure 8.1(b). It is subjected to forces $\overline{F}_1$ and $\overline{F}_2$ In this figure, the directions and magnitudes of these forces have been

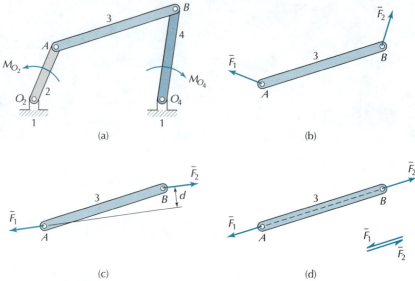

Figure 8.1 Two-force member analysis.

arbitrarily depicted. However, in order for link 3 to remain in equilibrium, the sum of the applied forces and sum of the applied moments must be zero (see Section 2.14). The sum of the applied forces is

$$\bar{F}_1 + \bar{F}_2 = \bar{0}$$

$$(8.2\text{-}1)$$

or

$$\bar{F}_1 = -\bar{F}_2$$

$$(8.2\text{-}2)$$

that is, the two applied forces are equal in magnitude and opposite in direction. Figure 8.1(c) illustrates such a condition. However, as shown, the two forces in Figure 8.1(c) provide a couple of magnitude $F_1 d$, where d is the perpendicular distance between the lines of action. For the net moment to be zero, the distance d must be zero; thus, the two forces must be collinear. Figure 8.1(d) shows the case where the two forces are applied such that the net force and the moment acting on the link are zero.

In summary, for a member to be in equilibrium when acted on by two forces, the two forces must be equal in magnitude, collinear, and opposite in direction.

8.3 THREE-FORCE MEMBER

A link acted on by only three forces is known as a *three-force member*. No moments can be applied. Consider the mechanism shown in Figure 8.2(a). Links 2 and 4 do not qualify as three-force members, because moments are applied to them. However, neglecting gravity, inertia, and friction in the kinematic pairs, link 3 is subjected to only three forces. A free-body diagram of the link is shown in Figure 8.2(b). One of the forces, $\bar{F}_3$, is assumed known and is applied through point C. The direction and magnitude of $\bar{F}_1$ and $\bar{F}_2$ have been arbitrarily chosen. For this body to remain in equilibrium, the sum of applied forces and the sum of applied moments must be zero. Summing the applied forces to zero gives

$$\bar{F}_1 + \bar{F}_2 + \bar{F}_3 = \bar{0}$$

$$(8.3\text{-}1)$$

or

$$\bar{F}_3 = -(\bar{F}_1 + \bar{F}_2)$$

$$(8.3\text{-}2)$$

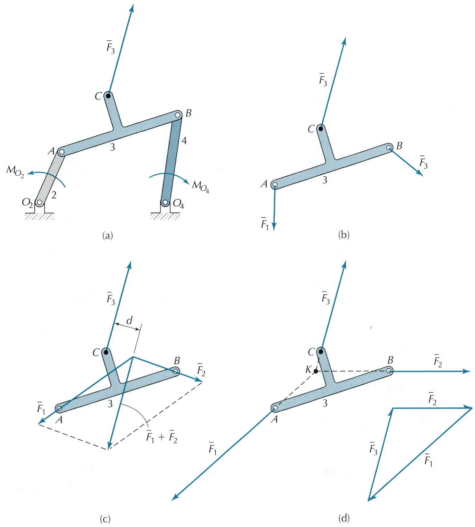

Figure 8.2 Three-force member analysis.

Figure 8.2(c) illustrates a case that satisfies Equation (8.3-2). However, as shown, the three forces provide a force couple of magnitude F_3d, and thus link 3 is not in equilibrium. In order for the force couple to vanish, distance d must equal zero, which is accomplished by having the three forces concurrent at some point. Figure 8.2(d) shows the three lines of action concurrent at point K. As shown in this example, the point of concurrency need not lie within the physical bounds of the link.

In summary, for a member acted on by three forces to be in equilibrium, the forces must add vectorially to zero and must have their lines of action concurrent at some point.

8.4 FORCE TRANSMISSION IN FRICTIONLESS KINEMATIC PAIRS

Forces and moments in a mechanism are transmitted through kinematic pairs of adjoining links. In this chapter, we restrict ourselves to frictionless kinematic pairs. In such instances, forces are transmitted perpendicular to the surfaces of members in contact. If friction and gravity are neglected for a sliding pair, the force transmitted between the members is perpendicular to the direction of the slide. Figure 8.3(a) illustrates a slider crank mechanism with a force, $\bar{P}$, applied to the slider and a moment, M, applied to the crank. A free body diagram of the slider is shown in Figure 8.3(b). The line of action of the force *from* link 3 *on* link 4, designated as $\bar{F}_{34}$, is known since link 3 is a two-force member. Force $\bar{F}_{14}$ is drawn perpendicular to the surfaces in contact, and thus is also perpendicular to the direction of the slide.

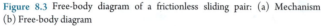

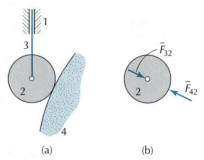

Figure 8.3 Free-body diagram of a frictionless sliding pair: (a) Mechanism (b) Free-body diagram

Figure 8.4 Free body diagram of a frictionless rolling pair

A frictionless rolling pair is shown in Figure 8.4(a). Here, the transmitted force is perpendicular to the surfaces of the links at the point of contact. A free body diagram of the roller is shown in Figure 8.4(b). Neglecting rolling friction, the roller is a two-force member.

For a turning pair, the direction of the force transmitted is dependent on the relative position of the other kinematic pairs in the kinematic chain, and cannot be determined by examining the turning pair alone. Considering the mechanism shown in Figure 8.4, there is a turning pair between links 2 and 3. The direction of the force transmitted between these two links depends on the direction of the force between links 2 and 4. As shown, the two forces acting on link 2 must be collinear for the link to remain in static equilibrium.

8.5 FORCE POLYGONS

For each three-force member within a mechanism, a *force polygon* may be constructed as an aid in determining either directions or magnitudes of forces. As an example, consider the three-force member shown in Figure 8.5(a). Let points A, B, and C represent kinematic pairs with adjacent links of the mechanism, or locations of externally applied loads to the mechanism. Let us say at B both the direction and magnitude of the applied force are known. Also, the line of action of the force at C is given. It is now required to determine the force magnitudes and directions at A and C.

We begin by extending the known lines of action of forces through B and C. These lines intersect at point K as shown in Figure 8.5(b). Since this is a three-force member, the line of action of the force at A must pass through the same intersection point. Now that we have all three lines of action, we can construct the force polygon by first drawing in the known force through point B using an appropriate scale. At the head and tail of this vector, we proceed to draw the other two lines of action. The intersection point of these lines of action dictates the shape of the force polygon. The polygon is completed, as shown in Figure 8.5(c), by labeling the vectors, all head to tail to ensure that

$$\bar{F_1} + \bar{F_1} + \bar{F_3} = \bar{0}$$

(8.5-1)

Note that for a force polygon, unlike velocity and acceleration polygons, there is no pole point.

A special case occurs when two of the forces acting on a three-force member are parallel, as shown in Figure 8.6, where their intersection occurs at infinity. It follows that the third force acting on the member must be parallel to the other two. In this instance, it is not possible to draw a force polygon to determine the unknown quantities. Instead, it is necessary to apply Equations (2.14-2) and (2.14-3) directly.

8.6 STATIC FORCE ANALYSIS USING FORCE POLYGON METHOD

Utilizing force polygons as presented in Section 8.5, it is possible to determine the static forces transmitted throughout an entire mechanism. The force polygon method is suitable for only one configuration.

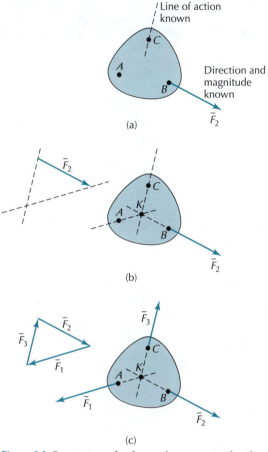

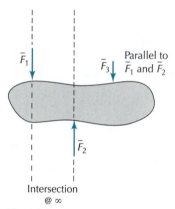

Figure 8.5 Construction of a force polygon associated with a three-force member: (a) Given information, (b) Construction of force polygon, (c) Final result.

Figure 8.6

The method consists of the following general steps:

1. Draw a diagram of the mechanism to scale, in the configuration for which an analysis is required.
2. Break the mechanism into a suitable number of subsystems (either individual links or groups of links) and, if required for clarification, draw a free body diagram for each.
3. Complete the force polygons for those subsystems which can be analyzed immediately (such as those subjected to known external forces).
4. Use results from step 3 to construct the remaining force polygons by examining the forces transmitted between subsystems.

EXAMPLE 8.1 STATIC FORCE ANALYSIS OF A THREE-FORCE MEMBER

Figure 8.7 illustrates a mechanism that incorporates a three-force member and is in static equilibrium. The tensile load in the rope attached to point A is 100 Newtons. Determine the forces applied to link 3 at B and O_3. Neglect the effect of gravity.

(Continued)

EXAMPLE 8.1 Continued

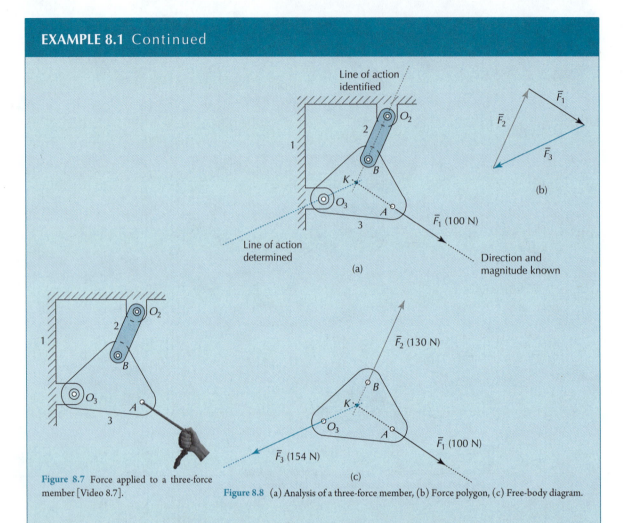

Figure 8.7 Force applied to a three-force member [Video 8.7].

Figure 8.8 (a) Analysis of a three-force member, (b) Force polygon, (c) Free-body diagram.

SOLUTION

Both the magnitude and direction of the force applied at A are given. Since link 2 is a two-force member, the line of action of the force through B is readily identified. The lines of actions of the forces through A and B intersect at K as shown in Figure 8.8(a). Since link 3 is a three-force member, the line of action of the force through O_3 must also pass through K to maintain static equilibrium of the link.

Using a selected scale, the known force vector at A of magnitude 100 Newtons is drawn as illustrated in Figure 8.8(b). Then, the lines of action of the forces through B and O_3 are placed at the head and tail of the vector. The force vectors are drawn head-to-tail so that their vector sum is zero. Using the selected scale, the magnitudes of the forces at B and O_3 are measured to be

$$\left|\bar{F}_2\right| = 130 \text{ N}; \quad \left|\bar{F}_3\right| = 154 \text{ N}$$

The free-body diagram of link 3 is illustrated in Figure 8.8(c).

[**Video 8.7**] provides an animated sequence of the construction of the force polygon presented in this example.

Video 8.7
Force Polygon

EXAMPLE 8.2 STATIC FORCE ANALYSIS OF A SLIDER CRANK MECHANISM

Consider a slider crank mechanism shown in Figure 8.9 having zero offset and with

$$\theta = 120°; \quad r_2 = 1.2 \text{ in}; \quad r_3 = 4.8 \text{ in}$$

A 250-lb force is applied to the slider as shown. Determine the moment required from the base link onto link 2 to keep the mechanism in static equilibrium.

SOLUTION
Free-body diagrams of links 3 and 4 are shown in Figure 8.9. We note that link 3 is a two-force member, and link 4 is a three-force member. Therefore, for link 4

$$\bar{P} + \bar{F}_{14} + \bar{F}_{34} = \bar{0} \tag{8.6-1}$$

Assuming frictionless kinematic pairs, $\bar{F}_{14}$ is perpendicular to the direction of the slide. Also, the force from link 3 is along the direction defined by points B and D. For link 4, we construct the force polygon, shown in Figure 8.9.

We now consider the analysis of link 3, and note that at point D, based on Newton's third law of mechanics (Appendix C),

$$\bar{F}_{43} = -\bar{F}_{34} \tag{8.6-2}$$

Since link 3 is a two-force member

$$\bar{F}_{23} = -\bar{F}_{43} \tag{8.6-3}$$

Now consider the force transmitted from link 3 onto link 2. We note that

$$\bar{F}_{32} = -\bar{F}_{23} \tag{8.6-4}$$

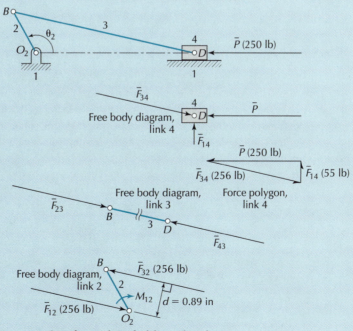

Figure 8.9 Static force analysis of a slider crank mechanism.

(Continued)

EXAMPLE 8.2 Continued

To keep link 2 in static equilibrium, the sum of the forces and sum of the moments acting on this link must be zero. The sum of the forces being zero is accomplished by setting

$$\overline{F}_{12} = -\overline{F}_{32} \tag{8.6-5}$$

However, $\overline{F}_{12}$ and $\overline{F}_{32}$ are not collinear. Therefore, link 2 is not a two-force member. Thus, in addition to the forces, an external moment must be applied to the link to counteract the couple produced by the forces. The magnitude of the moment is

$$M_{12} = \left|\overline{F}_{32}\right| d \tag{8.6-6}$$

where d is the perpendicular distance between the two lines of action. This moment must be applied in the clockwise direction to counteract the counterclockwise moment created by the force couple. Therefore

$$M_{12} = \left|\overline{F}_{32}\right| d = 256\ \mathrm{lb} \times 0.89\ \mathrm{in} = 227\ \mathrm{in\text{-}lb\ CW}$$

EXAMPLE 8.3 STATIC FORCE ANALYSIS OF A LEVER MECHANISM

For the mechanism shown in Figure 8.10(a), determine the magnitude and sense of the torque to be applied to link 2 by the base link to keep the mechanism in static equilibrium while being subjected to a force of magnitude 50 N.

$$r_{O_2 B} = 3.0\ \mathrm{cm}$$

SOLUTION

By inspection of Figure 8.10(a), links 4 and 6 are two–force members, and links 3 and 5 are three-force members. Link 2 is not a two–force member because a torque is to applied to this link. The lines of action of the forces are shown in Figure 8.10(b). For link 5, the line of action of force $\overline{F}$ and the line of action of the force from link 6 intersect at K_1. Then the line of action of the third force on link 5, through the turning pair at C, is drawn through this intersection point. For link 3, the line of action of the force through C intersects the line of action of the force acting on link 3 from link 4 at K_2. The line of action of the third force acting on link 3, through the turning pair at B, is drawn through this intersection point.

Force polygons of links 5 and 3 are also shown in Figure 8.10(b). Since

$$\overline{F}_{32} = -\overline{F}_{23}$$

then we construct a free body diagram of link 2 as shown in Figure 8.10(b). The moment that must be supplied to link 2 is

Video 8.11
Scissor Jack

$$M_{12} = \left|\overline{F}_{32}\right| d = 26\ \mathrm{N} \times 2.8\ \mathrm{cm} = 73\,\mathrm{N\text{-}cm\ CW}$$

(Continued)

EXAMPLE 8.3 Continued

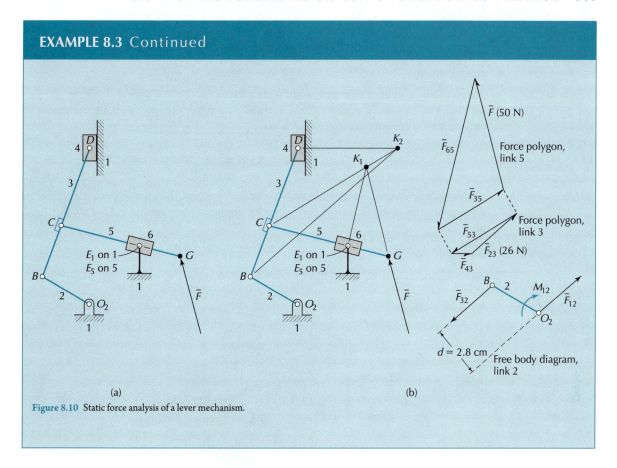

(a)   (b)

Figure 8.10 Static force analysis of a lever mechanism.

EXAMPLE 8.4 STATIC FORCE ANALYSIS OF A SCISSOR JACK

Figure 8.11(a) illustrates a scissor jack mechanism that is used for applications requiring the lift of heavy loads. Typically, a scissor jack is used to lift a vehicle axle during a tire change or used for vertical lift platforms. This mechanism is able to convert rotational input motion at the input crank into vertical linear output motion at the top bracket. The design permits rotational input motion requiring relatively low applied torque to generate a high-force vertical translational output. Hence, a human arm can provide the low-torque to turn the crank, resulting in forces strong enough to lift a vehicle. Figures 8.11(b) and 8.11(c) show the side view of the scissor jack in collapsed and extended positions, respectively. Figure 8.11(d) shows a schematic of a similar system. Gear sectors on links 4 and 6, and on links 2 and 8 are in mesh. They ensure link 5 remains parallel to itself as the mechanism moves. However, if the scissor jack is on a horizontal surface, and vertically applied load, $\overline{P}$, is through centerline of the scissor jack, then the interactive forces between the gear teeth are negligible.

Starting from the configuration with $\alpha = \alpha_0 = 25°$, determine the axial forces in links 4 and 9 as a function of the number of turns of the input crank, up to 100 turns.

$$r_{O_2A} = r_{AB} = r_{CD} = r_{O_8D} = L = 18.0 \text{ cm}$$

$$pitch = 0.3 \text{ cm}; \quad |\overline{P}| = 4000 \text{ N}$$

(*Continued*)

EXAMPLE 8.4 Continued

SOLUTION

If we neglect the interactive forces between the gear sectors that are in mesh, then links 2, 4, 6, 8, and 9 are two-force members. Links 3, 5, and 7 are three-force members. By inspection, because of the symmetry of the system, we have

$$\left|\bar{F}_{45}\right| = \left|\bar{F}_{43}\right| = \left|\bar{F}_{23}\right| = \left|\bar{F}_{12}\right|$$

For each rotation of the input crank, the distance between the turning pairs at A and D is reduced by the pitch. Therefore, the horizontal distance, h, as a function of the number of turns, n, is

$$h = h_0 - \frac{pitch}{2}n, \quad \text{where } h_0 = L\cos\alpha_0 = 16.3 \text{ cm}$$

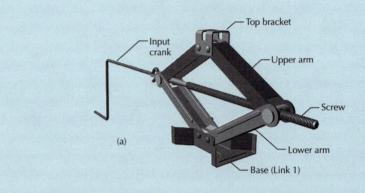

(a)

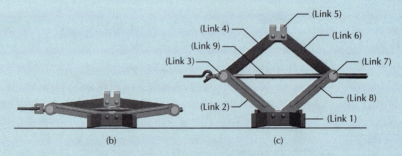

(b)   (c)

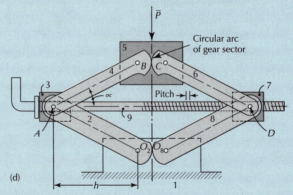

(d)

Figure 8.11 Scissor jack [Video 8.11]: (a) Isometric view, (b) Collapsed position, (c) Extended position, (d) Schematic.

(*Continued*)

EXAMPLE 8.4 Continued

From Figure 8.11(d) we have

$$\alpha = \cos^{-1}\left(\frac{h}{L}\right) = \cos^{-1}\left(\frac{2\,h_0 - pitch\,n}{2\,L}\right)$$

For an arbitrary configuration, the force polygons for links 5 and 3 are illustrated in Figure 8.12, from which we obtain

$$\left|\bar{F}_{45}\right| = \frac{\left|\bar{P}\right|}{2\sin\alpha}; \quad \left|\bar{F}_{93}\right| = 2\left|\bar{F}_{43}\right|\cos\alpha = 2\left|\bar{F}_{45}\right|\cos\alpha$$

By inspection we have

$$\text{compressive load in link } 4 = \left|\bar{F}_{45}\right|; \quad \text{tensile load in link } 9 = \left|\bar{F}_{93}\right|$$

Substituting given values in the above equations for the starting configuration, we obtain

$$\left|\bar{F}_{45}\right|_{\alpha=25°} = 4730 \text{ N}; \quad \left|\bar{F}_{93}\right|_{\alpha=25°} = 8580 \text{ N}$$

Figure 8.13 illustrates plots of the angle α and magnitudes of the compressive load in link 4 and tensile load in link 9 as a function of the number of turns. Because of the symmetry of the system, as α approaches 90° the compressive load in link 4 approaches one-half the magnitude of $\bar{P}$, while the tensile load in link 9 vanishes.

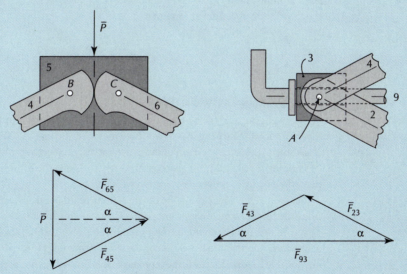

Figure 8.12 Force polygons for links 5 and 3.

(Continued)

EXAMPLE 8.4 Continued

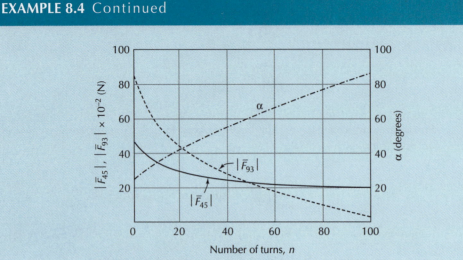

Figure 8.13 Analysis of a scissor jack.

EXAMPLE 8.5 STATIC FORCE ANALYSIS OF A FRONT-END LOADER

Consider the front-end loader mechanism shown in Figure 8.14. The mobility of this mechanism is equal to two. The two inputs are provided by the linear actuators, links 5 and 7. For static analysis, it is permissible to designate the linear actuators as two-force members.

For the position shown, a load of 5000 lb is applied to the bucket, link 6. Determine

(a) the magnitude of the force in the actuators, links 5 and 7
(b) the magnitude of bending moment at B

$$r_{AB} = 32 \text{ in}$$

SOLUTION

(a) Link 2 is a two-force member. Links 4 and 6 are three-force members.

We begin the analysis by considering link 6. Both the magnitude and direction of the load are known. Since link 7 is a two-force member, the line of action of the force at point F is known. The lines of action of these two forces intersect at point K_1. The line of action of the force through D must also pass through K_1. Using the three lines of action, we complete the first force polygon associated with link 6.

Because of the equal and opposite directions of interactive forces, we can easily determine the forces acting on link 3 at points C and D. In addition, there are two other forces acting on link 3 through the turning pairs located at A and B. Link 3 can be referred to as a four-force member because it is connected to other links through four turning pairs. We are not able to solve for the forces applied to this link using the cases presented in Sections 8.2 and 8.3. Instead, it is necessary to consider a subsystem consisting of a combination of links 3, 6, and 7. This is permissible because under static conditions there is no relative motion between links.

The subsystem is subjected to three external forces. Therefore the properties of a three-force member may be applied. The line of action of the external load and that of the two-force member of link 2 intersect at point K_2. The line of action of the force through the turning pair at B

(Continued)

EXAMPLE 8.5 Continued

must also pass through the same intersection point. Having determined the three lines of action, we construct a force polygon corresponding to the subsystem. Included in this polygon is force $\bar{F}_{43}$.

We now consider the other three-force member, link 4. We employ the result from the previous force polygon and realize that

$$\bar{F}_{34} = -\bar{F}_{43} \tag{8.6-7}$$

The line of action of the force through B has already been determined. This line of action intersects that coming from link 5 at K_3 The third line of action through point O_4 is determined because it must also go through K_3. From the three lines of action, we determine the force polygon corresponding to link 4.

The magnitudes of the forces in links 5 and 7 are

$$|\bar{F}_{54}| = 20,400 \text{ lb} \quad \text{and} \quad |\bar{F}_{76}| = 10,500 \text{ lb}$$

respectively.

(b) From the force polygon analysis, we determine the force $\bar{F}_{23}$ which acts through the turning pair at A. To determine the bending moment at B, we take the component of this force that is perpendicular to the direction of link 3. The result is

$$(F_{23})_{\text{L}} = 10,400 \text{ lb}$$

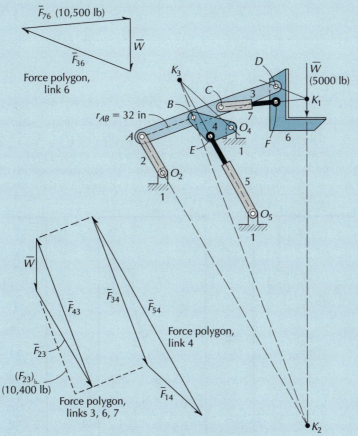

Figure 8.14 Static force analysis of a front-end loader mechanism.

(Continued)

EXAMPLE 8.5 Continued

Video 8.15
Front End Loader
Mechanism

By inspection, there is no bending moment in link 3 at A. The magnitude of the bending moment at B is then

$$M_B = r_{AB} \times (F_{23})_{\perp} = 32 \text{ in} \times 10{,}400 \text{ lb} = 3.33 \times 10^5 \text{ in-1b}$$

Figure 8.15 shows two positions of the front-end loader mechanism. The speeds of the animation shown through [**Video 8.15**] are so slow that for all positions the mechanism is essentially in static equilibrium, that is, for all configurations, lines of action of forces for each three-force member intersect at a common point.

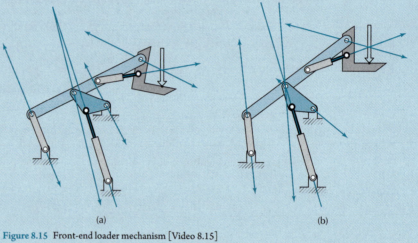

(a)   (b)

Figure 8.15   Front-end loader mechanism [Video 8.15]

8.7 PRINCIPLE OF SUPERPOSITION

When a mechanism is subjected to more than one force simultaneously, it is not possible to immediately determine their combined effect in one graphical force analysis. Instead, we may perform multiple analyses where each has only one of the forces being applied. The combined result may be found by superimposing individual results. This procedure is referred to as the *Principle of Superposition*. This principle is illustrated in Figure 8.16, and in the following examples.

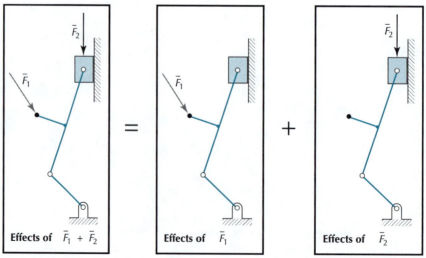

Effects of $\bar{F}_1 + \bar{F}_2$ **Effects of** $\bar{F}_1$ **Effects of** $\bar{F}_2$

Figure 8.16   Example of Principle of Superposition

EXAMPLE 8.6 STATIC FORCE ANALYSIS OF A PARKING BRAKE MECHANISM

Figure 8.17 illustrates two views of a parking brake mechanism that is used to either engage or disengage the brakes of an automobile. This mechanism uses the principle of mechanical advantage to magnify a low-force input motion, into a high-force output motion. Normally, a parking brake is a fail-safe system that is required to operate in the absence of automobile power. Therefore, human power alone must be sufficient to engage or disengage the brakes of an automobile. Since brakes require a high-force to engage them, the parking brake mechanism serves as a means to magnify the relatively low-power exerted by a human leg. When a person steps and pushes on the brake pedal, the pedal rotates about the pivot. The distance from the foot pedal to the pivot is much greater than the brake cable to the pivot, resulting in the increase of the magnitude of the force. As the foot pedal is depressed further, a pawl and ratchet engages in discrete steps, preventing retraction of the pedal, and hence maintaining the tension developed in the brake cable. When it is desired to release the brakes, the person depresses the foot pedal again, causing the pawl to disengage from the ratchet, and allowing the mechanism to return to its initial position (due to the spring).

Video 8.17
Parking Brake
Mechanism

For the position shown in Figure 8.17(b), determine the magnitude of the force required to be applied to the pedal and the resultant force at the pivot to overcome

(a) the tension in the brake cable
(b) the tensile force in the spring
(c) the superimposed effects of the cable and the spring.

$$\left|\overline{F}_{cable}\right| = 60 \text{ N}; \quad \left|\overline{F}_{spring}\right| = 80 \text{ N}$$

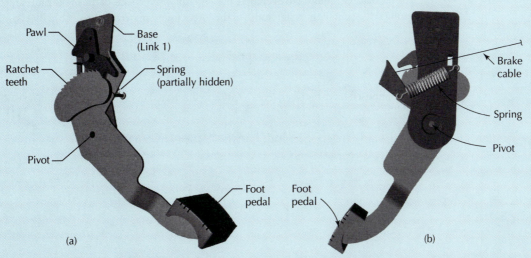

Figure 8.17 Parking brake mechanism [Video 8.17]

SOLUTION

(a) Referring to Figure 8.18, we consider only the force in the brake cable for this part of the problem. The cable can be considered as a two-force member in tension. Link 2 is a three-force member. The force polygon associated with link 2 is constructed from which we measure

$$\left|\overline{F}_{pedal, cable}\right| = 28 \text{ N}; \quad \left|\overline{F}_{12,cable}\right| = 82 \text{ N}$$

(Continued)

EXAMPLE 8.6 Continued

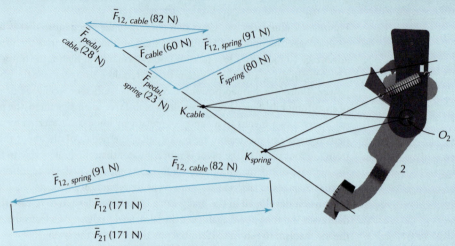

Figure 8.18 Analysis of a parking brake mechanism.

(b) For this part of the problem we consider only the force on the spring. The spring may be treated as a two-force member. From the associated force polygon of link 2 we measure

$$\left|\bar{F}_{pedal,spring}\right| = 23 \text{ N}; \quad \left|\bar{F}_{12,spring}\right| = 91 \text{ N}$$

(c) The pedal force required to overcome the superimposed effects of the brake cable and the spring is the sum of the magnitudes of the forces found in parts (a) and (b). That is,

$$\left|\bar{F}_{pedal}\right| = \left|\bar{F}_{pedal,cable}\right| + \left|\bar{F}_{pedal,spring}\right| = 28 \text{ N} + 23 \text{ N} = 51 \text{ N}$$

The superimposed force between links 1 and 2, as illustrated in Figure 8.18, has a magnitude

$$\left|\bar{F}_{12}\right| = \left|\bar{F}_{12,cable} + \bar{F}_{12,spring}\right| = 171 \text{ N}$$

EXAMPLE 8.7 STATIC FORCE ANALYSIS OF A FOUR-BAR MECHANISM

Consider the four-bar mechanism shown in Figure 8.19(a). The mechanism is subjected to two forces, one on link 3, and the other on link 4. Determine the torque required to be supplied from the base link onto link 2 in order to keep the mechanism in equilibrium. Neglect the effect of gravity.

$$r_{O_2B} = 10 \text{ in}$$

(Continued)

EXAMPLE 8.7 Continued

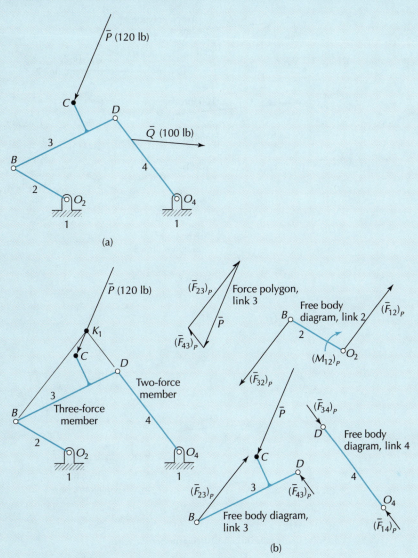

Figure 8.19 (a) Mechanism, (b) Load applied to link 3.

SOLUTION

We begin by considering only load $\overline{P}$ on link 3 and neglecting load $\overline{Q}$ on link 4, as shown in Figure 8.19(b). In this instance, link 3 may be considered as a three-force member, whereas link 4 is a two-force member. Through the analysis shown, the force $(\overline{F}_{32})_P$ is determined.

Next, we remove load $\overline{P}$ from link 3 and apply load $\overline{Q}$ to link 4, as shown in Figure 8.19(c). In this instance, link 3 may be considered as a two-force member, and link 4 is now a three-force member. Through the analysis shown, the force $(\overline{F}_{32})_Q$ is determined.

The total load between links is found by superimposing the above results, that is,

$$\overline{F}_{32} = (\overline{F}_{32})_P + (\overline{F}_{32})_Q \qquad (8.7\text{-}1)$$

(Continued)

EXAMPLE 8.7 Continued

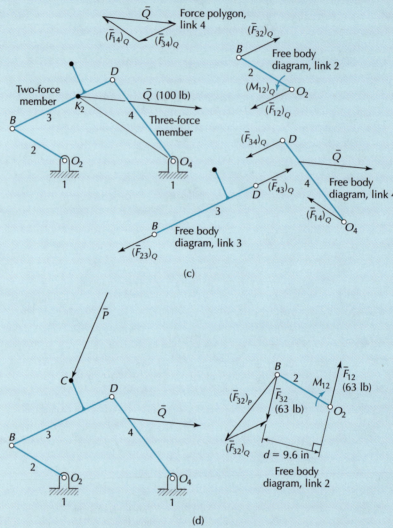

Figure 8.19 (c) Load applied to link 3, (d) Superimposed results.

The result is shown in Figure 8.19(d), from which

$$\left|\bar{F}_{32}\right| = 63 \text{ lb}$$

To maintain static equilibrium of link 2, we require

$$\bar{F}_{12} = -\bar{F}_{32} \tag{8.7-1}$$

In addition, a moment is supplied from the base link onto link 2. Its magnitude and direction are

$$M_{12} = \left|\bar{F}_{32}\right| \times d = 63 \text{ lb} \times 9.6 \text{ in} = 605 \text{ in-lb CW}$$

where $d = 9.6$ in is the perpendicular distance between the lines of action of the forces acting on link 2. This moment must be supplied in the clockwise direction to counteract the counter clockwise moment created by the force couple.

8.8 GRAPHICAL DYNAMIC FORCE ANALYSIS OF A MECHANISM LINK INERTIA CIRCLE

We now consider the analysis of a mechanism link under dynamic conditions. Through the use of inertia forces and inertia moments introduced in Chapter 2, dynamic problems can be treated in the same manner as static problems. Therefore, concepts introduced for static analyses are employed for dynamic problems.

Consider the mechanism link shown in Figure 8.20(a). The center of mass is indicated as point G. The mass of the link is m, and the polar mass moment of inertia about its mass center is I_G. In addition, for the given position, suppose we know the linear and angular accelerations of the link. For the time being, we isolate the link from the rest of the mechanism.

On the link shown in Figure 8.20(a) are drawn the vector of linear acceleration of the center of mass, $\bar{a}_G$, and the angular acceleration, $\ddot{\theta}$. We may generate the linear acceleration of the mass center by applying an external force that is parallel to $\bar{a}_G$. The force (see Section 2.12.1) required is

$$\bar{F} = m\bar{a}_G$$

(8.8-1)

Equation (8.8-1) is independent of the location of application of $\bar{F}$. Therefore, regardless of where the force is applied, we always obtain the same linear acceleration of G. However, different angular accelerations of the link will be generated by varying the location of $\bar{F}$. If $\bar{F}$ is applied at the center of mass (Figure 8.20(b)), then it will not provide a moment about G, and based on Equation (2.12-4), the link will have zero angular acceleration. However, by applying the force at a position other than the mass center, the force will cause a moment about G. Using Equation (2.12-4), the applied moment will cause the link to undergo an angular acceleration. By applying the force as shown in Figure 8.20(c), $\bar{F}$ also causes a moment, M_G, about the mass center in the

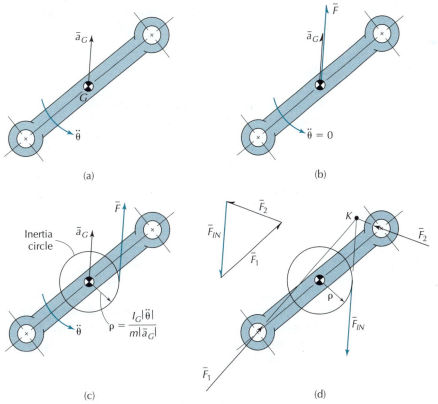

(a)

(b)

(c)

(d)

Figure 8.20 Dynamic force analysis of a mechanism link

counterclockwise direction and as a result causes the link to have on angular acceleration in the same direction. We recognize that

$$|M_G| = |\bar{F}|\rho \tag{8.8-2}$$

where ρ is the perpendicular distance of the line of action of the force from G. Combining Equations (2.12-4), (8.8-1), and (8.8-2), we obtain

$$|M_G| = |\bar{F}|\rho = m\,|\bar{a}_G|\rho = I_G|\ddot{\theta}| \tag{8.8-3}$$

Combining Equations (2.14-15), (2.14-18), and (8.8-3), we get

$$\rho = \frac{|M_G|}{|\bar{F}|} = \frac{I_G|\ddot{\theta}|}{m|\bar{a}_G|} = \frac{|M_{IN}|}{|\bar{F}_{IN}|} \tag{8.8-4}$$

Substituting given motions in Equation (8.8-4), we may calculate ρ, which indicates how far distant $\bar{F}$ must be placed from G in order to produce both the given linear and angular accelerations. As a guide, we draw an *inertia circle* of radius ρ, centered at G, as shown in Figure 8.20(c). Placing $\bar{F}$ tangent to the inertia circle ensures that the perpendicular distance is ρ.

It must be recognized, however, that acceleration of the link is not caused by $\bar{F}$. Rather, for the current illustration, forces are actually applied through turning pairs of adjacent links. Figure 8.20(d) illustrates forces $\bar{F}_1$ and $\bar{F}_2$ at the two turning pairs of the link, for which it is required that

$$\bar{F} = \bar{F}_1 + \bar{F}_2 \tag{8.8-5}$$

Therefore, instead of $\bar{F}$, we employ an inertia force that has the same line of action as $\bar{F}$ but acting in the opposite direction, as shown in Figure 8.20(d). This inertia force when combined with forces $\bar{F}_1$ and $\bar{F}_2$ must sum to zero, that is,

$$\bar{F} + \bar{F}_{IN} = \bar{0} \tag{8.8-6}$$

Also, the inertia force produces an inertia moment in a direction opposite to that produced by the actual forces, such that

$$M_{IN} = -M_G \tag{8.8-7}$$

or

$$M_G + M_{IN} = 0 \tag{8.8-8}$$

The form of Equations (8.8-6) and (8.8-8) is similar to that of a static problem when we consider one of the forces as an inertia force, and one of the moments as an inertia moment. This link may be considered as a three-force member in which one of the forces is the inertia force.

The Principle of Superposition, presented in Section 8.7, may be used for dynamic analyses, where inertia forces would be considered one at a time, and the results could then be superimposed to find their combined effect.

8.9 DYNAMIC FORCE ANALYSIS USING FORCE POLYGON METHOD

To perform a dynamic force analysis of a mechanism, the following general steps are required:

1. Draw a diagram of the mechanism to scale, in the configuration for which the analysis is required.
2. Using prescribed input motion(s), construct an acceleration polygon and determine the linear accelerations of link mass centers, the link angular accelerations, the inertia forces, and the inertia moments.
3. Indicate on the diagram of the mechanism all externally applied and inertia forces (offset each inertia force from the mass center to provide the required inertia moment).
4. Treat each inertia force and externally applied load individually, draw force polygons for each, and determine the effect of each throughout the mechanism.
5. If the combined effect of more than one force is to be determined, employ the Principle of Superposition.

EXAMPLE 8.8 DYNAMIC FORCE ANALYSIS OF A SLIDER CRANK MECHANISM

For the slider crank mechanism shown in Figure 8.21(a), the link lengths are

$$r_{O_2B} = r_2 = 3.5 \text{ in}; \quad r_{BD} = r_3 = 9.0 \text{ in}; \quad r_{BG_3} = 4.5 \text{ in}$$

and the constant input rotational speed is

$$\dot{\theta}_2 = 500 \text{ rpm CCW} = 52.4 \text{ rad/sec CCW}$$

An outline of the acceleration polygon is given in Figure 8.21(b).
    The coupler, link 3, has a uniform cross section and the following properties:

$$m_3^* = 3.0 \text{ lbm}; \quad I_{G_3}^* = 47.0 \text{ lbm-in}^2$$

For the position shown, determine the forces on link 3 at the turning pairs due to the inertia of link 3.

SOLUTION

Since the coupler has a uniform cross section, we locate the position of its center of mass midway between B and D on the acceleration polygon, as shown in Figure 8.21(b). The magnitude of the acceleration of the center of mass is

$$|\bar{a}_{G_3}| = 6350 \text{ in/sec}^2$$

and the tangential acceleration component between points B and D is

$$a_{DB}^T = 7330 \text{ in/sec}^2$$

(Continued)

EXAMPLE 8.8 Continued

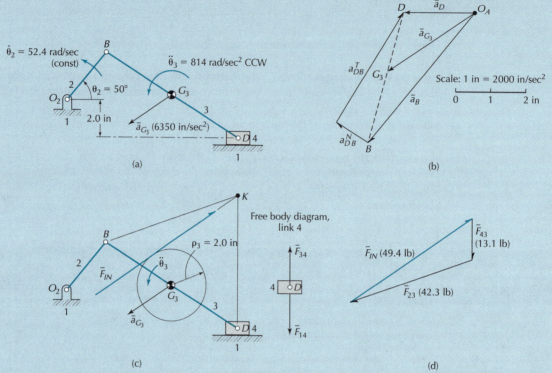

Figure 8.21 Dynamic force analysis of a slider crank mechanism: (a) Mechanism, (b) Acceleration polygon, (c) Mechanism with inertia circle, (d) Force polygon.

The angular acceleration of link 3 is then

$$\ddot{\theta}_3 = \frac{a_{DB}^T}{r_{DB}} = \frac{7330}{9.0} = 814 \text{ rad} / \sec^2 \text{ CCW}$$

The units of the quantities employed in this example correspond to the third set listed in Table 2.3. In the calculation of the inertia force, we substitute the corresponding value of the gravitational constant, g_c, in Equation (2.14-15) to calculate the magnitude of the inertia force:

$$\left| \overline{F}_{IN} \right|_3 = m_3 \left| \overline{a}_{G_3} \right| = \frac{m_3^*}{g_c} \left| \overline{a}_{G_3} \right|$$

$$= \frac{3.0 \text{ lbm}}{\left(386 \dfrac{\text{lbm-in}}{\text{lb-sec}^2} \right)} \times 6350 \text{ in} / \sec^2 = 49.4 \text{ lb}$$

From Equation (2.14-18), the magnitude of the inertia moment is

$$\left| M_{IN} \right|_3 = I_{G_3} \left| \ddot{\theta}_3 \right| = \frac{I_{G_3}^*}{g_c} \left| \ddot{\theta}_3 \right|$$

$$= \frac{47.0 \text{lbm-in}^2}{\left(386 \dfrac{\text{lbm-in}}{\text{lb-sec}^2} \right)} \times 814 \text{ rad} / \sec^2 = 99.1 \text{ in- lb}$$

EXAMPLE 8.8 Continued

Using Equation (8.8-4), the radius of the inertia circle is

$$\rho_3 = \frac{|M_{IN}|_3}{|\overline{F}_{IN}|_3} = \frac{99.1 \text{ in-lb}}{49.4 \text{ lb}} = 2.0 \text{ in}$$

The inertia force, $\overline{F}_{IN}$, is added to the drawing of the mechanism, offset by the radius of the inertia circle. The direction of $\overline{F}_{IN}$ is opposite to that of $\overline{a}_{G_3}$. In this example, the link has an angular acceleration in the counterclockwise direction. Thus, the forces at the turning pairs are providing a moment in the counterclockwise direction. Therefore, we place the inertia force as shown in Figure 8.21(c) so that it applies an inertia moment in the clockwise direction, opposite to that of the actual moment. The sum of the moments is then zero. Since the mass of the slider is not considered in the analysis, the slider (link 4) is a two-force member. The direction of the force transmitted from the slider onto link 3 is perpendicular to the direction of the slide. The lines of action of the inertia force and that from the slider intersect at K. The third line of action, through the turning pair at B, must pass through the same intersection point. Using the inertia force and the other two lines of action, the force polygon is completed, as shown in Figure 8.21(d). The magnitudes of the forces located at B and D as a result of the inertia in link 3 are

$$|\overline{F}_{43}| = 13.1 \text{ lb}; \quad |\overline{F}_{23}| = 42.3 \text{ lb}$$

The directions of these forces are indicated in the force polygon.

EXAMPLE 8.9 DYNAMIC FORCE ANALYSIS OF A FOUR-BAR MECHANISM

Figure 8.22(a) shows a four-bar mechanism. Link 2 rotates at a constant rotational speed of 120 rad/sec in the counterclockwise direction. Also indicated in the figure are the results of the acceleration analysis. Pertinent properties of links 3 and 4 are listed in Table 8.1.

Determine the magnitude and sense of the moment to be applied to link 2 from base link 1 to overcome the inertias of links 3 and 4.

$$r_1 = 9.0 \text{ cm}; \quad r_2 = 2.0 \text{ cm}; \quad r_3 = 13.0 \text{ cm}; \quad r_4 = 7.0 \text{ cm}$$

SOLUTION

We will solve this problem using the Principle of Superposition. For the first part of the solution, only the inertia of link 3 will be considered. In the second part, only the effects of the inertia of link 4 will be evaluated. Finally, the results will be superimposed. The units of the quantities employed in this example correspond to the second set listed in Table 2.2, for which $g_c = 1$.

For the first part of the problem, using the acceleration parameters given in Figure 8.22(a), the magnitude of the inertia force of link 3 is

$$|\overline{F}_{IN}|_3 = m_3 |\overline{a}_{G_3}| = \frac{m_3^*}{g_c} |\overline{a}_{G_3}|$$

$$= \frac{0.50 \text{ kg}}{1} \times 236 \text{ m/sec}^2 = 118 \text{ N}$$

(Continued)

EXAMPLE 8.9 Continued

The magnitude of the inertia moment of link 3 is

$$\left|M_{IN}\right|_3 = I_{G_3}\left|\ddot{\theta}_3\right| = \frac{I_{G_3}^*}{g_c}\left|\ddot{\theta}_3\right|$$

$$= \frac{2.52 \times 10^3 \, \text{kg-m}^2}{1} \times 725 \ \text{rad/sec}^2 = 1.83 \, \text{N-m}$$

The radius of the inertia circle of link 3 is

$$\rho_3 = \frac{\left|M_{IN}\right|_3}{\left|\overline{F}_{IN}\right|_3} = \frac{1.83 \ \text{N-m}}{118 \ \text{N}} = 0.0155 \ \text{m} = 1.55 \ \text{cm}$$

Figure 8.22(b) shows the corresponding analysis in which only the inertia associated with link 3 is considered. This link is treated as a three-force member. Link 4 is assumed to have no mass and is treated as a two-force member. From this analysis, we determine

$$\left(\overline{F}_{32}\right)_3 = -\left(\overline{F}_{23}\right)_3$$

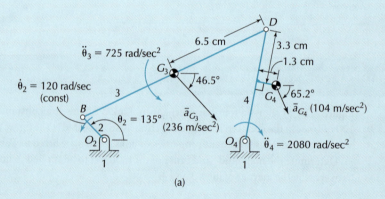

(a)

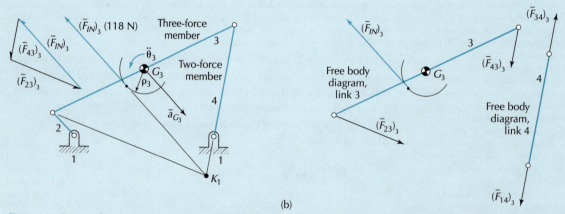

(b)

Figure 8.22 Dynamic force analysis of a four-bar mechanism: (a) Input parameters and results of acceleration analysis, (b) Effects of inertia of link 3.

(*Continued*)

EXAMPLE 8.9 Continued

For the second part of the problem, only the inertia of link 4 is considered. In this instance, link 4 is treated as a three-force member, and link 3 is treated as a two-force member. Carrying out analyses similar to those completed for link 3 gives

$$|\bar{F}_{IN}|_4 = 83.2 \text{ N}; \quad |M_{IN}|_4 = 1.66 \text{ N-m}; \quad \rho_4 = 2.00 \text{ cm}$$

Figure 8.22(c) shows the force analysis for this part of the problem. The analysis provides

$$(\bar{F}_{32})_4 = -(\bar{F}_{23})_4$$

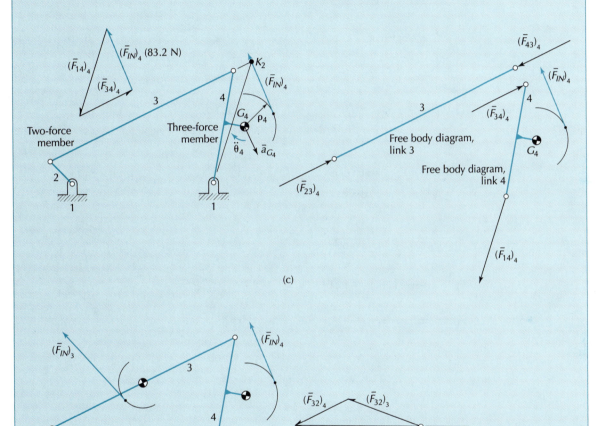

Figure 8.22 (c) Effects of inertia of link 4 analysis, (d) Superimposed result.

TABLE 8.1 PARAMETERS OF THE FOUR-BAR MECHANISM SHOWN IN FIGURE 8.22

	Link 3	Link 4
m^* (kg)	0.5	0.8
I_G^* (kg-m^2)	2.52×10^{-3}	8.00×10^{-4}

(Continued)

EXAMPLE 8.9 Continued

The superimposed results are illustrated in Figure 8.22(d). They include

$$\left|\overline{F}_{32}\right| = \left|\left(\overline{F}_{32}\right)_3 + \left(\overline{F}_{32}\right)_4\right| = 160\ N; \quad d = 0.014\ m$$

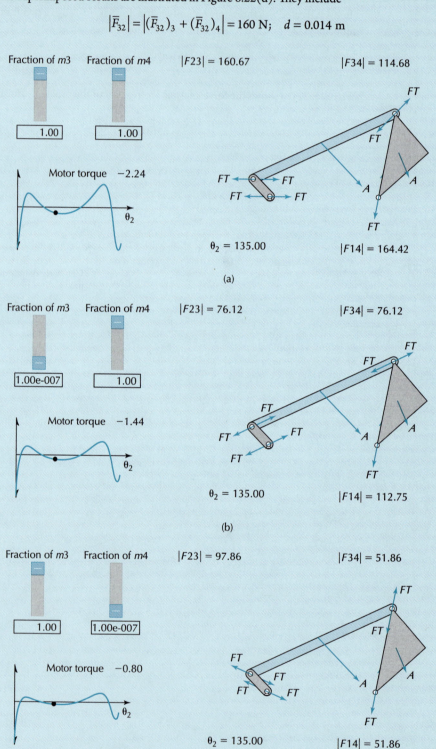

Figure 8.23 Dynamic analysis of a four-bar mechanism [Model 8.23]: (a) Links 3 and 4 have mass, (b) Only link 4 has mass, (c) Only link 3 has mass.

(Continued)

EXAMPLE 8.9 Continued

where d is the perpendicular distance between the lines of action of forces $\bar{F}_{32}$ and $\bar{F}_{12}$. The moment required from the base link on link 2 is then

$$M_{12} = |\bar{F}_{32}|d = 160 \text{ N} \times 0.014 \text{ m} = 2.24 \text{ N-m CW}$$

Results of this example show that the moment to be supplied from the base link onto link 2 is in the clockwise direction, opposite to its direction of rotational speed. A dynamic force analysis of the same mechanism throughout a complete cycle of motion is provided through [Video 8.23]. During the cycle, the driving moment takes on positive and negative values. Positive values correspond to the counterclockwise direction. With this model it is possible to adjust values of the masses and, consequently, values of polar mass moments of inertia about the mass centers. Figure 8.23(a) shows the torque curve when both links 3 and 4 have mass. Figure 8.23(b) shows the case where the mass of link 3 has been set to zero, and only the mass of link 4 remains. In this instance, link 3 acts as a two-force member. Figure 8.23(c) shows the case where the mass of link 4 has been set to zero, and only the mass of link 3 remains.

Video 8.23
Model 8.23
Dynamic Analysis
of a Four-Bar
Mechanism

PROBLEMS

Static Force Analysis

P8.1 Assuming the mechanism shown in Figure P8.1 is in static equilibrium, determine forces $\bar{Q}$ and $\bar{F}_{12}$.

$r_{O_2B} = 4.0 \text{ cm}$

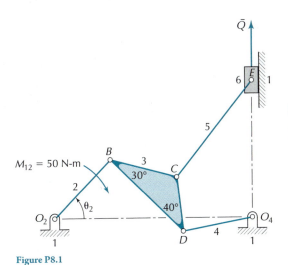

Figure P8.1

P8.2 For the vise grip mechanism shown in Figure P8.2, use force polygons to determine

  (a) the applied force $\bar{P}$ necessary to produce a force $\bar{Q}$ of magnitude 120 lb

  (b) the force $\bar{Q}$ that can be produced by an applied force $\bar{P}$ of magnitude 15 N

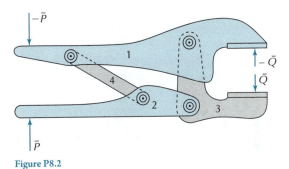

Figure P8.2

P8.3 For the quick-return mechanism in static equilibrium shown in Figure P8.3, the cutting force $\bar{F}$ has a magnitude of 30 N. Neglecting inertia, determine

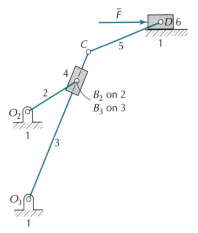

Figure P8.3

(a) the driving torque required on link 2

(b) the bearing forces at O_2 and O_3

$$r_{O_2B_2} = 4.0 \text{ cm}$$

P8.4 Construct a complete set of force polygons for the mechanism in static equilibrium shown in Figure P8.4, and if $\bar{Q} = 150$ lb, determine:

(a) $\bar{P}$

(b) $\bar{F}_{41}$

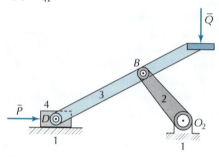

Figure P8.4

P8.5 Figure P8.5 shows a scaled diagram of a hand-operated toggle clamp. In the position shown, draw the static force polygons. Determine the clamping force, $\bar{F}$, when the hand force, $\bar{P}$, of 25 lb is applied as shown.

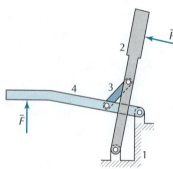

Figure P8.5

P8.6 Determine the torque required from the base link on link 2 for static equilibrium of the mechanism shown in Figure P8.6. Force $\bar{F}$ has a magnitude of 20 lb.

$$r_{O_2B_2} = 2.0 \text{ i}$$

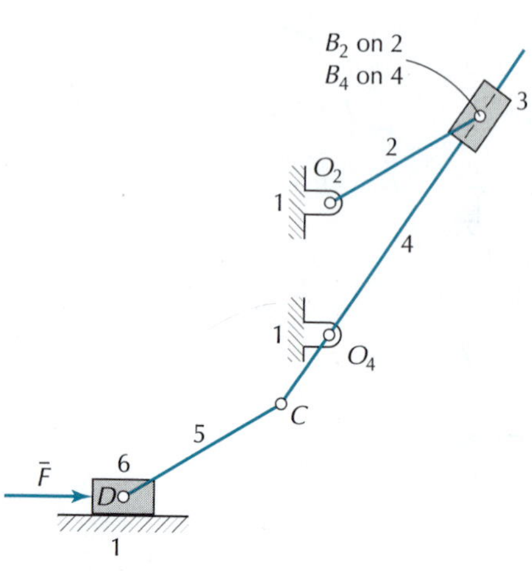

P8.7 Determine the required cylinder pressure for static equilibrium of the mechanism shown in Figure P8.7. Torque M_{14} has a magnitude of 100 N-m. The diameter of the piston is 1.0 cm.

$$r_{O_4D} = 6.0 \text{ cm}$$

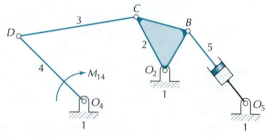

Figure P8.7

P8.8 Determine the required input torque M_{12} for static equilibrium of the mechanism shown in Figure P8.8. Force $\bar{F}$ has a magnitude of 1000 N.

$$r_{O_2B} = 3.0 \text{ cm}$$

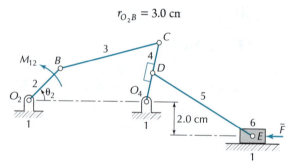

Figure P8.8

P8.9 Determine the pressure required in the 2.0 cm diameter cylinder shown in Figure P8.9 to maintain static equilibrium. The image shown is to scale.

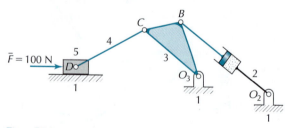

Figure P8.9

P8.10 For the given position of the mechanism shown in Figure P8.10, on the next page, determine the magnitude and sense of the torque required to be applied to crank O_6E by the base link, to maintain static equilibrium.

$$r_{O_2B} = 7.0 \text{ cm}; \quad r_{O_6E} = 3.0 \text{ cm}; \quad M_{12} = 5.0 \text{ N-m CW}$$

Figure P8.6

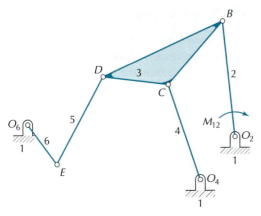

Figure P8.10

P8.11 For the mechanism illustrated in Figure P8.11, determine the magnitude and direction of the force $\bar{P}$ required to be applied to link 2, to maintain static equilibrium while being subjected to the external force shown on link 6.

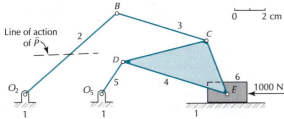

Figure P8.11

P8.12 For the mechanism illustrated in Figure P8.12, determine the magnitude of force $\bar{P}$, applied to link 1, required to produce a crimping force $\bar{Q}$, of magnitude 80 N.

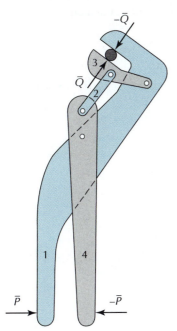

Figure P8.12

P8.13 For the mechanism illustrated in Figure P8.13, determine the magnitude and direction of moment M_{12}, applied to link 2, required to maintain static equilibrium while being subjected to force $\bar{F}$ and moment M_{14}.

$$r_{O_2B} = 5.0 \text{ cm}; \quad |\bar{F}| = 800 \text{ N}; \quad M_{14} = 50 \text{ N-m CW}$$

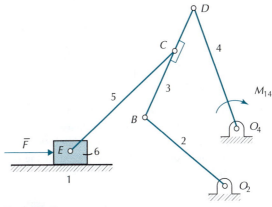

Figure P8.13

P8.14 Figure P8.14 shows a clamping mechanism. Determine the clamping force $\bar{Q}$ when the mechanism is subjected to the applied load $\bar{P}$ shown.

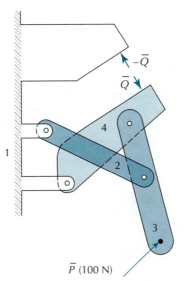

Figure P8.14

P8.15 Figure P8.15, on the next page, shows a clamping mechanism. Determine the clamping force $\bar{Q}$ when the mechanism is subjected to the applied load $\bar{P}$ of magnitude 100 N.

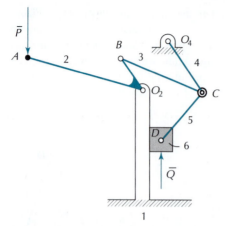

Figure P8.15

P8.16 Figure P8.16 shows a truck with a lifting mechanism. Determine the loads in the two actuators 2 and 4 to hold the container weighing 1600 N in the position shown. Neglect the weight of the other links.

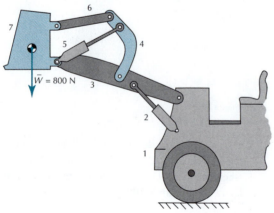

Figure P8.16

P8.17 Figure P8.17 shows a front-end loader mechanism. Determine the loads required in the actuators 2 and 5 to maintain equilibrium while holding the container weighing 800 N. Neglect the weight of the other links.

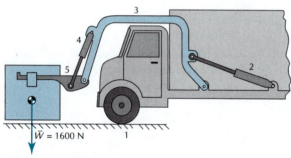

Figure P8.17

Dynamic Force Analysis

P8.18 For the mechanism shown in Figure P8.18, determine the magnitude and sense of the torque to be applied to link 2 from the base link to overcome the inertia of link 3.

$$r_{O_2B} = 5.0 \text{ cm}; \quad r_{BD} = 10.0 \text{ cm}; \quad r_{BG_3} = 5.0 \text{ cm}$$
$$m_3 = 800 \text{ gr}; \quad I_{G_3} = 6.4 \times 10^{-3} \text{ kg-m}^2$$
$$\dot{\theta}_2 = 600 \text{ rpm CCW(constant)}$$

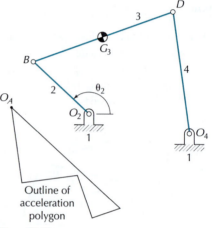

Figure P8.18

P8.19 For the mechanism shown in Figure P8.19, link 3 has a mass of 3.0 lbm and polar mass moment of inertia of 0.04 lb-in-sec² about its mass center G_3. Sliding block 2 has a constant velocity of 10 ft/sec upward. Determine the instantaneous force $\bar{F}$ required to produce this motion, assuming that the sliders are massless.

$$r_{BD} = 10.0 \text{ in}; \quad r_{BG_3} = 4.0 \text{ in}$$
$$\bar{a}_{G_3} = 2360 \text{ in/sec}^2 @180°; \quad \ddot{\theta}_3 = 392 \text{rad/sec}^2 \text{ CW}$$

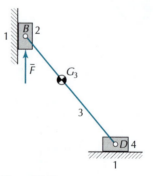

Figure P8.19

P8.20 For the mechanism shown in Figure P8.20, a gas force of 5000 lb acts to the left on link 4. Determine the magnitude and sense of the torque to be applied to link 2 from base link 1 to overcome

(a) the gas force
(b) the inertia of link 3
(c) the inertia of link 4
(d) the combined effects of parts (a), (b), and (c)

$$r_{O_2B} = 1.5 \text{ in}; \quad \dot{\theta}_2 = 1800 \text{ rpm CCW (constant)}$$
$$m_3 = 0.90 \text{ lbm}; \quad I_{G_3} = 0.010 \text{ lb-in-sec}^2$$
$$m_4 = 0.60 \text{ lbm}; \quad I_{G_4} = 5.0 \times 10^{-3} \text{ lb-in-sec}^2$$

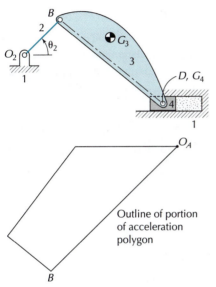

Figure P8.20

P8.21 For the given position of the mechanism shown in Figure P8.21, determine the magnitude and sense of the torque required to be applied to link 2 by the base link, to overcome the inertia of link 3.

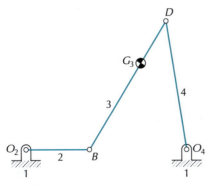

Figure P8.21

$$r_{O_2O_4} = 10.0 \text{ cm}; \quad r_{O_2B} = 4.0 \text{ cm};$$
$$r_{BD} = 9.0 \text{ cm}; \quad r_{BG_3} = 6.0 \text{ cm}$$
$$r_{O_4D} = 7.8 \text{ cm}; \quad m_3 = 600 \text{ gr};$$
$$I_{G_3} = 1.5 \times 10^{-3} \text{ kg-m}^2$$
$$\dot{\theta}_2 = 200 \text{ rpm CCW};$$
$$\ddot{\theta}_2 = 100 \text{ rad/sec}^2 \text{ CCW};$$
$$\bar{a}_{G_3} = 19.4 \text{ m/sec}^2 \text{ @} 210°;$$
$$\ddot{\theta}_3 = 122 \text{ rad/sec}^2 \text{ CW}$$

P8.22 For the given position of the mechanism shown in Figure P8.22, determine the magnitude and sense of the torque required to be applied to link 2 by the base link, to overcome the inertia of link 5.

$$r_{O_2B} = 3.0 \text{ cm}; \quad r_{CE} = 6.0 \text{ cm}; \quad r_{CG_5} = 3.0 \text{ cm}$$
$$m_5 = 5.0 \text{ gr}; \quad I_{G_5} = 5.0 \times 10^{-6} \text{ kg-m}^2$$
$$\dot{\theta}_2 = 3000 \text{ rpm CW (constant)}$$

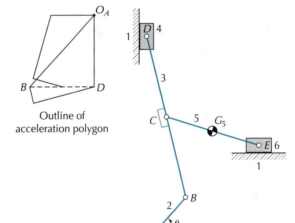

Figure P8.22

P8.23 For the given position of the mechanism shown in Figure P8.23, determine the magnitude and sense of the torque required to be applied to link 2 by the base link, to overcome

(a) the inertia of link 3
(b) the inertia of link 4
(c) the inertia of links 3 and 4

$$r_{O_2B} = 1.0 \text{ in}; \quad \dot{\theta}_2 = 475 \text{rpm CW (constant)}$$
$$m_3 = 9.0 \text{ lbm}; \quad I_{G_3} = 1.5 \times 10^{-2} \text{ lb-in-sec}^2$$
$$m_4 = 11.5 \text{ lbm}; \quad I_{G_4} = 8.0 \times 10^{-2} \text{ lb-in-sec}^2$$

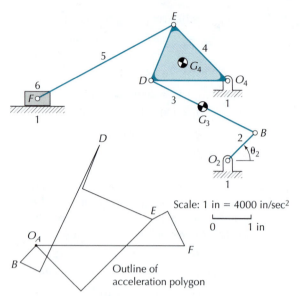

Scale: 1 in = 4000 in/sec²

Figure P8.23

P8.24 For the given position of the mechanism shown in Figure P8.24, determine the magnitude and sense of the torque required to be applied to link 2 by the base link, to overcome

(a) the inertia of link 3

(b) the inertia of link 4

(c) the inertia of links 3 and 4

$$r_{O_2B} = 5.0 \text{ cm}; \quad r_{BD} = 10.0 \text{ cm}; \quad r_{BG_3} = 3.0 \text{ cm}$$
$$m_3 = 400 \text{ gr}; \quad m_4 = 600 \text{ gr}; \quad I_{G_3} = 6.3 \times 10^{-4} \text{ kg-m}^2$$
$$\theta_2 = 40°; \quad \dot{\theta}_2 = 200 \text{ rpm CW}; \quad \ddot{\theta}_2 = 200 \text{ rad/sec}^2 \text{ CCW}$$
$$\bar{a}_{G_3} = 25.2 \text{ m/sec}^2 @ 192°; \quad \ddot{\theta}_3 = 122 \text{ rad/sec}^2 \text{ CW}$$
$$\bar{a}_{G_4} = 27.8 \text{ m/sec}^2 @ 180°$$

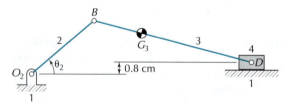

Figure P8.24

P8.25 For the given position of the mechanism shown in Figure P8.25, determine the magnitude and sense of the torque required to be applied to link 2 by the base link, to overcome the inertia of link 4.

$$r_{O_2B} = 5.0 \text{ cm}; \quad r_{O_4D} = 8.0 \text{ cm}; \quad r_{O_4G_4} = 4.0 \text{ cm}$$
$$m_4 = 0.50 \text{ kg}; \quad I_{G_4} = 5.0 \times 10^{-4} \text{ kg-m}^2$$
$$\ddot{\theta}_2 = 1200 \text{ rpm CW (constant)}$$

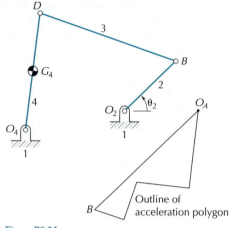

Figure P8.25

P8.26 For the given position of the mechanism shown in Figure P8.26, determine the magnitude and sense of the torque required to be applied to link 2 by the base link, to overcome the inertia of link 6.

$$r_{O_2B} = 5.0 \text{ cm}; \quad m_6 = 50 \text{ gr}; \quad I_{G_6} = 4.0 \times 10^{-5} \text{ kg-m}^2$$
$$\dot{\theta}_2 = 100 \text{ rpm CW}; \quad \ddot{\theta}_2 = 80 \text{ rad/sec}^2 \text{ CW}$$
$$\bar{a}_{G_6} = 380 \text{ cm/sec}^2 @253°; \quad \ddot{\theta}_6 = 90 \text{ rad/sec}^2 \text{ CCW}$$

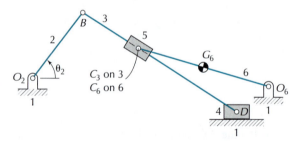

Figure P8.26

P8.27 For the given position of the mechanism shown in Figure P8.27, determine the magnitude and sense of the torque required to be applied to crank O_2B by the base link, to overcome the inertia of link 5. (*continued on next page*)

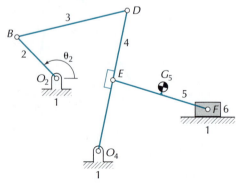

Figure P8.27

..

$\theta_2 = 135°; \quad \dot{\theta}_2 = 150$ rpm CCW (constant)

$r_{O_2B} = 4.0$ cm; $\quad m_5 = 49.2$ gr; $\quad I_{G_5} = 6.0 \times 10^{-3}$ kg-m^2

$\bar{a}_{G_5} = 145$ m/sec^2 @ 222°; $\quad \ddot{\theta}_5 = 29.7$ rad/sec^2 CCW

P8.28 For the given position of the mechanism shown in Figure P8.28, determine the magnitude and sense of the torque required to be applied to crank O_2B by the base link, to overcome

(a) the inertia of link 4
(b) the inertia of link 5
(c) the inertia of links 4 and 5

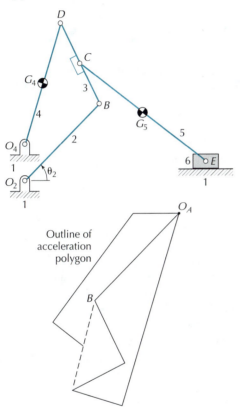

Figure P8.28

$r_{O_2B} = 6.0$ cm; $\quad \dot{\theta}_2 = 12.0$ rad/sec CCW (constant)

$m_4 = 30$ gr; $\quad I_{G_4} = 6.0 \times 10^{-5}$ kg-m^2

$m_5 = 50$ gr; $\quad I_{G_5} = 6.5 \times 10^{-5}$ kg-m^2

P8.29 For the given position of the mechanism shown in Figure P8.29, determine the magnitude and sense of the torque required to be applied to crank O_2B by the base link, to overcome the inertia of link 5.

$r_{O_2B} = 6.0$ cm; $\quad \dot{\theta}_2 = 12.0$ rad/sec CW (constant)

$m_5 = 50.0$ gr; $\quad I_{G_5} = 6.5 \times 10^{-5}$ kg-m^2

$\bar{a}_{G_5} = 690$ cm/sec^2 @ 318°; $\quad \ddot{\theta}_5 = 108$ rad/sec^2 CW

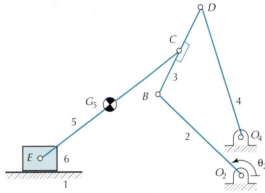

Figure P8.29

CHAPTER 9

ANALYTICAL FORCE ANALYSIS AND BALANCING OF PLANAR MECHANISMS

9.1 INTRODUCTION

In this chapter, analytical equations are presented for use in completing kinetic analyses of four-bar mechanisms and slider crank mechanisms. They are based on the governing equations of motion (Section 2.12) and the analytical expressions of kinematics of the links (Chapter 4). With use of a computer, multiple analyses throughout cycles of motion may be readily completed. Also presented are methods for either reducing or eliminating the net force imparted to the base link, called the *shaking force*, created by the unbalanced inertias of the moving links.

In all of these analyses, we restrict ourselves to cases where the rotational speed of input link 2 is constant. We will prescribe

$$\dot{\theta}_2 = \omega = \text{constant} \tag{9.1-1}$$

Therefore, we can set

$$\theta_2 = \omega t \tag{9.1-2}$$

Unless otherwise noted, we will assume that motions of all links of planar mechanisms take place in a single plane.

9.2 FORCE ANALYSIS OF A FOUR-BAR MECHANISM

Consider the four-bar mechanism illustrated in Figure 9.1(a). Locations of the centers of mass of links 2, 3, and 4 with respect to a turning pair are specified by distances b_i ($i = 2, 3, 4$) and angles φ_i ($i = 2, 3, 4$). Free-body diagrams of individual links are shown in Figure 9.1(b). The driving torque, M_{12}, is one of the quantities that must be determined.

To complete a kinetic analysis, we must determine accelerations of the centers of mass of the moving links. Referring to Figure 9.1(a) and implementing Equation (9.1-1), the results for link 2 are

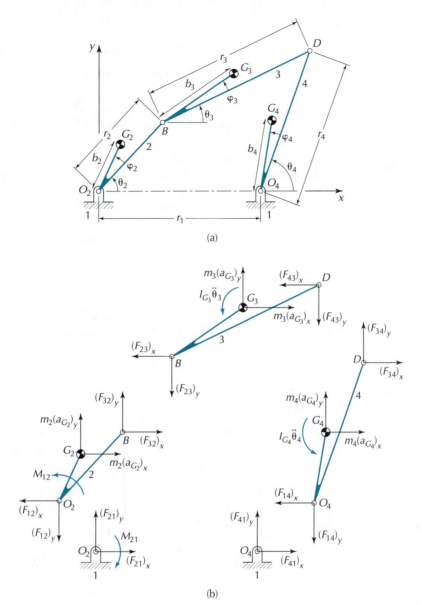

Figure 9.1 Dynamic analysis of a four-bar mechanism.

$$(r_{G_2})_x = b_2 \cos(\theta_2 + \varphi_2)$$

$$(a_{G_2})_x = (\ddot{r}_{G_2})_x = -b_2 \omega^2 \cos(\theta_2 + \varphi_2)$$

$$(r_{G_2})_y = b_2 \sin(\theta_2 + \varphi_2)$$

$$(a_{G_2}) = (\ddot{r}_{G_2})_y = -b_2 \omega^2 \sin(\theta_2 + \varphi_2) \qquad (9.2\text{-}1)$$

The results for link 3 are

$$(r_{G_3})_x = r_2 \cos\theta_2 + b_3 \cos(\theta_3 + \varphi_3)$$

$$(a_{G_3})_x = -r_2 \omega^2 \cos\theta_2 - b_3 \left[\dot{\theta}_3^2 \cos(\theta_3 + \varphi_3) + \ddot{\theta}_3 \sin(\theta_3 + \varphi_3) \right]$$

$$(r_{G_3})_y = r_2 \sin\theta_2 + b_3 \sin(\theta_3 + \varphi_3)$$

$$(a_{G_3})_y = -r_2 \omega^2 \sin\theta_2 - b_3 \left[\dot{\theta}_3^2 \sin(\theta_3 + \varphi_3) - \ddot{\theta}_3 \cos(\theta_3 + \varphi_3) \right] \qquad (9.2\text{-}2)$$

The results for link 4 are

$$(r_{G_4})_x = r_1 + b_4\cos(\theta_4 + \varphi_4)$$

$$(a_{G_4})_x = -b_4\left[\dot{\theta}_4^2\cos(\theta_4 + \varphi_4) + \ddot{\theta}_4\sin(\theta_4 + \varphi_4)\right]$$

$$(r_{G_4})_y = b_4\sin(\theta_4 + \varphi_4)$$

$$(a_{G_4})_y = -b_4\left[\dot{\theta}_4^2\sin(\theta_4 + \varphi_4) - \ddot{\theta}_4\cos(\theta_4 + \varphi_4)\right] \tag{9.2-3}$$

The governing equations of motion for link 2 are determined by considering the x and y components of Equation (2.12-1), as well as Equation (2.12-6). Referring to Figure 9.1(b), we obtain

$$(F_{32})_x - (F_{12})_x = m_2(a_{G_2})_x \tag{9.2-4}$$

$$(F_{32})_y - (F_{12})_y = m_2(a_{G_2})_y \tag{9.2-5}$$

$$-(F_{32})_x r_2\sin\theta_2 + (F_{32})_y r_2\cos\theta_2 + M_{12} = 0 \tag{9.2-6}$$

Similar equations may be generated for links 3 and 4. The governing equations of motion of all moving links may then be combined in matrix form and expressed as

$$[A]\{x\} = \{B\} \tag{9.2-7}$$

where

$$[A] = \begin{bmatrix} -1 & 0 & 1 & 0 & 0 & 0 & 0 & 0 & 0 \\ 0 & -1 & 0 & 1 & 0 & 0 & 0 & 0 & 0 \\ 0 & 0 & A_{3,3} & A_{3,4} & 0 & 0 & 0 & 0 & 1 \\ 0 & 0 & -1 & 0 & -1 & 0 & 0 & 0 & 0 \\ 0 & 0 & 0 & -1 & 0 & -1 & 0 & 0 & 0 \\ 0 & 0 & A_{6,3} & A_{6,4} & A_{6,5} & A_{6,6} & 0 & 0 & 0 \\ 0 & 0 & 0 & 0 & 1 & 0 & -1 & 0 & 0 \\ 0 & 0 & 0 & 0 & 0 & 1 & 0 & -1 & 0 \\ 0 & 0 & 0 & 0 & A_{9,5} & A_{9,6} & 0 & 0 & 0 \end{bmatrix} \tag{9.2-8a}$$

$$A_{3,3} = -r_2\sin\theta_2; \qquad A_{3,4} = r_2\cos\theta_2$$

$$A_{6,3} = -b_3(\sin\theta_3 + \varphi_3); \qquad A_{6,4} = b_3\cos(\theta_3 + \varphi_3)$$

$$A_{6,5} = r_3\sin\theta_3 - b_3\sin(\theta_3 + \varphi_3); \qquad A_{6,6} = -r_3\cos\theta_3 + b_3\cos(\theta_3 + \varphi_3)$$

$$A_{9,5} = -r_4\sin\theta_4; \qquad A_{9,6} = r_4\cos\theta_4$$

$$\{x\} = \begin{Bmatrix} (F_{12})_x \\ (F_{12})_y \\ (F_{23})_x \\ (F_{23})_y \\ (F_{34})_x \\ (F_{34})_y \\ (F_{14})_x \\ (F_{14})_y \\ M_{12} \end{Bmatrix}; \qquad \{B\} = \begin{Bmatrix} m_2(a_{G_2})_x \\ m_2(a_{G_2})_y \\ 0 \\ m_3(a_{G3})_x \\ m_3(a_{G3})_y \\ I_{G_3}\ddot{\theta}_3 \\ m_4(a_{G4})_x \\ m_4(a_{G4})_y \\ (I_{G_4} + m_4 b_4^2)\ddot{\theta}_4 \end{Bmatrix} \tag{9.2-8b; 9.2-8c}$$

The above equations are functions of the angular displacements, angular velocities, and angular accelerations of links 3 and 4. These quantities may be determined using equations provided in Section 4.3.3.

Solving Equation (9.2-7) permits other results to be determined. For instance, the magnitude of the force transmitted through a turning pair between links *i* and *j* is

$$| \bar{F}_{ij} | = [(F_{ij})_x^2 + (F_{ij})_y^2]^{1/2}$$

(9.2-9)

The direction of the force is

$$\alpha_{ij} = \tan 2^{-1} [(F_{ij})_x , (F_{ij})_y], \qquad -\pi < \alpha_{ij} < \pi$$

(9.2-10)

The *shaking force* is transmitted through the base pivots onto the base link and is expressed as

$$\bar{F}_S = \bar{F}_{21} + \bar{F}_{41}$$

(9.2-11)

The magnitude of the shaking force is

$$| \bar{F}_S | = \{ [(F_{21})_x + (F_{41})_x]^2 + [(F_{21})_y + (F_{41})_y]^2 \}^{1/2}$$

(9.2-12)

and its direction is

$$\alpha_S = \tan 2^{-1} \{ [(F_{21})_x + (F_{41})_x], [(F_{21})_y + (F_{41})_y] \}, \qquad -\pi < \alpha_S < \pi$$

(9.2-13)

The *shaking moment* about the base pivot O_2 is due to the reaction of the driving torque onto the base link, and pin force at base pivot O_4, and is expressed as

$$M_S = - M_{21} + (F_{41})_y r_1$$

(9.2-14)

As an illustration of the application of the above equations, consider the four-bar mechanism shown in Figure 9.2. Link 2 is driven at $\omega = 50$ rad/sec CCW. The parameters of the links are given in Table 9.1.

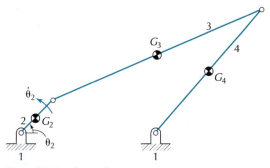

Figure 9.2 Four-bar mechanism.

TABLE 9.1 PARAMETERS OF THE FOUR-BAR MECHANISM SHOWN IN FIGURE 9.2

	Link 1	Link 2	Link 3	Link 4
r_i (cm)	6.00	2.00	10.0	7.00
b_i (cm)	—	1.00	5.00	3.50
m (gr)	—	6.24	31.2	21.8
φ_i (degrees)	—	0	0	0
I_{G_i} (gr-cm^2)	—	—	260	89.0

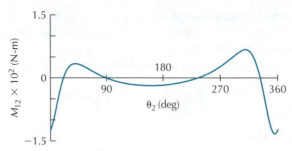

Figure 9.3 Driving torque.

Employing Equations (9.2-7) and (9.2-8), a plot of the driving torque is shown in Figure 9.3. Employing Equations (9.2-12) and (9.2-13), Figure 9.4(a) shows a polar plot of the shaking force. Each point on the curve defines the magnitude and direction of the force for each value of θ_2. A typical shaking force is depicted for $\theta_2 = 30°$. The related calculated values are

$$(F_{12})_x = 7.47 \text{ N}; \quad (F_{12})_y = 2.60 \text{ N}$$
$$(F_{14})_x = -3.93 \text{ N}; \quad (F_{14})_y = -4.76 \text{ N}$$
$$\bar{F}_S = [(F_{21})_x + (F_{41})_x]\bar{i} + [(F_{21})_y + (F_{41})_y]\bar{j}$$
$$= (7.47 - 3.93)\,\bar{i} + (2.60 - 4.76)\bar{j}$$
$$= 3.54\,\bar{i} - 2.16\,\bar{j} \text{ N}$$
$$\alpha_S = \tan 2^{-1}\{[(F_{21})_x + (F_{41})_x], [(F_{21})_y + (F_{41})_y]\}$$
$$= \tan 2^{-1}(3.54, -2.16) = -31.4°$$

Figure 9.4(b) shows the three moving links and the interactive forces acting on the base link. The shaking force for this configuration is also illustrated.

Employing Equations (9.2-9) and (9.2-10), Figure 9.5 is a polar plot of the force at the turning pair between links 2 and 3. Employing Equation (9.2-14), Figure 9.6 is a plot of the shaking moment.

Undue shaking force transmitted to the base link generally creates an undesirable condition. In Section 9.6, a procedure is presented whereby the net shaking force transmitted through the base pivots can be eliminated.

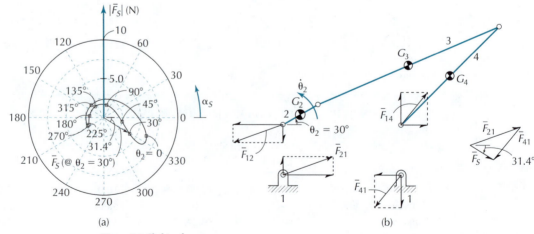

(a)   (b)

Figure 9.4 Shaking force.

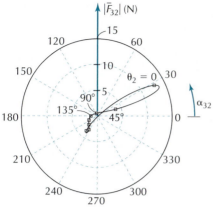

Figure 9.5 Force at turning pair.

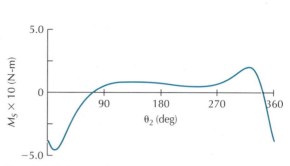

Figure 9.6 Shaking moment.

9.3 FORCE ANALYSIS OF A SLIDER CRANK MECHANISM

Figure 9.7(a) shows a slider crank mechanism. The rotational speed, ω, of the crank is constant. Locations of the centers of mass of links 2 and 3 with respect to a turning pair are specified by distances $b_i\,(i=2,3)$ and angles $\varphi_i\,(i=2,3)$. Free body diagrams of the links are shown in Figure 9.7(b). Based on these diagrams, we can generate a set of governing equations of motion expressed as

$$[A]\{x\} = \{B\} \tag{9.3-1}$$

where

$$[A] = \begin{bmatrix} -1 & 0 & 1 & 0 & 0 & 0 & 0 & 0 \\ 0 & -1 & 0 & 1 & 0 & 0 & 0 & 0 \\ 0 & 0 & A_{3,3} & A_{3,4} & 0 & 0 & 0 & 1 \\ 0 & 0 & -1 & 0 & -1 & 0 & 0 & 0 \\ 0 & 0 & 0 & -1 & 0 & -1 & 0 & 0 \\ 0 & 0 & A_{6,3} & A_{6,4} & A_{6,5} & A_{6,6} & 0 & 0 \\ 0 & 0 & 0 & 0 & 1 & 0 & 0 & 0 \\ 0 & 0 & 0 & 0 & 0 & 1 & -1 & 0 \end{bmatrix} \tag{9.3-2a}$$

$$A_{3,3} = -r_2 \sin\theta_2; \quad A_{3,4} = r_2 \cos\theta_2$$

$$A_{6,3} = r_3 \sin\theta_3 - b_3 \sin(\theta_3 + \varphi_3); \quad A_{6,4} = -r_3 \cos\theta_3 + b_3 \cos(\theta_3 + \varphi_3)$$

$$A_{6,5} = -b_3 \sin(\theta_3 + \varphi_3); \quad A_{6,6} = b_3 \cos(\theta_3 + \varphi_3)$$

$$\{x\} = \begin{Bmatrix} (F_{12})_x \\ (F_{12})_y \\ (F_{23})_x \\ (F_{23})_y \\ (F_{34})_x \\ (F_{34})_y \\ (F_{14})_y \\ M_{12} \end{Bmatrix}; \quad \{B\} = \begin{Bmatrix} m_2(a_{G_2})_x \\ m_2(a_{G_2})_y \\ 0 \\ m_3(a_{G_3})_x \\ m_3(a_{G_3})_y \\ I_{G_3}\ddot{\theta}_3 \\ m_4(a_{G_4})_x \\ 0 \end{Bmatrix} \tag{9.3-2b, 9.3-2c}$$

Equations (9.3-2) contain expressions for the angular displacement, angular velocity, and angular acceleration of link 3. They may be determined using Equations (4.3-77), (4.3-82), and (4.3-84).

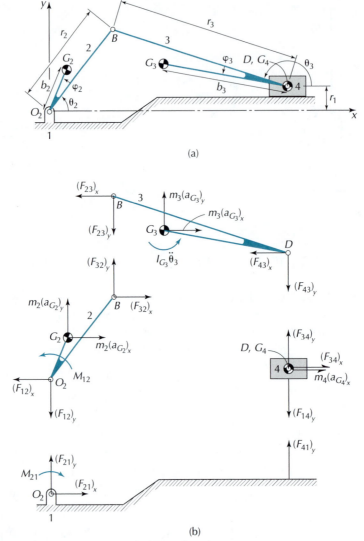

Figure 9.7 Dynamic analysis of a slider crank mechanism.

Equation (9.3-2c) is a function of the linear accelerations of the centers of mass of links 2, 3, and 4. Expressions for these accelerations may be determined using the procedure presented in Section 4.4. The results for link 2 are

$$(a_{G_2})_x = -b_2\,\omega^2 \cos(\theta_2 + \varphi_2); \qquad (a_{G_2})_y = -b_2\omega^2 \sin(\theta_2 + \varphi_2) \qquad (9.3\text{-}3)$$

The results for link 3 are

$$(a_{G_3})_x = -r_2\omega^2 \cos\theta_2 + \dot{\theta}_3^2[r_3 \cos\theta_3 - b_3 \cos(\theta_3 + \varphi_3)]$$
$$+ \ddot{\theta}_3[r_3 \sin\theta_3 - b_3 \sin(\theta_3 + \varphi_3)]$$
$$(a_{G_3})_y = -r_2\omega^2 \sin\theta_2 + \dot{\theta}_3^2[r_3 \sin\theta_3 - b_3 \sin(\theta_3 + \varphi_3)]$$
$$- \ddot{\theta}_3[r_3 \cos\theta_3 - b_3 \cos(\theta_3 + \varphi_3)]$$

$$(9.3\text{-}4)$$

The results for link 4 are

$$(a_{G_4})_x = \frac{-r_2 \omega^2 \cos(\theta_3 - \theta_2) + r_3 \dot{\theta}_3^2}{\cos\theta_3} \tag{9.3-5}$$

The shaking force is

$$\overline{F}_S = \overline{F}_{21} + \overline{F}_{41} \tag{9.3-6}$$

The magnitude and direction of the shaking force are

$$|\overline{F}_S| = \{(F_{21})_x^2 + [(F_{21})_y + (F_{21})_y]^2\}^{1/2}$$
$$\alpha_S = \tan 2^{-1}\{(F_{21})_x, [(F_{21})_y + (F_{41})_y]\}, \quad -\pi < \alpha_S < \pi \tag{9.3-7}$$

The magnitude and direction of a force transmitted through a turning pair may be found using Equations (9.2-9) and (9.2-10). The shaking moment is

$$M_S = -M_{21} + (F_{41})_y (r_2 \cos\theta_2 - r_3 \cos\theta_3) \tag{9.3-8}$$

EXAMPLE 9.1 DYNAMIC FORCE ANALYSIS OF A SLIDER CRANK MECHANISM

Figure 9.8 illustrates a slider crank mechanism. Link 2 is driven at $\omega = 55$ rad/sec CCW. The associated parameters are listed in Table 9.2. For one revolution of the crank, determine

(a)   the plot of the driving toque
(b)   the polar plot of the shaking force
(c)   the polar plot of the force at the turning pair between links 2 and 3
(d)   the plot of the shaking moment

TABLE 9.2 PARAMETERS OF THE SLIDER CRANK MECHANISM SHOWN IN FIGURE 9.8

	Link 1	Link 2	Link 3	Link 4
r_i (cm)	0	2.5	10	—
b_i (cm)	—	1.25	6.67	—
m (gr)	—	2.46	9.84	10
φ_i (degrees)	—	0	0	—
I_{G_i} (gr-cm^2)	—	—	82	—

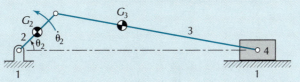

Figure 9.8 Slider crank mechanism.

(*Continued*)

EXAMPLE 9.1 Continued

SOLUTION
(a)   Employing Equations (9.3-1)–(9.3-5), the driving torque is shown in Figure 9.9.
(b)   Employing Equations (9.3-6) and (9.3-7), a polar plot of the shaking force is shown in Figure 9.10.
(c)   Employing Equations (9.2-9) and (9.2-10), a polar plot of the force between links 2 and 3 is shown in Figure 9.11.
(d)   Employing Equation (9.3-8), the shaking moment is shown in Figure 9.12.

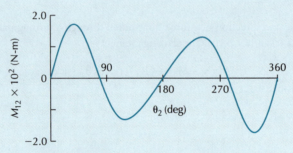

Figure 9.9 Driving torque.

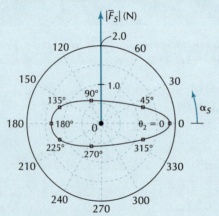

Figure 9.10 Shaking force.

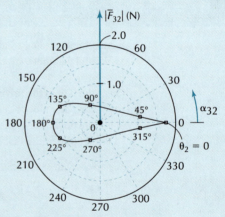

Figure 9.11 Force at turning pair.

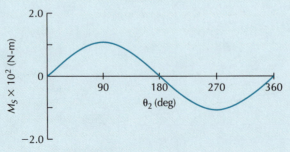

Figure 9.12 Shaking moment.

9.4 UNBALANCE AND BALANCING

Virtually all machines contain rotating components. If the center of mass of a single rotating component does not coincide with its axis of rotation, then there is an *unbalance*. A rotating component having unbalance will produce inertia forces that are transmitted to connected links. If the component is connected to the base link, a shaking force is created.

In most instances, it is desirable to have machines operate smoothly with minimum shaking force. *Balancing* is the process of designing or modifying machinery to reduce unbalance to an acceptable level. The most common approach to achieve balancing is by selectively adding mass to, or removing mass from, links of the machine.

Figure 9.13(a) shows a link turning at rotational velocity ω about base pivot O. It is composed of two lumped masses, m_A and m_B. The center of mass, G, is also illustrated, located distance e from the base pivot. The system shown in Figure 9.13(a) is equivalent to that given in Figure 9.13(b), where $M = m_A + m_B$. This system will produce a shaking force with magnitude $Me\omega^2$, as shown in Figure 9.13(c). The direction of this force rotates at rate ω, and is transmitted to the base link. To remove this shaking force, a *balance mass*, m_c, is added as shown, at distance r_c from the base pivot. The mass is added diagonally opposite from the base pivot, as illustrated in Figure 9.13(d). For balancing, it is required that

$$m_c r_c = Me \tag{9.4-1}$$

After balancing, the revised center of mass coincides with the axis of rotation.

The above is an illustration of *rotating unbalance.* It can occur as a result of pure rotational motion of a body about a fixed axis. Besides rotating unbalance, many mechanisms simultaneously generate *reciprocating unbalance.* This is caused by the inertial forces associated with a translating mass. Reciprocating unbalance may be generated by the translational motion of the slider of a slider crank mechanism. The coupler of a four-bar mechanism may also produce reciprocating unbalance.

In the following sections, we will investigate conventional means of eliminating or reducing the shaking forces produced by four-bar and slider crank mechanisms.

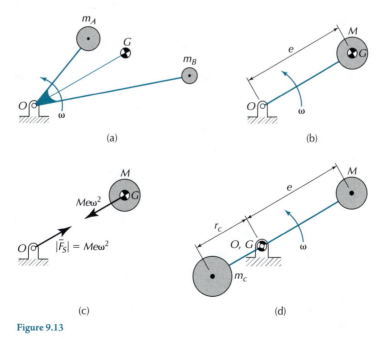

Figure 9.13

9.5 FORCE BALANCING OF A FOUR-BAR MECHANISM

This section presents a method of achieving a four-bar mechanism that is completely force balanced, developed by Berkof and Loewen [10]. This method involves determining an expression for the center of mass of the entire mechanism in terms of the link properties and angular displacement of the input link. This expression will be used to determine values of the link properties under which the center of mass of the *entire mechanism* remains stationary for all configurations of the mechanism. As a result, there will be no acceleration of the center of mass of the mechanism, and thus the net inertia force transmitted through the two base pivots to the base link will be zero.

Figure 9.14 illustrates a four-bar mechanism. The real axis of a complex plane is drawn through the base pivots. Vectors R_{sj} ($j = 2, 3, 4$) point to the centers of mass of the moving links. As illustrated in Figure 9.1(a), parameters b_i and φ_i define the positions of the mass centers of the links with respect to the appropriate turning pair.

The position of the center of mass, G, of the moving links of the mechanism is

$$R_s = \frac{1}{M}(m_2 R_{s2} + m_3 R_{s3} + m_4 R_{s4}) \tag{9.5-1}$$

where

$$M = m_2 + m_3 + m_4 \tag{9.5-2}$$

From Figures 9.1(a) and 9.14

$$R_{s2} = b_2 e^{i(\theta_2 + \varphi_2)}$$
$$R_{s3} = r_2 e^{i\theta_2} + b_3 e^{i(\theta_3 + \varphi_3)} \tag{9.5-3}$$
$$R_{s4} = r_1 + b_4 e^{i(\theta_4 + \varphi_4)}$$

Substituting Equations (9.5-3) in Equation (9.5-1) and rearranging, we obtain

$$MR_s = (m_2 b_2 e^{i\varphi_2} + m_3 r_2)e^{i\theta_2} + (m_3 b_3 e^{i\varphi_3})e^{i\theta_3} + (m_4 b_4 e^{i\varphi_4})e^{i\theta_4} + m_4 r_1 \tag{9.5-4}$$

Equation (9.5-4) is a function of the angular displacement of the three moving links. One of the three angular displacements can be removed by substituting the loop closure equation (refer to Section 4.2), which in this case is

$$r_2 e^{i\theta_2} + r_3 e^{i\theta_3} - r_4 e^{i\theta_4} - r_1 = 0 \tag{9.5-5}$$

Solving Equation (9.5-5) for the unit vector associated with link 3, we obtain

$$e^{i\theta_3} = \frac{1}{r_3}(r_1 - r_2 e^{i\theta_2} + r_4 e^{i\theta_4}) \tag{9.5-6}$$

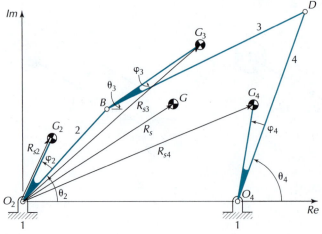

Figure 9.14

Substituting Equation (9.5-6) in Equation (9.5-4), and rearranging, we obtain

$$MR_s = \left(m_2 b_2 e^{i\varphi_2} + m_3 r_2 - m_3 b_3 \frac{r_2}{r_3} e^{i\varphi_3} \right) e^{i\theta_2} + \left(m_4 b_4 e^{i\varphi_4} + m_3 b_3 \frac{r_4}{r_3} e^{i\varphi_3} \right) e^{i\theta_4}$$

$$+ m_4 r_1 + m_3 b_3 \frac{r_1}{r_3} e^{i\varphi_3} \tag{9.5-7}$$

In Equation (9.5-7), only θ_2 and θ_4 are time dependent. Therefore, the equation is independent of time if the coefficients that multiply $e^{i\theta_2}$ and $e^{i\theta_4}$ are zero. That is, if

$$m_2 b_2 e^{i\varphi_2} + m_3 r_2 - m_3 b_3 \frac{r_2}{r_3} e^{i\varphi_3} = 0 \tag{9.5-8}$$

and

$$m_4 b_4 e^{i\varphi_4} + m_3 b_3 \frac{r_4}{r_3} e^{i\varphi_3} = 0 \tag{9.5-9}$$

then vector R_s must be constant. We can employ Equations (9.5-8) and (9.5-9) to determine the properties of the moving links that will provide force balance. These equations involve parameters associated with the three moving links. They may be satisfied by first specifying the parameters of any one of the moving links, say, link 3, and then by solving for the parameters of links 2 and 4. In this case, we rearrange Equations (9.5-8) and (9.5-9) and isolate the parameters of link 3 onto the right-hand sides as

$$m_2 b_2 e^{i\varphi_2} = m_3 \left(b_3 \frac{r_2}{r_3} e^{i\varphi_3} - r_2 \right) \tag{9.5-10}$$

$$m_4 b_4 e^{i\varphi_4} = -m_3 b_3 \frac{r_4}{r_3} e^{i\varphi_3} \tag{9.5-11}$$

To solve for the parameters of links 2 and 4, it is convenient to employ coordinate systems relative to each link (see Figure 9.15). The components of Equations (9.5-10) and (9.5-11) are

$$\boxed{\begin{aligned} m_2 (b_2)_{\xi_2} &= m_3 \left(b_3 \frac{r_2}{r_3} \cos\varphi_3 - r_3 \right) \\ m_2 (b_2)_{\eta_2} &= m_3 b_3 \frac{r_2}{r_3} \sin\varphi_3 \end{aligned}} \tag{9.5-12}$$

$$\boxed{\begin{aligned} m_4 (b_4)_{\xi_4} &= -m_3 b_3 \frac{r_4}{r_3} \cos\varphi_3 \\ m_4 (b_4)_{\eta_4} &= -m_3 b_3 \frac{r_4}{r_3} \sin\varphi_3 \end{aligned}} \tag{9.5-13}$$

Equations (9.5-12) and (9.5-13) give the net required values to provide force balance. If a mechanism already exists, the magnitude and locations of the balance masses required may be determined to provide force balance. This is illustrated in the example below.

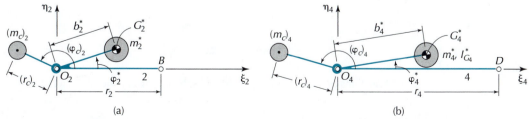

Figure 9.15 Local coordinates for placement of balance masses on a four-bar mechanism: (a) link 2, (b) link 4.

EXAMPLE 9.2 FORCE BALANCING OF A FOUR-BAR MECHANISM

Figure 9.16(a) illustrates a four-bar mechanism having the following parameters:

$$r_1 = 6.0 \text{ cm}$$

$$r_2 = 2.0 \text{ cm}; \quad m_2^* = 250 \text{ gr}; \quad b_2^* = 0.70 \text{ cm}; \quad \varphi_2^* = 5°$$

$$r_3 = 10.0 \text{ cm}; \quad m_3 = 500 \text{ gr}; \quad b_3 = 3.0 \text{ cm}; \quad \varphi_3 = 10°$$

$$r_4 = 7.0 \text{ cm}; \quad m_4^* = 250 \text{ gr}; \quad b_4^* = 2.5 \text{ cm}; \quad \varphi_4^* = 15°$$

$$I_{G_3} = 250 \text{ gr-cm}^2; \quad I_{O_4}^* = 500 \text{ gr-cm}^2$$

The constant rotational speed of link 2 is $\omega = 120$ rad/sec CCW. Figure 9.16(b) shows a polar plot of the shaking force.

Determine the magnitudes and locations of balance masses to be added to links 2 and 4 that will provide complete force balance. Add the masses at the following radial distances from the base pivots:

$$(r_c)_2 = 1.0 \text{ cm}; \quad (r_c)_4 = 3.5 \text{ cm}$$

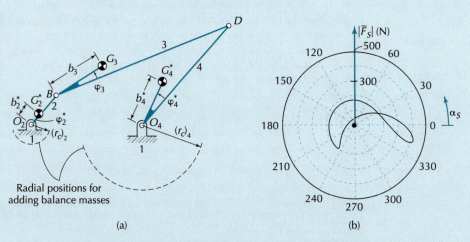

Figure 9.16 Four-bar mechanism prior to force balancing: (a) mechanism, (b) shaking force.

SOLUTION

From Equations (9.5-12) and (9.5-13)

$$m_2 (b_2)_{\xi_2} = -705 \text{ gr-cm}; \quad m_2 (b_2)_{\eta_2} = 52.1 \text{ gr-cm}$$

$$m_4 (b_4)_{\xi_4} = -1000 \text{ gr-cm}; \quad m_4 (b_4)_{\eta_4} = -182 \text{ gr-cm}$$

To provide force balance of the mechanism, balance masses $(m_c)_2$ and $(m_c)_4$ are added to links 2 and 4, respectively. The combined effect caused by the given links, and that of the added balance masses, must provide the above calculated values. That is, for link 4

$$m_4 (b_4)_{\xi_4} = m_4^* (b_4^*)_{\xi_4} + (m_c)_4 ((r_c)_4)_{\xi_4} \tag{9.5-14a}$$

$$m_4 (b_4)_{\eta_4} = m_4^* (b_4^*)_{\eta_4} + (m_c)_4 ((r_c)_4)_{\eta_4} \tag{9.5-14b}$$

where

$$(b_4^*)_{\xi_4} = b_4^* \cos \varphi_4^* = 2.42 \text{ cm}; \quad (b_4^*)_{\eta_4} = b_4^* \sin \varphi_4^* = 0.65 \text{ cm}$$

Using Equations (9.5-14a) and (9.5-14b) and substituting known values, we obtain

(Continued)

EXAMPLE 9.2 Continued

$$(\varphi_c)_4 = \tan 2^{-1}[(m_c)_4((r_c)_4)_{\xi_4}, (m_c)_4((r_c)_4)_{\eta_4}]$$
$$= \tan 2^{-1}[m_4(b_4)_{\xi_4} - m_4^*(b_4^*)_{\xi_4}, m_4(b_4)_{\eta_4} - m_4^*(b_4^*)_{\eta_4}]$$
$$= -168.1°$$

The coordinates of the balance mass for link 4 are

$$((r_c)_4)_{\xi_4} = (r_c)_4 \cos(\varphi_c)_4 = -3.4 \text{ cm}$$
$$((r_c)_4)_{\eta_4} = (r_c)_4 \sin(\varphi_c)_4 = -0.72 \text{ cm}$$

From Equation (9.5-14a), the balance mass required on link 4 is

$$(m_c)_4 = \frac{m_4(b_4)_{\xi_4} - m_4^*(b_4^*)_{\xi_4}}{((r_c)_4)_{\xi_4}} = 478 \text{ gr}$$

The mass of link 4 after adding the balance mass is

$$m_4 = m_4^* + (m_c)_4 = 728 \text{ gr}$$

The polar mass moment of inertia of link 4 about its base pivot after addition of the balance mass is

$$I_{O_4} = I_{O_4}^* + (m_c)_4(r_c)_4^2 = 6.360 \text{ gr-cm}^2$$

A similar analysis for link 2 provides

$$(\varphi_c) = 177.6°; \quad (m_c)_2 = 880 \text{ gr}; \quad m_2 = 1130 \text{ gr}$$

The force-balanced mechanism is shown in Figure 9.17. After force balancing, the polar plot of the shaking force is shown as a dot at the origin.

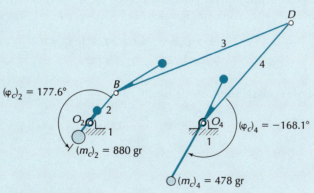

Figure 9.17 Force balanced mechanism.

Force balancing has an effect on the forces at the turning pairs as well as the shaking moment of the mechanism. Considering Example 9.2, Figure 9.18 shows a polar plot of the force between links 2 and 3. The solid line indicates the result using the mechanism prior to force balancing, and the dashed line corresponds to the force-balanced mechanism. The maximum magnitude of the force in the turning pair is greater for the force balanced mechanism. Figure 9.19 shows a plot of the shaking moment before and after force balancing. The maximum magnitude is greater after force balancing.

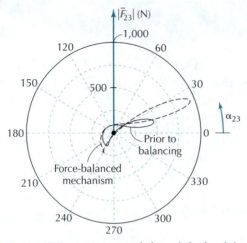

Figure 9.18 Force at turning pair, before and after force balancing.

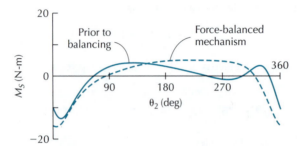

Figure 9.19 Shaking moment, before and after force balancing.

9.6 FORCE BALANCING OF A SLIDER CRANK MECHANISM

Figure 9.20 illustrates a typical slider crank mechanism that will be considered in this section. The slider has zero offset, and the centers of mass of the crank and coupler are on lines connecting the turning pairs associated with each link. The location of the center of mass of link 3 is specified by dimensions r_B and r_D. Comparing Figures 9.7 and 9.20, we have

$$r_1 = 0; \quad \varphi_2 = 0 \left(\text{or } 180° \right); \varphi_3 = 0$$
$$b_3 = r_D; \quad r_B + r_D = r_3 \tag{9.6-1}$$

In addition, we consider geometries for which

$$\frac{r_2}{r_3} < 0.25 \tag{9.6-2}$$

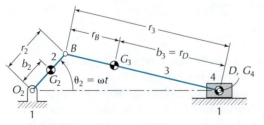

Figure 9.20 Slider crank mechanism.

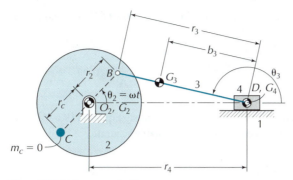

Figure 9.21 Reference mechanism.

The conditions described by Equations (9.6-1) and (9.6-2) are commonly used in piston engines.

Based on the above, a simplified model will be developed for the kinetic analysis of a slider crank mechanism. This model includes development of a simplified expression of the linear acceleration of link 4. Also, we will replace the distributed mass of link 3 with a *lumped-mass model*. These modifications greatly reduce the computational effort required to complete an analysis, while still providing acceptably accurate results.

Figure 9.21 shows a mechanism having its center of mass of link 2 coinciding with the base pivot O_2, that is,

$$b_2 = 0 \qquad (9.6\text{-}3)$$

This is regarded as the reference mechanism. With respect to this mechanism, the shaking force can be adjusted by adding a balance mass, m_c. As shown in Figure 9.21, a balance mass is added to link 2 at distance r_c from the base pivot and at angular position $\theta_2 + 180°$.

9.6.1 Lumped-Mass Model of the Coupler

A simplified model of the coupler is obtained by replacing the distributed mass of the coupler shown in Figure 9.22(a) with two lumped masses, m_B and m_D. The lumped masses are located at the *crank pin* and the *wrist pin* and are connected by a massless rigid rod, as shown in Figure 9.22(b).

The two couplers shown in Figure 9.22 would generate equivalent results in a kinetic analysis, provided they have the same total mass, the same location of the center of mass, and the same polar mass moment of inertia with respect to the center of mass. These conditions are expressed as

$$m_B + m_D = m_3 \qquad (9.6\text{-}4)$$

$$m_B r_B = m_D r_D \qquad (9.6\text{-}5)$$

$$m_B r_B^2 + m_D r_D^2 = I_{G_3} \qquad (9.6\text{-}6)$$

Since there are only two adjustable quantities (m_B and m_D) in Equations (9.6-4)-(9.6-6), not all three conditions can be satisfied. Thus, the lumped-mass coupler is an approximate representation

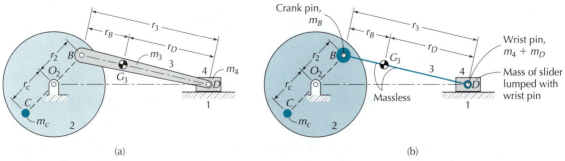

(a)   (b)

Figure 9.22 Slider crank mechanism employing alternative models of coupler: (a) distributed mass model, (b) lumped mass model.

of the distributed mass coupler. Solving Equations (9.64) and (9.6-5) for m_B and m_D, and implementing Equation (9.6-1)

$$m_B = \left(\frac{r_D}{r_B + r_D}\right)m_3 = \frac{r_D}{r_3}m_3; \quad m_D = \frac{r_B}{r_3}m_3$$

(9.6-7)

Although this lumped-mass model does not satisfy Equation (9.6–6), it will be shown that related values of calculated shaking force are acceptably accurate.

9.6.2 Accelerations of Lumped Masses

In this subsection we will determine the accelerations at the crank pin (point B), the wrist pin (point D), and the location at which the balance mass is added (point C). Points B and C execute circular motion. Expressions of their accelerations can be readily determined to be

$$\bar{a}_B = -r_c\omega^2(\cos\omega t\,\bar{i} + \sin\omega t\,\bar{j})$$

(9.6-8)

and

$$\bar{a}_C = r_c\omega^2(\cos\omega t\,\bar{i} + \sin\omega t\,\bar{j})$$

(9.6-9)

The wrist pin undergoes rectilinear motion. An expression of its acceleration is given in Equation (9.3-5), being a function of the lengths, angular displacements, and angular velocities of links 2 and 3. To develop a simplified expression for this acceleration, we begin by employing Equation (4.3-77) for the angular displacement of link 3. When the offset, r_1, is zero, and substituting Equation (9.1-2),

$$\theta_3 = 180° - \sin^{-1}\left(\frac{r_2}{r_3}\sin\omega t\right)$$

(9.6-10)

Employing the identity

$$\sin^{-1}u = \cos^{-1}[1 - u^2]^{\frac{1}{2}}$$

(9.6-11)

for which, in this case

$$u = \frac{r_2}{r_3}\sin\omega\,t$$

(9.6-12)

then Equation (9.6-10) becomes

$$\theta_3 = 180° - \cos^{-1}\left[1 - \left(\frac{r_2}{r_3}\sin\omega t\right)^2\right]^{\frac{1}{2}}$$

(9.6-13)

Using the infinite series

$$[1 - u^2]^{\frac{1}{2}} = 1 - \frac{1}{2}u^2 - \frac{1}{8}u^4 - \ldots$$

Equation (9.6-13) is expressed as

$$\theta_3 = 180° - \cos^{-1}\left[1 - \frac{1}{2}\left(\frac{r_2}{r_3}\right)^2\sin^2\omega t - \frac{1}{8}\left(\frac{r_2}{r_3}\right)^4\sin^4\omega t - \ldots\right]$$

(9.6-14)

For the condition described in Equation (9.6-2), the maximum value of the third term inside the square brackets of Equation (9.6-14) is less than 1/64 that of the second term. It is therefore reasonable to drop the third term and all other higher-order terms of the infinite series. We also employ the identity

$$\sin^2 \omega t = \tfrac{1}{2} - \tfrac{1}{2}\cos 2\omega t$$

and Equation (9.6-14) becomes

$$\theta_3 \approx 180° - \cos^{-1}\left[1 - \frac{1}{4}\left(\frac{r_2}{r_3}\right)^2(1 - \cos 2\omega t)\right] \tag{9.6-15}$$

Employing Figure 9.21, the position of the link 4 is

$$r_4 = r_2 \cos\theta_2 - r_3 \cos\theta_3 \tag{9.6-16}$$

Substituting Equations (9.1-2) and (9.6-15) in Equation (9.6-16), we obtain

$$r_4 = r_2 \cos\omega t - r_3 \cos\left\{180° - \cos^{-1}\left[1 - \frac{1}{4}\left(\frac{r_2}{r_3}\right)^2(1 - \cos 2\omega t)\right]\right\} \tag{9.6-17}$$

Employing the identity

$$\cos(a-b) = \cos a \cos b + \sin a \sin b$$

in Equation (9.6-17) for which

$$a = 180°; \quad b = \cos^{-1}\left[1 - \frac{1}{4}\left(\frac{r_2}{r_3}\right)^2(1 - \cos 2\omega t)\right] \tag{9.6-18}$$

and rearranging and simplifying, we get

$$r_4 = r_2 \cos\omega t + r_3\left[1 - \frac{1}{4}\left(\frac{r_2}{r_3}\right)^2(1 - \cos 2\omega t)\right] \tag{9.6-19}$$

Differentiating Equation (9.6-19) twice with respect to time, we obtain

$$\ddot{r}_4 = -r_2\omega^2\left(\cos\omega t + \frac{r_2}{r_3}\cos 2\omega t\right) \tag{9.6-20}$$

Therefore, the acceleration of link 4 is

$$\overline{a}_D = -r_2\omega^2\left(\cos\omega t + \frac{r_2}{r_3}\cos 2\omega t\right)\overline{i} \tag{9.6-21}$$

Equation (9.6-21) is an explicit function of only the link dimensions and the rotational speed of link 2.

9.6.3 Force Analysis and Balancing

We now implement the lumped-mass model of the coupler introduced in Section 9.6.1, and expressions for accelerations presented in Section 9.6.2, as the basis of completing a force analysis. Figure 9.23 illustrates a slider crank mechanism along with the accelerations of points B, C, and D. Also shown are the lumped masses at each of these locations. At the crank pin is the lumped mass m_B of the coupler as given by Equation (9.6-7). Point C is the location where a balancing

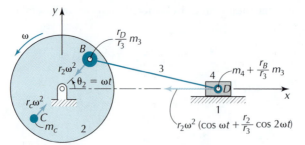

Figure 9.23 Accelerations in a lumped mass model.

mass may be added. At the wrist pin we combine the mass of the slider, m_4, with lumped mass m_D (Equation (9.6-7)).

The shaking force is imparted to the base link as a result of the inertia of the links. For this model, it is the negative of the sum of the masses at B, C, and D multiplied by their respective accelerations. It is expressed as

$$\overline{F}_S = (F_S)_x \overline{i} + (F_S)_y \overline{j} \tag{9.6-22}$$

where after simplification

$$(F_S)_x = (F_S)_{x,\text{primary}} + (F_S)_{x,\text{secondary}} \tag{9.6-23a}$$

$$(F_S)_{x,\text{primary}} = [(m_3 + m_4)r_2 - m_c r_c]\omega^2 \cos\omega t \tag{9.6-23b}$$

$$(F_S)_{x,\text{secondary}} = \left(m_4 + \frac{r_B}{r_3}m_3\right)\frac{r_2^2\omega^2}{r_3}\cos 2\omega t \tag{9.6-23c}$$

and

$$(F_S)_y = -\left(\frac{r_D}{r_3}m_3 r_2 - m_c r_c\right)\omega^2 \sin\omega t \tag{9.6-24}$$

The first term on the right-hand side of Equation (9.6-23a) is called the *primary component*. By inspection of Equation (9.6-23b), its frequency of oscillation equals the rotational speed of the crank. The second term on the right-hand side of Equation (9.6-23a) is the *secondary component*. From Equation (9.6-23c), we see that its frequency of oscillation is twice the rotational speed of the crank.

An examination of Equations (9.6-23) and (9.6-24) reveals that the value of $m_c r_c$ has no influence on the secondary component. Also, it is impossible to find a value of $m_c r_c$ that will simultaneously eliminate the y component of the shaking force and the primary component in the x direction. It is thus impossible to achieve complete force balance by adding a balance mass to the crank. However, by properly sizing the balance mass, the magnitude of the shaking force can be considerably reduced from that generated by the reference mechanism.

Consider the slider crank mechanism illustrated in Figure 9.24 with

$$r_2 = 2.5 \text{ cm} = r_c; \quad r_3 = 10.0 \text{ cm}; \quad m_3 = 8.0 \text{ gr}; \quad m_4 = 12.0 \text{ gr}$$

$$r_B = 4.0 \text{ cm}; \quad r_D = 6.0 \text{ cm}; \quad \omega = 55.0 \text{ rad/sec CCW}$$

Figure 9.25(a) shows a polar plot of the shaking force of the reference mechanism (i.e., $m_c = 0$). Results are illustrated using both the lumped-mass model introduced in this section and the distributed-mass model presented in Section 9.3. There is no discernible difference between the two. The percentage differences obtained using the two models are shown in Figure 9.25(b).

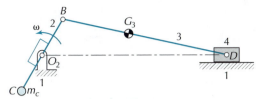

Figure 9.24 Slider crank mechanism.

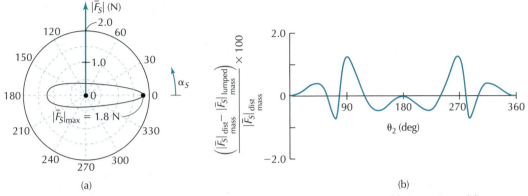

Figure 9.25 Comparison of results using a distributed mass model and a lumped mass model: (a) shaking forces, (b) percentage differences.

The maximum absolute difference is less than 1.5 percent. This result justifies utilization of a lumped-mass model to obtain acceptable results. The maximum magnitude of the shaking force is 1.80 N.

Figure 9.26 shows the shaking force for three levels of balance mass, m_c. All have a balance mass added according to the relation

$$m_c = \frac{\dfrac{r_D}{r_3}m_3 r_2 + \gamma\left(\dfrac{r_B}{r_3}m_3 + m_4\right)r_2}{r_c}, \quad 0 \leq \gamma \leq 1.0 \tag{9.6-25}$$

Figure 9.26(a) corresponds to case for which $\gamma = 0$, and from Equation (9.6-25)

$$m_c = \frac{r_D}{r_3}\frac{r_2}{r_c}m_3 = 4.8 \text{ gr} \tag{9.6-26}$$

In this case, using Equation (9.6-24), the component of shaking force in the y direction is eliminated. The dashed line in Figure 9.26(a) corresponds to the shaking force when $m_c = 0$. The maximum magnitude of the shaking force is now 1.44 N. Its value has been reduced 20 percent compared to that produced by the reference mechanism.

We now consider increasing the balance mass, using values of γ in the range $0 < \gamma < 1.0$. By introducing various values of γ, we obtain the result illustrated in Figure 9.26(b) showing the maximum magnitude of the shaking force during a cycle has been minimized. This is the condition of optimum balancing. Under this condition, the same maximum magnitude of shaking force occurs three times during each cycle. For this illustration, they occur when $\theta_2 = 0°$, 95°, and 265°. The maximum shaking force has magnitude of 0.754 N, which is a 58 percent reduction, compared to that generated by the reference mechanism.

For this illustration, the condition of optimum balance is achieved using $\gamma = 0.595$, which corresponds to $m_c = 13.8$ gr. In general, the value of γ that produces optimum balancing depends on the relative masses of the coupler and slider, and the relative lengths of links 2 and 3. Optimum force balancing is typically obtained by searching in the range $0.50 < \gamma < 0.75$.

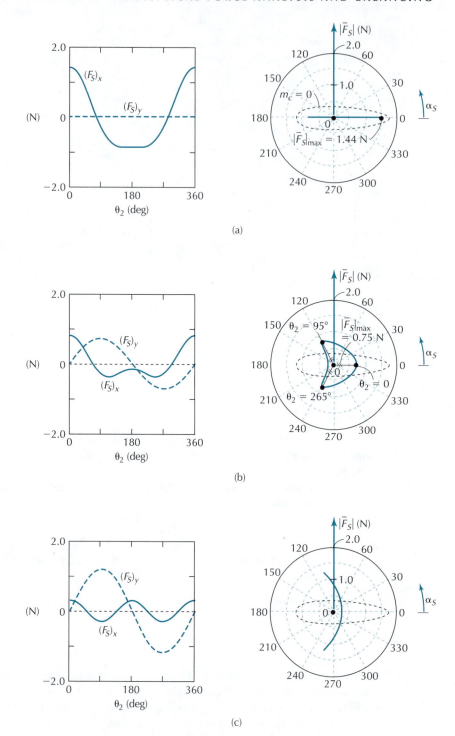

Figure 9.26 Illustrative example of employing different levels of balance mass in a slider crank mechanism: (a) $m_c = 4.8$ gr, (b) $m_c = 13.8$ gr, (c) $m_c = 20.0$ gr [Video 9.26].

Figure 9.26(c) corresponds to case for which $\gamma = 1.0$. Using Equations (9.6-1) and (9.6-25),

$$m_c = \frac{r_2}{r_c}(m_3 + m_4) = 20.0 \text{ gr} \qquad (9.6\text{-}27)$$

Comparing Equations (9.6-23b), (9.6-23c), and (9.6-27), it can be shown in this instance that the primary component of the shaking force in the x direction has been counterbalanced, and only the secondary component remains. Although the shaking force in the x direction now has a relatively small value, the maximum magnitude of the shaking force is greater than that produced for the condition of optimum balancing. Employing $\gamma = 1.0$ in Equation (9.6-25) yields a calculated value of balance mass which, if used, will cause a shaking force higher than the optimum condition.

[Video 9.26] illustrates the levels of balance mass and associated polar plots shown in Figure 9.26.

Video 9.26
Shaking Force
Analysis

PROBLEMS

Shaking Force

P9.1 For the parameters of the four-bar mechanism (see Figure 9.1) given in Table P9.1, determine

(a) the shaking force throughout a cycle of motion as a result of the inertia of links 3 and 4

(b) the shaking moment throughout a cycle of motion as a result of the inertia of links 3 and 4

(Mathcad program: fourbarforce)

$$\omega = 50 \text{ rad/sec CCW}$$

TABLE P9.1

	Link 1	Link 2	Link 3	Link 4
r_i (cm)	6.0	1.5	9.0	8.0
b_i (cm)	—	1.0	4.0	3.0
m_i (gr)	—	2.0	10.0	7.0
φ_i (degrees)	—	0	5.0	0
I_{G_i} (gr-cm²)	—	—	80.0	30.0

P9.2 For the parameters of the slider crank mechanism (see Figure 9.7) given in Table P9.2, determine

(a) the shaking force throughout a cycle of motion as a result of the inertia of links 3 and 4

(b) the shaking moment throughout a cycle of motion as a result of the inertia of links 3 and 4

(Mathcad program: slidercrankforce)

$$\omega = 50 \text{ rad/sec CCW}$$

TABLE P9.2

	Link 1	Link 1	Link 1	Link 1
r_i (cm)	0.5	2.0	9.0	—
b_i (cm)	—	1.0	7.0	—
m_i (gr)	—	2.5	10.0	15.0
φ_i (degrees)	—	0	5.0	—
I_{G_i} (gr-cm²)	—	—	80.0	—

Balancing of Four-Bar and Slider Crank Mechanisms

P9.3 For the given parameters of a four-bar mechanism (see Figure 9.1), determine the masses and the locations of the centers of mass of links 2 and 4 so

that the mechanism is completely force balanced. **(Mathcad program: fourbarbalance)**

$$r_1 = 6.0 \text{ cm}; \quad r_2 = 2.0 \text{ cm}; \quad r_4 = 7.0 \text{ cm}$$

$$r_3 = 10.0 \text{ cm}; \quad m_3 = 600 \text{ gr}; \quad b_3 = 3.0 \text{ cm}$$

$$\varphi_3 = 10°; \quad I_{G_3} = 200 \text{ gr-cm}^2$$

$$(r_c)_2 = 2.0 \text{ cm}; \quad (r_c)_4 = 3.5 \text{ cm}$$

P9.4 For the given parameters of a four-bar mechanism (see Figure 9.16), determine the magnitudes and locations of the balance masses to be added to links 2 and 4 so that the mechanism is completely force balanced. **(Mathcad program: fourbarbalance)**

$$r_1 = 60 \text{ cm}$$

$$r_2 = 2.0 \text{ cm}; \quad m_2^* = 200 \text{ gr}$$

$$b_2^* = 0.70 \text{ cm}; \quad \varphi_2^* = -5°$$

$$r_3 = 10.0 \text{ cm}; \quad m_3 = 450 \text{ gr}$$

$$b_3 = 3.0 \text{ cm}; \quad \varphi_3 = -10°$$

$$r_4 = 7.0 \text{ cm}; \quad m_4^* = 250 \text{ gr}; \quad b_4^* = 2.5 \text{ cm}; \quad \varphi_4^* = -15°$$

$$(r_c)_2 = 2.0 \text{ cm}; \quad (r_c)_4 = 3.5 \text{ cm}$$

$$I_{G_3} = 300 \text{ gr-cm}^2; \quad I_{O_4}^* = 800 \text{ gr-cm}^2$$

P9.5 For the given parameters of a slider crank mechanism (see Figure 9.22), determine the balance mass to be added to link 2 so that the mechanism has optimum force balancing. **(Mathcad program: slidercrankbalance)**

$$r_2 = 2.0 \text{ cm} = r_c; \quad r_3 = 9.0 \text{ cm}; \quad m_3 = 10.0 \text{ gr}; \quad m_4 = 12.0 \text{ gr}$$

$$r_B = 4.0 \text{ cm}; \quad r_D = 5.0 \text{ cm}; \quad \omega = 50.0 \text{ rad/sec CCW}$$

P9.6 For the given parameters of a slider crank mechanism (see Figure 9.22), determine the balance mass to be added to link 2 so that the mechanism has:

(a) balanced shaking force in the y direction

(b) optimum balancing

(c) balanced primary shaking force component in the x direction

(Mathcad program: slidercrankbalance)

$$r_2 = 2.0 \text{ cm} = r_c; \quad r_3 = 10.0 \text{ cm}; \quad m_3 = 6.0 \text{ gr}; \quad m_4 = 15.0 \text{ gr}$$

$$r_B = 4.0 \text{ cm}; \quad r_D = 6.0 \text{ cm}; \quad \omega = 55.0 \text{ rad/sec CCW}$$

CHAPTER 10

FLYWHEELS

10.1 INTRODUCTION

A *flywheel* is a mechanical component designed to store and release kinetic energy. It is mounted on a rotating shaft in a machine, and its mass is symmetrical about the axis of rotation so that shaking forces are reduced. Some flywheels can simply be a solid metal disc.

Flywheels are commonly used for reducing periodic speed fluctuations that occur in internal combustion engines under steady-state operation. Figure 10.1(a) illustrates a single-cylinder engine (see Chapter 13, Section 13.12.1), with a flywheel attached to the crankshaft. A free body diagram of the flywheel is illustrated in Figure 10.1(b), indicating the directions of the applied torques. The drive torque is supplied by the engine, and the load torque is the reaction on the flywheel from the externally driven system.

A plot of the drive torque generated by a single-cylinder engine is shown in Figure 10.1(c). This torque is a superposition of the effects of combustion of the mixture of fuel and air during the power stroke and the inertia of the links. Under steady-state operating conditions, this torque repeats itself every two revolutions (4π radians). The load torque may also fluctuate depending on the application. When the torques vary, there will be fluctuations in the rotational speed, ω. When the magnitude of the drive torque is greater than the magnitude of the load torque, the net torque on the flywheel will cause an angular acceleration of the crankshaft and an increase in rotational speed. Alternatively, when the magnitude of the load torque is greater than the magnitude of the drive torque, the rotational speed will decrease. Adding a flywheel to the crankshaft increases the polar mass moment of inertia of the rotating elements. For a given set of drive and load torques, the addition of a flywheel will reduce the speed fluctuations. The greater the polar mass moment of inertia of the flywheel, the lower these speed fluctuations.

A gasoline-powered lawn mower is an example of the system illustrated in Figure 10.1. The blade of the lawn mower is rigidly attached to the crankshaft and provides a significant portion of the polar mass moment of inertia about the axis of rotation, and therefore it also acts as a form of flywheel. A properly designed system prevents excessive fluctuations of speed when there are modest variations in the load torque requirements.

A single-cylinder engine produces a greater variation in drive torque compared to that produced by a multi-cylinder engine (see Chapter 13, Section 13.12.2). Flywheels, in some form, are used in the design of virtually all internal combustion engines.

Flywheels are also employed in systems for which the drive torque can be considered constant, as in electric motors, when load torque is variable. A punch press is a typical example. The punching operation is performed during only that portion of the cycle when the load torque on the flywheel is larger than the drive torque. The electric motor drives the flywheel to increase its rotational speed and thus provide the source of energy for the sudden surge of power needed in the punching

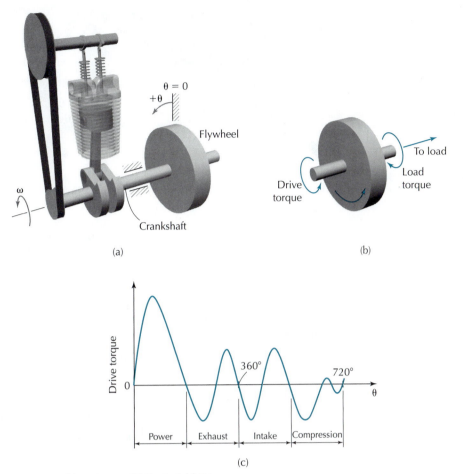

Figure 10.1 (a) Mechanism. (b) Flywheel. (c) Drive torque curve.

operation. The power rating of an electric motor used in a punch press can be significantly reduced by use of a flywheel.

Flywheels have recently become the subject of extensive research as energy storage devices. A typical system consists of a flywheel suspended by nearly frictionless magnetic bearings inside a vacuum chamber and can be connected to either an electric motor or an electric generator. These flywheels are made from composite materials that permit greater rotational speed and increased capacity for energy storage. Here, the flywheel stores kinetic energy by driving it with an electric motor. The energy may later be retrieved by engaging the spinning flywheel with an electric generator.

This chapter will be restricted to providing a mathematical model for determining the required size of a flywheel to keep the cyclical rotational speed within stated limits, based on applied torques under cyclical steady-state operating conditions. First, we will consider systems in which all the components attached to a rotating shaft have the same rotational speed. Then, we will examine systems that involve more than one rotational speed.

10.2 MATHEMATICAL FORMULATION

The governing equation for rotational motion of a body (see Equation (2.12-4)) is

$$M_G = I_G \ddot{\theta} \tag{10.2-1}$$

A free-body diagram of a flywheel is shown in Figure 10.2 indicating positive directions of rotational speed, ω, the *drive torque*, T_D, and *load torque*, T_L, applied to the flywheel. Employing Equation (10.2-1), and dropping subscript G, we obtain

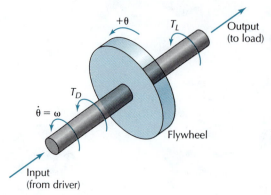

Figure 10.2 Flywheel: free-body diagram.

$$M = T_D + T_L = I\frac{d^2\theta}{dt^2} \tag{10.2-2}$$

or

$$M = I\frac{d}{dt}\left(\frac{d\theta}{dt}\right) \tag{10.2-3}$$

When we substitute

$$\omega = \frac{d\theta}{dt} \tag{10.2-4}$$

for the rotational speed, Equation (10.2-3) becomes

$$M = I\frac{d\theta}{dt}\frac{d}{d\theta}\left(\frac{d\theta}{dt}\right) = I\omega\frac{d\omega}{d\theta} \tag{10.2-5}$$

or

$$Md\theta = I\omega\,d\omega \tag{10.2-6}$$

Integrating both sides of Equation (10.2-6) produces

$$\int_{\theta=\theta_1}^{\theta=\theta_2} M\,d\theta = \int_{\omega=\omega_1}^{\omega=\omega_2} I\omega\,d\omega = \tfrac{1}{2}I(\omega_2^2 - \omega_1^2) \tag{10.2-7}$$

The left-hand side of Equation (10.2-7) represents the work done on the flywheel by the torques and will be designated as ΔE. The right-hand side represents the corresponding change of kinetic energy stored in the flywheel (see Appendix C, Section C.10.1).

We are interested in the maximum speed fluctuation during each cycle of motion. This requires that we find those values of θ which generate the maximum value of ΔE, designated as ΔE_{max}. Under these circumstances, from Equation (10.2-7) we have

$$\Delta E_{max} = \tfrac{1}{2}I(\omega_{max}^2 - \omega_{min}^2) = \tfrac{1}{2}I(\omega_{max} + \omega_{min})(\omega_{max} - \omega_{min}) \tag{10.2-8}$$

We introduce the average rotational speed

$$\omega_0 = \frac{(\omega_{max} + \omega_{min})}{2} \tag{10.2-9}$$

and Equation (10.2-8) becomes

$$\Delta E_{max} = I\omega_0\left(\omega_{max} - \omega_{min}\right) \tag{10.2-10}$$

Further, we introduce the *coefficient of speed fluctuation*:

$$\boxed{C_S = \frac{\omega_{max} - \omega_{min}}{\omega_0}} \tag{10.2-11}$$

Application	Typical Value of C_S
Alternators, generators	0.005
Punch press	0.10
Rock crusher	0.20

and Equation (10.2-10) becomes

$$\Delta E_{\max} = I\omega_0^2 C_S \qquad\qquad (10.2\text{-}12)$$

or

$$I = \frac{\Delta E_{\max}}{\omega_0^2 C_S} \qquad\qquad (10.2\text{-}13)$$

Equation (10.2-13) is an expression for the required polar mass moment of inertia in terms of the average rotational speed, the maximum change of kinetic energy in the flywheel throughout one cycle, and the coefficient of speed fluctuation.

Equation (10.2-13) represents the total polar mass moment of inertia of not only the flywheel but for the whole machine. If the machine has a polar mass moment of inertia of I_{machine}, then the polar mass moment of inertia required for the flywheel is

$$I_{\text{flywheel}} = I - I_{\text{machine}} \qquad\qquad (10.2\text{-}14)$$

The coefficient of speed fluctuation is a dimensionless quantity. In the analysis presented, we will consider situations in which there are small speed fluctuations about the average value ω_0; that is, the coefficient of speed fluctuation will be much less than unity. Typical allowable values of the coefficient of speed fluctuation are given in Table 10.1.

If steady-state operating conditions are assumed, the system will repeat itself after an integer number of revolutions. For instance, speed fluctuation of the reciprocating engine shown in Figure 10.1(a) will repeat itself every two revolutions (4π radians) of the crankshaft.

A convenient method for analyzing and determining the requirements of a flywheel is to plot the drive torque and load torque curves on the same graph. Using the notation presented in this section, it is recommended that the *positive* of the drive torque and the *negative* of the load torque be drawn onto the same graph. An illustrative graph is shown in Figure 10.3.

Under steady-state conditions, during each cycle the energy input to the flywheel provided by the drive torque must equal the energy removed from the flywheel by the load torque. Therefore, throughout a cycle, the average absolute values of the drive and load torques must be equal.

The net torque on the flywheel, M, is the sum of the drive and load torques (Equation (10.2-2)). For the graph in Figure 10.3(a), since the negative of the load torque is plotted, the net torque on the flywheel for a given rotation is the *difference* between the two curves. A plot of the net torque on the flywheel is given in Figure 10.3(b). Points of intersection of the curves in Figure 10.3(a) correspond to there being no net torque applied to the flywheel. At these locations, there is no acceleration or deceleration, and rotational speed is at a local extremum. A plot of the rotational energy throughout a cycle is shown in Figure 10.3(c). At each point of intersection, the energy is also at a local extremum.

When the driving torque has a magnitude greater than the load torque, there is a net torque in the direction of rotation, and the rotational speed (and energy in the flywheel) increases. Alternatively, when the magnitude of the load torque is greater than that of the drive torque, the net torque is in a direction opposite to that of the direction of rotation, and the rotational speed decreases.

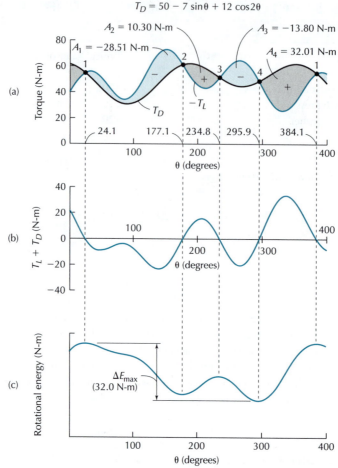

$$T_L = -50 + 10\cos\theta - 15\sin 3\theta$$
$$T_D = 50 - 7\sin\theta + 12\cos 2\theta$$

Figure 10.3 Typical load and drive torque curves [Video 10.3].

Since these graphs are in terms of torque versus rotation, areas between the drive torque and load torque curves represent changes of energy in the flywheel. Also, since intersection points of the curves correspond to local extrema of energy, it is possible to compare the relative levels of energy at all intersection points. The value of ΔE_{max} can then be determined by comparing all of the energies, selecting the maximum and minimum values, and taking the difference.

[Video 10.3]
provides an animated sequence of the construction of the plot of the curve of rotational energy illustrated in Figure 10.3.

Video 10.3
Flywheel Analysis

Figure 10.4 shows a model of a single-cylinder engine operating under steady-state conditions. The drive and load torque curves are also given. The load torque is constant, and the negative of its value is plotted. The drive torque curve is highly simplified and only deviates from a constant value during the power and compression strokes. In order for this system to operate under steady-state conditions, average values of the drive torque and of the negative of the load torque must be equal.

Each cycle in this demonstration lasts two revolutions (4π radians) of crank rotation. In any given cycle, there are only two points of intersection between the torque curves; each point corresponds to a local extremum of rotational energy.

Intersection point $1A$ is set to coincide with zero rotation. It corresponds to a local minimum. During the interval of rotation between intersection points $1A$ and $2A$, since the drive torque is greater in magnitude than the load torque, both energy and rotational speed increase. This increase will continue until intersection point $2A$, where both energy and rotational speed will have reached maximums. Beyond point $2A$, and up to point $1B$, both energy and rotational speed decrease. Between points $1A$ and $2A$, the area bounded by the torque curves equals that bounded by the torque curves between points $2A$ and $1B$. In either case, the area is

$$\Delta E_{max} = 1905 \text{ N-m}$$

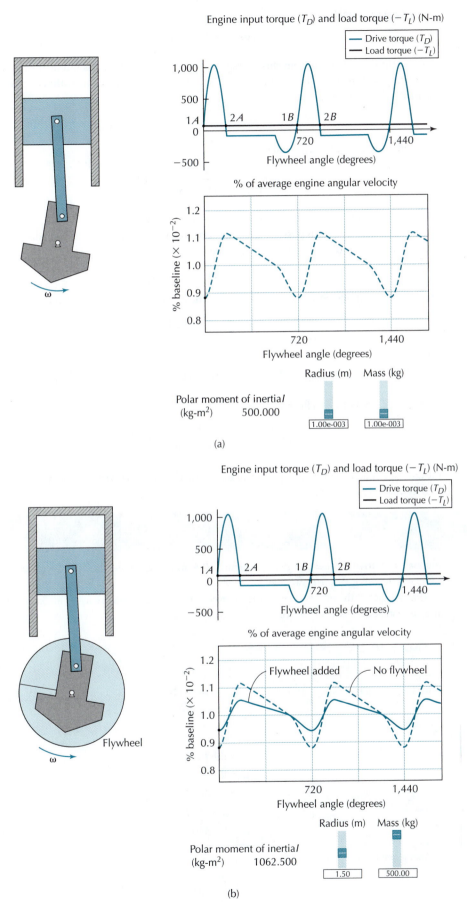

Figure 10.4 Flywheel in a single-cylinder engine [Video 10.4].

The average rotational speed is

$$\omega_0 = 4.0 \, \text{rad/sec}$$

The amount of fluctuation from this average depends on the polar mass moment of inertia of the rotating components. The polar mass moment of inertia of the machine without a flywheel is assumed to have a constant value of

$$I = I_{\text{machine}} = 500 \, \text{kg-m}$$

The coefficient of speed fluctuation is

$$C_S = \frac{\Delta E_{\text{max}}}{\omega_0^2 I} = \frac{1905}{4.0^2 \times 500} = 0.238$$

The corresponding speed curve in this case is provided in Figure 10.4(a).

Using [Model 10.4], it is possible to adjust both the mass and radius of the solid disc flywheel. For instance, Figure 10.4(b) shows the case where

Video 10.4
Model 10.4
Engine Flywheel

$$m = 500 \, \text{kg}; \quad r = 1.5 \, \text{m}$$

and the corresponding polar mass moment of inertia of the flywheel (see Section C.9) is

$$I_{\text{flywheel}} = \tfrac{1}{2} m r^2 = 560 \, \text{kg-m}^2$$

Thus

$$I = I_{\text{machine}} + I_{\text{flywheel}} = (500 + 560) \, \text{kg-m}^2 = 1060 \, \text{kg-m}^2$$

and the coefficient of speed fluctuation has reduced to

$$C_S = \frac{\Delta E_{\text{max}}}{\omega_0^2 I} = \frac{1905}{4.0^2 \times 1060} = 0.112$$

The corresponding speed curve is plotted in Figure 10.4(b). For comparison, the case where there is no flywheel is also drawn.

EXAMPLE 10.1

The load and drive torques are given in Figure 10.5. In this example, the load torque is constant, and the drive torque is variable.

The energies at the intersection points between the torque curves are listed in Table 10.2. Intersection point 1 is given an arbitrary energy level of E.

TABLE 10.2 ENERGY VALUES AT THE INTERSECTION POINTS IN FIGURE 10.5

Point	Energy Level (N-m)
1	E_1
2	$E_2 = E - 1600$
3	$E_3 = E + 3200$
4	$E_4 = E - 800$
5	$E_5 = E + 4800$
6	$E_6 = E + 1500$
7	$E_7 = E + 3400$

Also given is

$$\omega_0 = 800 \, \text{rpm}; \quad C_S = 0.02$$

(Continued)

EXAMPLE 10.1 Continued

Determine the required polar mass moment of inertia.

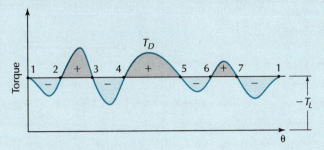

Figure 10.5 Torque curves.

SOLUTION

Examination of Table 10.2 shows that the maximum energy level corresponds to point 5, and the minimum energy level is at point 2. Thus

$$\Delta E_{max} = E_5 - E_2 = [(E + 4800) - (E - 1600)]\,\text{N-m} = 6400\,\text{N-m}$$

Substituting in Equation (10.2-13), we obtain

$$I = \frac{\Delta E_{max}}{\omega_0^2 C_S} = \frac{6400\,\text{N-m}}{\left(800 \times \dfrac{2\pi}{60}\right)^2 \dfrac{\text{rad}^2}{\text{sec}^2} \times 0.02} = 45.5\,\text{N-m-sec}^2 = 45.5\,\text{kg-m}^2$$

EXAMPLE 10.2

Figure 10.6 illustrates a graph that has a variable load torque and constant drive torque. Each cycle lasts three revolutions (6π radians). In addition,

$$\omega_0 = 180\ \text{rpm} = 18.85\ \text{rad/sec}; \qquad I_{machine} = 125\ \text{kg-m}^2$$

Assuming steady-state conditions, determine

(a) the average power required
(b) the maximum and minimum rotational speeds throughout a cycle
(c) the mass of a 0.6-meter-diameter solid disc flywheel to produce $C_S = 0.025$

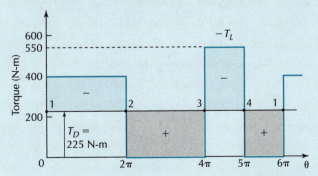

Figure 10.6 Torque curves.

(*Continued*)

EXAMPLE 10.2 Continued

SOLUTION

Since the averages of the absolute values of the drive and load torques must be equal, the constant drive torque is

$$T_D = -\frac{\text{area under load torque curve}}{\text{cycle of operation}}$$

$$= \left(\frac{400 \times 2\pi + 550 \times \pi}{6\pi}\right) \text{N-m} = 225 \text{ N-m}$$

TABLE 10.3 ENERGY VALUES AT THE INTERSECTION POINTS IN FIGURE 10.6

Point	Energy Level (N-m)
1	E_1
2	$E_2 = E_1 - (400 - 225) \times 2\pi = E_1 - 350\pi$
3	$E_3 = E_2 + 225 \times 2\pi = E_1 + 100\pi$
4	$E_4 = E_3 - (550 - 225) \times \pi = E_1 - 225\pi$

Energies at the intersection points between the load and drive torque curves are listed in Table 10.3.

Examination of Table 10.3 shows that the maximum energy level corresponds to point 3, and the minimum energy level is at point 2. Thus, ΔE_{max} is calculated as

$$\Delta E_{max} = E_3 - E_2 = 450\pi \text{ N-m}$$

Based on the above, the solutions are

(a)

$$T_D \omega_0 = 225 \text{ N-m} \times 18.85 \text{ rad/sec} = 4240 \text{ N-m/sec} = 4.24 \text{ kW}$$

(b)

$$I = \frac{\Delta E_{max}}{\omega_0^2 C_S}; \quad I = I_{machine}$$

and therefore

$$C_S = \frac{\Delta E_{max}}{\omega_0^2 I} = \frac{450\pi}{18.85^2 \times 125} = 0.0318$$

However,

$$C_S = \frac{\omega_{max} - \omega_{min}}{\omega_0}$$

Therefore

$$\omega_{max} - \omega_{min} = C_S \times \omega_0 = (0.0318 \times 18.85) \text{ rad/sec} = 0.600 \text{ rad/sec}$$

and

$$\omega_{min} = \omega_0 - \frac{\omega_{max} - \omega_{min}}{2} = (18.85 - 0.300) \text{ rad/sec} = 18.55 \text{ rad/sec}$$

(*Continued*)

EXAMPLE 10.2 Continued

$$\omega_{max} = (18.85 + 0.300)\ \text{rad/sec} = 19.15\ \text{rad/sec}$$

(c)

$$I = \frac{\Delta E_{max}}{\omega_0^2 C_S} = \left(\frac{450\pi}{18.85^2 \times 0.025}\right)\ \text{kg-m}^2 = 159\ \text{kg-m}^2$$

Therefore, using Equation (10.2-14), we obtain

$$I_{\text{flywheel}} = I - I_{\text{machine}} = (159 - 125)\ \text{kg-m}^2 = 34\ \text{kg-m}^2$$

But, using Equation (C.9.9)

$$I_{\text{flywheel}} = \tfrac{1}{2}\,\text{mass}\,r^2 = 34\ \text{kg-m}^2 = \tfrac{1}{2}\,\text{mass}\,0.30\ \text{m}^2$$

$$\text{mass} = 756\ \text{kg}$$

EXAMPLE 10.3

The load torque and drive torque are

$$T_L = (-100 - 20\sin 3\theta)\ \text{N-m}$$

$$T_D = (100 + 80\sin\theta)\ \text{N-m}$$

The torque curves are plotted in Figure 10.7. Also

$$I = 4.0\ \text{kg-m}^2$$

Determine

(a) ΔE_{max}
(b) the maximum angular acceleration during a cycle of motion

As required for steady-state conditions, the absolute average values of the drive and load torques are equal.

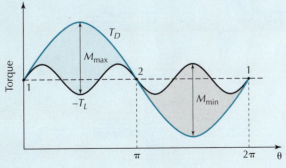

Figure 10.7 Torque curves.

(*Continued*)

EXAMPLE 10.3 Continued

SOLUTION

(a) By inspection, rotational energy is gained in the range from $\theta = 0$ to $\theta = \pi$. The minimum energy level is at point 1, and the maximum energy level is at point 2. Therefore

$$\Delta E_{max} = \int_{\theta=0}^{\theta=\pi} (T_D + T_L)\, d\theta$$

$$= \left[\int_{\theta=0}^{\theta=\pi} (80 \sin\theta - 20 \sin 3\theta)\, d\theta \right] \text{N-m} = 147 \text{ N-m}$$

(b) The net moment, M, on the flywheel corresponds to the difference between the two torque curves. By inspection of Figure 10.7, the maximum differences occur when $\theta = \pi/2$ and $3\pi/2$. The maximum positive moment is

$$M_{max} = (T_D + T_L)\big|_{\theta = \pi/2} = (80 + 20) \text{ N-m} = 100 \text{ N-m}$$

Using Equation (10.2-1), we obtain

$$\ddot\theta_{max} = \frac{M_{max}}{I} = \frac{100 \text{ N-m}}{4.0 \text{ kg-m}^2} = 25 \text{ rad/sec}^2$$

10.3 BRANCHED SYSTEMS

In the operation of a machine, components usually rotate at different speeds. Consider the geared system shown in Figure 10.8. Gear A is rotating at ω_2. Then, the magnitude of the rotational speed of gear B is

$$|\omega_3| = \frac{r_A}{r_B} |\omega_2| \tag{10.3-1}$$

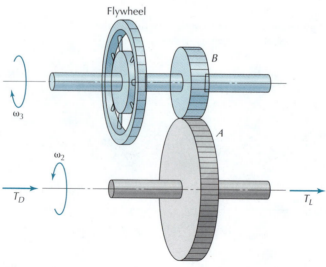

Figure 10.8 Branched system.

Drive and load torques are applied to the shaft on which gear A is mounted. A flywheel is mounted on the same shaft as gear B. If the polar mass moments of inertia of the two gears and flywheel are I_A, I_B, and I_{flywheel}, respectively, then the total rotational energy may be expressed as

$$\text{energy} = \tfrac{1}{2} I_A \omega_2^2 + \tfrac{1}{2}(I_B + I_{\text{flywheel}})\omega_3^2$$

$$= \tfrac{1}{2} I_A \omega_2^2 + \tfrac{1}{2}(I_B + I_{\text{flywheel}})\left(\frac{r_A}{r_B}\omega_2\right)^2 = \tfrac{1}{2} I_{\text{eff}} \omega_2^2 \quad (10.3\text{-}2)$$

where

$$\boxed{I_{\text{eff}} = I_A + (I_B + I_{\text{flywheel}})\left(\frac{r_A}{r_B}\right)^2} \qquad (10.3\text{-}3)$$

can be considered to be the effective polar mass moment of inertia with respect to shaft A.

EXAMPLE 10.4 FLYWHEEL ANALYSIS OF A BRANCHED SYSTEM

For the branched system of Figure 10.8, the drive torque is illustrated in Figure 10.9. The load torque is constant. Gear A has an average speed of 300 rpm, and its diameter is two times that of gear B. The polar mass moments of inertia of the gears are

$$I_A = 8.0 \text{ kg-m}^2; \; I_B = 1.0 \text{ kg-m}^2$$

Determine the polar mass moment of inertia of a flywheel on the same shaft as gear B that is required to have a coefficient of speed fluctuation of 0.005.

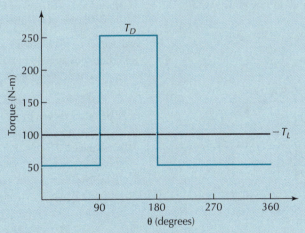

Figure 10.9 Torque curves.

SOLUTION

From the drive torque curve, it may be determined that

$$T_L = -100 \text{ N-m}; \quad \Delta E_{\text{max}} = 75\pi \text{ N-m}$$

From Equation (10.2-13) we have

$$I = I_{\text{eff}} = \frac{\Delta E_{\text{max}}}{\omega_0^2 C_S} = \frac{75\pi \text{ N-m}}{\left(300 \times \dfrac{2\pi}{60}\right)^2 \dfrac{\text{rad}^2}{\text{sec}^2} \times 0.005} = 47.6 \text{ kg-m}^2$$

(Continued)

EXAMPLE 10.4 Continued

Rearranging Equation (10.3-3) gives

$$I_{\text{flywheel}} = \left(I_{\text{eff}} - I_A\right)\left(\frac{r_B}{r_A}\right)^2 - I_B$$

$$= \left[(47.6 - 8.0)\left(\frac{1}{2}\right)^2 - 1.0\right] \text{kg-m}^2 = 8.90 \text{ kg-m}^2$$

PROBLEMS

Sizing of Flywheels, Energy Analysis

P10.1 Figure P10.1 shows the load torque diagram of an engine that operates at an average speed of 2000 rpm. Determine the value of the constant drive torque necessary to move the crankshaft at this speed, and then determine the mass of a solid disc flywheel required to limit speed fluctuation to 0.8 percent of the nominal speed, given that the outside diameter of the flywheel is 1.2 m. Compare this with the mass of a rim disc flywheel (i.e., assume all mass at the radius of the rim) of the same diameter that is required for the same speed fluctuation.

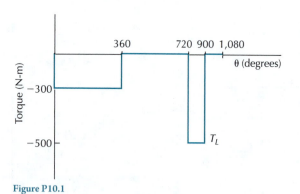

Figure P10.1

P10.2 The crankshaft of a punch press rotates at a maximum speed of 100 rpm, which falls by 10 percent with each punching operation. The flywheel must be able to provide 10 kJ of energy during the 20° while the holes are punched.
Determine
(a) the coefficient of speed fluctuation
(b) the mass of a rim disc flywheel of diameter 2.8 m necessary for the press

P10.3 The maximum variation in stored energy of a flywheel is 3000 N-m over each cycle of operation. For a coefficient of speed fluctuation of 0.05 at an average speed of 300 rpm, determine the radius of gyration of the 100-kg rim disc flywheel. Also determine the maximum and minimum angular velocities of the machine.

P10.4. A steel (7000 kg/m^3) solid disc flywheel has a diameter of 1.2 m and a thickness of 20 mm. Determine
(a) the difference in stored kinetic energy of the flywheel for a speed increase from 200 to 210 rpm
(b) the coefficient of speed fluctuation

P10.5. A flywheel of mass 600 kg and radius of gyration 1.05 m rotates at 3000 rpm. Determine the kinetic energy that it delivers for a 2 percent drop in speed.

P10.6 Figure P10.6 shows the load torque variation for 1 cycle of a shaft driven by a constant torque.
(a) Determine the energy delivered in each cycle for a flywheel mounted on the shaft so that the coefficient of speed fluctuation is 0.05 at an average speed of 200 rpm.
(b) If the flywheel is a solid steel (7000 kg/m^3) disc of thickness 10 mm, determine its diameter.

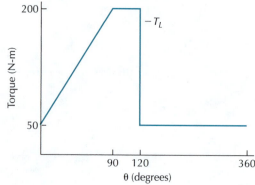

Figure P10.6

P10.7 A solid disc flywheel has a maximum variation in stored kinetic energy amounting to 1000 J. Given that its diameter is 0.9 m and the velocity of the edge varies between 8 and 9 m/sec, determine its mass.

P10.8 A 1.0 m diameter solid disc flywheel of mass 100 kg is designed to handle 1500 N-m of kinetic energy change at an average speed of 200 rpm. Determine the coefficient of speed fluctuation.

P10.9 A machine performs one complete cycle of operation in three revolutions, and the load torque curve is shown in Figure P10.1. The driving torque is constant, and the polar mass moment of inertia of the machine is 100 kg-m². The average speed is 200 rpm. Determine

(a) the coefficient of speed fluctuation

(b) the maximum acceleration and deceleration

(c) the mass of a 60-cm-diameter solid disc flywheel required to reduce the coefficient of speed fluctuation to 0.020

P10.10 The main shaft of a rotating machine has a constant load torque and is driven by the torque shown in Figure P10.10. The machine has a moment of inertia of 8.0 kg-m² and an average operating speed of 400 rpm. Determine

(a) the minimum and maximum angular speeds in each cycle

(b) the greatest acceleration and deceleration in each cycle

(c) the thickness of a steel (7000 kg/m³) 30 cm diameter solid disc flywheel, which when attached to the shaft will prevent the maximum rotational speed from exceeding 410 rpm

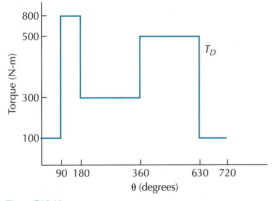

Figure P10.10

P10.11. A rotating engine with constant drive torque has a load of –5000 N-m for half a turn, 250 N-m for the next 180°, and –500 N-m for the final 90° of its cycle. The minimum and maximum speeds are 98 and 104 rpm, respectively.

(a) Determine the constant input torque and sketch the input and load torque curves against shaft rotation.

(b) Sketch the output speed curve below the torque curves, showing roughly the maximum, minimum, and average values.

(c) Determine the polar mass moment of inertia from the given data.

(d) Determine the thickness of a steel (7000 kg/m³) solid disc flywheel, 100 cm in diameter, which when mounted on the shaft reduces the coefficient of speed fluctuation to 0.052.

P10.12 Figure P10.12 is the negative of the load torque curve of an engine for one cycle of operation. Given that the average operating speed of the engine is 400 rpm, determine

(a) the required constant drive torque

(b) the crank angles at which the angular velocity is a maximum or minimum in every cycle

(c) the coefficient of speed fluctuation of the engine given that its polar mass moment of inertia is 8.0 kg-m²

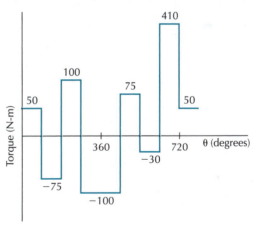

Figure P10.12

P10.13 An engine operates at an average speed of 3500 rpm with its speed fluctuating 20 rpm about this mean value. Given that the engine is fitted with a 2.5-m-diameter rim disc flywheel with mass of 35 kg, determine the kinetic energy consumed per cycle.

P10.14 The torque exerted on the crankshaft of an engine is given by (20,000 + 1500 sin2θ − 1320 cos2θ) N-m. For a constant load torque, determine

(a) the average power of the engine operating at 1000 rpm

(b) the mass of a 1.2-m-diameter solid disc flywheel so that the speed variation is less than 0.3 percent about the mean

(c) the rotational acceleration of the flywheel when θ = 45°

P10.15 For an engine with a drive torque given by $(12{,}000 + 2366 \sin 3\theta)$ N-m and a load torque given by $(-12{,}000 - 1200) \sin \theta$ N-m operating at 200 rpm, determine the coefficient of speed fluctuation, given that the polar mass moment of inertia of the flywheel is 120 kg-m^2.

Branched System

P10.16 In the system of Figure 10.8, the load torque of the system is $T_L = -100 \sin^2 (\theta/2)$ N-m. Gear A turns at a nominal speed of 300 rpm, driven by a constant torque, and its diameter is three times that of gear B. Determine the diameter of a 100 kg rim disc flywheel that is needed to keep speed fluctuations within 0.9 percent of the average speed. If the flywheel had been mounted on the main shaft, what would its diameter have been? Neglect the effect of the gears.

CHAPTER 11
SYNTHESIS OF MECHANISMS

11.1 INTRODUCTION

Most of the problems presented so far have involved the *analysis of mechanisms*. Some dealt with the kinematics of given mechanisms. An input motion was prescribed, and the task was to determine the resultant motions. Consider the mechanism given in Figure 11.1(a), where through analysis we obtain the function graph shown in Figure 11.1(b). Other analyses included the determination of loads between links for a given input motion, or determination of the torque required to maintain a given input motion.

Synthesis of mechanisms can be regarded as the reverse task of analysis. A typical problem starts with a desired motion of at least one of the links. It is then necessary to select the most suitable type of mechanism for that particular situation and determine its dimensions such that the motion comes as close as possible to that desired. Given the function graph in Figure 11.1(b), the problem of synthesis would be to select the appropriate type of mechanism for the task and determine its dimensions. The study of cams, presented in Chapter 7, is another example involving synthesis of mechanisms. Examples started with the desired motion of the follower, as set out by a prescribed displacement diagram, and it was then required to determine the cam profile that would produce that motion.

Video 11.2
Corkscrew
Mechanisms

More than one type of mechanism may be designed to provide similar output motion. Figure 11.2 illustrates four varieties of corkscrew mechanisms [11]. All produce vertical motion of the link connected to the cork. These mechanisms may be compared based on their cost, number of links, input motion required, etc.

Type synthesis involves selection of the kind of mechanism to be used for a particular application. Designers generally rely on their experience as a guide to carry out this task, and therefore should be familiar with the capabilities and typical applications of a variety of mechanisms.

Dimension synthesis entails determination of the proportions of the selected type of mechanism. Determining the shape of a disc cam is an example of dimension synthesis. Numerous graphical and analytical methods have been developed for dimension synthesis of planar four-bar and slider crank mechanisms. Some common algorithms are presented in this chapter.

Still other studies of synthesis involve the *kinetics* of mechanisms. A typical problem is to design a particular type of mechanism that will operate in a desired manner when subjected to a specified input force or input torque. This could involve varying either the distribution of mass in the links or the sizing of flywheels attached to rotating cranks. Henceforth in this chapter, unless otherwise noted, *synthesis* will imply dimension synthesis.

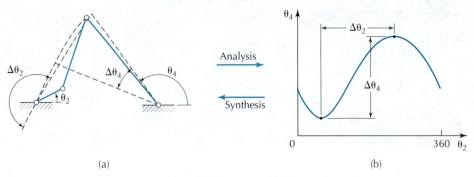

Figure 11.1 Analysis and synthesis of a mechanism: (a) Mechanism, (b) Function graph.

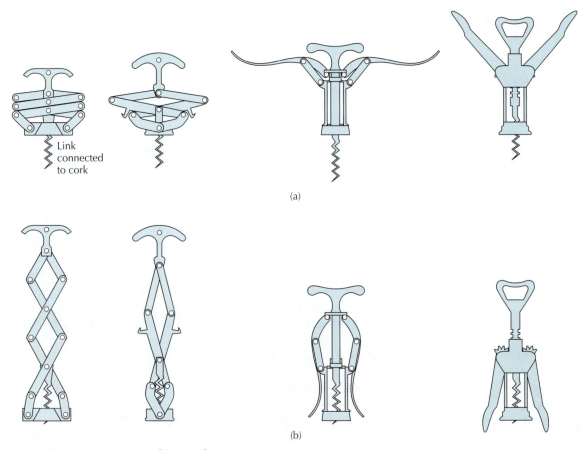

Figure 11.2 Corkscrew mechanisms [Video 11.2].

11.2 CLASSIFICATION OF SYNTHESIS PROBLEMS

Problems of synthesis can generally be assigned to one of three categories. Each is described in a subsection.

11.2.1 Function Generation

A synthesis problem of *function generation* involves determining the dimensions of a mechanism such that it will coordinate either linear or angular motions of two links in a desired manner. For example, Figure 11.3 shows a disc cam mechanism for which the output motion of the follower is

Absolute Maximum Values

Function	Absolute Maximum	Maximum at (deg)
Follower Displacement	8.000e−001	100.0
Follower Velocity	1.833e+001	50.0
Follower Acceleration	5.184e+002	260.0
Follower Jerk	1.866e+004	310.0
Pressure Angle	2.544e+001	318.0

Output Graphs (normalized with the maximum values)

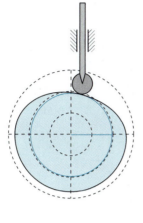

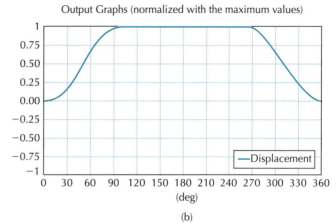

(a)   (b)

Figure 11.3 Cam mechanism.

a function of the rotation of the cam. For this mechanism, it is often possible to design a cam for which the desired motion of the follower is obtained precisely for every cam rotation.

A four-bar mechanism may also be employed for function generation. For the mechanism shown in Figure 11.1(a), we consider θ_4 as a function of θ_2. The form of the functional relation generated by a four-bar mechanism was covered in Chapter 4. Combining Equations (4.3-54) and (4.3-56)–(4.3-59), the relationship may be expressed as

$$\theta_4 = g(\theta_2, r_1, r_2, r_3, r_4) \tag{11.2-1}$$

As indicated, function g depends on the input angular displacement, θ_2, and the four link lengths.

Suppose that a four-bar mechanism is to be designed that will coordinate angular displacements according to the relation

$$\theta_4 = f_{\theta_2}(\theta_2) \tag{11.2-2}$$

In selecting a four-bar mechanism to generate a function, there are only a finite number of link dimensions which may be adjusted to achieve a desired result. Therefore, it is generally not possible to obtain the desired value of a function for every value of θ_2. Instead, we must settle for the likelihood of a difference between the desired function, f_{θ_2}, and that actually produced, g. This difference is known as *structural error, e,* which is expressed as

$$\boxed{e(\theta_2, r_1, r_2, r_3, r_4) = g(\theta_2, r_1, r_2, r_3, r_4) - f_{\theta_2}(\theta_2)} \tag{11.2-3}$$

Figure 11.4(b) shows typical plots of a desired function and an approximation of the function generated by a four-bar mechanism illustrated in Figure 11.4(a). The link lengths were determined using a procedure presented in this chapter. For the plots shown, we do not consider full rotation of links 2 and 4. Rather, link 2 rotates in the range starting from $(\theta_2)_s$ and finishing at $(\theta_2)_f$, while link 4 moves between $(\theta_4)_s$ and $(\theta_4)_f$. The structural error is also shown in a separate plot. An intersection of the two functions is called a *precision point*. Procedures have been developed for the design of a four-bar mechanism where it is possible to specify four or five precision points [12]. However, they are quite complicated compared to an algorithm that requires only three precision

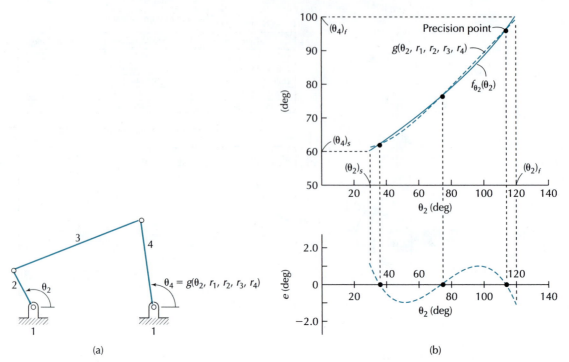

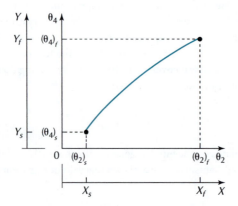

Figure 11.4 Desired and approximate functions.

Figure 11.5 Function generation curve.

points. A procedure requiring only specification of three precision points, which is sufficient for many design problems, is presented in this chapter.

It is often convenient to express a functional relation in the dimensionless form of

$$Y = f_X(X) \tag{11.2-4}$$

where X is proportional to the input motion and Y is proportional to the output motion. As illustrated in Figure 11.5, we may relate the values of X and Y to the rotational movements of links 2 and 4 by recognizing that

$$X = X_S \quad \text{when} \quad \theta_2 = (\theta_2)_s; \quad X = X_f \quad \text{when} \quad \theta_2 = (\theta_2)_f \tag{11.2-5}$$
$$Y = Y_S \quad \text{when} \quad \theta_4 = (\theta_4)_s; \quad Y = Y_f \quad \text{when} \quad \theta_4 = (\theta_4)_f$$

Furthermore, if we define

$$\Delta X = X_f - X_s; \quad \Delta\theta_2 = (\theta_2)_f - (\theta_2)_s \tag{11.2-6}$$
$$\Delta Y = Y_f - Y_s; \quad \Delta\theta_4 = (\theta_4)_f - (\theta_4)_s$$

then for any value between the starting and finishing positions of link 2

$$\frac{\theta_2 - (\theta_2)_s}{X - X_s} = \frac{\Delta\theta_2}{\Delta X} = r_X \tag{11.2-7}$$

or

$$\theta_2 = r_X(X - X_s) + (\theta_2)_s \tag{11.2-8}$$

Similarly, for the rotation of link 4 we have

$$\theta_4 = f_{\theta_2}(\theta_2) = r_Y(Y - Y_s) + (\theta_4)_s \tag{11.2-9}$$

where

$$r_Y = \frac{\Delta\theta_4}{\Delta Y} \tag{11.2-10}$$

The positions of precision points play a role in the distribution of structural error. Consider function f_X shown in Figure 11.6(a), where two different approximations of this function are shown. Each approximation has its distinct locations of the precision points. The structural error for each is shown in Figure 11.6(b).

   Chebyshev spacing is commonly employed to locate the precision points for a given range of input motion [3]. Figure 11.7 illustrates the method of graphically locating three precision points. A semicircle is drawn on the X axis with diameter ΔX and with center at $(X_s + X_f)/2$. Half of a regular hexagon is then inscribed in the semicircle with two of its sides perpendicular to the X axis. Lines drawn perpendicular to the X axis from the vertices of the half polygon determine the locations of the precision points. The positions may be expressed as

$$X_1 = \frac{X_f + X_s}{2} - \frac{X_f - X_s}{2}\cos 30° = 0.933X_s + 0.0670X_f$$

$$X_2 = \frac{X_f + X_s}{2} = 0.500X_s + 0.500X_f$$

$$\tag{11.2-11}$$

$$X_3 = \frac{X_f + X_s}{2} + \frac{X_f - X_s}{2}\cos 30° = 0.0670X_s + 0.933X_f$$

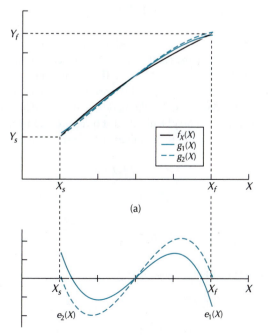

Figure 11.6

(a)

(b)

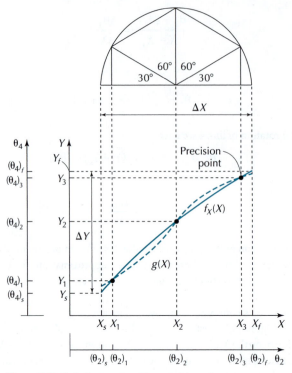

Figure 11.7 Chebyshev spacing of three position points.

Based on Equations (11.2-8) and (11.2-9), we can relate the positions of links 2 and 4 to the values of X and Y at the precision points using

$$(\theta_2)_j = r_X (X_j - X_s) + (\theta_2)_s, \quad j = 1, 2, 3 \tag{11.2-12}$$

$$(\theta_4)_j = r_Y (Y_j - Y_s) + (\theta_4)_s, \quad j = 1, 2, 3 \tag{11.2-13}$$

A slider crank mechanism may be designed to generate prescribed linear motion of the slider as a function of the rotational motion of the crank. For the mechanism shown in Figure 11.8(a), the corresponding function is shown in Figure 11.8(b). The function produced will depend on the offset and the lengths of the crank and coupler. Therefore, the generated function will likely be an approximation of that which is desired. Using the algorithm presented in this chapter, the desired function and that actually produced may be matched at three precision points.

We may relate positions of the slider to the values of Y of a function using

$$s = r_Y (Y - Y_s) + s_s \tag{11.2-14}$$

where

$$r_Y = \frac{\Delta s}{\Delta Y} \tag{11.2-15}$$

Also, corresponding to the precision points,

$$s_j = r_Y (Y_j - Y_s) + s_s, \quad j = 1, 2, 3 \tag{11.2-16}$$

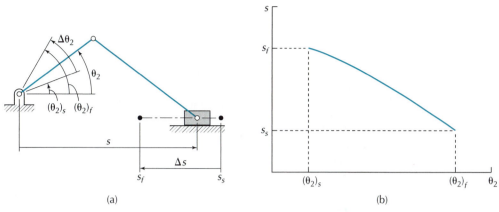

Figure 11.8 Slider crank mechanism and function generation curve.

11.2.2 Path Generation

A synthesis problem of *path generation* involves determining dimensions of a mechanism that will guide a point on a link along a specified path.

Figure 1.16 shows an application of path generation. A point on the coupler is utilized in a film drive mechanism.

A wide variety of shapes of paths may be generated. Figure 1.14 illustrates a small sampling of the path of the trace point from the coupler of a four-bar mechanism. One of the curves illustrated includes a *cusp*: a point on the coupler path for which there are multiple tangents to the curve.

An extensive catalog of coupler curves that are generated by four-bar mechanisms was prepared by Hrones and Nelson [13]. Using a catalog of coupler curves, we may select mechanism dimensions to perform a specific function.

11.2.3 Rigid-Body Guidance

A synthesis problem of *rigid-body guidance* involves determining the dimensions of a mechanism so that during its motion a point on one of the links (i.e., the rigid body) passes through prescribed positions, while at the same time the link is constrained to undergo desired rotations. Figure 11.9(a) shows an example where it is required to design a mechanism in which one link starts from position and orientation 1, and moves to final position and orientation 3. An intermediate position and orientation 2 are also specified. For each position, point A on the link has prescribed coordinates (x_i, y_i), $i = 1, 2, 3$. Also, rotation of the link between positions 1 and 2, and positions 1 and 3, are prescribed as θ_{12} and θ_{13}, respectively. This synthesis problem is called *three-position*

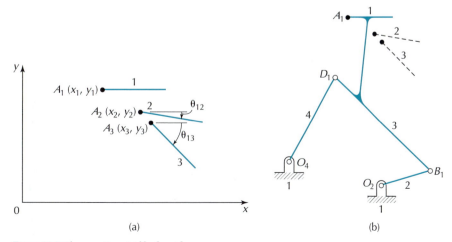

Figure 11.9 Three position rigid body guidance.

rigid-body guidance. Figure 11.9(b) shows a four-bar mechanism that provides the required motion. It was designed using a synthesis procedure presented in this chapter.

It is possible to design a four-bar mechanism to satisfy a finite number of positions of rigid-body guidance of the coupler. A procedure involving four-position rigid-body guidance is suited to the use of a computer [14], but impractical for hand calculation. However, synthesis procedures for two- or three-position rigid-body guidance are relatively easy to complete either graphically or analytically, using a calculator. These methods are sufficient to solve a wide class of practical design problems and are presented in this chapter.

In many instances it will be possible to generate more than one solution to a particular synthesis problem. The solutions may be compared for suitability based on the dimensions of the links, the extent of motions, and the transmission angles.

11.3 ANALYTICAL DESIGN OF A FOUR-BAR MECHANISM AS A FUNCTION GENERATOR

In this section, a method is presented for the design of a four-bar mechanism to generate a function having three precision points. Employing equation (4.3-53) along with the identity

$$\cos(\theta_i - \theta_j) = \cos\theta_i \cos\theta_j + \sin\theta_i \sin\theta_j$$

gives the *Freudenstein equation*

$$h_1 \cos\theta_4 - h_3 \cos\theta_2 + h_5 = \cos(\theta_2 - \theta_4) \qquad (11.3\text{-}1)$$

Three distinct equations may be generated from Equation (11.3-1) by inserting three pairs of angles for links 2 and 4 corresponding to the precision points, that is,

$$\boxed{h_1 \cos(\theta_4)_i - h_3 \cos(\theta_2)_i + h_5 = \cos((\theta_2)_i - (\theta_4)_i), \quad i = 1, 2, 3} \qquad (11.3\text{-}2)$$

Quantities h_1, h_3, and h_5 can be determined from the solution of these three equations. If the length of the base link is selected, then the lengths of the three moving links may be determined. We employ the following equations, which were found by rearranging Equations (4.3-54):

$$\boxed{r_2 = \frac{r_1}{h_1}; \quad r_4 = \frac{r_1}{h_3}; \quad r_3 = (r_1^2 + r_2^2 + r_4^2 - 2r_2 r_4 h_5)^{1/2}} \qquad (11.3\text{-}3)$$

In determining configurations of the mechanism corresponding to a precision point, a negative value obtained from Equations (11.3-3) must be interpreted in a vector sense. That is, the link is drawn in the opposite direction to that defined in Figure 4.3(c).

EXAMPLE 11.1 SYNTHESIS OF A FOUR-BAR GENERATING MECHANISM WITH THREE PRECISION POINTS

Design a four-bar mechanism that will approximately generate the function

$$Y = X^{1/2}, \qquad X_s = 1.0 \le X \le 5.0 = X_f \tag{11.3-4}$$

Also

$$r_1 = 1.00 \text{ cm}; \quad \Delta\theta_2 = -90°; \quad \Delta\theta_4 = -40° \tag{11.3-5}$$

Use Chebyshev spacing of three precision points.

SOLUTION

This example specifies the motions of links 2 and 4. However, there is no stipulation of the starting angular configurations of these links. Therefore, we are free to select

$$(\theta_2)_s = 120°; \quad (\theta_4)_s = 100° \tag{11.3-6}$$

Using Equations (11.2-11), we obtain

$$X_1 = 0.933X_s + 0.0670X_f$$
$$= 0.933 \times 1.0 + 0.0670 \times 5.0 = 1.268$$
$$X_2 = 3.000; \quad X_3 = 4.732 \tag{11.3-7}$$

Also, from Equation (11.2-6) we get

$$\Delta X = X_f - X_s = 4.000 \tag{11.3-8}$$

Then from Equations (11.2-7), (11.2-12), (11.3-7), and (11.3-8) we have

$$(\theta_2)_1 = (\theta_2)_s + \frac{X_1 - X_s}{\Delta X}\Delta\theta_2$$

$$= 120° + \frac{1.268 - 1.000}{4.000}(-90°) = 113.97°$$

$$(\theta_2)_2 = 75.00°; \quad (\theta_2)_3 = 36.03° \tag{11.3-9}$$

Using Equation (11.3-4), the values of Y at the precision points are

$$Y_j = X_j^{1/2}, \quad j = 1,2,3$$
$$Y_1 = X_1^{1/2} = 1.126; \quad Y_2 = 1.732; \quad Y_3 = 2.175 \tag{11.3-10}$$

Also

$$\Delta Y = Y_f - Y_s = X_f^{1/2} - X_s^{1/2} = 1.236 \tag{11.3-11}$$

Combining Equations (11.2-13), (11.3-10), and (11.3-11), we obtain

$$(\theta_4)_1 = (\theta_4)_s + \frac{Y_1 - Y_s}{\Delta Y}\Delta\theta_4$$

$$= 100° + \frac{1.126 - 1.000}{1.236}(-40°) = 95.92°$$

$$(\theta_4)_2 = 76.31°; \quad (\theta_4)_3 = 61.97° \tag{11.3-12}$$

(Continued)

EXAMPLE 11.1 Continued

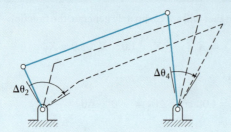

Figure 11.10 Four-bar mechanism.

Substituting the values from Equations (11.3-9) and (11.3-12) in Equation (11.3-2), we get

$$
\begin{bmatrix} -0.103 & 0.406 & 1.000 \\ 0.237 & -0.259 & 1.000 \\ 0.470 & -0.809 & 1.000 \end{bmatrix} \begin{Bmatrix} h_1 \\ h_3 \\ h_5 \end{Bmatrix} = \begin{Bmatrix} 0.951 \\ 1.000 \\ 0.899 \end{Bmatrix}
\tag{11.3-13}
$$

Solving for the unknowns, we obtain

$$
h_1 = 2.959; \quad h_3 = 1.438; \quad h_5 = 0.672
\tag{11.3-14}
$$

Substituting Equations (11.3-14) in Equations (11.3-3), the lengths of the links are

$$
r_2 = 0.338 \text{ cm}; \quad r_3 = 1.132 \text{ cm}; \quad r_4 = 0.695 \text{ cm}
$$

Figure 11.10 shows the mechanism in three configurations corresponding to the precision points.

Having the link lengths of the four-bar mechanism, we may now evaluate the function that would actually be produced by the mechanism and then determine the structural error as a function of the rotation of link 2. For instance, consider the mechanism with

$$
\theta_2 = (\theta_2)_s = 120°
$$

Using Equations (4.3-56)–(4.3-59), gives

$$
\left(\theta_4\right)_{@\theta_2 = 120°} = 98.902°
$$

which does not match the prescribed functional value of

$$
(\theta_4)_s = 100°
$$

That is, in this configuration, using Equation (11.2-3), the structural error of the angular displacement of link 4 is

$$
e_{@\theta_2 = 120°}\left(\theta_4 - (\theta_4)_s\right)_{@\theta_2 = 120°} = 98.902° - 100° = -1.098°
$$

The non-dimensional structural error at $X = X_s = 1.0$ is then

$$
(e(X))_{@X=1} = \left(\theta_4 - (\theta_4)_s\right)_{@\theta_2 = 120°} \frac{\Delta Y}{\Delta \theta_4} = -1.098° \frac{1.236}{-40°} = 0.034
$$

Similar calculations may be completed for other configurations of the mechanism. Combining these results, Figure 11.11(a) shows plots of the nondimensional desired function and that produced by the mechanism in the required range of motion. A plot of the nondimensional structural error is provided in Figure 11.11(b).

(Continued)

EXAMPLE 11.1 Continued

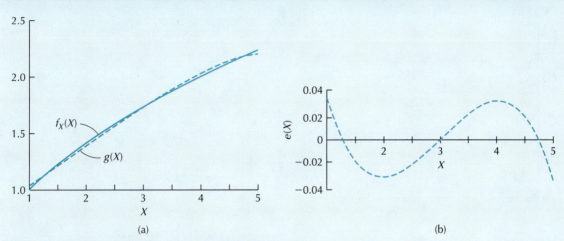

(a)   (b)

Figure 11.11   Function generating four-bar mechanism—first solution.

Other synthesized mechanisms can be produced by selecting alternative starting angular configurations of links 2 and 4. If we select

$$(\theta_2)_s = 160°; \quad (\theta_4)_s = 90° \tag{11.3-15}$$

and reproduce the above procedure, the results are

$$r_2 = 1.685 \text{ cm}; \quad r_3 = 6.375 \text{ cm}; \quad r_4 = -5.279 \text{ cm}$$

The corresponding mechanism and function graph are shown in Figures 11.12 and 11.13. Note in this case that the value of structural error has changed. Although the structural error is reduced compared to the first solution, a designer may not wish to incorporate the second solution because the link lengths are greater and the transmission angle is smaller.

Figure 11.12   Four-bar mechanism—second solution.

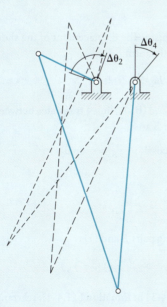

(*Continued*)

EXAMPLE 11.1 Continued

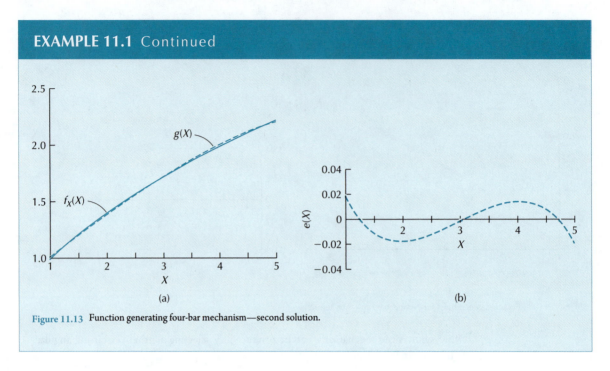

Figure 11.13 Function generating four-bar mechanism—second solution.

There is no guarantee the calculated results will produce an acceptable mechanism as far as link dimensions and transmission angle are concerned. In general, a solution must be checked relative to its suitability in completing a required task. The use of Working Model 2D software is very effective in carrying out such checks.

11.4 ANALYTICAL DESIGN OF A SLIDER CRANK MECHANISM AS A FUNCTION GENERATOR

Figure 11.14 shows a slider crank mechanism. The displacement of the slider is a function of the rotation of the crank. This function takes the form

$$s = f_{\theta_2}(\theta_2, r_1, r_2, r_3) \tag{11.4-1}$$

As link 2 moves between positions $(\theta_2)_s$ and $(\theta_2)_f$, link 4 translates between s_s and s_f.

The x and y coordinates of turning pair B are

$$x_B = r_2 \cos\theta_2; \quad y_B = r_2 \sin\theta_2 \tag{11.4-2}$$

and those of turning pair D are

$$x_D = s; \quad y_D = r_1 \tag{11.4-3}$$

The square of the distance between turning pairs B and D is

$$r_3^2 = (x_D - x_B)^2 + (y_D - y_B)^2 \tag{11.4-4}$$

Substituting Equations (11.4-2) and (11.4-3) in Equation (11.4-4), rearranging, and simplifying, we obtain

$$s^2 = k_1 s \cos\theta_2 + k_2 \sin\theta_2 - k_3 \tag{11.4-5}$$

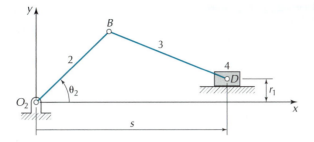

Figure 11.14 Slider crank mechanism.

where

$$k_1 = 2r_2; \quad k_2 = 2r_1 r_2; \quad k_3 = r_1^2 + r_2^2 - r_3^2 \tag{11.4-6}$$

Three distinct equations may be generated from Equation (11.4-5) by inserting three pairs of angles of link 2 and the positions of the slider, that is,

$$s_i^2 = k_1 s_i \cos(\theta_2)_i + k_2 \sin(\theta_2)_i - k_3, \quad i = 1, 2, 3 \tag{11.4-7}$$

Quantities k_1, k_2, and k_3 can be determined from the solution of these equations. Employing Equations (11.4-6), the mechanism geometry is determined using

$$r_2 = \frac{k_1}{2}; \quad r_1 = \frac{k_2}{2r_2}; \quad r_3 = (r_1^2 + r_2^2 - k_3)^{\frac{1}{2}} \tag{11.4-8}$$

Chebyshev spacing of precision points (Equations (11.2-11)) may also be used for synthesizing a slider crank mechanism.

EXAMPLE 11.2 SYNTHESIS OF A SLIDER CRANK GENERATING MECHANISM WITH THREE PRECISION POINTS

Design a slider crank mechanism that will approximate the function

$$Y = X^{3/2}, \quad X_s = 1.0 \leq X \leq 4.0 = X_f$$

by utilizing three precision points with Chebyshev spacing. Also

$$\Delta\theta_2 = -120°; \quad \Delta s = 5.0 \text{ cm}$$

SOLUTION
We make the selection

$$(\theta_2)_s = 150°; \quad s_s = 2.000 \text{ cm}$$

Then from the function graph shown in Figure 11.15

$$X_1 = 1.201; \quad Y_1 = 1.316; \quad (\theta_2)_1 = 141.96°; \quad s_1 = 2.226 \text{ cm}$$
$$X_2 = 2.500; \quad Y_2 = 3.953; \quad (\theta_2)_2 = 90.00°; \quad s_2 = 4.109 \text{ cm}$$
$$X_3 = 3.799; \quad Y_3 = 7.405; \quad (\theta_2)_3 = 38.04°; \quad s_3 = 6.575 \text{ cm}$$

Substituting the above values in Equation (11.4-7), and solving for the unknowns,

$$k_1 = 5.522 \text{ cm}; \quad k_2 = 5.865 \text{ cm}^2; \quad k_3 = -11.02 \text{ cm}^2$$

Using Equations (11.4-8), the dimensions of the mechanism are

$$r_1 = 1.062 \text{ cm}; \quad r_2 = 2.761 \text{ cm}; \quad r_3 = 4.446 \text{ cm}$$

(Continued)

EXAMPLE 11.2 Continued

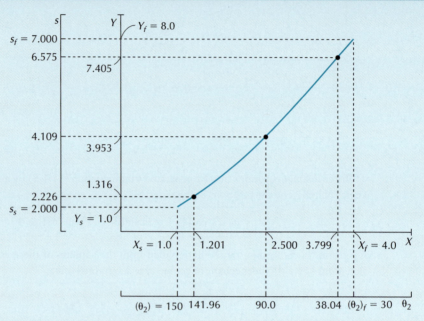

Figure 11.15 Slider crank mechanism.

Figure 11.16 shows the mechanism in three configurations corresponding to the precision points.

Having the dimensions of the slider crank mechanism, we then evaluate the function that would actually be produced by the mechanism and compare it with the desired function to determine the structural error as a function of the rotation of link 2. The actual positions of the slider are evaluated using Equations (4.3-77) and (4.3-78). Figure 11.17 provides the results in a nondimensional form, using a procedure similar to that employed in Example 11.1.

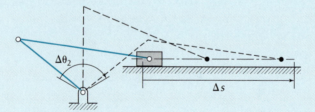

Figure 11.16 Function generating slider crank mechanism.

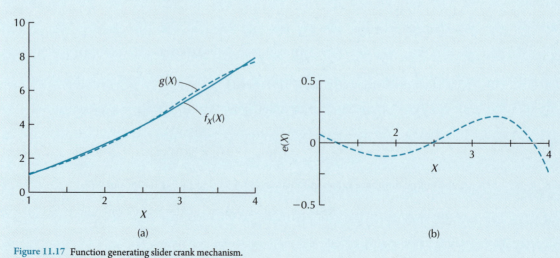

(a)   (b)

Figure 11.17 Function generating slider crank mechanism.

11.5 GRAPHICAL DESIGN OF MECHANISMS FOR TWO-POSITION RIGID-BODY GUIDANCE

Suppose that it is necessary to move the rigid body shown in Figure 11.18(a) from position 1 to position 2. Two points on the body are identified as A and B, and subscripts indicate the position number. Figure 11.18(b) illustrates one method of achieving this, by connecting a slider at A and B through a turning pair and having each slider move along a straight and stationary slide. If an actuator is attached to the slider at A and drives it from A_1 to A_2, then point B will follow in the required manner. In this instance, points A and B move along straight lines. Alternatively, as shown in Figure 11.18(c), a point on the body may be driven along a circular path by implementing a crank arm that is pinned to the rigid body. Point A on the body will move from A_1 to A_2 provided that the base pivot is located on the perpendicular bisector of line segment $A_1 A_2$. Either or both points A and B may be connected to a crank arm through a turning pair. Figure 11.18(c) shows the case where one crank guides point A from position 1 to position 2. For point B, we employ a straight slide and slider, as was used in Figure 11.18(b). This construction yields a slider crank mechanism. Note that there are an infinite number of locations where the base pivot could be located on the perpendicular bisector. Still another mechanism is shown in Figure 11.18(d). Here, we use one crank to guide point A and another to guide point B. In this instance we have a four-bar mechanism.

In this section, three different mechanisms were identified to carry out the same task, where the motion between the positions is distinct. A designer must select the most appropriate type of mechanism for each particular application (i.e., carry out type synthesis).

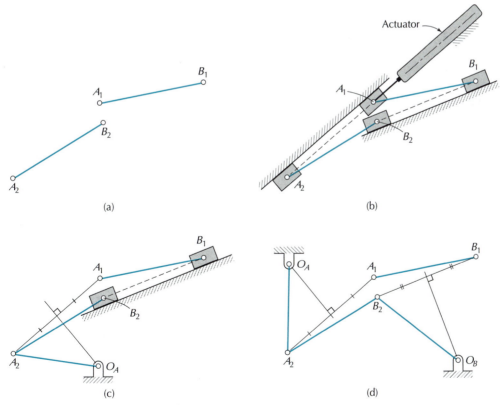

Figure 11.18 Two-position rigid-body guidance.

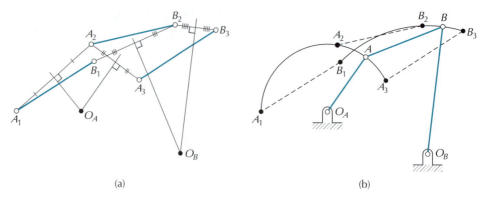

(a)   (b)

Figure 11.19 Three-position rigid body guidance.

11.6 GRAPHICAL DESIGN OF A FOUR-BAR MECHANISM FOR THREE-POSITION RIGID-BODY GUIDANCE

Consider the three positions of a rigid body containing points A and B, as shown in Figure 11.19(a). The three positions of point A are labeled A_1, A_2 and A_3. These points define a circle. The center of this circle is located at the intersection of the perpendicular bisectors of line segments $A_1 A_2$ and $A_2 A_3$ and is labeled O_A. A link pinned to the body at point A and to the base link at O_A can guide point A through its three positions. Likewise, the three positions of point B, labeled B_1, B_2, and B_3 define a circle centered at O_B. A rigid link pinned to the body at B and pinned to a base pivot at O_B will guide point B through its three positions. This construction has formed a four-bar mechanism $O_A A B O_B$, which guides the body through the three specified positions. Figure 11.19(b) illustrates the mechanism.

The procedure presented in this section started with selecting the location of the turning pairs on the coupler, and ended up by determining the base pivots. A trial and error procedure may be needed to determine a desirable location of the base pivots. In Section 11.7, an analytical procedure is presented whereby the base pivots maybe specified, and locations of the moving pivots are determined.

11.7 ANALYTICAL DESIGN OF A FOUR-BAR MECHANISM FOR THREE-POSITION RIGID-BODY GUIDANCE

An analytical procedure is presented for the synthesis of a four-bar mechanism to carry out three-position rigid-body guidance (see Figure 11.9). It employs the concept of displacement matrices presented in Appendix C, and was originally presented by Suh and Radcliffe [52]. We consider the case where we have a rotation about a fixed origin followed by a translation. The coordinates of the point before the motion are (x_{B_1}, y_{B_1}). The coordinates after the motion are (x_{B_n}, y_{B_n}), $n = 2, 3$. Using Equations (C.4-1), (C.4-15), and (C.4-16), the relationships between the coordinates are

$$x_{B_n} = C_{1n} x_{B_1} - S_{1n} y_{B_1} + A_{13_n}$$
$$y_{B_n} = S_{1n} x_{B_1} + C_{1n} y_{B_1} + A_{23_n} \tag{11.7-1}$$

where

$$A_{13n} = x_{B_n} - x_{B_1} C_{1n} + y_{B_1} S_{1n}$$
$$A_{23n} = y_{B_n} - x_{B_1} S_{1n} - y_{B_1} C_{1n}$$
$$C_{1n} = \cos\theta_{1n}; \quad S_{1n} = \sin\theta_{1n} \tag{11.7-2}$$

Employing a crank to guide the body through the positions, the coordinates of the base pivot are (x_{O_2}, y_{O_2}) and the coordinates of the moving turning pair in the starting position are (x_{B_1}, y_{B_1}).

Since the length of the crank is constant, the distance between the base pivot O_2 and the moving turning pair is the same for all positions. Then

$$[(x_{B_1} - x_{O_2})^2 + (y_{B_1} - y_{O_2})^2]^{1/2} = [(x_{B_n} - x_{O_2})^2 + (y_{B_n} - y_{O_2})^2]^{1/2} \qquad (11.7\text{-}3)$$

Squaring both sides of Equation (11.7-3), substituting Equations (11.7-1), and simplifying, we obtain

$$
\boxed{
\begin{aligned}
& x_{B_1}(A_{13n}C_{1n} + A_{23n}S_{1n} - x_{O_2}C_{1n} - y_{O_2}S_{1n} + x_{O_2}) \\
& + y_{B_1}(A_{23n}C_{1n} - A_{13n}S_{1n} + x_{O_2}S_{1n} - y_{O_2}C_{1n} + y_{O_2}) \\
& = A_{13n}x_{O_2} + A_{23n}y_{O_2} - \tfrac{1}{2}(A_{13n}^2 + A_{23n}^2), \qquad n = 2, 3
\end{aligned}
}
\qquad (11.7\text{-}4)
$$

Two distinct equations may be generated from Equation (11.7-4), one for $n = 2$ and the other for $n = 3$. If we specify the coordinates of the base pivot, then the equations may be solved for x_{B1} and y_{B1}. To determine the coordinates of D_1 (see Figure 11.9), we start with Equation (11.7-4) and then replace O_2 with O_4 and replace the coordinates of B_1 with those of point D_1.

The link lengths of the four-bar mechanism are determined using

$$
\boxed{
\begin{aligned}
r_1 &= \left[(x_{O_4} - x_{O_2})^2 + (y_{O_4} - y_{O_2})^2\right]^{1/2} \\
r_2 &= \left[(x_{B_1} - x_{O_2})^2 + (y_{B_1} - y_{O_2})^2\right]^{1/2} \\
r_3 &= \left[(x_{D_1} - x_{B_1}) + (y_{D_1} - y_{B_1})^2\right]^{1/2} \\
r_4 &= \left[(x_{D_1} - x_{O_4})^2 + (y_{D_1} - y_{O_4})^2\right]^{1/2}
\end{aligned}
}
\qquad (11.7\text{-}5)
$$

Care must be taken to ensure that the calculated values of link lengths produce a suitable mechanism. This is illustrated in the following example.

EXAMPLE 11.3 DESIGN OF A FOUR-BAR MECHANISM FOR THREE-POSITION RIGID-BODY GUIDANCE

It is required to move a headlight cover from the open position to the closed position as shown in Figure 11.20. Design a four-bar mechanism to perform this task by specifying three positions of rigid-body guidance. Employ the given locations of base pivots O_2 and O_4. During its motion, the headlight cover must not cross the fender of the automobile.

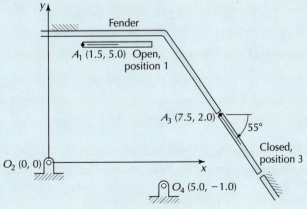

Figure 11.20 Three-position rigid-body guidance.

(*Continued*)

EXAMPLE 11.3 Continued

SOLUTION

Corresponding to the given information provided in Figure 11.20, we have

$$x_{O_2} = 0.0; \quad y_{O_2} = 0.0; \quad x_{O_4} = 5.0 \text{ cm}; \quad y_{O_4} = -1.0 \text{ cm} \tag{11.7-6}$$

The open and closed positions of the headlight cover are numbered 1 and 3, respectively. Thus

$$\begin{aligned} x_1 &= 1.5 \text{ cm}; \quad y_1 = 5.0 \text{ cm} \\ x_3 &= 7.5 \text{ cm}; \quad y_3 = 2.0 \text{ cm}; \quad \theta_{13} = -55° \end{aligned} \tag{11.7-7}$$

For the second position, we have a choice for the position and orientation of the headlight cover. If we select the following intermediate values between positions 1 and 3:

$$x_2 = 4.5 \text{ cm}; \quad y_2 = 3.5 \text{ cm}; \quad \theta_{12} = -27.5° \tag{11.7-8}$$

Substituting values from Equations (11.7-7) and (11.7-8) in Equations (C.4-16) we obtain

$$C_{12} = \cos\theta_{12} = 0.887; \quad S_{12} = -0.462$$

$$A_{132} = x_2 - x_1 C_{12} + y_1 S_{12} = 0.861 \text{ cm}; \quad A_{232} = -0.242 \text{ cm}$$

$$C_{13} = 0.574; \quad S_{13} = -0.819; \quad A_{133} = 2.544 \text{ cm}; \quad A_{233} = 0.361 \text{ cm}$$

Substituting the above values in Equations (11.7-4), one for $n = 2$ and the other for $n = 3$, and placing in matrix form, we get

$$\begin{bmatrix} 0.875 & 0.182 \\ 1.164 & 2.291 \end{bmatrix} \begin{Bmatrix} x_{B_1} \\ y_{B_1} \end{Bmatrix} = \begin{Bmatrix} -0.400 \\ -3.301 \end{Bmatrix} \tag{11.7-9}$$

Solving Equation (11.7-9) gives

$$x_{B_1} = -0.175 \text{ cm}; \quad y_{B_1} = -1.352 \text{ cm} \tag{11.7-10}$$

Employing equations similar to (11.7-4) and solving for the coordinates of D_1 gives

$$x_{D_1} = 3.281 \text{cm}; \quad y_{D_1} = -0.418 \text{cm} \tag{11.7-11}$$

and using Equations (11.7-5) gives

$$r_1 = 5.099 \text{cm}; \quad r_2 = 1.363 \text{cm}$$

$$r_3 = 3.580 \text{cm}; \quad r_4 = 1.815 \text{cm} \tag{11.7-12}$$

Video 11.21A
Headlight Cover, Undesirable Design

Figure 11.21(a) shows the resulting mechanism in its starting position. Although the mechanism can be assembled in each of the three positions, links 2 and 4 do not have continuous rotational motion in one direction in order to move through the three positions (see [Video 11.21A]). For this reason, it is not a desirable solution.

(Continued)

EXAMPLE 11.3 Continued

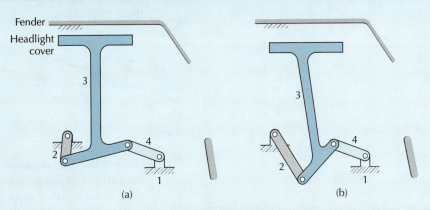

Figure 11.21 Headlight cover: (a) Undesirable design [Video 11.21A]. (b) Desirable design [Video 11.21B].

To obtain a desirable solution, one may try altering the location and orientation of the headlight cover for the second position. If we select

$$x_2 = 6.2 \text{ cm}; \quad y_2 = 2.8 \text{ cm}; \quad \theta_{12} = -44° \qquad (11.7\text{-}13)$$

then the calculated link lengths are

$$r_1 = 5.099 \text{ cm}; \quad r_2 = 2.665 \text{ cm}$$

$$r_3 = 2.622 \text{ cm}; \quad r_4 = 1.777 \text{ cm} \qquad (11.7\text{-}14)$$

Video 11.21B
Headlight Cover,
Desirable Design

The result is illustrated in Figure 11.21(b), and in this instance link 4 moves continuously in one direction while the headlight cover moves from the open position to the closed position (see [Video 11.21B]).

PROBLEMS

P11.1 Design a four-bar mechanism that will approximately generate the function

$$Y = X^{1/2}, \quad 1 \le X \le 4$$

Also

$$r_1 = 1.00 \text{ cm}; \quad (\theta_2)_s = 120°; \quad (\theta_2)_f = 30°$$
$$(\theta_4)_s = 100°; \quad (\theta_4)_f = 50°$$

Use Chebyshev spacing of three precision points.
(**Mathcad program: fourbarfuncsyn**)

P11.2 Design a slider crank mechanism that will approximately generate the function

$$Y = X^{3/2}, \quad 1 \le X \le 3$$

Also

$$(\theta_2)_s = 150°; \quad (\theta_2)_f = 90°$$
$$s_s = 2.0 \text{ cm}; \quad s_f = 4.0 \text{ cm}$$

Use Chebyshev spacing of three precision points.
(**Mathcad program: slidercrankfuncsyn**)

P11.3 Figure P11.3, on the next page, illustrates a function graph. Determine a slider crank mechanism that will generate the function with three precision points. Employ Chebyshev spacing of the precision points. (**Mathcad program: slidercrankfuncsyn**)

$$\Delta\theta_2 = -50°; \quad \Delta s = 4.0 \text{ cm}$$

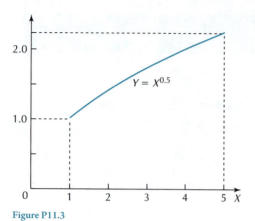

Figure P11.3

P11.4 Figure P11.3 illustrates a function graph. Determine a four-bar mechanism that will generate the function with three precision points. Employ Chebyshev spacing of the precision points. (**Mathcad program: fourbarfuncsyn**)

$$r_1 = 1.00 \text{ cm}; \quad \Delta\theta_2 = -30°; \quad \Delta\theta_4 = -30°$$

P11.5 Given the three positions of the rigid body shown in Figure P11.5, graphically design a four-bar mechanism that will guide the mechanism through the three positions.

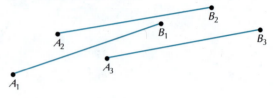

Figure P11.5

P11.6 Design a four-bar mechanism, with the given locations of the base pivots, that will guide a rigid body through the specified three positions. (**Mathcad program: fourbarrbg**)

$$x_{O_2} = 0.0; \quad y_{O_2} = 0.0; \quad x_{O_4} = 5.0 \text{ cm}; \quad y_{O_4} = -1.0 \text{ cm}$$

$$x_1 = 1.5 \text{ cm}; \quad y_1 = 5.0 \text{ cm}$$

$$x_2 = 6.3 \text{ cm}; \quad y_2 = 3.0 \text{ cm}; \quad \theta_{12} = -40°$$

$$x_3 = 7.0 \text{ cm}; \quad y_3 = 2.5 \text{ cm}; \quad \theta_{13} = -50°$$

CHAPTER 12
DESIGN OF MECHANISMS AND MACHINES

12.1 INTRODUCTION

When a device is needed to perform a mechanical function, the first step is to survey the many existing examples of mechanisms and machines and to adapt one of those solutions to satisfy that purpose. This book and other books provide many examples of existing mechanisms and machines. In some cases, however, there are no preexisting solutions, and a new mechanism must be created or designed. The design of mechanisms and machines is now discussed. The general engineering design process is presented, along with a detailed description of the various design phases. To facilitate this description, some examples of the design process involved for simple mechanisms are presented. In Chapter 13, numerous case studies on the design of various mechanisms and machines are presented, which emphasize different aspects of the design process. Also, exercise problems for various aspects of the design process are available at the end of this chapter, at the end of Chapter 13, and in Appendix A. These provide opportunities for the student of design to practice the materials described herein.

12.2 INTRODUCTION TO DESIGN

Engineering design is a process to create solutions to engineering problems. The solution may take the form of a system, a device, or a process that meets a set of specified design goals and objectives. Design is initially a creative process that generates multiple ideas and concepts. Those concepts are then evaluated and decisions are made on how to proceed with a solution that best matches the design goals and objectives. The analysis phase involves mathematics and engineering sciences. Design can be complex and open-ended, involving multiple iterations before a satisfactory solution is achieved. Generally, there is no unique design solution, only alternate solutions each with their own merits, which are evaluated against the design objectives.

It is important to emphasize the differences between engineering design activity and engineering science analysis. Many of the examples and problems presented in Chapter 1 to Chapter 11 are examples of mathematical and engineering science analysis (unless specifically indicated as design problems). In mathematics and engineering science, the problems are well-posed and are presented in a compact form. They require application of specialized areas of knowledge, such as kinematics, dynamics, the law of toothed gearing, and other such knowledge. The solution to each problem is generally unique, and these problems have a readily identifiable closure. In contrast, engineering design problems must be formulated or interpreted by the designer from a list of facts, needs, or client statement, typically known as a problem description. The problem description is usually incomplete due to missing information, or there may be uncertainty regarding the nature of the problem itself. The solution often requires consideration of several subject areas, beyond just mathematics and engineering science. There is generally no unique solution, only alternate solutions each with their own merits. When a final design solution is prepared and submitted, it should be scrutinized, and there are always recommendations for future work and revisions.

Mechanism and machine design involves the creation and synthesis of new ideas. Hence, one may ask "How do I learn to design or create things?" There are many opinions and textbooks on how to learn to design. Some say that design can be learned by studying and looking at many existing designs, analyzing them, and developing an aesthetic appreciation of design. This approach is often used by architects, artists, or even engineers. Others say that design can be learned by doing "design over and over again," taking the time to reflect on the process, reflecting on success and failure, and learning by hands-on experience. This approach is also used by some designers, whereby design experience is gradually built up over many years. Others will say that design can be learned/done by following a set of general design methodologies and processes, consisting of techniques, tools, and steps. This approach can be helpful with design projects of high complexity and many considerations. It is the authors' opinion that learning to design involves a balance of all three of the above approaches.

Individuals and novice designers seeking to develop their design skills are encouraged to use all three approaches. First, perform periodic reviews of the numerous examples of existing mechanisms and machines throughout this book and other books. These existing examples should be studied in depth, and where possible, the physical models or computer models of these mechanisms should be explored. Through the understanding of existing devices comes the knowledge necessary to develop new designs. Second, individuals should practice the process of creation and design often, by working on design-based problems such as those provided within Chapters 12 and 13 and also in Appendix A. Third, individuals should become familiar with formal design methodology and process, found in textbooks such as [15, 16, 17]. Design methodology and process is very useful when combined with the other two approaches, because it helps to organize and focus the design effort. Following a design process can greatly increase the likelihood that a good design solution will be achieved, in a reasonable period of time. The next section provides an explanation of formal design methodology and process, with emphasis on techniques that are useful for the design of mechanisms and machines.

12.3 DESIGN METHODOLOGY AND PROCESS

Using a formal design process can help organize and focus the design effort. Figure 12.1 lists the seven suggested phases of the design process, as well as specific activities within each phase. The phases and activities shown in the design process are generally depicted as a linear progression. However, the actual design process is nonlinear and iterative in nature as depicted by the oval dashed arrows. The design process can involve multiple iterations, reformulation of the problem, backtracking, re-design, etc.

12.3.1 Problem Formulation

The first phase of the design process is *Problem Formulation*. Here, the designer seeks to define the purpose and scope of the design activity, the problem to be solved, and the boundaries of the problem solution. The activities to be completed in the problem formulation phase from a checklist that ensures the design activity gets off to a good start. These tasks include the following:

Task 1: Create a *Design Goal Statement* for the design project, based on the information provided in the *Problem Description*. The problem description is usually presented by a client, an employer, or another colleague. The problem description arises from the recognition that there is a problem with the current situation, a deficiency with a system, or a lack of an existing design. It is very likely that the problem description will have insufficient information, which will require the designer to search for and gather more information from various sources. The *Design Goal Statement* attempts to summarize (a) the problem with the existing situation and (b) the general approach and methods to solve that problem. This statement should clearly express the need for the

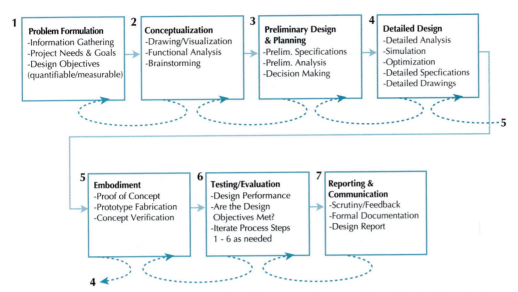

Figure 12.1 The seven phases of the design process. Although the process is depicted as a linear progression, it is an iterative process in practice.

design activity. Two examples of problem descriptions with their corresponding design goal statements are provided next. One example describes the need for a design to transmit torque between two nonparallel shafts, while the second example describes the need for a rotational device design subjected to clockwise (CW) and counterclockwise (CCW) torque.

EXAMPLE 12.1-A DESIGN OF TORQUE TRANSMISSION MECHANISM

PROBLEM DESCRIPTION
Connecting two shafts to allow for transfer of torque between them is difficult when they constantly change their relative alignment. In this case, torque needs to be transmitted from an input shaft to an output shaft, while they are spinning, where the alignment will vary as they spin. There is already a mounting plate built onto each of the shafts, to which a new coupling mechanism design could be mounted. Therefore, any new proposed design should accommodate those existing plates. It is important for this application that the mechanism be lightweight. It is also suggested that metal parts may be too heavy, so other materials should be considered. A diagram has been provided by the client, as shown in Figure 12.2, which illustrates the configuration and location of the potential device as "??".

PROPOSED DESIGN GOAL STATEMENT
Design a mechanism to connect two shafts together, such that the shafts may change their relative angle with each other and can transmit at least 3 N-m of torque. The two shafts are initially co-axial, but the output shaft may develop an angle of up to 30° off axis, during normal operation. To facilitate low weight, the mechanism must be constructed entirely out of ABS (acrylonitrile butadiene styrene) plastic sheet. Each shaft has a mounting plate attached perpendicular on its end, to which the new mechanism must attach.

Figure 12.2 Problem description illustration for a torque transmission mechanism.

EXAMPLE 12.1-B DESIGN OF ONE-DIRECTION ROTATIONAL DEVICE

PROBLEM DESCRIPTION

A machine has rotating components, but cannot operate properly if one of those components rotates in the wrong direction. At present, the said component is subjected to both clockwise (CW) and counterclockwise (CCW) torques, and it rotates in either direction. Upon examination of the machine, it is established that the component output should only rotate in a CW direction. Additionally, the device needs to transmit a reasonable amount of output torque in the CW direction for the machine to operate properly. Figure 12.3 illustrates the configuration and location of the potential device as "??".

PROPOSED DESIGN GOAL STATEMENT

Design a rotating device where its output can rotate only in the CW direction, where that device will be incorporated into a machine. During normal operation this rotating device will be subjected to CW and CCW torques. If torque is applied in the CCW direction, the output rotation must stop. If torque is applied in the CW direction, the output rotation must be able to transmit up to 100 N-m of output torque.

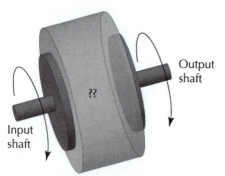

Figure 12.3 Problem description illustration for a one-direction rotational device.

Task 2: Conduct an extensive *Information Gathering* search to review existing devices that may address the stated design goal statement. This ensures there are no existing design solutions that can already satisfy the goal. It also exposes the designer to solutions that partially satisfy the goal. Suggested sources of information include:

- **Library:** Textbooks, journal papers, or conference papers in the subject area.
- **Internet Sources:** There are numerous online search-engines such as Google, Yahoo, Bing, and many others where appropriate keywords relating to the design can be entered, and relevant documents can be found.

- **Internet-Based Image Search:** Image-based searches are particularly useful for mechanism and machine design. Keywords will provide dozens of images that can be visually scanned for potential design relevance. Such visual-based scanning is highly recommended for designers who are new to mechanism design. New designers often do not know the technical names for specific mechanisms, and hence it is difficult to search with text-based approaches.
- **Internet-Based Technology Magazines:** These have extensive coverage of many technical areas, including magazines or articles from the ASME (American Society of Mechanical Engineers), SAE (Society of Automotive Engineers), or other organizations that address the subject area.
- **Patents:** Millions of patents are catalogued and listed online. These are an excellent source of information for mechanism and machine design. In particular, patents that are over 20 years old are no longer protected, and their contents are free for all to use. They often have excellent diagrams and descriptions of how the mechanisms work. One suggested online source is the United States Patent and Trademark Office.
- **Stakeholders:** These are the people who are involved in, or affected by, the design. This may include the clients who have commissioned the design.
- **Experts/professionals:** These are people who have worked on similar designs in the past.

Task 3: Develop a set of *Design Objectives and Constraints.* These are the specific performance targets that will dictate the direction of the design activities. Sometimes these are referred to as "Design Requirements" in other books; however, they are referred to as *Design Objectives* in this textbook. The design objectives are specific engineering performance targets describing how your design will perform, and they should also include a quantifiable/numeric value. As a designer, you will aspire to meet or exceed your design objectives. *Design Constraints* are similar to design objectives, however they are items/descriptions that restrain or restrict the design activity. To best illustrate the formulation of design objectives and constraints, Example 12.1-A and Example 12.1-B are further developed by Example 12.2-A and Example 12.2-B, respectively. Design objectives and constraints are best listed in a table format to consolidate the information.

EXAMPLE 12.2-A DESIGN OF TORQUE TRANSMISSION MECHANISM

DESIGN OBJECTIVES AND CONSTRAINTS

Table 12.1-A lists a set of proposed Design Objectives and Constraints for the design of a torque transmission mechanism to transfer torque between two shafts.

In this example, a number of design objectives can be developed. Note that the design objectives for torque, angular alignment, material, and shaft coupling were all derived from the information provided in the design goal statement. However, much more information is needed. As the designer contemplates the problem further, a number of questions will arise. The problem definition states that the design should be lightweight, but no value was specified. Further, the designer will realize that there is no mention of operational speed, life cycle, or other requirements in the description, yet these two parameters would likely have a large role in the design. Therefore, more information gathering must be done by the designer to learn more about the intended application, using the suggestions of Task 2 in Section 12.3.1. Presuming that such additional information gathering takes place, the designer may reach the conclusion that an operating speed of 60 rpm, a life cycle of three years, and a maximum mass of 1 kg are reasonable values for the application. Based on this information, the design objectives and constraints listed in Table 12.1-A have been formulated.

TABLE 12.1-A DESIGN OBJECTIVES AND CONSTRAINTS FOR THE TORQUE TRANSMISSION MECHANISM

Design Objective	Description	Value (Units)
Torque	Mechanism must transmit a minimum torque.	3 (N-m)
Angular alignment	Mechanism must operate properly with angular misalignments of up to 30 degrees.	30 (degrees)
Mass	Must be lightweight with a maximum value 1 kg.	1 (kg)
Material	Must be constructed from only ABS plastic sheet of 3 mm thickness.	ABS plastic sheet
Shaft coupling	Must attach to existing shaft mounting plates, as illustrated in Figure 12.2.	Via mounting plate
Operating speed	Must operate at rotational speeds at least 60 rpm	60 (rpm)
Cycle life	Must operate for at least 3 years	3 (years)

EXAMPLE 12.2-B DESIGN OF ONE-DIRECTION ROTATIONAL DEVICE

DESIGN OBJECTIVES AND CONSTRAINTS

Table 12.1-B lists a set of proposed Design Objectives and Constraints for the design of a one-direction rotational device.

For this example, the problem description provides very few details regarding the quantities or performance targets needed for the objectives. In many cases, a mechanism designer will be given a similar level of limited information, and it is up to them to investigate further. This requires further information gathering from the client, other designers that worked on similar designs, textbooks, or other written materials.

For this example, assume that the designer proceeds to gather more information and develops a number of design objectives including: reverse torque capability, the mounting system, the materials, and the useful life of the mechanism. The reverse torque capability objective states that the proposed mechanism should not permit any CCW output torque. The mounting system objective states that the support structure that holds the mechanism should allow it to rotate with low friction in the CW direction. The material objective specifies that the material used to build the component should be strong and wear resistant. The useful life objective specifies the minimum number of rotations that the component should endure during its useful life prior to failure.

From the information gathered, the designer concludes that there are some restrictions to the scope of the design. These constraints include a limited volume within the machine, into which the new device must fit. Also, the new device must spin in the desired direction with a rotational speed of at least 10 radians/second for the machine to operate.

In addition, during the information gathering the designer learns that there are no weight limitations for the design. Also, that similar design concepts in other written works exhibit noise during operation. Upon further consideration and consultation with the client, it is decided that the design objectives of weight and noise generation will not be considered as influential factors in the design activity. Based on this information, the design objectives and constraints listed in Table 12.1-B have been established.

TABLE 12.1-B DESIGN OBJECTIVES AND CONSTRAINTS FOR ONE-DIRECTION ROTATIONAL DEVICE

Design Objective	Description	Value (Units)
CW torque	Transmission of output torque in desired direction	100 (N-m)
CCW torque	Component output torque must stop. Should resist specified torque without rotation.	Ø (N-m)
Mounting	Rotational mounting system for low friction during rotation	Ball bearings
Material	Must be strong and wear resistant	Steel
Cycle Life	Must operate for at least 1,000,000 rotations	1 million (rotations)
Weight	Not considered as influential for this application	—
Noise generation	Not considered as influential for this application	—
Constraint	**Description**	**Value (Units)**
Maximum size	Mechanism must not exceed size range	$30 \times 30 \times 15$ $(L \times W \times H)$ (mm)
Minimum operating speed	Device must operate smoothly at rotational speeds of at least 10 radians/sec	10 (rad/sec)

It is important to establish the design objectives and constraints at the beginning of the design activity. There are a few important reasons for this:

(i) The design objectives and constraints should generally be established before the conceptualization phase and should provide the ideal performance values which the design solution should aim to achieve. Further, each objective should be focused on best satisfying the design goal statement and should be relatively independent of specific design concepts. Later, after the conceptualization phase, the different proposed concepts can be judged against these original design objectives. This process will be described in Section 12.3.2.

(ii) The design objectives and constraints serve as an important method to document the development of a design. As the design evolves, new information may be uncovered (extra requirements or obsolete ideas) which will require the update or modification of the original design objectives. This is part of the iterative nature of the design process. Additionally, if a design project is undertaken by a team of designers, a central document that lists the past and current design objectives serves to keep all team members' efforts coordinated.

(iii) A very important reason to establish clear design objectives and constraints, is to ensure that the expectations of the client are met. By writing down the design objectives and their associated numeric targets/values at the outset of a project, both the designer and the client can generally agree on the performance expected. If the design objectives require modification later during the design phase, both designer and client should meet to discuss the impact on the project. Consider the problem description in Example 12.1-A. Here the weight of the design is listed as "lightweight." Without specifying a quantity, the term "lightweight" is open to interpretation, opinion, and ultimately misunderstanding between the designer and the client. Not all design objectives can be precisely quantified, but an effort to specify their performance as a written description should be done. Proper documentation of design objectives serves to create effective communication and to establish measurable performance metrics against which a design solution can be judged by all involved.

12.3.2 Conceptualization

The next phase of the design process is the *Conceptualization* phase. This is a creative phase, where several ideas or concepts are generated. It is the divergent phase of the design process, where many possibilities are considered without too much judgment. Each concept should attempt to solve a number of the design objectives, but not necessarily all of them. For a typical design project, an effort should be made to generate at least three *Conceptual Design Alternatives*. These should be documented, along with sketches and descriptions of the conceptual designs.

To say it simply, the generation of creative ideas is one of the wonders of being human. A writer by the name of William Plomer once wrote, "Creativity is the power to connect the seemingly unconnected." There is no formula or set of rules that can be applied to generate ideas out of thin air. Rather, what is presented here are a set of practical strategies to help individual designers to enhance their inherent creativity. In particular, these strategies are presented since they have high relevance to mechanism and machine design.

Creative Strategy 1: The designer should study and review existing design examples on a periodic basis. Creativity cannot occur in a vacuum, but is often the result of exposure to many other ideas, previous experience, and practical lessons. There are many great books that summarize hundreds of mechanisms and machines [18, 19]. Such books are organized by subject chapters and are helpful to read before undertaking a major mechanism design effort.

Creative Strategy 2: *Morphological Analysis* is a technique that encourages a designer to consider the combination of two seemingly unrelated concepts. The technique is best implemented by creating a *Morphological Chart*, with two axes of information. Each axis describes an attribute, design objective, or some function of the design. The information in the two axes should be relatively independent, to help with the generation of interesting ideas. The use of morphological analysis often leads to impractical ideas, however, those ideas may themselves eventually lead to practical ones. An example of morphological analysis is described in Example 12.3, for the generation of novel ways to transmit torque between two shafts.

Creative Strategy 3: *Illustration and Drawing* is an important way to record ideas and generate new ideas. Some designers are abstract thinkers while others are visual and spatial thinkers. By preparing a drawing, limitations can be revealed, or ideas can be built upon. The authors strongly recommend that students of design learn to prepare hand-drawn sketches, using pencil and paper. This should always be the first step. Example 12.4-B provides examples of concept sketches for the one-direction rotation device. As a note, many students rush too quickly to use computer aided drawing tools. For idea generation, computer tools are often too restrictive for quick sketching. It has been the experience of the authors, that the drawing function/tool-set available with many computer aided drawing softwares will heavily influence the idea generation process. This occurs because some ideas cannot be quickly and easily drawn with the predefined software functions and toolsets. Computer aided drawing tools should be reserved for the later phases of *Preliminary Design* and *Detailed Design* of the design process.

Creative Strategy 4: Allow ample time for *Reflection* on ideas, and allow for iteration. Creativity cannot be rushed, and setting a strict timeline for the creative phase of the process may limit the best solutions from emerging. Iteration is important, since often the original concept can lead to a better idea for the concept. Sometimes an idea suddenly emerges for a design solution, while the designer is in the midst of another unrelated activity. This is sometimes referred to as an "Ah-ha moment," and such moments are more likely to occur when the creative process is given ample time.

Creative Strategy 5: *Functional Decomposition* is a method that has been developed to assist designers in the concept generation process [20–22]. It is especially useful in situations with complex designs, and it is well suited for mechanism and machine design. Functional decomposition can be used for the preliminary design phase; or during the concept generation phase if the designer becomes "stuck" during the process. Two examples of functional decomposition are provided in the next section, in Examples 12.5-A and 12.5-B.

Creative Strategy 6: The previous five creative strategies can be done by an individual designer. Sometimes, it is useful to engage a team of persons in the creative process. In this case, the *Brainstorming* method can be an effective group activity for the generation of new ideas. To best employ the brainstorming method, a set of guidelines [23] should be followed.

EXAMPLE 12.3 USE OF MORPHOLOGICAL CHARTS

Table 12.3 provides an example of using a morphological chart to generate new ideas for the transmission of torque between two parallel shafts.

The chart is constructed by specifying one independent attribute on each axis. In this example, the horizontal axis contains an attribute called *Coupling Device Type*, of which five possible types are suggested. The vertical axis contains an attribute called *Attachment System*, of which six possible types are suggested. The chart is then systematically evaluated, to generate ideas. The purpose is to force the designer to consider combinations of seemingly unrelated mechanical components. Combinations that might otherwise be subconsciously dismissed, or ignored, are now required to be given at least 1–2 minutes of thought.

In this example, the column with the heading *Chain* is evaluated first, to see if ideas can emerge when comparing it to the headings listed in the rows. We first compare the *Chain* column to the *Keyed Slot* row. In other words, the designer will ask "Is it possible to transmit torque between two shafts using a *chain*, when a *keyed slot* is the only attachment system used on the shaft"? After some thought, there does not seem to be a valid combination, since a *Keyed Slot* cannot grip a *Chain*, so the box is left blank. The next combination is then evaluated: "Is it possible to transmit torque between two shafts with a *Chain*, when *Sprockets* are the attachment system on the shaft"? Certainly, because this is commonly done, a YES is placed in the box. However, that idea isn't particularly innovative. The next combinations are then evaluated: "Is it possible to transmit a torque with a *Chain* when the shaft only has *Holes*, *Springs*, *Pins*, etc., as attachment points?" When all the possible combinations using a *Chain* are considered, the next column is evaluated in the same way.

In the case of the column considering *Belt*-based torque transmission between shafts, the obvious combination is a *Pulley* attachment system. In the case of the column with *Gear*-based torque transmission, the obvious combination is a *Keyed Slot*. Beyond this, an interesting yet plausible answer is to use *Glue* (perhaps Super Glue™) to fasten the *Gears* to the shaft, and hence that combination is marked with YES.

Upon deeper evaluation, other more interesting ideas emerge as combinations. These ideas are denoted as Ideas A, B, C, and D. In the case of Idea A, it may be possible to use a flat *Belt* to couple the shafts, if the belt is dotted with holes. These holes would have a specific pitch distance between them, that would engage with the *Pins* that are protruding from the shaft. In the case of Idea B, torsional *Springs* could connect the *Gears* to each shaft. When the two gears are brought into mesh, the springs could be pre-loaded to create an antibacklash system. In the case of Idea C, round *Magnets* could be inserted into the *Holes* in the shafts. In this way, as one shaft spins, it could exert magnetic force onto the other, which may transmit small amounts of torque for some applications. In the case of Idea D, perhaps pressurized *Fluid* may exist inside the shafts, which would spray out of the *Holes* perpendicular to the shaft axis. In this way, if the two fluid jets are impinging, they may cause transfer of torque between shafts.

Note that Ideas A, B, C, and D may range from interesting to impractical, but with further development they may eventually lead to a useful torque transfer system. Certainly, without using a Morphological Chart, Ideas A, B, C, and D would likely have never emerged for consideration.

There are other noteworthy methods and practical strategies to help a designer during concept generation. These additional methods and strategies include: Mind Maps [23], the 6-3-5 Method [24], and recognition of creative and perceptual barriers. Detailed descriptions and examples of these methods are beyond the scope of this book, but references are provided. The student of design is encouraged to read about these methods and go beyond to explore other methods.

TABLE 12.3 MORPHOLOGICAL CHART TO GENERATE VARIOUS IDEAS FOR TORQUE TRANSMISSION BETWEEN TWO SHAFTS

		Coupling Device Type				
		Chain	**Belt**	**Gears**	**Magnets**	**Fluids**
Attachment System	Keyed slot			YES		
	Sprocket	YES				
	Holes				Idea C	Idea D
	Springs			Idea B		
	Pins		Idea A			
	Pulley		YES			
	Glue			YES		

EXAMPLE 12.4-B DESIGN OF ONE-DIRECTION ROTATIONAL DEVICE

CONCEPTUALIZATION: ILLUSTRATIONS OF CONCEPTUAL DESIGN ALTERNATIVES

Conceptual design sketches should be hand-drawn, and must illustrate the operational principle of the idea. The various components within the diagrams should be labeled. In this example, four different conceptual design alternatives for the one-direction rotational device have been sketched, and are provided in Figure 12.4.

Concept A shows a traditional toothed ratchet and pawl idea. The pawl is lightly pressed against the teeth using a spring. In this design, there are discrete positions where the pawl is engaged with the teeth, where no CCW motion can occur. However, in the CW direction, the ratchet will simply slide over the top of the teeth, dipping down between tooth positions.

Concept B shows a friction-based idea with two off-center pivot followers. These pivot followers are lightly pressed against the central disc using springs. If the disc rotates CW, it easily slides past the followers since they are offset. However, if the disc attempts to rotate CCW, the friction between the followers and the disc will cause the followers to wedge themselves into the disc, thereby preventing CCW motion.

Concept C shows a friction-based idea with two off-center rollers. This design also makes use of the "wedge principle." In this case, the rollers ride within the disc, and they are pressed outward via springs. When the disc rotates CW, the rollers are pushed inward. However, if the disc attempts to rotate CCW, the rollers roll out and wedge themselves into the housing, thereby preventing rotation.

Concept D shows a friction-based idea with two elliptic followers that can pivot. This design employs the "wedge principle" and is similar to Concept B. However, the followers are pinned to the disc allowing for a simple circular housing. The springs are used to force the follower into contact with the circular housing, but are not shown for clarity.

12.3.3 Preliminary Design, Analysis, and Decision Making

After the generation of conceptual design alternatives, the *Preliminary Design* phase will begin. During this phase, the conceptual designs must be developed further to the point where predictions can be made regarding their expected performance. Tools to further develop the designs may include *Functional Decomposition*, engineering science analysis, and mathematical analysis.

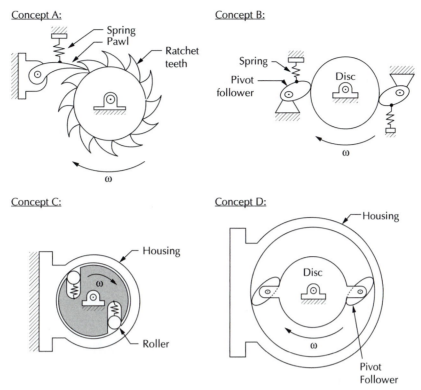

Figure 12.4 Sketches of conceptual design alternatives for a one-direction rotational device.

In some cases, preliminary computer modeling and simulation may also be useful to predict performance. When the conceptual design alternatives are sufficiently developed, a *Decision Selection Methodology* should be used to choose the best design concept. There are various possible ways to select from different design concepts, such as analytical methods or discussions with the design team, clients, or other experts. In this work, an analytical method called *Selection Tables* will be presented, since it is well suited to the design of mechanisms and machines. Using this approach, the developed design concepts will be evaluated numerically against the design objectives and assigned scores according to how well they satisfy the objectives. Generally, the design concept with the best total score will be selected and carried forward to the *Detailed Design* phase.

In Example 12.4-B, a number of conceptual designs for a one-direction rotational device were illustrated. Although these concepts were drawn nicely and briefly described, they are not sufficiently developed and not detailed enough to make predictions about their performance. Therefore, preliminary design is about further development and design of these concepts, in accordance with the design objectives established for the one-direction rotational device.

12.3.3.1 Functional Decomposition within Preliminary Design

Functional Decomposition is a useful and systematic technique to develop the details of a design [20–22]. It is especially useful for complex designs consisting of many parts, such as mechanism and machine designs. Functional decomposition (sometimes referred to as Functional Analysis in other textbooks [22]) can be used to specify the desired functions of a mechanism. The first step in functional decomposition is the creation of a *Functional Block Diagram* to represent the system functions. Note that this diagram is not a component/part layout chart. Rather, it is a flowchart representing the desired functions/actions that the system must perform. By drawing the flowchart in terms of functions/actions, the operation of the mechanism/machine can be created and developed, without worrying about mechanical details. After a functional block diagram is complete, the designer can refer back to the function blocks when creating the mechanical components needed to achieve those functions.

The basic component of a functional block diagram is a block, where each block is a function/action, and has inputs and outputs, as shown in Figure 12.5(a). The output of one block serves as the input(s) for the next block(s), as shown in Figure 12.5(b). For example, a simplified representation of a fluid pump is depicted in Figure 12.6. Note that the blocks represent functions or actions, whereas the input/output arrows represent physical objects or energy flows. In this simple example, the block with a dotted line represents the overall system boundaries The inputs to the system are energy and fluid on the left, and the outputs of the system are pressurized fluid and waste heat on the right. Internally, there are three functional blocks shown. The first block is the function to change the input energy into a more useable form such as mechanical torque. The second block describes the function to clean the fluid entering the system. The third block describes the function to compress the fluid, having two required inputs which are mechanical torque and clean fluid. The outputs of the third block are pressurized fluid and waste heat, which exit the overall system.

The functional block diagram of Figure 12.6 is simple in form, and it requires more functional detail to be useful. For example, more detail is needed for the block describing the conversion of energy into mechanical torque, or the block describing the compression of fluid. The major advantage of using functional decomposition with block diagrams is that hierarchical levels can be used, where each subsequent (lower-level) diagram provides more and more detail. For example, Figure 12.7 provides such detail, by illustrating a lower-level functional diagram of the "convert energy into torque" block. Note that the inputs and outputs of this block must be identical to those of the higher-level block in Figure 12.6. When creating the diagram of Figure 12.7, the designer must consider the inputs and outputs at the higher-level diagram and must create lower-level details to match with those. Here, three new functional blocks are proposed and added. The first block describes the conversion of the input energy into rotation, and it provides an output of high rotational speed. This provides more detail for the function, and it allows the designer to start thinking about ways to accomplish these details. Also, this level of detail precludes other possible conversion methods, by specifying that the energy is to be converted into high-speed rotation.

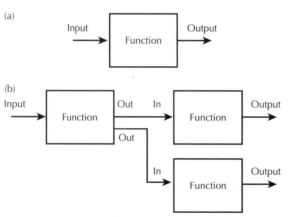

Figure 12.5 Basic elements of functional block diagrams.

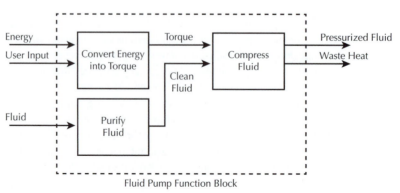

Fluid Pump Function Block

Figure 12.6 High-level functional block diagram of a fluid pump.

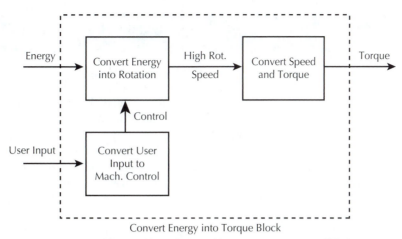

Figure 12.7 Lower-level functional block diagram of "convert energy into torque" block.

Clearly, the designer may jump to the conclusion that an electric motor can perform this function, however, the designer should resist such specific choices at this stage. Only the function regarding the conversion of energy into rotation should be shown. This is done since there may be other possible ways to convert energy into rotation, so the diagram is drawn in this general way to allow for various ideas to be considered later. The second block describes the regulation/conversion of the rotational speed into a useful speed and torque. Again, the designer might jump to the conclusion of using a gearbox to perform this regulation/conversion function. However, by leaving the block written as a conversion function, other ideas such as the use of belts and pulleys, or friction drives, or other mechanisms could be considered later. The third block describes user input that is required to turn the pump on and off. It describes conversion of user input into a useful mechanical control action, which has an output indicated as control.

The creation of a functional block diagram is gradual and iterative. The diagram will start out simple as drawn in Figure 12.6 and will become more detailed by the addition of lower-levels that represent each of the high-level blocks. Very often when drawing the lower-level diagram details, inputs or outputs will emerge that were not considered at the higher-level diagrams. Some of these may need to be transferred up into the high-level diagrams, which may go on to influence other high-level blocks and then their own sub-blocks, and so on. In this way, the process is iterative, but is very useful in designing the functional details of the mechanism. Note that there is no single correct version of a functional block diagram, and many variations are possible. The end result of the functional block diagram may be of less importance than the act of creating the diagram. The act of creating it allows the designer to critically examine the design objectives and the functions to achieve them, the function inputs and outputs, and their relations to each other and determine if additional functions are needed. This is a great way to engage the creative process of design.

The best way to further demonstrate functional decomposition and the generation of functional block diagrams is through Examples 12.5-A and 12.5-B. These demonstrate functional decomposition as applied to the design of the torque transmission device and to the one-direction rotational device. There are a number of possible ways to prepare these diagrams, based on the assignment of various functions.

EXAMPLE 12.5-A DESIGN OF TORQUE TRANSMISSION MECHANISM

FUNCTIONAL DECOMPOSITION
The creation of a functional block diagram for the torque transmission mechanism is presented here. The first step prior to drawing the diagram is to review the Problem Description, the Design Goal Statement, and the Design Objectives listed in Table 12.1-A. Some of these will

(Continued)

EXAMPLE 12.5-A Continued

be useful when creating the diagram, while others will not apply. Also, a review of any conceptual sketches (if available) or other concept generation results should be done.

The diagram should start with the specification of the bounding box drawn as a dotted line, as shown in Figure 12.8. All the external inputs such as energy (forces) and materials entering the system are specified, as well as all output energy and materials exiting the system. These inputs and outputs are shown as arrows that cross the bounding box. In this example, the inputs are specified as torque and shaft misalignment, while the output is specified as torque. Next, the designer must create the practical/necessary functions that need to take place within this mechanism, to match up with its inputs and outputs. Although there is no shape or form to the proposed mechanism yet, the designer may realize that the mechanism will need to be connected to the external shaft that supplies the torque, in some way. Also, some way to connect the mechanism output to an external receiving shaft is needed. These two functions are added to the diagram, labeled as "Connect to Input Shaft" and "Connect to Output Shaft." Upon further reflection, the designer may realize that the mechanism will be rotating, and its rotating parts must be supported on both ends. However, the output end may be misaligned with respect to the input. This information is added to the diagram with the function blocks labeled as "Support Input Structure" and "Stabilize Output Structure." The word stabilize is used since it is known that the output will be misaligned, yet it must still be stable enough to allow for the transmission of torque. Finally, there must be some way to connect the input side of the mechanism to the output side. This function is labeled as "Couple Input to Output," which is a rather broad definition, but will be refined further in the next step. As a note, the actual process to create Figure 12.8 took a few iterations, until there was a smooth progression of functional steps, separated by arrows of force or material flows.

Although Figure 12.8 is helpful, it lacks sufficient detail to be immediately useful.

Therefore the next step is to decompose each functional block, into a set of lower-level functional blocks, with more detail. Figure 12.9 shows a lower-level functional block diagram, based on the information from Figure 12.8. Each of the original blocks from the higher-level diagram now appear as dotted bounded blocks, with the proposed lower-level functional blocks within. Also note that a new external input has been added as an arrow, listed as "Ground Link." This input was necessary since the mechanism must be supported in some way, with respect to ground. Since the ground link arrow enters the system as a whole, it should be added to the higher-level diagram of Figure 12.8. However, this has been purposefully left as is, to demonstrate the iterative nature of drawing functional block diagrams, and the relationship between high-level to low-level diagram creation.

The blocks of Figure 12.9 provide more detailed functional information. For example, functions such as alignment, fastening, allowing motion or support, constraint, stabilization, and so on, have been added. It is likely that some blocks may still lack sufficient detail, and those functional blocks can be decomposed further into an even lower-level diagram, as needed. For brevity, this is not shown here. The remaining details of Figure 12.9 are left to the reader to explore.

When the functional decomposition of the Torque Transmission Mechanism is completed, the resulting block diagram can be used for other purposes. For example, it can be used to communicate the functional details to other designers or customers for feedback, or it could be used for mechanical component selection or design.

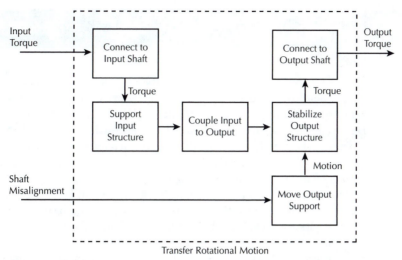

Figure 12.8 High-level functional block diagram of torque transmission mechanism.

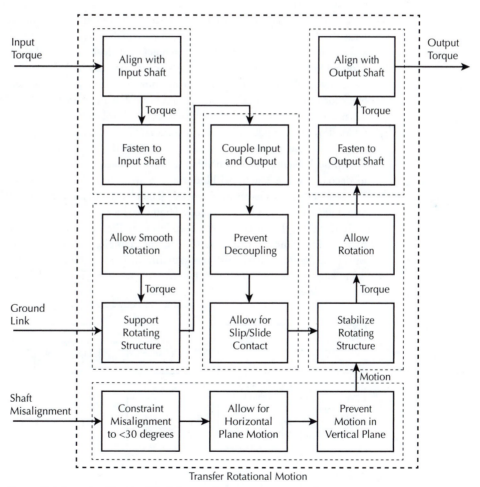

Figure 12.9 Lower-level functional block diagram of torque transmission mechanism.

EXAMPLE 12.5-B DESIGN OF ONE-DIRECTION ROTATIONAL DEVICE

FUNCTIONAL DECOMPOSITION

The creation of the functional block diagram for the one-direction rotational device is presented here. Prior to drawing the diagram, the designer should review the Problem Description, the Design Goal Statement, and the Design Objectives listed in Table 12.1-B. Also, there are some conceptual sketches developed for the one-direction rotational device in Example 12.4-B, as shown in Figure 12.4. These sketches will be particularly useful, since they can help to define the block functions in a more specific way.

The diagram creation begins with the specification of the bounding box drawn as a dotted line, as shown in Figure 12.10. All the external inputs entering the system, as well as all outputs exiting the system, should be defined first. In this example, the inputs are CW input torque and CCW input torque, while the output is only CW torque. Next, the designer must create the practical/necessary functions needed within this mechanism, to match up with its inputs and outputs. Some hypothetical steps that the designer may consider when developing the functional blocks are now described. The designer may consider the way that the mechanism will need to be connected to the external CW and CCW input torque, in some way. Also, some way to connect the mechanism output to an external receiving link is needed. These two functions are added to the diagram, labeled as "Connect to Input Link" and "Connect to Output Link." Next, the designer may consider that the purpose of the mechanism is to transmit CW rotation and to prevent CCW rotation, hence two functional blocks are added as "Allow Rotation" and "Prevent Rotation." Note that the inputs to these blocks are "CW Torque" and "CCW Torque," respectively, which helps to qualify the meaning of the subsequent blocks. Note that only the block for "Allow Rotation" has an output torque. Finally, there must be some way to support this mechanism with respect to the ground plane. This is added by the function "Support from Ground" with the input arrow being labeled as "Ground Link." Since the manner of support is not yet clear, the output arrow is dashed and is to be defined in a lower-level diagram.

Although Figure 12.10 is helpful, it requires more detail to be useful. The next step is to decompose the functional blocks into a set of detailed lower-level functional blocks. Figure 12.11 shows a lower-level functional block diagram, where some of the original blocks from Figure 12.10 now appear as dotted bounded blocks.

In this example, the conceptual sketches from Figure 12.4 are available and can be helpful for defining the lower-level functions. For example, Figure 12.4(a) illustrates a rotating wheel and Figure 12.4(b) illustrates a rotating disc, where both spin CW. Additionally, there is a linkage that is pushed into contact with the wheel or disc, labeled as a pawl or follower,

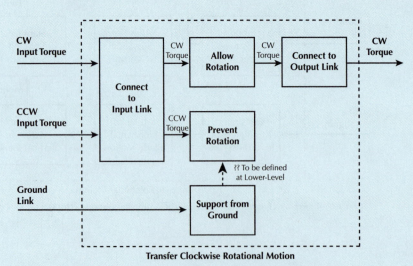

Figure 12.10 High-level functional block diagram of one-direction rotational device.

(*Continued*)

EXAMPLE 12.5-B Continued

respectively. This information may be incorporated into Figure 12.11. The designer may begin by expanding the level of detail of the "Support from Ground" block from Figure 12.10. In this case, some function to support the pawl (or follower) with respect to ground is needed, which has been added as the "Support Pawl Link" block. In addition, some function to press the pawl against the rotating wheel is needed. This is added as the "Push Pawl Against Wheel" block. In association with this, a block labeled "Apply Force on Pawl" is added. These three blocks are bounded by the dotted block that represents a higher level function "Support from Ground."

Next, the designer may consider the decomposition of the high-level blocks called "Allow Rotation" and "Prevent Rotation." From Figure 12.4, it is known that there is a rotating wheel, and hence the block "Apply CW Torque to Wheel" has been added. The output of this block must lead to a function that allows rotation to take place. Since it is known that the Pawl from Figure 12.4 will be involved, this can be written more specifically as "Allow Pawl to Slip over Wheel," thereby allowing the wheel to rotate. In the other case, the "Apply CCW Torque to Wheel" block leads to the function that states "Jam Pawl into Wheel," thereby preventing the wheel from rotating. In this way, the functional diagram is made more specific, by incorporating information from the conceptual diagram.

Some blocks in Figure 12.11 may still lack sufficient detail, and those functional blocks can be decomposed further into an even lower-level diagram, as needed.

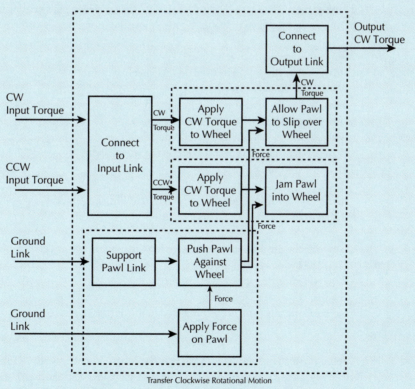

Figure 12.11 Low-level functional block diagram of one-direction rotational device.

Functional decomposition is a great method to create and develop the functions or actions needed for a mechanism to achieve its design objectives. The design student is encouraged to practice this method further, using some of the examples provided later in this chapter. More examples and advanced application of the functional decomposition method can also be found in books such as references 20–22.

12.3.3.2 Parameter Specification within Preliminary Design

After performing concept generation activities, and also functional decomposition of the concept, a number of new details, points of interest, and questions about the concept will likely arise. These details may include questions regarding the forces exerted, the speeds achieved, the range of motion, and so on. Also, questions regarding the materials to be used, the shape and strength of the components, use of standard parts such as bolts and springs, and other points will arise. These specific design details are referred to as *Design Parameters*. The main purpose of the preliminary design process is to develop the design parameters with sufficient level of detail to allow for estimates on the ability of the concept to satisfy the design objectives. Example 12.6-B illustrates one way to develop the preliminary design of the one-direction rotational device, in the context of component design, selection, and configuration. For brevity, only development of Concept A is provided.

EXAMPLE 12.6-B DESIGN OF ONE-DIRECTION ROTATIONAL DEVICE

PRELIMINARY DESIGN: PARAMETER DEVELOPMENT FOR CONCEPT A, INVOLVING COMPONENT DESIGN, SELECTION AND CONFIGURATION

Some preliminary design details of Concept A can be developed, using the conceptual sketch of Figure 12.4(a) and the functional block diagram of Figure 12.11. The process begins by considering all the outstanding questions that arise when reviewing this existing information. In the case of Concept A shown in Figure 12.4(a), these questions may be (in no particular order):

- What is the diameter of the wheel?
- How many teeth should be used on the wheel?
- What angle should the teeth have?
- How thick is the wheel?
- How long is the pawl?
- Where is the pawl located relative to the wheel?
- How much force will the pawl exert on the wheel teeth?
- Etc.

In the case of the functional block diagram of Figure 12.11, these questions may be (in no particular order):

- How will the mechanism be connected to the input link?
- How will the pawl be supported?
- How will force be applied to the pawl?
- How will the wheel be supported?
- How will the mechanism be connected to the output?
- Etc.

Some information is already available, from the design objectives listed in Table 12.1-B: It is known that the wheel and pawl should be made from steel. It is known that the overall size of the mechanism must fit within 30 × 30 × 15 mm. It is known that the mechanism will employ ball bearings, which provide for low friction rotation, and also satisfy the design objective for a long life cycle. It is known the mechanism should apply at least 100 N-m of output torque in the CW direction, and so on.

The next step is to sort through all these design questions/details and prioritize them in terms of precedence relationships and importance. Precedence means that some design parameters must be specified or established, before other design parameters can be dealt with. For example, the diameter of the wheel should be established prior to establishing how many teeth to use, or what the tooth angle should be. Also, the overall size of the mechanism should be established first (this is provided by the design objectives), before establishing the diameter

EXAMPLE 12.6-B Continued

of the wheel. In addition, the pawl support should be established before determining how long the pawl will be. The most important and highest precedence design parameters are known as *Design Drivers*. These are early-stage design choices and details that heavily influence the remainder of the design decisions. A list of suggested design parameters are presented next in relative priority/importance. The design drivers are indicated as well.

1. Wheel diameter (design driver)
2. Connection of mechanism with input link (design driver)
3. Wheel support system (design driver)
4. Connection of mechanism with output link
5. Pawl support system
6. Location of the pawl relative to the wheel
7. Length of the pawl
8. Angle of the teeth
9. Force of pawl exerted on the wheel
10. Number of teeth on the wheel
11. Thickness of the wheel
12. Etc.

Once the prioritized list of design parameters has been prepared, the next step is to establish a set of values or component choices for these parameters. This can be done in the form of a table, which lists the prioritized design parameters, along with the proposed values and specific details. Table 12.4-B illustrates this as applied to this example. This table consists of three columns, where the third column is titled Rationale for Choice, to help illustrate how the proposed values were developed or chosen. In some cases, the parameter values must be developed from scratch, such as the number of teeth, or the location of the pawl. In other cases, pre-existing mechanical components can be selected for parameter values, such as ball bearings and shaft diameters for the wheel support system. Developing the parameter values begins at the top of the table, since those parameters have highest priority. Lower priority parameter values must be developed so that they are compatible with the higher level values.

TABLE 12.4-B PRELIMINARY DESIGN PARAMETERS FOR CONCEPT A: ONE-DIRECTION ROTATIONAL DEVICE

Preliminary Design Parameters	Proposed Value (Units)	Rationale for Choice
Wheel diameter	10 mm	Fits within maximum permitted volume and still allows space for pawl and spring.
Connection to input link	Cotter pin connection	Selected after review of various attachment methods in mechanical design books [25].
Wheel support system	Ball bearing and shaft system	Specified in design objectives. Selected after review of mechanical design books.
Connection output link	Cotter pin connection	Selected since cotter pin is used for input connection, to keep system common.
Pawl support system	Bushing and shaft	Selected after review of various methods for simplicity and strength.

(Continued)

EXAMPLE 12.6-B Continued

TABLE 12.4-B Continued

Location of pawl	4 mm to left and 2 mm above wheel	Fits within maximum permitted volume and accommodates wheel and support size.
Length of pawl	6 mm	Fits within maximum permitted volume and accommodates wheel and support size.
Angle of teeth	20 degrees	Low angle allows easy slip of pawl over teeth during CW motion.
Pawl force on wheel	5 N	Keeps pawl firmly in place against teeth for CCW, but still allows for easy CW slip.
Number of teeth	16	Maintains 20 degree angle and allows enough height for good pawl engagement.
Thickness of wheel	3 mm	Standard material thickness available from supplier, allows for quick machining.
Etc.		

After the preparation of Table 12.4-B, a new "preliminary design sketch" should be prepared that consolidates all the new design parameters and choices. Ideally, this new preliminary design should now embody a sufficient level of detail, that some amount of engineering analysis can be done to assess the performance of the design concept.

When the preliminary design activity is completed, a new set of detailed design parameters should emerge. These parameters should be of sufficient number and detail that some level of engineering analysis can be done to assess the design concept properties or performance. Specifically, to assess if the proposed preliminary design will satisfy the design objectives for the project. In the case of Example 12.6-B, since the wheel diameter, thickness, teeth dimensions, and material have been specified, as well as the parameters of the pawl, there is now sufficient information to perform engineering stress analysis. This stress analysis can be directly used to determine if the teeth or pawl are strong enough to meet the design objective of 100 N-m of output torque in the CW direction, without damage. Or, these design parameters can be used to determine if the mechanism meets the design objective of maximum size to fit within the specified volume. The main purpose of preliminary design is to (a) develop and specify the design parameters needed to analyze the design, and (b) determine to what degree it can satisfy the design objectives. If it is determined that some design objectives are not met, the corresponding parameters should be changed in an attempt to shift the design to meet those objectives. As is the nature of design, iteration will occur when parameter choices are conflicting, but priorities should be used to resolve these conflicts to achieve the best balance.

12.3.3.3 Decision Making within Preliminary Design

After performing the preliminary design and analysis for each of the concepts, a great deal of information will be available about each of the possible design solutions. Also, the amount of work needed to further develop the design details will begin to increase exponentially, as the level of detail increases. This is a critical phase of the design process. The designer or design team has finite

TABLE 12.5 BLANK DESIGN SELECTION TABLE OUTLINE

Design Objectives	Value (Units)	Concept A	Concept B	Concept C
Torque	(N-m)			
Angular alignment	(degrees)			
Mass	(kg)			
Material	(type)			
Shaft coupling	(yes/no)			
Operating speed	(rpm)			
Cycle life	(years)			
TOTAL SCORE:	—			

time and resources and cannot continue toward the detailed design of each and every possible design concept. However, the designer must ensure that no concepts are abandoned prematurely without proper consideration. At this stage, there is likely sufficient information on each conceptual design alternative to make estimates on the concepts' ability to satisfy the design objectives. This information must be consolidated and reviewed.

Since not all design alternatives can be pursed for further development, a *Design Selection* must take place, where the designer will choose a single design concept for further development, while the remaining concepts will be shelved. There are many possible ways to choose one design concept from among the alternatives. For example, it could be done by listing the pros and cons of each concept. It could be done through a committee of experts who review each concept and provide their opinion. It could also be done by using the designer's best guess and judgement. Each of those selection methods has its merits and disadvantages. However, the purpose of any selection method is to choose the best conceptual design alternative based on the factual information available. Therefore, the authors of this text present the use of a logical *Design Selection Methodology* to select a single design concept, from among all the developed design alternatives.

The *Design Selection Methodology* presented here makes use of a *Selection Table* that scores each design concept, against the design objectives and constraints for the project. Table 12.5 provides an example of a blank design selection table for the torque transmission device.

Upon examination of Table 12.5 we can see the first column (left) contains all the design objectives related to the design. Their units of measure are listed in the next column. The next three columns (3rd, 4th, and 5th) are each devoted to a specific design concept. Each design concept must be sufficiently developed, so that a value may be estimated corresponding to each objective. Along the bottom of this table (last row), a tally of scores is to be computed.

Although Table 12.5 appears reasonable at first glance, it cannot be used directly as shown. The reason for this is illustrated in Table 12.6. Assume that Concept A, Concept B, and Concept C have been developed to the point where values for the design objectives can be estimated and have been entered into Table 12.6. To determine which concept best (i.e., more closely) satisfies all the design objectives, it would be desirable to add all the values, to obtain the best score. However, there are different units for the values for each design objective, therefore they cannot be added together. Further, it has not been declared whether a high value or low value for the design objective is a good or bad characteristic. For example, it is not possible to get a total score by adding torque values with alignment values, with mass values, with material values, and so on.

In order to make proper use of a design selection table, the design objective values for each concept must first be *mapped* with the use of *evaluation scales*. Since each design objective has different units of measurement with a different measurement range, direct addition is not possible. Therefore, mapping design objective values onto a common numeric evaluation scale becomes essential to process the information. Table 12.7 provides an example of an evaluation scale for the design objective of Torque. The left column of the scale provides the design objective with its

TABLE 12.6 EXAMPLE OF DESIGN SELECTION TABLE RENDERED UNUSABLE DUE TO MIXED UNITS

Design Objectives	Value (Units)	Concept A	Concept B	Concept C
Torque	(N-m)	2.5	3.5	8.0
Angular alignment	(degrees)	45	35	25
Mass	(kg)	0.5	1.6	2.0
Material	(type)	ABS	ABS	Acrylic
Shaft coupling	(yes/no)	Yes	Yes	Yes
Operating speed	(rpm)	60	50	55
Cycle life	(years)	2.5	3.5	2.5
TOTAL SCORE:	—	???	???	???

TABLE 12.7(A) EVALUATION SCALE FOR TORQUE DESIGN OBJECTIVE

Design Objective Torque (N-m) Value Lists	Numeric Score
>4.1	10
3.1-4.0	9
2.7-3.0	7
2.5-2.6	5
2.0-2.4	3
<2.0	0

units of measure, along with the proposed range of values. These ranges of values represent performance characteristics against which the design concepts are judged. The right column contains the associated scores given, if the design objective falls within a particular range. In this way, an evaluation scale is used to "map" quantitative "scores" onto design objectives. This is needed, since a numeric analysis will be performed later, where these numeric evaluation scales are used. In the example of Table 12.7, the scores have a numeric range from 10 to 0 to describe the design objective as being: 10 = excellent, 9 = great, 7 = good, 5 = adequate, 0 = poor, or some other variation. Any numeric score range would be acceptable, however a range from 10 to 0 is recommended, as will be described later.

An evaluation scale is needed for each of the design objectives. Tables 12.7(a)–12.7(c) provide examples for all the evaluation scales needed to map the design objectives for the torque transmission device, into a common selection table. By creating all of these scales, the three design concepts can now be compared in a meaningful and numeric way.

A few more comments are made about the evaluation scales. First, the scales can have ranges that are linear or nonlinear. The only important aspect is that the ranges can be mapped onto scores that are deemed excellent (Score = 10) through to poor (Score = 0). Also, different design objectives can have different rows of ranges, where Table 12.7(b) has six rows while Table 12.7(c) has four rows. Yet both tables have numeric ranges from 10 to 0. Also, high values for design objectives may be a good or bad trait. For example, note the scale for mass in Table 12.7(b). Here, a high value for mass is a bad trait, so higher mass receives a lower score, whereas a high value for operational speed is a good trait, so higher operational speeds receive a higher score.

A proper selection table can now be created with the corresponding evaluation scales for each of the design objectives. Table 12.8 shows a proper selection table for the torque transmission

TABLE 12.7(B) EVALUATION SCALE FOR THE DESIGN OBJECTIVES OF ANGULAR ALIGNMENT (DEGREES), MASS (KG), AND OPERATING SPEED (RPM)[A]

Design Objective Angular Alignment (degrees)	Design Objective Mass (kg)	Design Objective Operating Speed (rpm)	Numeric Score
>40	< 0.8	> 90	10
40-30	0.8–1.0	60–90	9
25-29	1.1–1.2	51–59	7
20–24	1.3–1.5	45–50	5
15–19	1.6–2.0	40–44	3
<15	>2.0	<40	0

[a] All performance characteristics are mapped to a common numeric score ranging from 10 to 0, where 10 is considered excellent and 0 is considered poor.

TABLE 12.7(C) EVALUATION SCALE FOR THE DESIGN OBJECTIVES OF MATERIAL (TYPE), SHAFT COUPLING (YES/NO), AND CYCLE LIFE (YEARS)[A]

Design Objective Material (type)	Design Objective Coupling (yes/no)	Design Objective Cycle Life (years)	Numeric Score
ABS	Yes	> 4	10
Polycarbonate	—	3–4	9
Acrylic	—	2.5–2.9	5
Other	No	< 2.5	0

[a] All performance characteristics are mapped to a common numeric score ranging from 10 to 0, where 10 is considered excellent and 0 is considered poor.

TABLE 12.8 DESIGN SELECTION TABLE FOR THE TORQUE TRANSMISSION DEVICE MAKING USE OF EVALUATION SCALES FROM TABLES 12.7(A)-(C)

Design Objectives	Evaluation Scale	Concept A	Concept B	Concept C
Torque	Table 12.7(a)	5	9	10
Angular alignment	Table 12.7(b)	10	9	7
Mass	Table 12.7(b)	10	3	3
Material	Table 12.7(c)	10	10	5
Shaft coupling	Table 12.7(c)	10	10	10
Operating speed	Table 12.7(b)	9	5	7
Cycle life	Table 12.7(c)	5	9	5
TOTAL SCORE:	—	59	55	47

device. Table 12.8 was constructed by mapping the information contained in Table 12.6, with the evaluation scales of Tables 12.7(a)–(c).

Based on Table 12.8, a numeric score can now be tallied for each design concept, as shown in the bottom row denoted as *Total Score*. On the basis of this total score, a decision can be made about which design concept generally satisfies most of the design objectives. In the example of Table 12.8,

TABLE 12.9 WEIGHTING TABLE FOR THE DESIGN OBJECTIVES OF THE TORQUE TRANSMISSION DEVICE

Design Objectives	Weighting Factor	Normalized Weight
Torque	3×	0.3
Angular alignment	2×	0.2
Mass	1×	0.1
Material	0.5×	0.05
Shaft coupling	0.5×	0.05
Operating speed	1×	0.1
Cycle life	2×	0.2

it can be seen that Concept A has the highest score with a value of 59. Since higher values in each evaluation scale correspond to a value of excellent, the highest total score generally corresponds to the concept that best satisfies the design objectives compared to the other concepts.

An advanced version of the design selection table is known as the *Weighted Design Selection Table*. In many cases of engineering design, the design objectives have different levels of importance. The relative levels of importance can be quantified by assigning weighting factors to the design objectives. For example, consider the torque transmission device. The designer or the design team may feel that the design objective of torque is more important than the design objective of mass. Additionally, cycle life may be deemed more important than mass, but less important than torque, and so on. For example, Table 12.9 lists a hypothetical weighting table.

The choice of numerical weights of design objectives is important. By assigning weights a ranked list is created, such that objectives with higher weight have higher importance over others, with that importance being numerically proportional to their weighted scores. Weighting factor values of 1.0× imply equal importance with other objectives. Assigning relative weights require experience and should be done sparingly. To help with the process of assigning weighting factors, a number of techniques have been developed such as the *Pairwise Comparison* method and the *Hierarchical Weighting Factor* method [26].

A new *Weighted Design Selection Table* is shown in Table 12.10. This table contains all the previous information from Table 12.8, but also incorporates the weighting factors for the design objectives from Table 12.9. The original concept scores are shown, but are no longer tallied. Instead, a new column has been placed to the right of the original concept scores, called "weighted score." This column contains the original score multiplied by the normalized weight factor. The weighted score is tallied for each concept. It is interesting to observe how the weight factors can influence the decision-making process. Note that Table 12.8 indicates that Concept A has the highest total score, while Table 12.10 indicates that Concept B has the highest total score with a value of 8.1. This indicates that Concept B best meets the design objectives, when taking into account the relative importance of those design objectives.

A design selection table is a useful tool to help a designer choose from among alternative designs. Its great value is to create some objectivity during concept selection, by numerically evaluating possible design concepts. However, it should be emphasized that design selection tables are tools that require careful use. Care must be taken when estimating the design objective values for each concept, creating the evaluation scales, and assigning of relative weighting factors. Design selection tables are best used along with additional input from other design team members, client opinion, and thorough review of all design concepts. Ultimately, a single design concept is chosen and is taken to the next phase of the design process, known as detailed design.

TABLE 12.10 WEIGHTED DESIGN SELECTION TABLE FOR THE TORQUE TRANSMISSION DEVICE

Design Objectives	Evaluation Scale	Normalized Weight	Concept A		Concept B		Concept C	
			Original Score	Weighted Score	Original Score	Weighted Score	Original Score	Weighted Score
Torque	12.7(a)	0.3	5	1.5	9	2.7	10	3.0
Angular alignment	12.7(b)	0.2	10	2.0	9	1.8	7	1.4
Mass	12.7(b)	0.1	10	1.0	3	0.3	3	0.3
Material	12.7(c)	0.05	10	0.5	10	0.5	5	0.25
Shaft coupling	12.7(c)	0.05	10	0.5	10	0.5	10	0.5
Operating speed	12.7(b)	0.1	9	0.9	5	0.5	7	0.7
Cycle life	12.7(c)	0.2	5	1.0	9	1.8	5	1.0
TOTAL SCORE:	—	—	—	7.4	—	8.1	—	7.15

12.3.4 Detailed Design

The *Detailed Design* phase involves further development of the best concept selected during the decision making phase. Detailed design can involve many aspects, which generally involves development of the selected preliminary design to the point where it can be built. Detailed design often involves activities such as: *Mathematical Modeling* or *Computer Simulation* to predict performance, creation of detailed *Engineering Drawings*, preparation of a *Work-Plan and Timeline*, and the preparation of a *Bill of Materials* to list the needed parts and estimate costs.

Detailed engineering analysis often begins with mathematical modeling and analysis to predict the performance of the design. Many undergraduate engineering courses provide the knowledge to make such models, such as courses in solid mechanics, dynamics, mechanisms and machines, and others. In this textbook, Chapters 1 through 11 present the detailed mathematical modeling and analysis of various mechanisms. In this way, mechanism performance can be predicted based on the design parameters, with simple hand calculations. In some cases, the mathematical models are too complex to solve by hand and require a computer for solution. This can be done by using mathematical software such as PTC MathCAD, Wolfram Mathematica, or Mathworks MATLAB, among others.

In some cases, the physical configuration of the mechanism may be very complex, with several linkages and numerous joints. Although mathematical modeling is possible, it may be more convenient to use special-purpose computer simulation software that is made to model mechanisms and machines. For example, motion simulation software can be used to model and analyze the kinematics and dynamics of complex mechanisms. Such software can also reveal the reaction forces between the various links in contact within the mechanism. Figure 12.12 shows an example of the motion simulation of the one-directional rotation device. This simulation was done with Working Model 2D software. In the simulation, in addition to the motion study, the reaction forces at the revolute joints, as well as the contact forces exerted between linkages, are shown. These reaction and contact forces are computed for each increment in time, as the simulation moves through its various positions.

After designing the system of linkages and their connections within a mechanism, it becomes necessary for the detailed design of each individual link in that mechanism. Each link must be strong enough to withstand the forces exerted on it, therefore, it is usually necessary to determine the internal stress. The motion simulation software described previously allows for the calculation of the external forces acting on a link, but does not compute the resulting internal stress. Determination of internal stress is important. If the internal stress becomes too high, the linkage

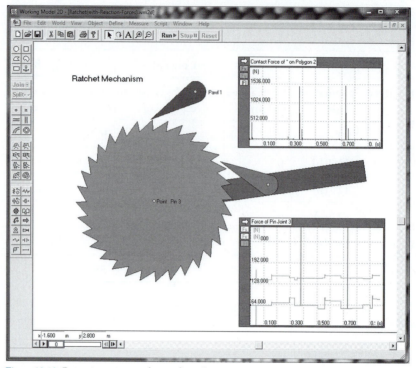

Figure 12.12 Dynamic motion simulation of one-direction rotational device.

may deform excessively, or even break. The subject of internal stress and deformation of objects is called Solid Mechanics [27, 28] and is outside the central scope of this book. Mechanism links are often complex in shape, and hence analysis of their internal stress is often done with computer simulation. A particularly useful simulation method to compute the internal stress of a link and its deformation is known as the finite element method [29]. Figure 12.13 provides an example of the application of the finite element method to the central linkage in a universal joint. Note that the central link has been discretized into hundreds of tetrahedral elements, and the shading of each element represents the magnitude of the stress at that location.

The physical shape of a linkage in a mechanism should be designed to minimise internal stress and also to satisfy the kinematic requirements. Presently, computer-assisted design (CAD) tools are available to create a computer model of the linkage. CAD tools offer considerable ease to quickly create drawings and, more importantly, allow for changes to be made easily, which helps during the iterative process of design. Sometimes, CAD tools can also be restrictive in their capabilities, making it difficult to draw certain shapes or forms. However, the advantages of CAD-based drawing have made it the standard method to produce linkage drawings. There are both 2D- and 3D-based CAD software tools available, with 3D tools becoming the norm. Examples of modern engineering-based CAD software include: AutoCAD, SolidWorks, CATIA, Pro/ENGINEER, Siemens NX, and other tools. Many of the 3D models of mechanisms and machines shown in this

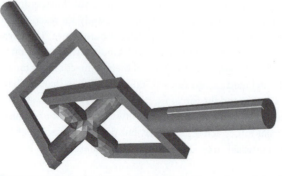

Figure 12.13 Finite element method simulation of torque transmission device.

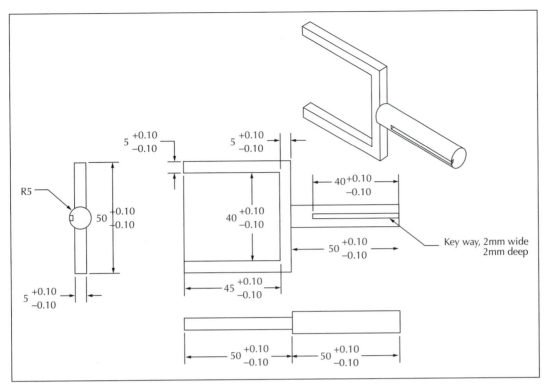

Figure 12.14 Detailed engineering drawings of the u-shaped link in the torque transmission device.

book were created with the Solidworks or Pro/ENGINEER CAD software. Examples include: the Clutch mechanism in Figure 13.11, the Synchronizer in Figure 13.13, and the Zero-max drive in Figure 13.48.

Once a 3D computer model of a mechanism is made using CAD software, a number of things can be done. The finite element method can be applied to predict the deformation and internal stress that a linkage of the mechanism will undergo. If the deformation or stress is found to be too high, the 3D linkage model can be re-designed and altered with the CAD software. The re-designed linkage can be analyzed again with the finite element method to determine its new performance. In this iterative way, a designer can converge toward a linkage design solution that satisfies the design objectives such as: geometry; strength, weight, stiffness (ability to withstand deformation), and others. This iterative-based design approach is often sufficient to create a good combination of performance characteristics for a linkage and the mechanism overall. Alternatively, there is a branch of engineering design that focuses on finding the mathematically optimum balance between the aforementioned design objectives, which is called *Optimization* [30]. Numerous optimization methods have been developed for design optimization, and mechanical designers are encouraged to explore optimization techniques for performance-critical applications such as aircraft structures, airfoils, turbine blades, and others.

After a 3D linkage model has been designed, simulated, and perhaps optimized, a set of detailed engineering drawings can be produced. Figure 12.14 provides an example of a typical detailed drawing for the central link in the torque transmission device. Normally, such a drawing is produced for each and every part within the mechanism, which results in a set of several or more drawings. Such drawings are typically provided to a machine shop for fabrication of the parts.

Upon preparing all of the engineering drawings, a production plan must be created to fabricate the design. *Project planning* [31] is important, since it serves to make sure all the resources needed to implement the design are available. It also serves to promote effective communication between the designer (or design team) and the client. Additionally, it serves to benchmark and measure progress on the design. At the most basic level, a project plan should include:

1. list of tasks (and components) needed to complete the project
2. An estimate of the time duration of each task
3. Sequential ordering and priority of those tasks
4. Cost and resource (equipment or people) estimates for each task

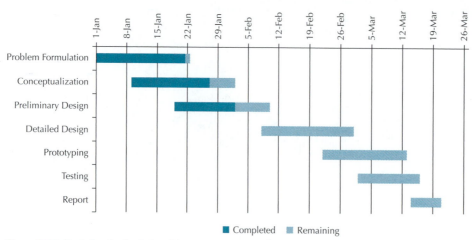

Figure 12.15 Gantt chart for the design of the torque transmission device.

A convenient and graphical format to consolidate project plan information is with the use of *Gantt charts* [31]. Figure 12.15 illustrates a simple Gantt chart for the design and fabrication of the torque transmission device.

12.3.5 Embodiment (Prototyping)

The *Embodiment* phase of the engineering design process involves the physical construction of a prototype to demonstrate the operation of a design. Prototyping is a highly important activity in the design process and serves a number of purposes. Translating a conceptual design idea into a physical prototype helps a designer learn and experience the physical limitations of construction, fabrication, and use of materials for a design. Early-stage prototypes often serve as a stepping stone for generating better designs. Also, early-stage prototypes can help communicate a design concept within a design team, or between a designer and the client. When presenting a physical prototype to a team or client, it is quite common to hear responses such as "OK, now I understand how it works" or even "Oh, I didn't realize it worked in that way." Such responses are quite common, even if the team or client had previously seen the design concept in the form of a sketch or engineering drawings. This is especially true for the design of mechanisms and machines, where the interplay of moving parts can often be difficult to comprehend or communicate with two-dimensional drawings. An additional value of prototyping is to demonstrate the functional performance or functional outcomes of a design concept. This is known as *testing and evaluation*, where a developed prototype is tested to see how well it meets the design objectives. This allows the designer to evaluate if the design has met, exceeded, or fell short of the intended design objectives, and it may warrant a re-design.

There are a number of classifications of prototypes, each with different purposes. An *alpha prototype* is a rough, first attempt at creating a physical implementation of a design concept. An alpha prototype may sometimes take the form of cardboard cutouts or plastic pieces glued together. They often serve to show the size or geometry of a design, or may even show limited mechanical function. *Functional prototypes* are prototypes that successfully demonstrate the intended function of a design concept, and can be tested to evaluate the design performance. In some cases a functional prototype may not have a finished or aesthetic appearance, since it only serves to demonstrate function. *Visual/scale prototypes* are intended to show form, finish and aesthetic appearance, but often do not function. For mechanism and machine design, it is often helpful to prepare an alpha prototype for complicated mechanisms. This may be done during the preliminary design phase, as depicted in Figure 12.1, to help with concept evaluation. A functional prototype should be developed during the embodiment phase, after the detailed design is complete. When constructing a functional prototype, an effort should be made to use the intended materials, parts, and manufacturing process to the best extent possible, as specified by the detailed design. This helps to validate many aspects of the design and its manufacturability.

There are many resources available to help an engineering designer to construct prototypes. This includes specialty parts suppliers [32–34], materials suppliers [32, 35], machine shops,

(a)   (b)

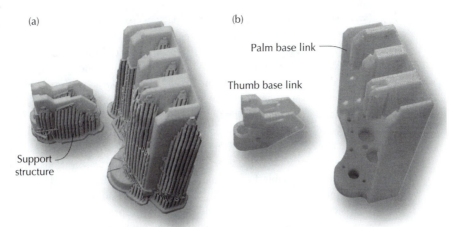

Palm base link

Thumb base link

Support
structure

Figure 12.16 Example of 3D printing of prototype parts for the palm and thumb pieces of new prosthetic hand: (a) with support structures, (b) released from supports.

and commercial businesses that specialize in various prototyping services. For mechanism and machine prototyping, a designer should keep up to date in the latest small-run machining and fabrication technologies. This includes traditional technologies such as CNC machining [36], laser machining [37], EDM machining [38], and other limited run technologies. For the purposes of prototype construction, avoid technologies that require expensive setup costs, such as injection molding or stamping, where the molds or dies can have high initial cost. A new and exciting technology for creating 3D prototypes of mechanisms and machines is called "3D printing" or "rapid prototyping" [39]. An example of a part manufactured by rapid prototyping at the University of Victoria using polycarbonate plastic is shown in Figure 12.16. This is a rapid prototype of the palm and thumb pieces for an experimental prosthetic hand. The original version of this hand is shown in Figure 1.30, which was manufactured by traditional machining, whereas the new version shown in Figure 12.16 is made up of hundreds of overlapped and fused layers of polycarbonate plastic. Figure 12.16(a) shows the rapid prototype with plastic support structures that are needed during the layering process, and Figure 12.16(b) shows the final structures after they are released from the supports.

   An example of an alpha prototype for the torque transmission device is shown in Figure 12.17. The prototype shown is for Concept B of Table 12.6. One of the design objectives was the material; therefore, Concept B was created using ABS plastic. The prototype is crudely constructed; however, it functionally represents the concept, and it works well enough to undergo the testing and evaluation process.

12.3.6 Testing and Evaluation

Testing and evaluation is an essential part of the engineering design process. Upon completion of the detailed design phase and the subsequent construction of a prototype, the design must be evaluated to determine if the design objectives have been satisfied.

   Testing and evaluation can take different forms, with specific types of tests used to determine various types of performance. For example, some tests may simply provide qualitative statements

Figure 12.17 Functional prototype for the torque transmission device.

(value statements) and observations. This may consist of yes/no statements, or statements that declare great, good, acceptable or poor performance. Other types of tests are quantitative (numeric, measurable) in nature and seek to establish specific units of measure such as: meters/second of speed, kilograms of mass, newtons of force, decibels of sound, and so on.

The design process described in this chapter is centered on the establishment of design objectives and the evaluation of those design objectives. Hence, design engineers should try to conduct their tests in such a way as to test for quantitative results. Quantification of test results (with proper testing) allows designers and clients to agree that the design objectives have been met.

The establishment of a test procedure is not trivial. In fact, there exists a formal method of creating a scientifically valid experiment, called *Design of Experiments.* Many books are dedicated specifically to the subject of design of engineering experiments, and formal approaches have been developed, such as those in [40–41]. The subject is outside the central scope of this book, however, the basic details of setting up a simple experiment are presented here. Processing the results of these experiments in a statistically meaningful way is left to the aforementioned books.

Beyond the design of the experiment is the design of the experimental apparatus itself—that is, the device which will provide the test forces or torques and also provide the support applied to the prototype, as well as the measurement tools that will collect data from the prototype. In fact, the design of the test apparatus may be just as involved and complex as that of the design of the prototype to be tested. For example, the test apparatus used to test and evaluate the torque transmission device of Figure 12.17 is shown in Figure 12.18.

A good *test plan* or *test procedure* must be developed prior to testing, or prior to designing a test apparatus. Such a test plan should include the following aspects:

1. Scope of the test, which includes:
   - The goal/purpose and justification for the test
   - The relevant parameters for the test
   - The expectations of the test (Hypothesis)
2. Administrative details:
   - Date and location of testing
   - Who is conducting the test
3. Design of the test (known as Design of Experiment)
   - Type of testing method and its relevance
   - List of test apparatus, measurement equipment, and model numbers
   - Identification of "dependent" and "independent" variables
   - Sampling procedure: Number of samples? Number of combinations?
4. Detailed, step-by-step procedure to conduct the test

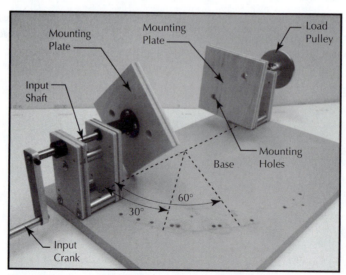

Figure 12.18 Test apparatus for torque transmission device.

(a)

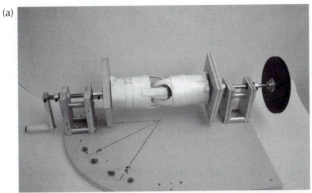

(b)

Figure 12.19 ABS plastic prototype of torque transmission device.

5. Conduct/Perform the test
   - Safety precautions
   - Data collection method (written, recorded, digital, etc.)
   - Observation of external factors (temperature, wind, noise, vibration, etc.)
6. Analysis of test data
7. Discussion of results and conclusions
   - Interpretation of data analysis, and other observations
   - Written test report

An example of a test being conducted is shown in Figure 12.19, where the prototype of the torque transmission mechanism is mounted in the test apparatus. With this test apparatus, different amounts of torque can be applied while the mechanism is rotated. Also, different relative angles between the input and output shaft can be tested. The main purpose of testing is to demonstrate that the design meets the performance expectations (design objectives) that were set out at the start of the design endeavor. A proper test plan, test procedure, and test report will provide clear documentation, for the design team and the client, that the design has been proven and demonstrated to work as intended.

12.3.7 Design Reporting and Documentation

Preparing a *Design Report* with supporting documentation is an essential part of the design process. This report should provide the answers to basic questions like why the design was undertaken and what purpose the design serves. It should allow the reader to follow the steps the designer used to go through the design process outlined in Figure 12.1. Additionally, it should provide enough supporting documentation and detail, so that the design can be reproduced by other engineers.

Generally, a single design report is provided for simple designs, once the design activity is completed. For more complex designs, a series of design reports are generally written. This may include a design proposal, a conceptual design report, and the final design report.

A *Design Proposal* identifies the design problem and outlines the design project to be undertaken. It should include an introduction describing the purpose of the work, background information for novice readers, and a literature review describing similar projects already done by others. Also, it should clearly indicate the Design Goal Statement, the proposed Design Objectives and Constraints for the project, rationale and justification for them, and the project work plan.

A *Conceptual Design Report* is a mid-phase design report intended to communicate the conceptual and preliminary design concepts developed for the design activity, for review by the client or other stakeholders. It should include refreshed and up-to-date contents from the design proposal report (i.e., purpose, background, literature, design goals, and design objectives). Also, it should present the *Conceptual Design Alternatives* that have been generated, their sketches and descriptions, preliminary engineering analysis of those generated design alternatives, and the *Design Selection Methodology* employed to select the best design approach. Additionally, it should also provide an updated project work plan that outlines the remaining steps to complete the project.

A *Final Design Report* is a professional-level engineering report providing a full description of the overall design process undertaken, from start to finish. The final design report should be a complete document that stands alone and that can be provided to clients or third parties with interest in the design. Therefore, it will include all the contents of the design proposal report, (with updates), and the conceptual design report (with updates). In addition, the final design report will include a detailed description of the final design, including: clear pictures and drawings of the design, a description of the features and functions, detailed engineering analysis (calculations of performance to demonstrate that design objectives have been met, for example: speed, weight, workspace, force, torque, stress, etc.), cost estimation for production, detailed engineering drawings, testing and evaluation data, discussion of test results, conclusion, and future work recommendations.

12.4 SUMMARY

A general engineering design process suitable for mechanism and machine design has been presented. This design process methodology suggests methods for information gathering, the formulation of design goals, and developing a set of clear design objectives. It presented creative strategies for generating and developing initial design concepts and then developing a preliminary design of those concepts. It presented a design selection methodology using selection tables to numerically evaluate various design concepts with each other. Methods to develop detailed designs were suggested, along with ideas on how to fabricate prototypes. Testing and evaluation procedures were suggested, along with suggestions on the writing of final design reports. In addition to the presented design methodologies, the reader is encouraged to continually seek and review other examples of mechanisms and machines, build their knowledge base of the state of the art, practice design on a frequent basis, and build up their experience level in design. In the next chapter, some case studies on the design of various mechanisms and machines are presented.

PROBLEMS

Problem Formulation

P12.1 As a design engineer, you meet with a potential client. After the meeting, the client provides you with a statement (see below) describing their thoughts and ideas on the subject. Read this statement and prepare the following:

(a) clear Design Goal Statement

(b) at least six Design Objectives along with their target values, which are of most relevance to this problem. Conduct some information gathering activities to find suitable target values for your design objectives.

Client Statement: Our company is in the business of making large highway advertisement signs. We seek a device that is able to trace out a specific shape in space and then cut shapes out of sheets of 1-mm-thick paper. The shapes we have in mind are simple circles, squares, triangles, stars, or similar. The cutout shapes need to be sized from 300 mm up to 1000 mm. The device should be manually controlled by a person, without the use of computers, and may make use of an assistive power source like electricity or compressed air. We have a set of master shapes that are 30 mm to 100 mm in size, and we would like the operator to trace out those master shapes with the device, which will amplify that trace 10× in order to cut it out from the paper stock. The machine should produce about 50 shapes per day and last for a reasonable period of time before it needs servicing.

P12.2 You are a design engineer on contract with a manufacturing company, to help design their products. The company is working on a new piece of equipment to harvest crops, but has encountered a technical problem. They provide you with a statement (see below) describing their thoughts and ideas on the problem. Read this statement and prepare the following:

(a) clear Design Goal Statement

(b) at least six Design Objectives along with their target values, which are of most relevance to this problem. Conduct some information gathering activities to find suitable target values for your design objectives.

Client Statement: We are working on a new piece of agricultural equipment, to be used to harvest crops. As part of this machine, we need a device that will transmit rotation and torque between two parallel shafts, where the shafts are separated by a large distance. It is very important that all of the rotational motion from one shaft be transmitted into the other shaft, with no relative slip between the two shafts. A speed ratio of one-to-one between the two shafts is desired. Previously, we had tried to use V-belts and pulleys to transmit the motion, but due to the high torque and moist environment, the belts kept slipping, causing relative rotation between the shafts. It is important to minimize the energy loss in transferring the torque and motion, so methods to achieve that should be implemented.

P12.3 You are an engineer employed at a heavy equipment manufacturing facility, and you meet with the assembly line workers. The assembly line workers describe their needs for a mechanism to assist their work, and they provide you with a written statement (see below). Read this statement and prepare the following:

(a) clear Design Goal Statement

(b) at least six Design Objectives along with their target values, that are of most relevance to this problem. Conduct some information gathering activities to find suitable target values for your design objectives.

Client Statement: The assembly line workers are currently assembling electric-powered forklifts. As part of the assembly process, we are required to pick up a tilt-cylinder part which weighs 40 kg, maneuver it into position, and bolt it into place. The final part position is such that it causes the workers to lean forward excessively while they manoeuver it into place. In addition, the part is difficult to hold, since it has no external shapes which are easy to grab with the hands. On a number of occasions, the part slipped and was dropped, but fortunately, no one has been hurt. A machine is needed that can help us complete this assembly task and overcome the existing problems.

Conceptualization

P12.4 One aspect of creativity is using conventional objects in unconventional ways. As an exercise to stimulate your creativity, list 10 different uses for a pendulum.

P12.5 As an exercise to stimulate your creativity, list 10 different uses for a four-bar crank-rocker mechanism.

P12.6 As an exercise to stimulate your creativity, list 10 different uses for a rack and pinion gear configuration

P12.7 As an exercise to stimulate your creativity, list 10 different uses for a slider-crank mechanism.

For problems **P12.8** to **P12.10**, create a 2D morphological chart to generate some novel design concepts, given the information provided below. For the 2D chart, begin by selecting a function/attribute for each dimension. Next, come up with 3–4 alternatives for each dimension. Then, construct a morphological chart to identify possible alternative design concepts. For those concepts that lead to interesting ideas, place the words "idea #" in the box, and describe the ideas below the chart in 2-3 sentences.

P12.8 You work for a manufacturer of renewable power generation equipment and have been asked to come up with a totally new way to harvest energy from the wind. Construct a 2D morphological chart to generate ideas, given the method described above.

P12.9 You work for a manufacturer of water pumps and have been asked to come up with a totally new way to pump water out of a reservoir tank. Construct a 2D morphological chart to generate ideas, given the method described above.

P12.10 You are working in an automobile design team, and the team leader has asked you to come up with a totally new type of braking system for the vehicle. Construct a 2D morphological chart to generate ideas, given the method described above.

P12.11 Create three different design concept sketches that represent possible solutions to the problem described in **P12.2**, given the design goal and the design objectives that you prepared. For each sketch, provide 2–3 sentences describing how the concept works.

P12.12 Create three different design concept sketches that represent possible solutions to the problem described in **P12.3**, given the design goal, and the design objectives that you prepared. For each sketch, provide 2-3 sentences describing how the concept works.

Preliminary Design: Functional Decomposition

For problems **P12.13** to **P12.15**, create a set of functional decomposition diagrams, given the information provided below. You may also conduct information gathering to learn more about the objectives and functions of these devices. Begin with a high-level diagram (single block) and then create a mid-level detail diagram (several blocks) and a low-level detail diagram (of any one of the mid-level blocks). Remember to clearly identify the inputs and outputs of this system, and those between the blocks.

P12.13 Create a functional decomposition diagram for a bicycle, given the method described above.

P12.14 Create a functional decomposition diagram for a portable cordless drill, given the method described above.

P12.15 Create a functional decomposition diagram for a hydraulic lift jack, given the method described above.

Preliminary Design: Decision Making

Problems **P12.16** to **P12.17** are exercises in the preparation and use of design selection tables, as part of the concept selection process. The first step is to prepare a blank selection table similar in format to that of Table 12.5. The design objectives in the first column of your selection table should be specific to your design problem. The next step is to populate your selection table with values that reflect the properties of each design concept A, B, and C, in a way similar to Table 12.6.

P12.16 Create a design selection table using the design objectives that were developed to answer question **P12.2**. The table should evaluate the three different design concept sketches that you prepared as an answer to question **P12.11**. The values used to populate your table should be based on estimates on the performance of your concept sketches. Use information gathering methods to assist with determining appropriate values for the table.

P12.17 Create a design selection table, using the design objectives that were developed to answer question **P12.3**. The table should evaluate the three different design concept sketches that you prepared as an answer to question **P12.12**. The values used to populate your table should be based on estimates on the performance of your concept sketches. Use information gathering methods to assist with determining appropriate values for the table.

Problems **P12.18** to **P12.19** are exercises in the preparation and use of evaluation scales. These evaluation scales should be similar in form to those shown in Tables 12.7(a), 12.7(b), or 12.7(c).

P12.18 Given the design selection table you created in problem **P12.16**, create a suitable evaluation scale for each of your design objectives.

P12.19 Given the design selection table you created in problem **P12.17**, create a suitable evaluation scale for each of your design objectives.

Problems **P12.20** to **P12.21** are exercises in the preparation of design selection tables that incorporate numeric scores as values for the design objectives, in a format similar to that of Table 12.8.

P12.20 Given the initial design selection table you created in problem **P12.16**, along with the evaluation scales you created in problem **P12.18**, create a final selection table that uses the scores to populate the table. At the bottom of the table, compute the total score for each design concept, A, B, and C. Does the concept with the highest score match your expectations for the best design concept? Explain your answer.

P12.21 Given the initial design selection table you created in problem **P12.17**, along with the evaluation scales you created in problem **P12.19**, create a final selection table that uses the scores to populate the table. At the bottom of the table, compute the total score for each design concept, A, B, and C. Does the concept with the highest score match your expectations for the best design concept? Explain your answer.

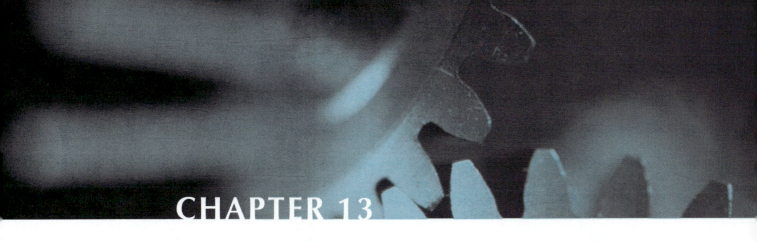

CHAPTER 13

DESIGN CASE STUDIES FOR MECHANISMS AND MACHINES

13.1 INTRODUCTION

This chapter describes an assortment of case studies in mechanism and machine design. Many of these mechanisms are in common use. By reviewing these case studies, the student of design will benefit in two ways. First, by following along with the problem formulation process whereby unstructured information is converted into a structured set of design goals and design objectives. Second, by being exposed to various practical solutions in mechanism and machine design.

Each case study begins with a problem description that provides general information normally available to a designer. Such information is typically unstructured, and is obtained from various sources and information gathering methods, as described in Section 12.3.1. From this, a design goal statement is formulated, which serves to guide the designer in the design process. Next, a set of design objectives are proposed, which specify performance values which the design must meet. Given the generality of the case studies provided here, the design objectives presented do not constitute a complete list, but serve as an example. Next, one or more possible design solutions that satisfy the design goal and design objectives are described, along with a diagram and video that graphically demonstrates their operation. Finally, a design summary and set of recommendations are provided to highlight the strengths and limitations of the design solutions. The reader is also encouraged to consider alternate solutions beyond those presented here. At the end of this chapter, several exercise problems that allow the reader to practice the various phases of the design process are presented. The exercise problems make reference to the case studies of this chapter, as well as to design problems listed in Appendix A.

13.2 SHAFT COUPLING

In mechanism and machine design, transmission of rotational motion between two shafts is frequently required. In Section 12.2, two examples are introduced involving the design of shaft coupling mechanisms. A few additional case studies of rotational motion transmission between two shafts are now presented.

13.2.1 Hooke's Coupling

Problem Description

It is necessary to transmit torque from one shaft to another, however, the two shafts may not be lined up along a straight line. Each shaft may become oriented in a way such that their axis centerlines will intersect, but may not be parallel. A device is needed that connects to each shaft, and will allow the output shaft to develop an angle of up to 30 degrees off axis, with respect to the input shaft. The input and output shafts cannot be modified, so a simple way is needed to connect this device to them. The device should transmit reasonable amounts of torque.

Design Goal Statement

Design a mechanism to transmit torque between two rotating shafts, where the input shaft and output shaft are not collinear, although the shaft centerlines must intersect. The mechanism must connect to the input and output shafts in a simple way, and allow both shafts to have a relative angle from 0 to 30 degrees with respect to each other, as they rotate. The mechanism must be simple by employing relatively few parts, and it should be strong enough to transmit a specified value of torque.

Proposed Design Objectives

Design objectives for Hooke's coupling mechanism are listed in Table 13.2.1.

Proposed Design Solution

There are many possible design solutions for the described problem and design objectives. One well known solution is a *Hooke's coupling*, also referred to as a *universal joint*, as illustrated in Figure 13.1(a). This coupling consists of an input and output shaft, denoted as links 2 and 4, respectively. They are connected by revolute joints to link 3. Link 3 is able to rotate within link 2, such that its rotation is perpendicular to the axis of link 2. Also link 3 is able to rotate within link 4 in a similar fashion. Figure 13.2 shows an application of a Hooke's coupling in the steering linkage of a classic automobile. The axis of the steering wheel shaft is misaligned with the axis of the steering gearbox shaft. A Hooke's coupling is employed to transmit motion between the shafts.

The Hooke's coupling satisfies the design objects presented in Table 13.2.1. However, it does present an interesting peculiarity, with regard to the relative speeds between the input shaft and output shaft, during its operation. This issue is described next, along with the implications for various applications. For the Hooke's coupling shown in Figure 13.1(a), the rotational speeds of the input and output are designated as $\dot{\theta}_2$ and $\dot{\theta}_4$. If the angle between the centerlines of the shafts is β, then starting from the configuration illustrated, the *speed ratio*, $\dot{\theta}_4/\dot{\theta}_2$, between links 4 and 2 is illustrated in Figure 13.1(b). The speed ratio is plotted for different values $\beta = 0$, $\beta = 20°$, and $\beta = 40°$. When $\beta = 0$, the shafts are collinear and the speed ratio is unity for all values of θ_2. However, when $\beta \neq 0$, and θ_2 is constant, $\dot{\theta}_4$ will fluctuate. Therefore, a Hooke's coupling cannot maintain

TABLE 13.2.1 DESIGN OBJECTIVES FOR HOOKE'S COUPLING MECHANISM

Design Objective	Description	Value (Units)
Torque	Mechanism must transmit a minimum torque	200 (N-m)
Angular alignment	Mechanism must operate with relative angles of 0 to 30 degrees between input and output shafts	30 (degrees)
Spacial alignment	Centerline of shaft axes will intersect	Shaft centerlines intersect
Shaft coupling	Input and output shafts will be connected by keys, collets, or other common means	Keys, collets
Operating speed	Low-speed operation	0–60 (rpm)

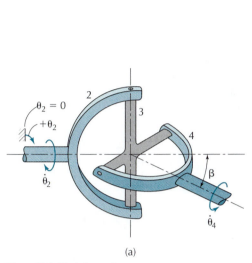

(a)

(b)

Figure 13.1 Hooke's coupling.

(a)   (b)

Figure 13.2 Hooke's coupling used in steering linkage.

<table>
<tr><td>
Video 13.3

Hooke's
Coupling
</td><td>
a constant speed ratio between the input and output, unless the shafts are collinear. For a single Hooke's coupling, as shown in Figure 13.3, the magnitude of variation of the speed ratio increases as β enlarges. An animation of this mechanism in operation is provided in [**Video 13.3**].
</td></tr>
</table>

a constant speed ratio between the input and output, unless the shafts are collinear. For a single Hooke's coupling, as shown in Figure 13.3, the magnitude of variation of the speed ratio increases as β enlarges. An animation of this mechanism in operation is provided in [**Video 13.3**].

To avoid the problem of variable speed ratio, two Hookes' couplings can be combined. In this way, the variable speed ratio of one coupling may be counteracted by that of the other coupling. Figure 13.4 shows such an arrangement. There will be a constant speed ratio between links 2 and 6 if the angles between the shafts of both Hooke's couplings are equal. In other words, the angle between link 2 and 4 must be equal to the angle between link 4 and 6. This scenario avoids output speed variation, but requires considerable spacial volume to implement, since two couplings are needed. An animation of the double Hooke's coupling in operation is provided in [**Video 13.4**].

Video 13.4

Double
Hooke's
Coupling

Summary and Suggestions

The Hooke's coupling is a design solution that meets the design objectives of Table 13.2.1. It can transmit torque at high angular misalignments, is simple in design, and can easily be implemented into a variety of mechanisms. A single Hooke's coupling will exhibit a variable speed ratio between input and output shafts. It will work well for low-speed applications, such as turning a steering wheel column. However, for high speed applications, this fluctuation of output speed can induce dynamic forces.

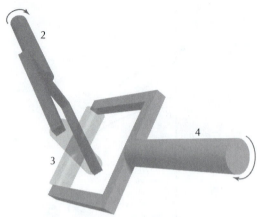

Figure 13.3 Hooke's coupling [Video 13.3].

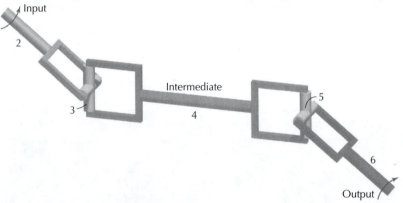

Figure 13.4 Double Hooke's coupling [Video 13.4].

As the speed increases, the magnitude of these dynamic forces can increase significantly, and lead to excessive vibration. To prevent dynamic forces leading to vibration, a double Hooke's coupling can be used, or other types of shaft couplings can be used for high-speed applications.

13.2.2 Constant-Velocity Coupling

Problem Description

A torque needs to be transmitted from one shaft to another, similar to case study 13.2.1. The orientation of the input and output shafts is such that their centerlines will intersect, but may not be parallel. The output shaft should be able to develop an angle of up to 30 degrees off axis, with respect to the input shaft. The primary need is to maintain a constant-velocity ratio between the input and output shafts, regardless of angular misalignment. The mechanism should be able to transmit high torques and also operate at high speeds.

Design Goal Statement

Design a mechanism to connect two rotating shafts that are not collinear, to allow for transfer of high torques at high speeds. The mechanism must connect to each shaft in a simple way, and allow both shafts to have a relative angle from 0 to 30 degrees with respect to each other, as they rotate. The mechanism design must be such that the velocity ratio between the input and output shafts remains constant at all times.

TABLE 13.2.2 DESIGN OBJECTIVES FOR CONTANT-VELOCITY COUPLING		
Design Objective	**Description**	**Value (Units)**
Torque	Must transmit a high torque	150 (N-m)
Angular alignment	Mechanism must operate with relative angles of 0 to 30 degrees between input and output shafts	30 (degrees)
Spacial alignment	Centerline of shaft axes will intersect	Shaft centerlines intersect
Shaft coupling	Input and output shafts will be connected by keys, collets, or other common means	Keys, collets
Velocity ratio	Ratio of output vs. input velocity must remain constant	Constant
Operating speed	High-speed operation	60–600 (rpm)

Proposed Design Objectives

Design objectives for constant-velocity coupling are listed in Table 13.2.2.

Proposed Design Solution

Video 13.5
Constant-
Velocity
Coupling

There are many possible design solutions for the described problem and design objectives. One well-known solution is a *constant-velocity coupling* that can transmit rotational motion between two shafts whose axes intersect, but are not necessarily parallel. An illustration of a constant-velocity coupling is shown in Figure 13.5. The complete mechanism is illustrated in Figure 13.5(a), while Figure 13.5(b) illustrates the major internal components. An animation of this mechanism in operation is provided in [**Video 13.5**]. As the animation progresses, a portion of the mechanism gradually disappears, revealing the inner components. A cross-section of the coupling is shown in Figure 13.6(a). The input and output shafts are connected to the inner and outer journals of the coupling. The angle between the shafts is β. There are six grooves in the *inner journal* and six corresponding grooves on the *outer journal*. Between each pair of grooves is a ball that transmits force from one journal to the other corresponding journal. All grooves are at an oblique angle with respect to the centerline axis of the shaft. Each pair of corresponding grooves have opposite oblique directions. As illustrated in Figure 13.6(b), each ball remains in the plane of symmetry between the two shafts, regardless of the value of β. The perpendicular distance *a* to the center of the ball is the same for both shafts. Since forces and motions are transmitted through the balls, the input and output shafts rotate at the same speed.

A constant-velocity joint is commonly used in front wheel drive vehicles. Smooth transmission of motion is achieved despite changes in alignment between the shafts. The animation provided in [**Video 13.5**] shows the case where the axes of rotation of the shafts remain stationary. However, if the angle between the centerlines of the shafts changes, the input and output rotational speeds would still remain equal.

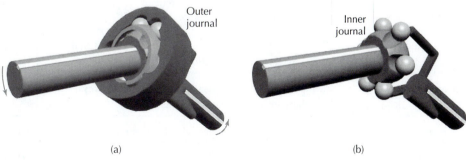

(a)   (b)

Figure 13.5 Constant-velocity coupling [Video 13.5].

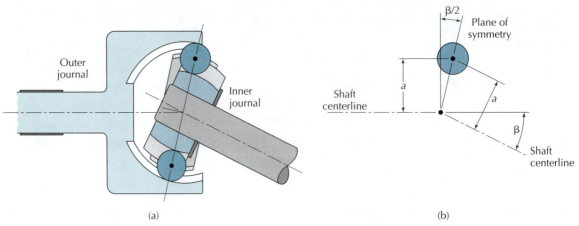

Figure 13.6 Constant-velocity coupling.

Summary and Suggestions

This type of joint is relatively compact, can transmit high torques at high speeds, and meets the design objectives of this problem. Constant-velocity joints rely on low rolling friction between the balls and their respective grooves. It is important that the grooves remain well lubricated and are free from foreign debris; otherwise wear will rapidly occur.

13.2.3 Oldham Coupling

Problem Description

A torque needs to be transmitted from an input shaft to an output shaft. The two shafts will be used in an application such that their centerlines are always parallel, but are not necessarily collinear. As part of the normal operation of the machine, the centerlines of the two shafts may become offset by a lateral distance that ranges from 0 to 3 shaft diameters. Also, the input and output shafts in this machine need to have a constant velocity ratio, regardless of offset misalignment.

Design Goal Statement

Design a coupling mechanism for use between two parallel, rotating shafts that are offset, to allow for transfer of torque. The mechanism must couple these rotating shafts, such that their relative lateral offset can vary from 0 to 3 shaft diameters. The velocity ratio between the input and output shaft must remain constant at all times. The mechanism must be simple and transmit moderate torques.

TABLE 13.2.3 DESIGN OBJECTIVES FOR OLDHAM COUPLING		
Design Objective	**Description**	**Value (Units)**
Torque	Must transmit a specified torque	50 (N-m)
Alignment	Input and output shafts must remain parallel throughout operation	Parallel
Spacial alignment	Input and output shafts may be offset by 0 to 3 shaft diameters during operation	0–3 (units of diameter)
Shaft coupling	Input and output shafts will be connected by keys, collets, or other common means	Keys, collets
Velocity ratio	Ratio of output vs. input velocity must remain constant	Constant
Operating speed	Low speed operation	0–60 (rpm)

Proposed Design Objectives

Design Objectives for Oldham coupling are listed in Table 13.2.3.

Proposed Design Solution

Video 13.8
Oldham
Coupling

There are many possible design solutions for the described problem and design objectives. One well-known solution is an *Oldham coupling*, as illustrated in Figure 13.7(a). Figure 13.7(b) illustrates an exploded view of the Oldham coupling, and Figure 13.7(c) illustrates a skeleton diagram of this coupling. An animation of this mechanism in operation is provided in [**Video 13.8**] and is illustrated in Figure 13.8. The input and output of this device are links 2 and 4, respectively. As the animation progresses, the output shaft is deliberately offset by a lateral amount to show how the mechanism transmits torque yet maintains a constant velocity ratio. The input link motion is transmitted to the output through an intermediate link 3, by means of a tongue-and-groove. Each tongue-and-groove pair allows relative sliding between the links. Since link 3 ensures the angular difference between the tongue-and-groove pairs remains constant as the links rotate, links 2 and 4 always have the same rotational speed.

Figure 13.9 shows an example application of an Oldham coupling employed in an automobile radio. A close-up view of the coupling is provided in Figure 13.9(b), where the centerlines of the

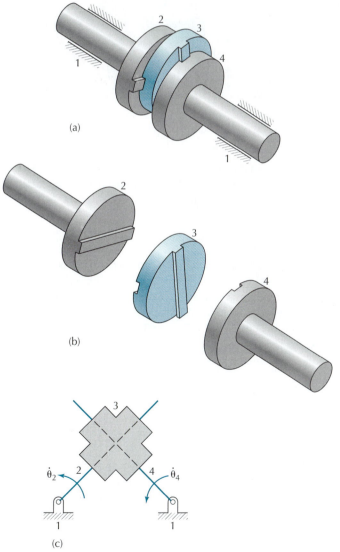

(a)

(b)

(c)

Figure 13.7 Oldham coupling.

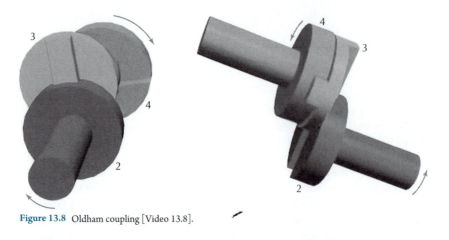

Figure 13.8 Oldham coupling [Video 13.8].

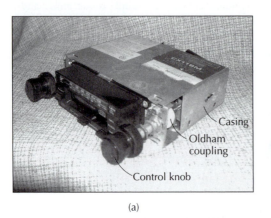

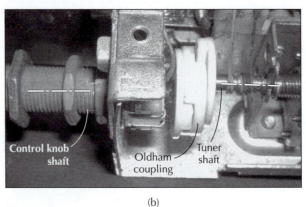

(a)   (b)

Figure 13.9 Application of an Oldham coupling: (a) Automobile radio, (b) Closeup of Oldham coupling.

control knob shaft and the tuner shaft are identified. The coupling is used to increase the lateral offset space between the control knobs to accommodate a tape cassette, and to also permit the tuner components to be moved inboard, leaving clearance within the casing.

Summary and Suggestions

This type of coupling is simple in design, is compact, and meets the design objectives of this problem. Since Oldham couplings rely on relative sliding motion, low friction between the tongue and groove is important. This is commonly achieved by using dissimilar materials for the components, such as Nylon or Teflon for link 3. Alternatively, lubricants may be used. Without proper lubrication, significant frictional losses will occur during torque transfer. Note that link 3 will radially oscillate when there is an angular offset between input and output shafts. This oscillation will create dynamic forces at high speeds, which makes Oldham couplings undesirable for high-speed applications.

13.2.4 Offset Drive

Problem Description

A torque needs to be transmitted from one shaft to another shaft, similar to case study 12.5.3. The two shafts will operate in an application where their centerlines are always parallel, but are not necessarily collinear. The centerlines of the two shafts will be laterally offset by an amount from 0 to 3 shaft diameters, however, this offset will not change during operation. The speed ratio between the input and output shafts should remain constant, regardless of offset. The application requires high efficiency to minimize power transmission loss. Also, the transmission of high torques is important.

Design Goal Statement

Design a coupling mechanism for use between two parallel, rotating shafts that are laterally offset, to allow for transfer of torque. The mechanism must couple these rotating shafts, such that their relative offset can vary from 0 to 3 shaft diameters. The speed ratio between the input and output shafts must remain constant at all times. The mechanism must transmit high torque and must have low frictional loss.

Proposed Design Objectives

Design objectives for an offset drive are listed in Table 13.2.4.

Proposed Design Solution

One possible, well-known solution for this design problem is an *offset drive*, as illustrated in Figure 13.10.

It can transmit rotational motion between two shafts that are parallel, yet offset by a specific amount. The offset drive is designed around the principle of the four-bar mechanism. In particular, the offset drive consists of three identical drag-link mechanisms, as described in Section 1.7.1. To function properly, the linkages must be designed such that they obey the rules of a drag-link

TABLE 13.2.4 DESIGN OBJECTIVES FOR AN OFFSET DRIVE

Design Objective	Description	Value (Units)
Torque	Must transmit a specified torque	100 (N-m)
Alignment	Input and output shafts must be parallel	Parallel
Spacial alignment	Mechanism must allow for input and output shaft offset from 0–3 shaft diameters	0–3 (units of diameter)
Shaft coupling	Input and output shafts will be connected by keys, collets, or other common means	Keys, collets
Velocity ratio	Ratio of output vs. input velocity must remain constant	Constant
Efficiency	Low frictional loss, defined as torque output divided by torque input	85%
Operating speed	Low-speed operation	0–60 (rpm)

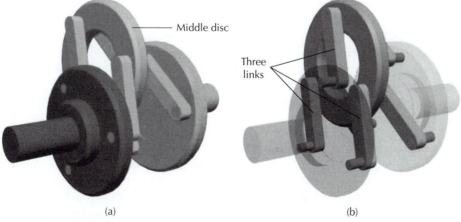

Middle disc

Three links

(a)   (b)

Figure 13.10 Offset drive [Video 13.10].

mechanism geometry. An animation of this mechanism in operation is provided in [**Video 13.10**], which shows three discs, connected by six links. On each side, one disc is connected to the *middle disc* by three links that remain parallel as the mechanism moves. There are effectively three drag-link mechanisms that work together, where the linkage lengths must be selected such that they can pass over one another. For any drag-link mechanism, the base link must have the shortest length. For the offset drive, the base link corresponds to the offset distance between the axes of the input and output shafts. A big advantage of the offset drive is that the rotational joints that comprise the four-bar mechanisms may use ball bearings. Since bearings are used, the resulting friction of the device is low, allowing for better efficiency when transmitting high torques.

Summary and Suggestions

The offset drive design is compact, is low friction, and meets the design objectives of this problem. It overcomes the frictional loss problems that occur with an Oldham coupling. However, the offset distance cannot be changed during operation, since this would effectively alter the base link length of the four-bar mechanism design, which would impact the operation. For situations where a constant offset distance is acceptable, the offset drive would work well and provide high efficiency.

13.3 SHAFT ENGAGEMENT MECHANISMS

13.3.1 Clutch Mechanism

Problem Description

An input shaft need to transmit torque to an output shaft, where the two shafts will be connected and disconnected from each other at various times. The amount of torque transmitted should be suitable for automotive drive-train or other high-torque applications. The two shafts should be aligned such that their centerlines are always collinear. During normal operation, the input shaft will be spinning, and it cannot be stopped during the connection/disconnection process. Also, the output shaft may be fully stopped or partially spinning prior to becoming connected to the input shaft. A mechanical action should be used to connect or disconnect the two shafts, using human input force. The connection/disconnection process will take place several dozen times per day, for a few years.

Design Goal Statement

Design a device that can connect/disconnect two collinear, rotating shafts, using human input force. While connected, the device must allow for the transmission of high torques suitable for automotive drive-trains. The device must allow the input shaft to continuously spin, while connecting/disconnecting the output shaft. The device must function for 100,000 connection/disconnection cycles.

Proposed Design Objectives

Design objectives for a clutch mechanism are listed in Table 13.3.1.

Proposed Design Solution

One possible solution for this design problem is a *clutch* mechanism, as shown in Figure 13.11. This mechanism is able to connect and disconnect two collinear shafts, by using a spring-loaded *friction pad*. Figure 13.12 shows a cross section of the well-known *plate clutch*, used in automobiles equipped with manual transmissions. The engaged and disengaged configurations are illustrated. The input shaft supplies torque from the engine and is rigidly connected to the *flywheel plate*. The output shaft supplies torque to the transmission and is rigidly connected to the friction pad. To disengage the clutch, the *pedal* is depressed by the foot of a human operator, as shown in Figure 13.12(a). In this configuration, it can be seen that the friction pad and flywheel are disconnected from each other. To engage the clutch, the pedal is released, thereby allowing the *diaphragm spring* to press the friction pad against the flywheel and thus providing a direct connection between the input and output shafts, as illustrated in Figure 13.12(b). In the first instant of contact the flywheel is spinning while the friction pad is stationary. However, the resulting friction between the two surfaces gradually transfers rotational motion and torque, until the output shaft reaches the same rotational speed as the input. An animation of the plate clutch in operation is provided in [**Video 13.11**].

TABLE 13.3.1 DESIGN OBJECTIVES FOR A CLUTCH MECHANISM

Design Objective	Description	Value (Units)
Torque	Transmit automotive drive-train level torque	400 (N-m)
Operating speed of output shaft	Output shaft speed same as input when connected. Drops to zero speed when disconnected	0–3000 (rpm)
Operating life	Number of connection/disconnection cycles before repair or replacement	100,000 (cycles)
Operation force	Force needed to engage/disengage device, must use human leg power	300 (N)
Constraint	**Description**	**Value (Units)**
Operating speed of input shaft	Input shaft always operates at speed and cannot stop during operation	1000–3000 (rpm)

Summary and Suggestions

The clutch mechanism meets the design objectives of this problem, by providing a way to connect/disconnect two collinear shafts to allow for the transfer of high torques. It employs a flywheel and friction pad system that is brought into contact when a spring-loaded force is removed via a foot pedal. In this sense, a plate clutch remains engaged until operator input is provided to disengage it. When engaging a clutch, some degree of operator skill is required, since the engagement must occur gradually. Sudden full engagement between the friction pad and flywheel causes a stop-start slipping, since the fast spinning input is not matched to the stationary output. The resulting reaction forces can slow the input shaft and feedback into the engine, thereby stalling the engine.

13.3.2 Synchronizer

Problem Description

An input shaft is engaged with an output shaft, via two gears in mesh. Torque should be transmitted from input to output, via the two gears, where the torque transfer can be engaged and disengaged at various times. The amount of torque transmitted should be suitable for automotive drive-train applications. During normal operation, the input shaft will be spinning during the

Figure 13.11 Clutch mechanism [Video 13.11].

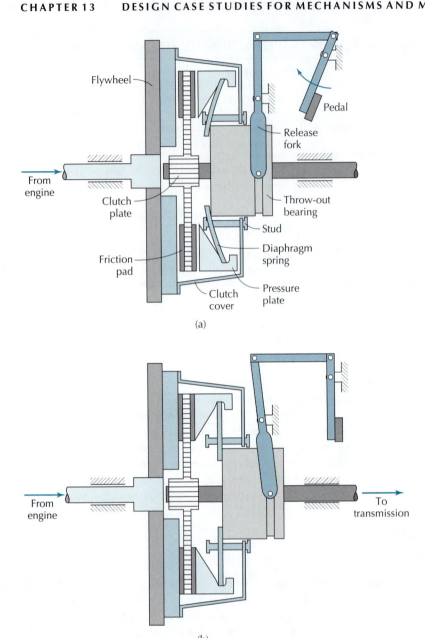

Flywheel

Pedal

Release
fork

From
engine

Clutch
plate

Throw-out
bearing

Stud

Diaphragm
spring

Friction
pad

Pressure
plate

Clutch
cover

(a)

From
engine

To
transmission

(b)

Figure 13.12 Clutch cross section: (a) Disengaged, (b) Engaged.

engage/disengage process. Also, the output shaft may be fully stopped or partially spinning prior to engagement with the input shaft. A mechanical action must be used to engage or disengage the two shafts, using human input force, and will take place several dozen times per day, for a few years.

Design Goal Statement

Design a device that can engage/disengage the torque transfer between two rotating shafts. The two shafts must employ gears in mesh for torque transfer. While engaged, the device must allow for the transmission of high torques suitable for automotive drive-trains. Human input force must supply the mechanical action to engage/disengage the shafts, while the input shaft continuously spins. The device must function for 100,000 engage/disengage cycles.

Proposed Design Objectives

Design objectives for a synchronizer are listed in Table 13.3.2.

TABLE 13.3.2 DESIGN OBJECTIVES FOR THE SYNCHRONIZER

Design Objective	Description	Value (Units)
Torque	Transmit automotive drive-train level torque	400 (N-m)
Operating speed of output shaft	Output shaft speed same as input when engaged. Drops to zero speed when disengaged	0–1000 (rpm)
Operating life	Number of connection/disconnection cycles before repair or replacement	100,000 (cycles)
Operation force	Force needed to engage/disengage device, must use human leg power	150 (N)
Constraint	**Description**	**Value (Units)**
Connection method	Mechanism used to couple the two shafts together	Gears
Maximum operating speed of input shaft	Maximum speed of input shaft during operation	3000 (rpm)

Proposed Design Solution

One possible solution for this design problem is a *synchronizer* mechanism, as illustrated in Figure 13.13. It is used to rigidly couple or uncouple a gear to/from a shaft, on which the gear is mounted. The synchronizer is commonly employed in manual transmissions of automobiles, as described in Section 6.2.2. When a gear and shaft are rigidly coupled, they act as one solid component capable of transmitting power through a meshing gear. When uncoupled, the gear will spin freely relative to the shaft, and no power may be transmitted. When used in the manual transmission application, the gear on the input shaft is always in mesh with the gear on the output shaft. The coupling process is carried out with no movement of the gear along the axis of the input shaft. Therefore, an operator can change transmission speed ratios without clashing the gear teeth.

To help with understanding the operation of a synchronizer, the operation of a manual transmission that incorporates them should be reviewed. In addition, the review of Figures 6.14, 6.15, and 6.16 is needed. As noted in Figure 6.14, gear 3 remains in mesh with gear 7 for all speed ratios. However, when the transmission is in "second gear" (Figure 6.15(d)), one of the synchronizers rigidly couples gear 7 to the output shaft, thereby allowing power to be transmitted through gears

(a)   (b)

Figure 13.13 Synchronizer [Video 13.13]: (a) Nuetral, (b) Engaged.

3 and 7. For all other speed ratios of the transmission, gear 7 is uncoupled from the output shaft, and alternative gears are used to transmit power.

Figure 13.14 shows cross-sectional views of a synchronizer in distinct configurations that occur during the engagement process. To commence engagement, the operator must first cut off torque being delivered to the transmission, by disengaging the clutch, as described in Section 13.3.1 and shown in Figure 6.14. Figure 13.14(c) illustrates the disengaged configuration in which the gear is free to spin on its shaft. Components of the synchronizer include the *hub*, which is splined to the shaft, and the *sleeve*, which is splined to the hub. By pushing the *shift fork* in the direction shown, the hub and sleeve move toward the gear. *Spring-loaded balls* located in detents in the sleeve deter relative movement between the hub and sleeve. Figure 13.14(d) illustrates the condition of initial contact between *synchronizing cones* on the gear and hub. As the cones are pressed together, frictional forces cause the gear, hub, and sleeve to rapidly achieve a common rotational speed. Upon further movement of the shift fork, the sleeve is forced to slide relative to the hub, toward the gear. At this point, the internal spline teeth on the sleeve slide over the external spline teeth on the gear. This is possible since the spline teeth on both the sleeve and gear have tappered end tips. Since the rotational speeds of the sleeve and gear are now the same, there is no clashing of the spline teeth. When the sleeve moves with respect to the hub, the balls move down against their springs. Now the gear is locked to the shaft through the sleeve, and the engagement is complete, as shown in Figure 13.14(e).

[**Video 13.14**] provides an animated sequence of the engagement of the synchronizer illustrated in Figure 13.14. Another illustration of a synchronizer is shown in Figure 13.13, in the uncoupled and rigidly coupled configurations. In the animation provided in [**Video 13.13**], when the synchronizer is uncoupled, the meshing gears stop rotating. However, in this instance, they are free to rotate at any speed.

[**Video 13.14**]
Cross
Section of
Synchronizer

[**Video 13.13**]
Synchronizer

Summary and Suggestions

The synchronizer mechanism meets the design objectives of this problem by providing a way to engage/disengage the torque between two shafts that are always coupled together by gears in

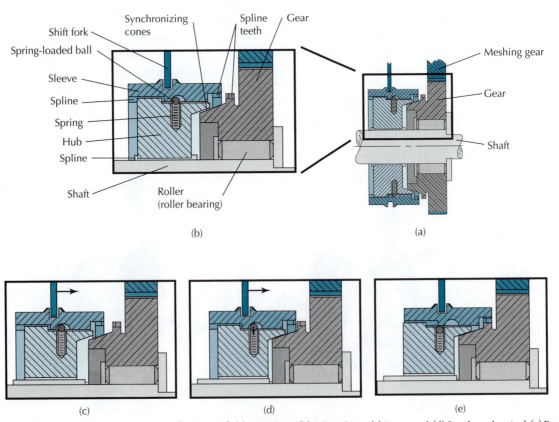

Figure 13.14 Cross section of a synchronizer [Video 13.14]: (a) Overall view, (b) Enlarged view, (c) Disengaged, (d) Speeds synchronized. (e) Engaged.

mesh. Since gears are employed, it allows for high-torque drive-train applications. It solves the engagement problem by employing friction-based cones that synchronize the speed between the input shaft and a splined gear, hence the name of the mechanism. After synchronization of speed, a sleeve is used to couple the input shaft to the splined gear, thereby permitting the transfer of torque. Disengagement is simply achieved by sliding the sleeve back and releasing the splined gear.

13.4 SHAFT/WORKPIECE CLAMPING MECHANISMS

13.4.1 Drill Chuck

Problem Description

Torque must be transferred from a motor shaft into a tool shaft, where the tool shaft is removable. The tool shaft has a smooth, round, steel surface at the connection interface. In this application, the tools will be drill bits, which have various diameters that should be accommodated for, when connected to the motor shaft. The tool and motor shafts will operate such that their centerlines are always collinear. High torques and high axial forces will exist in this application. A mechanical action should be used to connect or disconnect the two shafts, using human input force.

Design Goal Statement

Design a mechanism to connect/disconnect drill bits to/from a motor shaft, in a collinear manner. The connection method must make use of human input force. The mechanism must be able to accommodate various drill bit diameters. While connected, the mechanism must allow for the transmission of high torques and axial forces, as needed in drilling operations.

Proposed Design Objectives

Design objectives for a drill which are listed in Table 13.4.1.

Proposed Design Solution

One possible solution for this design problem is a *drill chuck* as shown in Figure 13.15(a). The chuck mechanism is located at the working end of an electric power drill, as shown in Figure 13.16. The chuck mechanism is able to clamp onto drill bits using a three fingered jaw. The clamping force is created when human hand power is applied via a *key* (not shown), which is used to turn the *scroll*

TABLE 13.4.1 DESIGN OBJECTIVES FOR DRILL CHUCK		
Design Objective	**Description**	**Value (Units)**
Torque	Transmit torque suitable for drilling operations in mild steel, using ½" diameter drill bit	1400 (N-m)
Axial force	Transmit axial force suitable for drilling operations in mild steel	900 (N)
Operating speed	Output shaft speed during operation	0–900 (rpm)
Alignment	Motor shaft and tool shaft alignment	Collinear
Connection force	Use human power	150 (N)
Constraint	**Description**	**Value (Units)**
Tool shaft size	Must accommodate various tool shaft diameters	2–3 (mm)

nut housing. The operation of the drill chuck is explained with the illustrations of Figure 13.15(a–e). Figure 13.15(a) illustrates the drill chuck in assembled form, and Figure 13.15(b) shows an exploded view of the same unit. When assembled, the scroll nut housing is press-fit over the *scroll nut*, and these components move as a rigid body. In order to clamp onto a drill bit, the bit is inserted into the *jaws*. When the scroll housing is rotated (via the key), meshing takes place between the internal threads on the two halves of the scroll nut and the scroll threads on each of the jaws. This meshing action causes the three jaws to move inward toward the centerline of the drill bit. Figure 13.15(c) shows the drill chuck without the scroll nut housing. In Figures 13.15(d) and 13.15(e), the scroll nut is also removed, and the jaws are shown in the open and closed positions with respect to a drill bit. An animation of the drill chuck is provided in [**Video 13.15**].

[**Video 13.15**]
Drill Chuck
Mechanism

Summary and Suggestions

The drill chuck mechanism meets the design objectives of this problem by providing a way to connect/disconnect drill bits onto the motor shaft of a drill. It employs helical threads that result in a large pressure angle between the meshing scroll threads on the jaws, with respect to the scroll nut

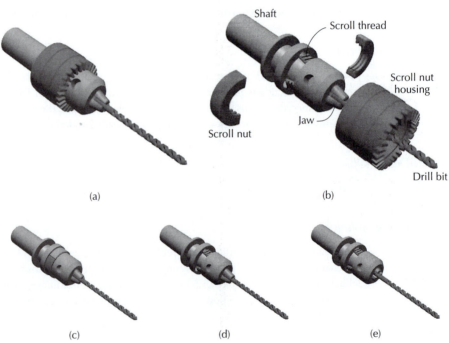

Figure 13.15 Drill chuck mechanism [Video 13.15].

Figure 13.16 Electric power drill.

housing. In this way, the design is not back-driveable, so that once a secure clamp is established, it will not self-loosen. This creates a secure clamp onto drill bits that can sustain high torques and axial forces, while they are used in drilling applications.

13.4.2 Lathe Chuck

Problem Description

Rotational motion and torque must be transferred from a motor shaft into a workpiece. The workpiece may need to be removed and repositioned from time to time. The workpiece will generally be round, may come in a variety of diameters, and may consist of different materials. The workpiece centerline and motor shaft will generally operate such that they are collinear. High torques, as well as high radial and axial forces, will be applied to the workpiece and should be accommodated in this application. A mechanical clamping action should be used to connect or disconnect the motor shaft from the workpiece. Either human input force or hydraulic input force should be considered to supply this clamping action.

Design Goal Statement

Design a mechanical device to clamp onto round workpieces, and thereby connect/disconnect them to/from a motor shaft. The device must clamp the workpiece in such a way that the workpiece centerline is collinear with the motor shaft centerline. The connection method must make use of either human input force or hydraulic input force. The connection device must be able to accommodate various workpiece diameters. While connected, the device must allow for the transmission of high torques and high radial and axial forces, as needed in metalwork machining applications.

Proposed Design Objectives

Design objectives for a lathe clutch are listed in Table 13.4.2.

Proposed Design Solution (Manual Lathe Chuck)[a]

One possible solution for this design problem, is a *manual lathe chuck* as shown in Figure 13.17. An illustration of the design is also provided in Figure 13.18. The lathe chuck

TABLE 13.4.2 DESIGN OBJECTIVES FOR LATHE CHUCK

Design Objective	Description	Value (Units)
Torque	Transmit torque suitable for metalwork operations on mild steel	150 (N-m)
Radial force	Transmit radial force suitable for metalwork operations on mild steel	400 (N)
Axial force	Transmit axial force suitable for metalwork operations on mild steel	2000 (N)
Operating speed	Output shaft speed during operation	0–2400 (rpm)
Alignment	Motor shaft and workpiece alignment	Collinear
Connection force	Method of supplying force to clamp workpiece	Human power[a] or hydraulic power[b]
Constraint	**Description**	**Value (Units)**
Workpiece size	Must accommodate various workpiece diameters	3–150 (mm)
Workpiece material	Must accommodate various workpiece materials	Metal, plastic, wood

Figure 13.17 Manual lathe chuck.

Figure 13.18 Manual lathe chuck [Video 13.18].

mechanism is able to clamp onto a workpiece using a three-piece *jaw*. This design solution employs human power as input for clamping the workpiece when it is inserted into the jaws. The clamping is accomplished by inserting a *wrench* into one of three *sockets* and rotating the wrench. Each socket is attached to a *pinion gear*, which is in mesh with a bevel *gear track* located on the edge of the *scroll*. As the socket is turned, the pinion gear rotates the scroll. The scroll consists of three spiral grooves that have been machined into the *spindle*. These grooves mesh with teeth on the three jaws. As the scroll rotates, all three jaws move radially inward to clamp onto the work-piece, or move radially outward to release the workpiece. An animation of the manual lathe chuck is provided in [**Video 13.18**].

Proposed Design Solution (Power Lathe Chuck)[b]

Another possible design solution is a *power lathe chuck*, which is actuated using hydraulic power and is illustrated in Figure 13.19(a). This alternate solution is provided, since there was an option in the design objectives for hydraulic power. In this design, a hydraulic cylinder (not shown) drives a *central piston* axially along the centerline of the chuck, as illustrated in Figure 13.19(b). The motion of the central piston is transferred into three *wedge blocks*, causing them to move tangentially along *tracks* located on the *spindle*, as shown in Figure 13.19(c). The wedge blocks have a set of *angled serrations*, which are in mesh with a matching set of angled serrations on the *jaws*. As the wedge blocks are forced to move tangentially, they in turn force the three jaws to move radially with respect to the centerline of the chuck. Figure 13.19(d) illustrates the jaws and internal components while the chuck is in the open position. An animation of the power lathe chuck is provided in [**Video 13.19**].

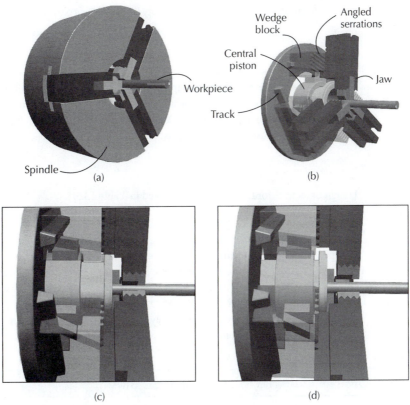

Figure 13.19 Power lathe chuck [Video 13.19].

Summary and Suggestions

Two different lathe chuck mechanisms are presented here, both of which meet the design objectives of this problem by providing a way to clamp/release workpieces onto the motor shaft of a lathe.

In the manual version, a human powered method of providing the clamping force was shown using a wrench and socket approach. That design employed a scroll (spiral grooves) in mesh with matching grooves on the jaws. This meshing arrangement creates a large pressure angle between the scroll grooves and jaw grooves, thereby creating non-back-driveable design. Once a secure clamp is established, it will not self-loosen. However, it can be easily opened by hand turning the pinion gear via the wrench.

In the power chuck version, hydraulic power is used to create the clamping forces. The hydraulic piston is arranged to move in the axial direction, for mechanism simplicity and compact size. To convert this axial motion into radial clamping motion, a set of tracks and serrations on the wedge blocks are used to slide jaws inward or outward. Here, the pressure angles between the tracks and serrations must be lower to allow the mechanism to function. For both designs, the three-jaw chuck provides a secure clamp for the workpiece and can sustain high torques and radial and axial forces for metalworking applications.

13.5 ROTATIONAL BRAKING SYSTEM

Problem Description

A device is needed to dissipate the angular kinetic energy of a rotating shaft, and thereby stop the rotation of that shaft. The device will be used for an automotive application, and hence the amount of angular kinetic energy to be dissipated is large. Additionally, the time to stop the rotating shaft should be relatively short. When the device is fully applied, the shaft must come to a complete stop, and the device should prevent any further rotation even if additional torque is applied. The stopping action should work regardless of the rotational direction of the shaft. The device will be subject to a dirty, wet, and cold environment while it functions. It will be employed several dozen times per day, for a period of two years. Mechanical simplicity (relatively few parts) has been identified as important.

Design Goal Statement

Design a mechanism to stop the rotation of shaft, by dissipating its angular kinetic energy in a short period of time. This is an automotive application where the rotating shaft has a high angular kinetic energy and high torque. The design must be mechanically simple, and must function regardless of the rotational direction of the shaft. The device must perform normally in a dirty (wet or dusty) environment. The device must function for at least 100,000 application cycles, before repair or replacement.

Proposed Design Objectives

Design objectives for the disc brake are listed in Table 13.5.1.

Proposed Design Solution (Disc Brake)

One possible solution for this design problem is a *disc brake,* as illustrated in Figure 13.20. A disc brake is able to dissipate the angular kinetic energy of a rotating shaft by converting it into heat, by employing the principle of friction. This is done by pressing two opposing *brake pads* onto a plate known as a *rotor.* The rotor is rigidly fixed to the rotating shaft and has a large contact area onto which the brake pads are pressed. To apply the brake mechanism, an input force is supplied by a human operator using foot motion. The foot input force is amplified using a hydraulic system, and

TABLE 13.5.1 DESIGN OBJECTIVES FOR THE DISC BRAKE

Design Objective	Description	Value (Units)
Shaft angular kinetic energy	Energy that the mechanism must dissipate per stop cycle	100 (kilojoules)
Shaft torque	Torque that the mechanism must withstand	3000 (N-m)
Stopping time	Amount of time to stop the shaft	2–3 (seconds)
Number of parts	Minimize the number of parts in mechanism	< 10 parts
Operating environment	Must operate as expected in dusty or wet environment	Dusty or wet
Operating life	Total number of cycles of operation before repair or replacement	100,000 (cycles)

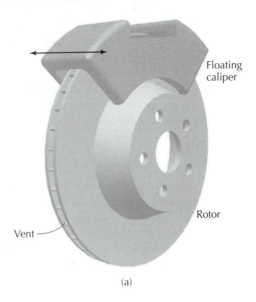

(a)

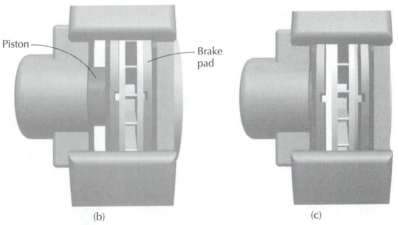

(b)   (c)

Figure 13.20 Disc brake [Video 13.20]: (a) Overall, (b) Brake applied, (c)Brake released.

hydraulic pressure is supplied to a *piston* which is mounted within a *floating caliper*, as illustrated in Figure 13.20(a). The caliper is held above the rotor on lubricated *slide pins* (not shown) allowing it to move perpendicular to the rotor surface and thereby pressing the brake pads onto the rotor. The slide pins are connected to the stationary portion of the wheel steering frame. When the brake is applied, the piston is extended, and it pushes one brake pad against the side of the rotor. Pressure in the piston cylinder also causes the floating caliper to shift to the left (Figure 13.20(b)).

This movement brings the other brake pad, mounted on the floating caliper, into contact on the opposite side of the rotor. As a result, the brake pads pinch the rotor tightly. Friction between the brake pads and rotor will reduce the rate of rotation by converting kinetic energy into heat. After the brake is released (Figure 13.20(c)), there is no spring to pull the brake pads away from the rotor. Either the brake pads stay in very light contact with the rotor or negligible wobble in the rotor will push the brake pads a small distance away from the rotor. In automotive applications, the amount of energy dissipated with each application of the brakes is considerable, and it results in significant heating of the rotors and brake pads. Most automobile rotors contain *ventilation slots* in the middle and contain air gaps/vents within the brake pads. This allows for air to move through the center of the rotor to efficiently dissipate heat energy. An animation of this mechanism in operation is provided in [**Video 13.20**].

[**Video 13.20**]
Disc Brake

Proposed Design Solution (Drum Brake)

Another possible solution for this design problem is a *drum brake*, as illustrated in Figure 13.21. Similar to a disc brake, a drum brake is able to dissipate angular kinetic energy into heat energy by employing the principle of friction. This is done with two *brake shoes* that press onto a *drum rotor* from within. The drum is rigidly fixed to the rotating shaft and has a large radial contact area onto which the brake shoes are pressed. Similar to a disc brake, human input force is supplied using foot motion and is converted to hydraulic pressure, which is supplied to a *piston*. When the brake is activated, the piston expands in both directions and pushes the brake shoes against the *drum* to provide the braking action. As the brake shoes contact the drum, there is a wedging action that increases the friction force (Figure 13.21(b)). After the brakes are released, the shoes are pulled away from the drum by springs. Other springs are used to hold the brake shoes in place. For drum brakes to function correctly, the brake shoes must remain close to the drum, to avoid excessive travel of the pistons. However, as the brake shoes wear down, a larger gap will form between the shoe and the drum, requiring adjustment to be made. This is why most drum brakes incorporate an *adjuster*. This adjustment action can only take place when the vehicle travels in reverse and when the brakes are applied. When the brakes are applied to stop the reversing vehicle, the *adjuster arm* swings upward and attempts to spin the *adjuster gear* (Figure 13.21(c)). Whenever the gap has enlarged sufficiently by wear, the adjuster lever will have enough upward swing motion to advance the adjuster gear by one tooth. This causes rotation in the threaded connection of the adjuster, causing it to slightly lengthen and thereby reducing the gap. An animation of this mechanism in operation is provided in [**Video 13.21**].

[**Video 13.21**]
Drum Brake

Summary and Suggestions

The disc brake and drum brake mechanisms meet the design objectives of this problem by providing a way to dissipate the angular kinetic energy of a rotating shaft, into heat. For automotive

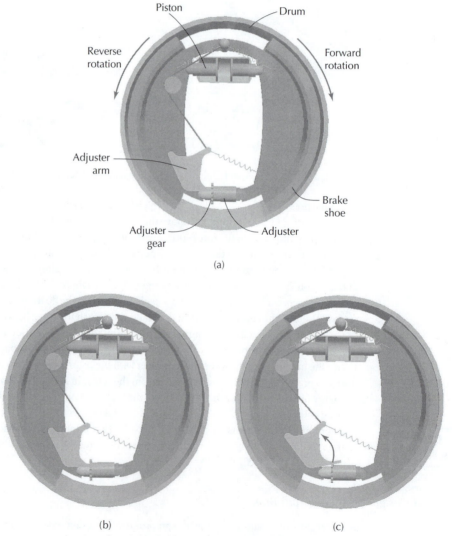

(a)

(b)   (c)

Figure 13.21 Drum brake [Video 13.21]: (a) Overall, (b) Brake applied, (c) Brake released.

applications, the rotational shaft is connected to the wheels of the vehicle. As such, the wheels are subject to the open environment along with water, dirt, and dust.

For the disc brake, the brake pad and rotor system described can function well in the open environment, since the squeezing action will dissipate water or dust and still allow for frictional contact to occur. An advantage of disc brakes is that they can cool quickly in the open environment and allow for prolonged braking in downhill applications or for frequent hard braking in performance applications. Older designs of disc brakes incorporated two pistons and a rigidly fixed caliper. That design has been largely eliminated because the floating caliper design is less expensive and more reliable. As with any frictional based system, eventually the brake pads, the rotor, or both will wear down and must be replaced. With proper selection of materials and size for specific energy dissipation levels, the disc brake mechanism can last for a 100,000 use cycles or more.

Drum brakes were the original braking system used in automobiles, and they pre-date disc brakes. However, their performance is less effective than disc brakes, since it is more difficult to dissipate the heat generated inside the drum. Brake pads are less effective at high temperatures, and drum brakes can have poorer performance during prolonged braking. The cost of producing drum brakes is lower than disc brakes, and hence they continue to be used in the rear wheels of automobiles. As a moving vehicle brakes, about 60–90% of the energy of that vehicle is dissipated by the front wheels, and hence manufacturers generally use disc brakes in the front wheels and drum brakes for the rear wheels to create a design compromise between cost and performance. High-performance vehicles employ disc brakes on all four wheels.

13.6 ONE-WAY ROTATIONAL MECHANISMS

It is often required to transmit rotational motion in only one direction while preventing rotation in the opposite direction. These so-called one-way mechanisms allow torque transmission in one direction only, while preventing torque transmission in the other direction. Some case studies of such mechanisms are presented, where each has different design features due to different design objectives.

13.6.1 Ratchet Mechanism

Problem Description

A device is needed to transmit torque from an input linkage to an output shaft in one direction only. If a reverse torque is applied, it must be mitigated in some way, to prevent transfer of the reverse torque to the output. The application will need to transmit high torques in the desired direction. Mechanical simplicity (relatively few parts) has been identified as important. The device will be subject to a dirty environment while it functions, and it should be compact in size.

Design Goal Statement

Design a mechanism to transfer torque from an input linkage to an output shaft, such that the output can only receive torque in one direction. The design must be mechanically simple and compact and must transmit high torques. The device must perform normally in a dirty (wet, greasy, or dusty) environment.

Proposed Design Objectives

Design objectives for the ratchet mechanism are listed in Table 13.6.1.

Proposed Design Solution

One possible solution for this design problem is a *ratchet mechanism*, as shown in Figure 13.22. For the arrangement shown, the ratchet *wheel* rotates about a central point and provides the output rotation. The wheel can only be rotated in the counterclockwise direction, since *pawl 1* is engaged with the *teeth* on the wheel. The input torque is supplied by the *handle*. Turning the handle in the

TABLE 13.6.1 DESIGN OBJECTIVES FOR THE RATCHET MECHANISM

Design Objective	Description	Value (Units)
CW torque	Allow applied torque to dissipate, without output shaft rotation	0 (N-m)
CCW torque	Transmit applied torque to rotate output shaft	200 (N-m)
Material	Must be strong and wear resistant	Steel
Number of parts	Minimize the number of parts in mechanism	< 10 parts
Operating environment	Must operate as expected in dusty and/or wet environment	Wet, greasy, or dusty
Overall size	Mechanism must not exceed size range	8 cm³ (volume)

[Video 13.22]
Ratchet
Mechanism

clockwise direction causes *pawl 2* to slip over the teeth, resulting in no torque applied to the wheel. Turning the handle in the counterclockwise direction causes pawl 2 to engage between two teeth, and it transmits the handle torque into the ratchet wheel. Springs (not shown) located at the turning pairs of the pawls keep the tips of the pawls in contact with the ratchet wheel. An animation of this mechanism in operation is provided in [**Video 13.22**].

[Video 13.23]
Ratchet
Wrench

A common application of this mechanism is in a *ratchet wrench*, where the driving torque can only be provided in one direction. Such a wrench is shown in Figure 13.23(a). In this mechanism, the output shaft corresponds to the square wrench head, which would convey the torque to the socket and onto the bolt. Figure 13.23(b) shows a close-up of the head of the wrench with the socket and bolt removed. The handle is rigidly connected to a multi-toothed ring which is like a pawl, as shown in Figure 13.23(c). The animation of [**Video 13.23**] illustrates how the wrench can provide torque in the clockwise direction and undergo slipping action in the counterclockwise direction. An additional feature of this mechanism is the ability to reverse the ratchet action. This can be done by re-setting the position of a cam, whereby the transmitted torque and ratcheting action are reversed (Figure 13.23(d)).

Summary and Suggestions

The ratchet wrench meets the design objectives of this problem. It employs the ratchet and pawl system to transmit torque in one direction while allowing the pawl to slip over the teeth in the other direction. An advantage of this system is that it can operate in wet or dusty conditions since the pawl can firmly engage into the teeth. However, due to the use of multiple teeth, there are a limited number of discrete positions where engagement can occur. As a result, it is sometimes possible for slip to occur, even in the desired direction, when the pawl is located between teeth. In this sense, the ratchet mechanism only works when the pawl becomes engaged at discrete locations.

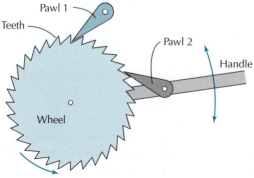

Figure 13.22 Ratchet mechanism [Video 13.22].

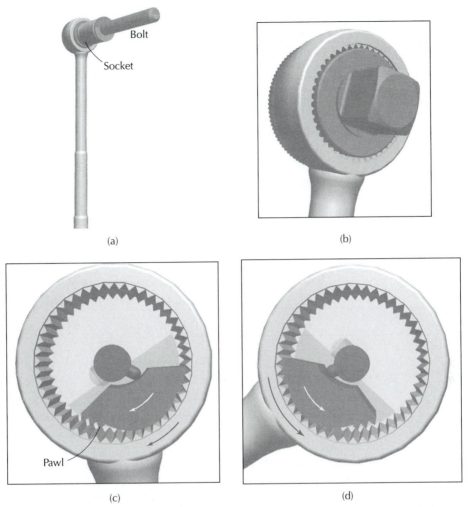

Figure 13.23 Ratchet wrench [Video 13.23].

13.6.2 Sprag Clutch

Problem Description

A device is needed to transmit torque from an input shaft to an output shaft in one direction only, similar to case study 13.6.1. If a reverse torque is applied, it should be mitigated in some way to prevent its transfer to the output shaft. In his application, the applied torque should be transferred to the output shaft at any position along the mechanism motion, in the desired direction. Also, it must operate with no noise or vibration, and will need to transmit high torques. Mechanical simplicity has been identified as important.

Design Goal Statement

Design a mechanism to transfer torque from an input shaft to an output shaft, such that the output can only receive torque in one direction. The mechanism must perform its operation with continuous engagement at any location, with minimal noise or vibration. The design must be mechanically simple and transmit high torques.

Proposed Design Objectives

Design objectives and constraints for a sprag clutch are listed in Table 13.6.2.

TABLE 13.6.2 DESIGN OBJECTIVES AND CONSTRAINTS FOR A SPRAG CLUTCH

Design Objective	Description	Value (Units)
CW torque	Transmit applied torque to rotate output shaft	250 (N-m)
CCW torque	Allow applied torque to dissipate, without output shaft rotation	0 (N-m)
Number of parts	Minimize the number of parts in mechanism	< 10 parts
Overall size	Mechanism must not exceed size range	50 cm^3 (volume)
Constraint	**Description**	**Value (Units)**
Continuous engagement	Device must be ready for engagement at any/all positions	–
Operating environment	Operating environment must be free of dust, debris, grease, or water.	Clean
Vibration	Device must operate smoothly, where vibration must remain below set value, measured as angular acceleration of inner race	< 1 (rad/s^2)
Noise	Device must operate with low noise	< 45 (dB)

Proposed Design Solution

One possible solution for this design problem is a *sprag clutch*, as shown in Figure 13.24. This mechanism consists of two concentric races, with a set of pivoting *sprag elements* between the *inner race* and *outer race*. In this example, the inner race supplies the input torque, while the outer race provides the output torque. The sprag elements are positioned in such a way that they cannot become tangent to the races, but are always angled in one direction. When the inner race is driven in the clockwise direction, the sprag elements pivot slightly and wedge themselves into both the inner and outer races. This holds the two races together, and both races turn in unison, thereby allowing for transfer of torque from the inner race to outer race. When the inner race is driven in the counterclockwise direction, the sprag elements pivot out of the way, and no motion is transmitted to the outer race. An animation of this mechanism in operation is provided in [**Video 13.24**].

Video 13.24
Sprag Clutch

Summary and Suggestions

The sprag clutch meets the design objectives of this problem, particularly the ability to transmit torque in one direction, with a smooth and continuous action. It employs the wedge principle to

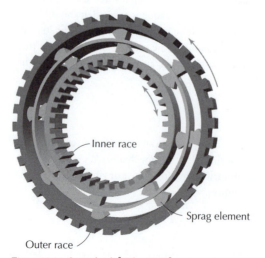

Figure 13.24 Sprag clutch [Video 13.24].

engage the sprag elements between two concentric races, to transmit torque. The advantage is that repeated engage/disengage action happens instantly, at any location along the race, unlike a ratchet where engage action can only occur at discrete locations. Another advantage is that the sprag clutch can run quietly in slipmode, since there is no clicking action over ratchet teeth. Also, it can operate at moderate to low speeds in slip mode. One disadvantage is that the sprag elements and raceways must be clean, lubricant-free surfaces to create the friction needed for the wedge action. If dirt or water enter the race, the sprag elements may have undesirable slip instead of wedging themselves, rendering the device ineffective.

13.7 ESCAPEMENT MECHANISM

Problem Description

A device is needed to regulate the output motion of a rotating wheel, such that it has intermittent motion with a precise period. The intermittent rotational output will be used for the indicator hands of a mechanical timepiece. As such, the period of the intermittent motion must be kept as constant as possible. Also, a mechanical-based energy source must be devised to provide power for the device to replenish energy lost to friction. Mechanical simplicity (relatively few parts) has been identified as important. Friction should be minimized, since it may affect the mechanism precision and use up valuable energy.

Design Goal Statement

Design a mechanism to provide periodic output motion to a rotating wheel, for use in turning the indicator hands of a mechanical timepiece. Also, design a mechanical energy source, to power the output motion. The design must be mechanically simple and must employ elements and features to minimize frictional loss.

Proposed Design Objectives

Design objectives for the escapement mechanism are listed in Table 13.7.1.

Proposed Design Solution

One possible solution for this design problem is an *escapement mechanism*, as illustrated in Figure 13.25. This mechanism provides a periodic release of stored energy, to create an intermittent

TABLE 13.7.1 DESIGN OBJECTIVES FOR THE ESCAPEMENT MECHANISM

Design Objective	Description	Value (Units)
Period of motion	Amount of time for mechanism to complete one cycle of motion	1/4 (second)
Accuracy	Ability to match the actual passage of time, as measured by error in ppm (parts per million)	10 (ppm)
Output motion	Nature of output motion must be intermittent, unidirectional rotation	Unidirectional rotation
Minimize friction	Employ materials or shapes to reduce revolute joint friction	Ruby jewel bearings
Number of parts	Minimize the number of parts in mechanism	< 10 parts
Mechanical energy source	Device used to supply energy to mechanism	–
Energy replenishment	Energy source must be easily recharged	Rechargeable

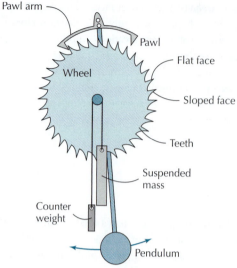

Figure 13.25 Escapement mechanism [Video 13.25].

output motion. The stored energy may be in the form of a wound spring, the potential energy of a suspended weight, or other. Escapement mechanisms were developed a few centuries ago and were commonly employed in mechanical timepieces. In all timepieces, an oscillator exists that is matched to the passing of time. Typical mechanical oscillators are pendulums with a small swing, or balance wheels, where they must maintain a precisely constant frequency of oscillation. However, mechanical oscillators are subject to frictional loss as they move. The escapement mechanism converts the stored energy source into pulses or "ticks" of intermittent output motion, where its output motion is precisely regulated by the mechanical oscillator. Simultaneously, the escapement mechanism adds energy to the oscillator to replace that which is lost to friction. In mechanical timepieces, the output motion is supplied to the hands of a clock, usually through a gearbox.

To understand the operation of an escapement mechanism, it is best to view it in motion. An animation is provided in [**Video 13.25**]. The escapement *wheel* has a few dozen *teeth* with nonsymmetric shape, as illustrated in Figure 13.25. Each tooth has two faces, where one is a *flat face* that is parallel with lines radial to the wheel while the other side of the tooth is a *sloped face* with respect to the radial lines. In the animation, the escapement wheel is driven by the stored energy in the *suspended mass* as it lowers. As the wheel turns, it operates against a double ended *pawl arm* that is rigidly attached to a swinging pendulum. For each oscillation cycle of the pendulum, the wheel is allowed to advance one tooth on the wheel.

Video 13.25
Escapement
Mechanism

Summary and Suggestions

The escapement mechanism meets the design objectives of this problem, by providing precise periodic output motion of a wheel for use in clocks. The output of the escapement wheel is used to turn the indicator hand for seconds, and through a gearbox for the indicator hands for minutes and hours. The energy source described here is a suspended weight that drops slowly to power the mechanism. To recharge the energy, the weight is lifted back to the top position by a person every day or two. The oscillator is the heart of any clock, and a pendulum is used in mechanical clocks together with the escapement mechanism, while the suspended mass replenishes energy at every cycle. In modern times, electric-based quartz oscillators together with electronics have replaced pendulums and escapement mechanisms, due to their very low cost and much higher accuracy.

13.8 VARIABLE APERTURE MECHANISM

Problem Description

A mechanism is needed to create a variable sized orifice, to regulate the passage of light through it. Further, the orifice shape should be circular regardless of the diameter of the orifice. The mechanism is to be used in photographic cameras. The orifice size should vary from a minimum of 4 cm in diameter, up to 20 cm in diameter. Also, the mechanism should be relatively compact in the direction of light passage, to allow it to fit among the optical lens system. The manner to adjust the orifice diameter should be a single control action.

Design Goal Statement

Design a mechanism to create a circular orifice that can vary in diameter from 4 cm to 20 cm. The orifice must maintain a circular shape within the diameter range specified. The design must be mechanically compact in the direction perpendicular to the orifice opening. The orfice diameter size must be adjusted using 1 degree of freedom of input actuation.

Proposed Design Objectives

Design objectives for the iris mechanism are listed in Table 13.8.10.

Proposed Design Solution

One possible solution for this design problem is an *iris mechanism*, as illustrated in Figure 13.26. An iris mechanism consists of a series of *vanes* equally spaced about the center of a circular *ring*, as illustrated in Figure 13.26(b). The vanes overlap each other and are able to rotate toward or away from the center of the ring. All the vanes are connected in such a way that they move in unison, with a single input motion. As all the vanes rotate toward the center, the opening in the middle of the mechanism is reduced, where the opening represents the *orifice (aperture)*, as shown in Figure 13.26(c). If all the vanes rotate away from the center, the orifice (aperture) is increased in size. Iris mechanisms can be formed with as few as three vanes, or up to a few dozen vanes. The more vanes that are used, the closer the central aperture will approximate a circular shape. For example, an iris mechanism with six vanes will have a hexagonal aperture shape, if the vanes have straight edges. The two extreme configurations of the iris mechanism are the fully open position (Figure 13.26(a)), and the reduced position (Figure 13.26(c)). In Figures 13.26(b) and 13.26(d), a single vane has been isolated to reveal its motion. One end of each vane is pinned to the base link. The other end of each vane has a pin that slides within a slot on a concentric ring. Rotation of the ring relative to the base link, provides motion for all vanes in unison. An animation of this mechanism in operation is provided in [**Video 13.26**].

Video 13.26
Iris Mechanism

Summary and Suggestions

The iris mechanism meets the design objectives of this problem, by providing a way create a variable-sized aperture that maintains an approximately circular shape. The more vanes that are used, the closer the approximation to a circular shape. Iris mechanisms are widely used in camera lenses to allow for passage of the desired amount of light. The shape of the opening is important for optical reasons. Also, iris mechanisms are used in other applications, such as the regulation of fluid flow. Figure 13.27 illustrates an iris mechanism that is employed to regulate airflow, and [**Video 13.27**] shows the same mechanism in motion.

Video 13.27
Flow Regulator

TABLE 13.8.1   DESIGN OBJECTIVES FOR THE IRIS MECHANISM

Design Objective	Description	Value (Units)
Variable diameter orifice	Orifice diameter is variable	4 –20 (cm)
Orifice shape	Shape of the opening regardless of size	Circular
Actuation method	Mechanical input motion with 1 degree of freedom to adjust orifice diameter	1 DOF input
Compact width	Must be compact in direction perpendicular to orifice opening	< 1 (cm)

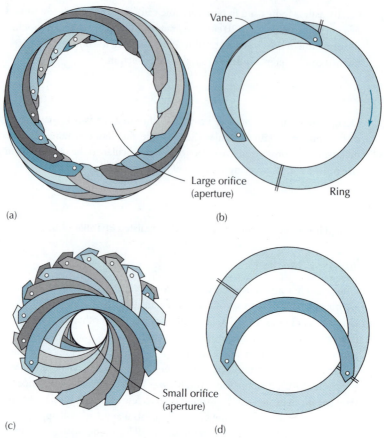

(a)

(b)

(c)

(d)

Figure 13.26 Iris mechanism [Video 13.26].

Figure 13.27 Flow regulator [Video 13.27].

13.9 ROTATIONAL TO LINEAR MOTION SYSTEMS

Converting mechanical motion from one form to another is an important aspect of mechanism and machine design. Often, motion and force are supplied in a form that is not immediately useful for the application, and a mechanism must convert the input motion into a usable form. One of the most common types of input motion is rotation, since it is readily produced by electric motors, internal combustion engines, turbines, or other sources. However, many mechanisms require linear motion and force. Hence, numerous systems have been devised to efficiently convert from one form to another. Case studies for three such mechanisms are presented next.

Problem Description

A device is needed to convert rotational input motion into linear output motion. The device should have a high efficiency in converting the input torque into linear force. The design should be scalable, to accommodate various linear travel lengths. It should allow for input rotation that is either clockwise or counterclockwise. The device should be mechanically simple and compact, for use in a wide variety of applications.

Design Goal Statement

Design a mechanism to convert rotational torque and motion into linear force and motion. The design must be mechanically simple with relatively few parts and must accommodate input rotation in either direction. It must employ components, materials, and features to minimize frictional loss to allow for efficient conversion of motion.

Proposed Design Objectives

Design objectives for linear motion system are listed in Table 13.9.1

Proposed Design Solution (Lead Screw and Nut)

One possible solution for this design problem is a *lead screw and nut*, as illustrated in Figure 13.28. The lead screw is a rod with a helical thread around its circumference. This thread has a *pitch* (distance between consecutive threads) which is the linear distance from one thread of the helix to the next. A *nut* with a matching internal thread is coupled to the screw. If the screw is rotated by 360 degrees, and the nut is prevented from rotating, the nut will move with linear motion along the screw equal to the pitch of the thread. In this way, the device can convert the rotary motion of the screw into the linear motion of the nut. Alternatively, the nut may be rotated while the screw is prevented from rotation, thereby causing linear motion of the screw. During operation, there is sliding contact between the screw and nut, hence lubrication with grease is essential for efficient operation. Also, plastic materials that have lower coefficients of friction on metal screws can be used for the nut. The efficiency of lead screws and nuts is generally between 50% and 65%. An interesting aspect of this mechanism, is that it is not back-driveable. In other words, linear force applied to the screw or nut will not result in rotational output of the nut or screw, respectively. This is due to a combination of friction and high pressure angle between the nut threads and screw threads. These mechanisms are inexpensive

TABLE 13.9.1 DESIGN OBJECTIVES FOR LINEAR MOTION SYSTEMS

Design Objective	Description	Value (Units)
Input torque	Mechanism must handle input torques within the prescribed range	0–1000 (N-m)
Output force	Mechanism must produce output forces within the prescribed range	0–10,000 (N)
Input motion	Accommodate either clockwise or counterclockwise input motion	CW or CCW
Output motion	The linear output stroke length should be within the prescribed range.	0.01–10 (m)
Efficiency	The amount of useful output energy, in comparison to the input energy	50–90 (%)
Minimize friction	Employ materials or shapes to reduce friction	Lubrication, different metals
Number of parts	Minimize the number of parts in mechanism	< 10 parts

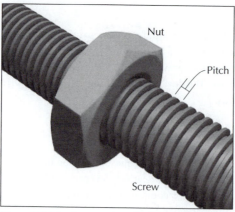

Figure 13.28 Lead screw and nut [Video 13.28].

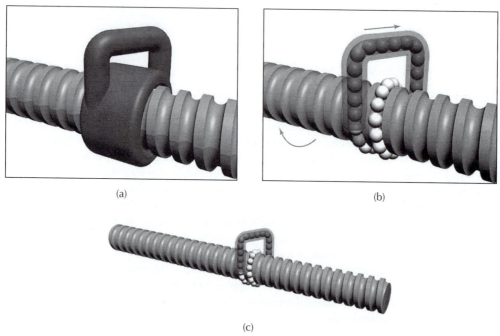

Figure 13.29 Recirculating ball thread [Video 13.29].

and are the most widely used method of converting rotational motion into linear motion, for a range of applied forces and linear travel lengths. An animation of this mechanism in operation is provided in [**Video 13.28**].

Proposed Design Solution (Recirculating Ball Screw and Nut)

Another possible solution for this design problem is a *recirculating ball screw* and *ball nut*, as illustrated in Figure 13.29. The kinematics of a recirculating ball screw are identical to those of a traditional lead screw, as described in the preceding section. Therefore, a distinction is not required when performing a kinematic analysis. However, a recirculating ball screw has drastically less friction than a traditional screw configuration. This is achieved by using ball bearings within grooves along the screw, to create rolling contact between the ball nut and ball screw. This results in a conversion efficiency of input torque to linear force, on the order of 85–90%. Unlike a lead screw, a recirculating ball screw is back-driveable. This means that linear force applied to the nut will result in rotational motion of the screw, although this is not a common mode of operation. This back-driveable operation is due to the high efficiency of these devices. Ball screw mechanisms have many parts in comparison to lead screws and are much more expensive. In this sense, they are in violation of the design objective for minimization of parts, however, they are vastly superior for handling higher forces, for efficiency, and for minimization of friction. Recirculating ball screws are usually used for high-precision applications with moderate loads, where precise amounts of linear travel for given rotations are required. This includes linear motion stages and some linear actuators. An animation of this mechanism in operation is provided in [**Video 13.29**].

Proposed Design Solution (Rack and Pinion)

A third possible solution for this design problem is a *rack-and-pinion* mechanism, as illustrated in Figures 1.24 and 1.25. The rack and pinion is a gear-train based system that employs a pinion gear in mesh with a rack. As described in Chapter 5, it is important to maintain the relative distance between the pinion and rack during operation, to minimize noise and backlash. As such, the extra parts needed (not shown) to provide this support add complexity to this design. Rack-and-pinion mechanisms are perfectly forward driven or back-driveable. In other words, they can efficiently convert rotational motion into linear motion, or linear motion back into rotational motion. They are moderately expensive, since racks are costly to produce in longer lengths. An animation of a rack and pinion in operation is provided in [**Video 1.24**] and [**Video 5.20**]. Rack-and-pinion systems are employed in automotive steering applications, and an animation of such operation is provided in [**Video 1.25**].

Summary and Suggestions

The lead screw mechanism, the recirculating ball screw mechanism, and the rack-and-pinion mechanism all meet the design objectives of this problem. They can all convert rotational motion and torque into linear motion and force in various ways. Each has its own advantages and disadvantages, and hence, is suitable for different applications.

13.10   DWELL MECHANISMS

Dwell mechanisms (also called *intermittent-motion mechanisms*) provide intervals of zero output motion while the input motion is continuous. Two such mechanisms are described below. Each employs a special shape of the coupler curve of a four-bar mechanism.

13.10.1 Coupler Curve Dwell Mechanisms
Problem Description

A device is needed to convert a continuous rotational input into an oscillatory output motion, where a portion of that output motion is stationary during the cycle. The device cannot incorporate electronic timers or components due to the operating environment, and must use mechanical linkages and joints alone. The application requires high efficiency to minimize power transmission loss. Also, the device should be capable of handling high torques and forces.

Design Goal Statement

Design a mechanism to convert a continuous input rotation into an output motion, where the output has periods of zero motion during the cycle. The device must use elements to minimize frictional loss in the mechanism. The design must be mechanically strong to handle high torques and forces.

Proposed Design Objectives

Design objectives for coupler curve dwell mechanism are listed in Table 13.10.1.

TABLE 13.10.1 DESIGN OBJECTIVES FOR COUPLER CURVE DWELL MECHANISMS

Design Objective	Description	Value (Units)
Intermittent output motion dwell	Motion must dwell for a period of time	2 (seconds)
Total cycle time	Total time for crank to complete one revolution	10 (seconds)
Link material	Link material	6061-T6 Aluminum
Stress in links	Stress must not exceed 25% of yield stress of linkage material	60 (MPa)
Trajectory of coupler linkage	Spatial trajectory (path through space) of coupler linkage that creates the dwell motion	Straight line[a] or circular arc[b]
Minimize friction	Employ elements to reduce rolling or sliding friction	Ball bearings or roller bearings
Operating speed of input	Low-speed operation	0–60 (rpm)

Proposed Design Solution (Straight-Line Dwell)[a]

Video 13.30
Straight-Line Dwell Mechanism

One possible design solution is a four-bar *straight-line dwell* mechanism for which a portion of the coupler curve is approximately a straight line[a], as illustrated in Figure 13.30. This mechanism is based on a crank-rocker mechanism design, as described in Section 1.7.1. Link 2 is the crank and link 4 is the rocker. Link 5 is a slider attached to coupler point C. As the mechanism operates, point C traces out a path in space (coupler curve) shown as the dashed *path of C*. Note that the dashed coupler curve has a shape like the letter D, for which there is an approximately straight section. Slider link 5 follows the path of C in space and eventually follows the straight section, where it slides along link 6, where link 6 will momentarily remain stationary. For the remainder of the cycle, link 6 will have a pulsed rotation.

Proposed Design Solution (Circular-Arc Dwell)[b]

Another possible design solution is a four-bar *circular-arc dwell* mechanism for which a portion of the coupler curve is approximately a circular arc[b], as illustrated in Figure 13.31. This alternate solution is provided, since there was an option shown in the design objectives for a circular arc[b].

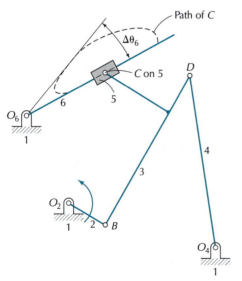

Figure 13.30 Straight-line dwell mechanism [Video 13.30].

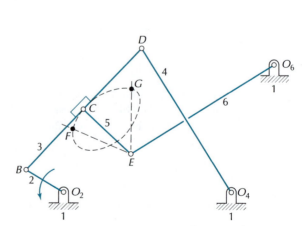

Figure 13.31 Circular-arc dwell mechanism [Video 13.31].

This mechanism is also a crank-rocker, where link 2 is the crank and link 4 is the rocker. As the mechanism operates, the point C traces out an egg-shaped coupler curve, shown as the dashed curve. Between points F and G, this curve is approximately a circular arc, centered at point E. Note that if a pivot point were to be located at the center of the circular arc (i.e., point E), that pivot point will not translate in space while point C moves along the circular arc. Therefore, Links 5 and 6 are added to the mechanism, sharing a moving pivot connection at point E. The other end of link 5 is at point C. As link 2 starts rotating from the position shown, link 6 remains stationary, as long as point C moves on the circular arc. Later in the cycle, when point C does not travel on the circular arc, link 6 will have a pulsed rotation. Also note that there are no sliding joints in this design, but only rotational joints.

Summary and Suggestions

Two different motion dwell mechanisms were presented that both satisfy the design objectives of this problem. One does this by using a straight-line portion of a coupler curve, and the other employs a circular-arc portion of a coupler curve. Depending on the rotational speed and length of input link 2, various cycle times can be achieved. Depending on how the link lengths are designed, various amounts of motion dwell can be achieved. Ball bearings can be used at all rotational joints to greatly reduce rolling friction, making these mechanisms very efficient. In addition, a high-precision mechanical motion can be achieved with the circular-arc design, by using high-accuracy bearings at its revolute joints. Using the principles described in Chapter 8 and 9, the forces within the linkages of this mechanism may be calculated.

13.10.2 Geneva Mechanism

Problem Description

A device is needed to convert a continuous rotational input into a stop-dwell-start oscillatory output motion. The application requires high-speed operation, so that the output shaft can perform the stop–dwell–start action at least 24 times per second, or faster. Due to the high inertial forces associated with the stop–start action, the device should have strong and lightweight moving parts. The application requires high efficiency to minimize power transmission loss and should operate for several million cycles in the intended application.

Design Goal Statement

Design a mechanism to convert a continuous input rotation, into a stop–dwell–start output motion cycle. The stop–dwell–start cycle must occur at least 24 times per second, and operate for several million cycles. The mechanism must use elements to minimize frictional loss in the components, and must be mechanically strong to handle high inertial forces.

Proposed Design Objectives

Design objectives for a Geneva mechanism are listed in Table 13.10.2.

Proposed Design Solution

One well-known solution for this design problem is known as a *Geneva mechanism*, as shown in Figure 13.32. The device consists of two rotating bodies that mesh with each other for only a portion of the rotational cycle. In this case, the input body is link 2 and the output is link 3. If link 2 is driven at constant rotational speed, then the output motion of link 3 stops and starts in regular intervals [42]. Figures 13.32(a) and 13.32(b) show two positions of the mechanism. In Figure 13.32(a), the *locking plate*, which is part of link 2, slides relative to a circular surface on link 3 and prevents any output motion of link 3. At a later stage of the cycle, the locking plate disengages from link 3 at precisely the same moment as *peg P* enters a slot on the *slotted wheel*, as shown in Figure 13.32(b). Since there are six equally spaced slots in link 3, the peg drives link 3 through one-sixth of a revolution (i.e., $60°$) each time it completes a cycle in a slot.

TABLE 13.10.2 DESIGN OBJECTIVES FOR A GENEVA MECHANISM

Design Objective	Description	Value (Units)
Output motion dwell	Motion of output must stop (dwell) for a percentage of total cycle time	66 (%)
Number of stop–dwell–start cycles/s	Number of motion cycles, for input crank to complete one revolution, per second	24 (cycles/sec)
Stress in links	Stress must not exceed 25% of yield stress of linkage material	60 (MPa)
Operating life	Total number of cycles of operation before repair or replacement	10 million (cycles)
Minimize friction	Employ elements to reduce rolling or sliding friction	Ball bearings or roller bearings
Operating speed of input	High-speed operation	240 (rpm)

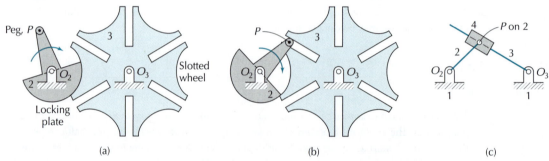

Figure 13.32 Geneva mechanism [Video 13.32]: (a) Wheel in locked position, (b) Wheel driven by peg, (c) Equivalent mechanism while peg is in slot.

The skeleton diagram representation of this mechanism for a one-sixth portion of its cycle is shown in Figure 13.32(c). This portion is equivalent to that shown in Figure 13.32(b). In this skeleton diagram, links 2 and 3 are connected to the input and output of the mechanism, respectively, and link 4 is a slider.

Summary and Suggestions

The Geneva mechanism presented meets the design objectives of this problem. It is able to convert a high-speed rotational input into an intermittent output rotation. Given an input rotational speed of 240 rpm (4 rev/sec), the six slot Geneva mechanism shown can achieve 24 start–dwell–stop cycles per second. For other applications, a Geneva mechanism may be designed with four, five, seven, eight or more equally spaced slots in link 3. If there are four slots in the slotted wheel, then each rotation of link 2, will rotate link 3 by 90°. One application that requires intermittent output motion to be produced from a constant input rotation is in movie cameras or projectors to periodically advance the film. This mechanism would be an alternative to the four-bar mechanism with a coupler point, as presented in Section 1.2.2 and illustrated in Figure 1.16.

13.10.3 Counting Mechanism

Problem Description

A device is needed to count the number of times an input shaft undergoes rotation. The count needs to be displayed in numerical form using alphanumeric digits. For this particular application, the device needs to be able to count up to 1000, where one turn of the input shaft corresponds to 10 counts. The mechanical design should be scalable, so that it can be easily expanded to count to

TABLE 13.10.3 DESIGN OBJECTIVES FOR A COUNTING MECHANISM

Design Objective	Description	Value (Units)
Numerical readout	Input shaft motion causes digits to advance, where advancement will be displayed as alphanumeric	Alphanumeric
Ratio of input shaft rotation to digit increment	One rotation of the input shaft will equal 10 counts of advancement of the numerical digits	1:10
Maximum count	Highest number that the mechanism will reach	1000
Expandability	Mechanism design must allow for easy expansion to count to higher numbers	–
Operating life	Total number of cycles of operation before repair or replacement	10 million (cycles)
Materials	Employ materials to reduce friction and wear on the mechanism components	Plastic and steel
Operating speed	Operating speed of input will be low speed	0–60 (rpm)

higher numbers for other applications. The device cannot incorporate electronic components due to the operating environment, and must use mechanical elements alone. The device will need to operate for several million cycles in the intended application.

Design Goal Statement

Design a mechanical counting mechanism to convert input shaft rotation, into a readable numeric output. The mechanism must be able to count up to 1000. The ratio between a single input rotation to number count must be 1 to 10. The design must allow for easy expansion for higher counts in other applications. The materials and component strengths must be appropriate for the device to operate for several million cycles.

Proposed Design Objectives

Design objectives for a counting mechanism are listed in Table 13.10.3.

Proposed Design Solution

Video 13.33
Counting
Mechanism

An interesting solution for this design problem is an intermittent-motion counting mechanism, as shown in Figure 13.33. This mechanism employs gears, linkages, and numbered wheels to achieve the design objectives. The *input shaft* is rigidly connected to the *first wheel*. For each rotation of the input shaft, the first wheel undergoes one full rotation. Each wheel has an *indexing tooth* on it, as shown in Figure 13.33(b). The indexing tooth on the first wheel will periodically engage with a *coupling gear*, which is in mesh with a *gear* attached to the *second wheel*. After each rotation of the first wheel, its index tooth engages the coupling gear and thereby advances the adjacent wheel by 1/10th of a turn corresponding to one digit, as shown in Figure 13.33(c). As the second wheel completes one full rotation, its own index tooth engages another coupling gear, which advances the third wheel by 1/10th of a turn. When a counting task has been completed, the counting mechanism must be reset to zero. This is done with a button (not shown in this example) that pushes against internal cams, to reset the count. An animation of this mechanism in operation is provided in [**Video 13.33**].

Summary and Suggestions

The counting mechanism presented meets the design objectives of this problem. It is able to convert input shaft rotation into discrete rotations of numbered wheels. As the first wheel completes one rotation, it engages the second wheel to give it a 1/10th turn. As the second wheel eventually

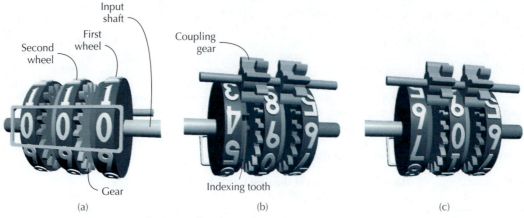

Figure 13.33 Counting mechanism [Video 13.33].

completes one turn, it engages the third wheel, and so on. By adding more wheels and coupling gears, this design is easily scalable to allow for higher counts. In this example, one input rotation corresponded to 10 counts. This can be changed to a 1:1 ratio by addition of a non-numbered wheel before the first wheel or by using a gearbox to reduce the input shaft speed by a factor of 10. This type of mechanism has many applications and is used in automobile odometers, revolution counters, and distance measuring wheels.

13.11 HARMONIC MOTION MECHANISM

13.11.1 Scotch Yoke Mechanism

Problem Description

A device is needed to convert shaft rotation into reciprocating linear motion of a component. The linear motion should have a velocity profile that is harmonic. The moving linear component should produce reasonable amounts of force, and must achieve a specific amount of linear displacement. The design should be simple, and use a small number of parts. The device will need to operate for many cycles, and should not wear out.

Design Goal Statement

Design a mechanism to convert shaft rotation into harmonic linear motion of a component. The mechanism must have a specific amount of linear displacement and must transmit reasonable force via its linear component. The mechanism must be simple (relatively few parts) and must operate for 100,000 cycles before repair or replacement. Elements or materials to minimize friction must be used.

Proposed Design Objectives

Design objectives for a scotch yoke mechanism are listed in Table 13.11.1.

Proposed Design Solution

One possible solution for this design problem is a *scotch yoke mechanism*, as shown in Figure 13.34. This mechanism transforms rotational motion into linear harmonic motion. The *crank* is rigidly attached to the *input shaft* at one end. The crank is attached to a *slider* at the other end via a revolute joint using a bearing. As the crank rotates, the slider moves back and forth within a track on a component referred to as a *yoke.* Angular displacement of the crank, denoted as θ_2, is measured with respect to a selected reference. By inspection, when $\theta_2 = 0$, as defined by the coordinate system, the yoke reaches one of its extremum positions with respect to the base link. This is also referred to as a *limit position*. The yoke has zero velocity when it is in a limit position. For this mechanism, if

TABLE 13.11.1 DESIGN OBJECTIVES FOR A SCOTCH YOKE MECHANISM

Design Objective	Description	Value (Units)
Input torque	Maximum torque permitted at shaft input	20 (N-m)
Output force	Desired output force of linear component	50 (N)
Linear displacement	Desired displacement of linear component	45 (cm)
Motion profile	Harmonic velocity profile	Harmonic
Number of parts	Minimize the number of parts in mechanism	< 8 parts
Operating speed	Low speed operation	0–60 (rpm)
Operating life	Total number of cycles of operation before repair or replacement	100,000 (cycles)
Minimize friction	Employ elements or materials to reduce rolling or sliding friction	Ball bearings or lubrication

Video 13.34
Scotch Yoke
Mechanism

the angular motion of the crank is constant, then the linear motion of the yoke is harmonic. A 2D animation of this mechanism in operation is provided in [**Video 13.34**]. The analysis of the limit positions of various mechanisms is presented in Section 2.8.

Summary and Suggestions

Video 13.35
Application of
Scotch Yoke
Mechanism

The scotch yoke mechanism presented meets the design objectives of this problem. It is able to convert input shaft rotation into harmonic linear motion of the yoke linkage. The scotch yoke mechanism is a basic mechanism, and it has use in many different applications. One example is its use in raising and lowering the windows of an automobile, as shown in Figure 13.35. An animation of this mechanism in operation is provided in [**Video 13.35**]. The glass window is attached to the yoke linkage, so that as the yoke moves up and down, so does the window. The regulator handle serves to provide the rotational input in this example. Also, a gear train is added between the handle and the crank, to reduce the crank rotation to 180 degrees, relative to 450 degrees for the regulator handle.

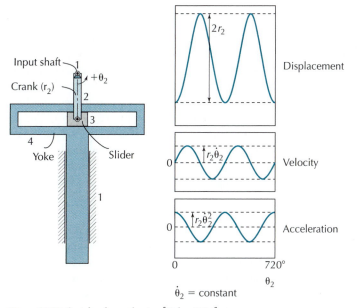

Figure 13.34 Scotch yoke mechanism [Video 13.34].

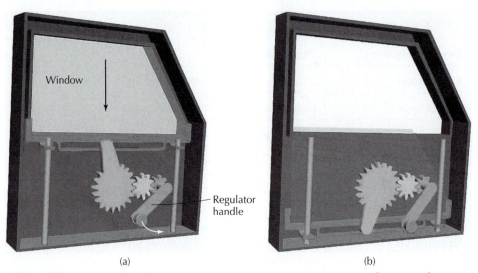

Figure 13.35 Application of a scotch yoke mechanism for raising/lowering window in car door [Video 13.35].

13.12 INTERNAL COMBUSTION ENGINES

Internal combustion engines (ICE) are machines that convert hydrocarbon fuels such as gasoline, diesel, or propane into useful mechanical energy. These fuels have a high energy density (total energy per unit volume) and allow them to be conveniently stored in fuel tanks. This allows a vehicle such as an automobile, boat, or plane to travel hundreds of kilometers before refueling. To convert these fuels into useful mechanical energy, they must be combusted (oxidized) with oxygen, which will produce large amounts of gas pressure and heat. One major aspect of mechanism design has been the development of machines that can harness the gas pressure produced by combustion and then convert it into useful mechanical work. The thermodynamic Otto cycle [43] defines a sequence of events that should take place to extract energy from the combustion of fuel and air, to produce certain gas pressures and temperatures. Many such *engines* have been devised over the past hundred years, and they vary depending on the type of fuel they combust. A few case studies of gasoline combustion engines that follow the Otto cycle are presented next.

13.12.1 Four-Stroke Piston Engines

Problem Description

A machine is needed to combust gasoline and convert the combustion gases into useful mechanical energy, in the form of shaft rotation. The combustion process should aspire to follow the thermodynamic Otto cycle. The device will be fed with a mixture of gasoline and oxygen (fuel–air mixture) from another machine. This fuel–air mixture should go into a combustion chamber, where it will be compressed and then ignited in some way. After ignition, the resulting high pressure produced by the combustion gases should be used to motivate the machine parts and produce the desired shaft rotation and torque. After harvesting the gas pressure, the machine should expel the combustion products, to ready itself for the next cycle that begins with fuel–air insertion. The device should repeat this cycle, a great many times. Since the combustion will produce hot gases, the machine should be able to withstand high temperatures and high pressures.

Design Goal Statement

Design a machine to combust gasoline and convert the combustion gases into shaft rotation and torque. The machine must have a combustion chamber made of materials that can sustain high temperatures and pressures. The machine must smoothly introduce the fuel, compress it, combust it, and expel the waste products, in a properly coordinated sequence following the Otto cycle. One part of the machine must regulate the fuel intake and coordinate this sequence. The machine must repeat this cycle millions of times before requiring repair or service. It must employ components, materials, and features to allow for efficient conversion of combustion energy into rotational motion.

Proposed Design Objectives

Design objectives and constraints for a four-stroke piston engine are listed in Table 13.12.1.

Proposed Design Solution (Overhead Valve Piston Engine)

One possible design solution is an *overhead valve piston engine*, as shown in Figure 13.36. The piston engine is also known as a reciprocating engine. The mechanical basis of the piston engine is the slider crank mechanism, as illustrated in Figures 1.5 and 1.6, whereby gas pressure exerts a force on the slider. This force is transferred to the rotating crank via the connecting rod. The piston-based design was originally developed for use in steam engines in 18th-century Europe. Steam engines incorporate a *piston* (slider), which moves inside of a *cylinder* (base link), whereby steam pressure pushes the piston to move. This design then led to the development of using the pressure from combustion gas to push the piston in the 19th-century. Combustion gas pressures are harnessed by the piston to create useful linear mechanical force that is converted into rotational motion of the output shaft (crankshaft). Figure 13.36 illustrates a cross-sectional view of a single cylinder piston engine that incorporates the Otto cycle with a four-stroke (four-phase) operating sequence. This particular engine design employs a *cam* mechanism to open and close valves located above (overhead) the *piston*, hence the name *overhead valve*. These overhead valves regulate (a) the fuel entry into the cylinder and (b) expulsion of waste gases from the cylinder. The engine sequence works as follows: A mixture of air and fuel is drawn into the cylinder through an open *intake valve*, as shown in Figure 13.36(a). The intake valve is then closed while the mixture is compressed as the piston is driven upwards, shown in Figure 13.36(b). Next, the fuel–air mixture is ignited while the valves remain closed, causing a rapid increase in gas pressure and hence downward force on the piston. This downward force is transferred by the connecting rod to the crankshaft in the direction of rotation (Figure 13.36(c)). As the piston reaches its bottom position, the *exhaust valve* opens, allowing the burned gas mixture to exit (Figure 13.36(d)). The cycle then repeats. An animation

TABLE 13.12.1 DESIGN OBJECTIVES AND CONSTRAINTS FOR A FOUR-STROKE PISTON ENGINE

Design Objective	Description	Value (Units)
Engine output power	Amount of useful energy/second that the engine will produce at the output shaft	20 (kilowatt)
Engine operating speed	Number of combustion cycles occurring per minute (Coupled to previous design objective)	1000 (rpm)
Output torque	Available torque at output shaft (coupled to previous two design objectives)	54 (N-m)
Operating life	Total number of cycles of operation before repair or replacement	50 million (cycles)
Materials	Employ materials to reduce friction and wear on the mechanism components	Steel, brass, silicone
Minimize friction	Employ elements to reduce rolling/sliding friction	Ball bearings/brass bearings
Constraint	**Description**	**Value (Units)**
Max. temperature	Maximum temperature of engine components within combustion area	300 (degrees Celsius)
Max. pressure	Maximum pressure within combustion area	5 (MPa)

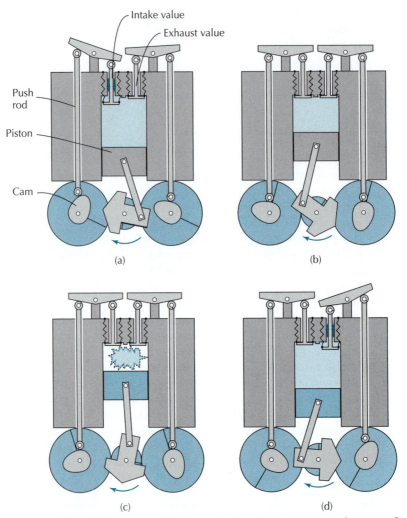

Figure 13.36 Overhead valve piston engine cross section throughout a cycle of motion [Video 13.36]:
(a) Intake, (b) Compression, (c) Power, (d) Exhaust.

<table>
<tr><td>**Video 13.36**
Overhead Valve
Piston Engine
Cross Section</td></tr>
</table>

of this mechanism in operation is provided in [**Video 13.36**]. A 3D animation of a variation of the overhead valve piston engine using a belt is also provided in [**Video 1.1**].

Proposed Design Solution (Sleeve Valve Piston Engine)

Another possible solution to this design problem is a *sleeve valve piston engine*, as shown in Figure 13.37. As with the previous proposed solution, this machine also makes use of a slider-crank-based *piston, sleeve, cylinder,* and *crankshaft*. It has a four-stroke operation and also works on the Otto cycle principle. However, it employs a different mechanism for addition of fuel and expulsion of waste combustion gases. This engine employs a concentric cylindrical sleeve located between the piston and the stationary cylinder wall. Two holes are located in the sleeve, one of which allows for the passage of the fuel mixture into the cylinder while the other allows waste gases to exit the cylinder. The motion of the sleeve is driven through a pivot and slider mechanism connected to the crankshaft via gears with a 2:1 ratio. An animation of this mechanism in operation is provided in [**Video 13.37**]. The engine sequence works as follows: The fuel inlet hole in the cylinder lines up with the inlet hole in the sleeve allowing for fuel to enter the cylinder. The inlet hole is then blocked by relative motion between the sleeve and cylinder, while the piston compresses the fuel–air mixture. The fuel is then ignited by a spark, creating high pressure to drive the piston down and thereby turn the crankshaft. Next, the exhaust hole in the cylinder lines up

<table>
<tr><td>**Video 13.37**
Sleeve Valve
Piston Engine</td></tr>
</table>

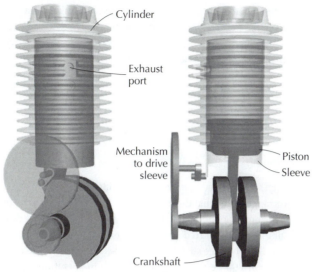

Figure 13.37 Sleeve valve piston engine [Video 13.37].

with the exhaust hole in the sleeve, while the piston moves upwards to push the exhaust gases out of the chamber. The exhaust hole is then blocked by relative motion, and the cycle repeats itself.

Summary and Suggestions

Two different piston engine designs are presented, that each meet the design objectives of this problem. They are able to harness the high-pressure gas energy produced when a fuel–air mixture is combusted in the cylindrical combustion chamber. The gas pressure is used to displace a piston, which turns a crankshaft, thereby creating rotational output motion and torque. The power produced by these designs is a function of several variables, such as fuel–air quantity per cycle, piston displacement distance, piston diameter, and number of cycles per second. Within the framework of the two designs presented, these variables can be adjusted, to produce different power designs for various applications. In both designs, steel components are used to handle the high heat and material stress created due to the combustion process and rapid motion. Another important aspect needed for both designs is the selection and use of materials for good seals and low friction between the piston and cylinder. Good seals are needed to prevent unwanted escape of gases, which can be achieved with piston rings that have low wear and good sliding properties. Lubrication of the piston and rotational joints is also achieved by use of oil. The two designs differed primarily by the mechanisms used to introduce the fuel–air mixture, or to expel the waste products. In practice, four-stroke, overhead valve engines are widely used in most applications, while the sleeve-valve engine is relatively uncommon.

13.12.2 Multi-Piston Engines

Problem Description

In some cases, a single cylinder piston engine cannot produce sufficient power for a desired application. Also, the nature of the four-stroke design is such that a power stroke occurs only one-quarter of the time, while the remaining strokes perform other functions that consume some of the power. This results in dynamic variation of the output torque. A design is needed to combine multiple four-stroke piston engines into one device. The device should produce more total power and should minimize variation of the output shaft torque. The device should allow for long duration operation and, if possible, should minimize engine noise.

Design Goal Statement

Design a mechanism to combine multiple four-stroke piston engines into an engine with a single output shaft. The engine must produce output power that is proportional to the number of pistons employed in the design. Additionally, the engine must minimize dynamic variation in the output shaft torque. The engine must endure tens-of-millions of cycles before requiring repair or service. It must employ components, materials, and features to allow for low-friction efficient operation, and to withstand high temperatures and pressures.

Proposed Design Objectives

Design objectives for multi-piston engines are listed in Table 13.12.

Proposed Design Solution (V-2 Piston Engine)

> **Video 13.38**
> V-2 Piston Engine

One possible design solution is a *V-2 piston engine*, as illustrated in Figure 13.38. The design consists of two cylinders arranged such that the cylinder axes have a relative angle between each other. This so-called V-shaped angle is typically 90 degrees, but different designs developed in the past have used angles ranging from 45 to 120 degrees. One reason for employing a two-cylinder design is to stagger the four-stroke sequence such that a power-stroke is available for half the time (50% duty cycle). An animation of a 90-degree, V-2 piston engine in operation is provided in [**Video 13.38**]. In the two-cylinder engine, the connecting rods of each piston are connected to a single crankshaft. The crankshaft employs a counterweight to balance its rotation. Inspection of the engine reveals that the power stroke (combustion and gas expansion) of one piston will occur after 180 degrees of crankshaft rotation of the power stroke of the other piston. This piston stroke arrangement along with the crankshaft counterweight helps to balance the output torque of the crankshaft and minimizes output torque variation. To improve this design further, a V-4 configuration could be employed, with an appropriate crankshaft design and firing sequence. Each piston would be connected to the crankshaft in such a way that its power stroke would occur 90 degrees out-of-phase with the next piston, resulting in a 100% duty cycle for power stroke for every crankshaft rotation.

Proposed Design Solution (Radial Piston Engine)

> **Video 13.39A**
> Radial Piston Engine
>
> **Video 13.39B**
> Radial Piston Engine with Propeller

Another possible design solution is a *radial piston engine*, as illustrated in Figure 13.39. In this arrangement, nine four-stroke pistons are employed, where all pistons are constrained to move radially from a common point, concentric with a crankshaft. Radial engines were developed as early as 1901 employing three cylinders, but typically use five, nine, or more cylinders. They were primarily used for early aircraft applications, due to their ability to be air-cooled, which provided good power versus weight performance. Figure 13.39(b) illustrates the same radial engine with a *propeller* added to the output shaft. The offset distance on the crankshaft connection point is carefully designed such that the displacement of all pistons is constant. For a four-stroke system, the timing of fuel inlet and exhaust outlet valves on each piston is very important. The open and

TABLE 13.12.2 DESIGN OBJECTIVES FOR MULTI-PISTON ENGINES

Design Objective	Description	Value (Units)
Engine output power	Amount of useful energy/second that the engine will produce at the output shaft	100 (kilowatt)
Output shaft speed	Output shaft speed	1000 (rpm)
Output shaft torque	Output shaft torque (coupled to previous two design objectives)	250 (N-m)
Operating life	Total number of cycles of operation before repair or replacement	50 million (cycles)
Minimize output torque oscillation	Employ design configuration to provide near constant output torque	–

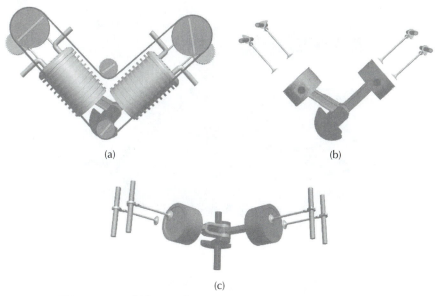

(a)

(b)

(c)

Figure 13.38 V-2 piston engine [Video 13.38].

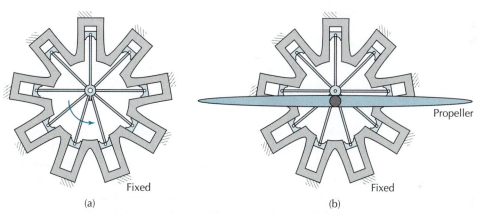

Propeller

Fixed

(a)

Fixed

(b)

Figure 13.39 Radial piston engine: (a) Radial engine alone, [Video 13.39A] (b) Radial engine with propeller [Video 13.39B].

Video 13.40
Airplane Radial
Piston Engine

close sequence of each valve must be such that the power-stroke on sequential cylinders do not conflict, but provide a smooth continuous application of power to the crankshaft. The use of an odd number of pistons ensures overlap of power-strokes, which provides smoother operation, as some power-stroke energy can be used for compression of the next cylinder. An animation of a radial piston engine in operation is provided in [**Video 13.39A**], and an animation of the same engine using a propellor is provided in [**Video 13.39B**]. Figure 13.40 shows a picture of a radial engine incorporated in a vintage single propeller airplane.

Summary and Suggestions

Two different multi-piston engine designs are presented that each meet the design objectives of this problem. They are able to combine piston engines in multiple configurations to increase the overall power of an engine. Also, they minimize dynamic torque variation in the output shaft. There are also numerous other arrangements of multi-piston engines. For example, automobile engines may employ four, six, or eight pistons (cylinders), in various arrangements. Figure 13.41 illustrates some typical arrangements of cylinders, which include: in a row (in-line) as illustrated in Figure 13.41(a), in two rows with a V-angle as illustrated in Figure 13.41(b), or in two rows opposing each other (flat), as illustrated in Figure 13.41(c).

Figure 13.40 Airplane with radial piston engine on front [Video 13.40].

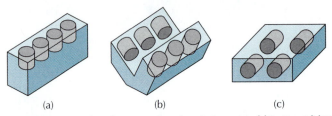

(a)   (b)   (c)

Figure 13.41 Typical configurations of a multi-cylinder engine: (a) In-Line 4 (b) V-6 (c) Flat 4.

13.12.3 Wankel Engine

Problem Description

A machine is needed to convert the combustion gases of gasoline and air into useful mechanical energy, similar to case study 13.12.1. The machine output should be shaft rotation. The designer is aware of the existing reciprocating piston engine design and its limitations such as: balance issues due to the reciprocating piston and connecting rod, intake and outlet valve complexity, and overall engine size and weight. A new engine that avoids those limitations is desired. The new engine should still employ a combustion process that follows the Otto cycle. It must still be fed with a fuel–air mixture that enters a combustion chamber, where it will be compressed and then ignited in some way. After ignition, the combustion gases should motivate the machine parts, and produce the desired shaft rotation and torque. After harvesting the gas pressure, the machine should expel the combustion products and then ready itself for another cycle, repeatedly. The machine should be able to withstand high-temperatures and high pressures, and operate for many millions of cycles.

Design Goal Statement

Design a machine to combust gasoline and convert the combustion gases into shaft rotation and torque. It must be a novel design that does not employ the reciprocating piston engine approach. It must have a combustion chamber made of materials that can sustain high temperatures and pressures and minimize frictional losses. The machine must introduce the fuel, compress it, combust it, and expel the waste products in a properly coordinated sequence following the Otto cycle. The machine must have a simple system to regulate the fuel intake, the exhaust gas expulsion, and the sequence coordination that is less complex than cam-based valves. The machine must be smaller and lighter than a piston engine of comparable output power. It must operate for millions of cycles before requiring repair or service.

Proposed Design Objectives

Design objectives for a Wankel engine are listed in Table 13.12.3.

TABLE 13.12.3 DESIGN OBJECTIVES FOR WANKEL ENGINE

Design Objective	Description	Value (Units)
Engine output power	Amount of useful energy/second that the engine will produce at the output shaft	20 (kilowatts)
Engine operating speed	Number of combustion cycles occurring per minute (coupled to previous design objective)	2000 (rpm)
Output torque	Available torque at output shaft (coupled to previous two design objectives)	27 (N-m)
Operating life	Total number of cycles of operation before repair or replacement	50 million (cycles)
Materials	Employ materials to reduce friction and wear on the mechanism components	Steel, brass, silicone
Minimize friction	Employ elements to reduce rolling/sliding friction	Ball bearings/brass bearings
Constraint	**Description**	**Value (Units)**
Avoid reciprocating components	Avoid reciprocating piston engine approach, for better output shaft balance	—
Size	Smaller size than piston engine of comparable output power	$25 \times 15 \times 35$ (cm) $[L \times W \times H]$ (volume)
Weight	Smaller weight than piston engine of comparable output power	125 (kg)
Simple fuel–air inlet and exhaust outlet	Simple design for fuel-air inlet and exhaust outlet, in comparison to piston engine valve system	—

Proposed Design Solution

Video 13.42
Wankel Engine

One possible design solution is a *Wankel engine*, as illustrated in Figure 13.42. The basic principle is a mechanism with an eccentric spinning *rotor* that rotates within a specially shaped *housing*. Three pockets of gas exist between the rotor and the housing, and as the rotor spins, these gas pockets move around the rotor and change their volume. At each of the three corners of the rotor is *a seal* that slides against the surface of the stationary housing, preventing gases from leaking into adjacent pockets. The eccentric motion of the rotor is created by an internal gear machined into the rotor that meshes with a fixed external gear. The rotor is connected to a central *output shaft*, thereby providing output rotation and torque. The operation of the mechanism is best understood by viewing the animation of a Wankel engine provided in [**Video 13.42**]. Figure 13.42 shows four images of the positions of the rotor during a cycle consisting of intake, compression, power (i.e., combustion of the fuel–air mixture), and exhaust the spent fuel. During intake, one of the gas pockets around the rotor becomes connected to the *intake port*, as shown in Figure 13.42(a). A fuel–air mixture is drawn into the increasing volume between the rotor and the internal wall of the housing. As the rotor continues to rotate, the mixture becomes closed off from the intake port and is compressed, as shown in Figure 13.42(b). Approaching maximum compression, the mixture is ignited by a spark plug, as shown in Figure 13.42(c). The combustion of the mixture generates a high gas pressure that creates force on the side of the rotor, corresponding to the power stroke. This force pushes the rotor and the volume of the cavity increases. Finally, the *exhaust port* becomes connected to the gas pocket, as shown in Figure 13.42(d). As the rotor continues to rotate, the volume of the pocket decreases, thereby expelling the burned mixture. The cycle is then repeated. Note that three overlapping cycles occur for a single rotation of the rotor. As such, there are three spark plug ignitions for each revolution of the rotor.

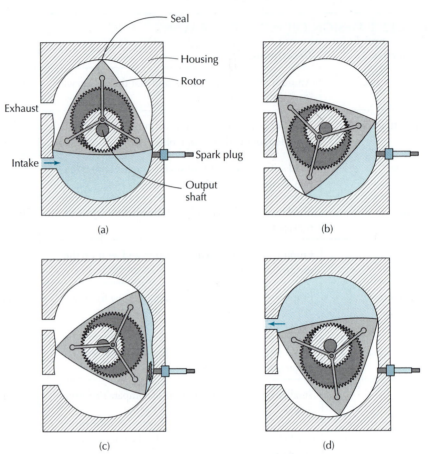

Figure 13.42 Wankel engine [Video 13.42]: (a) Intake, (b) Compression, (c) Power, (d) Exhaust.

Summary and Suggestions

The Wankel engine presented meets the design objectives of this problem. It is able to harness the gas pressure produced by combusting a fuel–air mixture to drive the rotation of a triangular-like rotor, which in turn rotates the output shaft. The Wankel engine has some interesting advantages. It is a very compact and lightweight engine that is much smaller than a piston engine of comparable output power. It has fewer moving parts than a piston engine, and its rotating parts are inherently balanced, providing for a much smoother operation and capability for higher rotational speeds. Additionally, it does not require cam-based valves, since it employs a simple inlet and outlet port system. The Wankel engine was used for motorcycles and cars throughout the 1970s, 1980s, and 1990s, most notably by the Mazda car company for a series of sports cars. Earlier versions of the engine suffered from problems with the seal system and with fuel consumption, which led to performance issues. Recently, the Wankel engine is seeing renewed development for small aircraft and UAV (unmanned aerial vehicles), since it has a high power-to-weight ratio.

13.13 PUMP MECHANISMS

13.13.1 Positive-Displacement Pump

Problem Description

A machine is needed to pump fluid from an inlet port to an outlet port. The pumping action should occur by moving specific amounts of fluid, from the inlet side to the outlet side, per unit time. Additionally, the pump action should occur such that all fluid entering the pump moves through in one direction only. There should be no leakage of fluid backward through the pump when it is on or off. Mechanical input power to the mechanism will be supplied by rotation of an input shaft.

The design should be simple, and use relatively few parts. The pump will need to operate for many cycles and should not wear out.

Design Goal Statement

Design a machine to pump fluid from an inlet port to an outlet port, making use of rotational power from an input shaft. The machine must move a specific amount of fluid in terms of volume per second. The fluid movement must occur in one direction only, and no fluid is permitted to leak backward through the pump, whether it is on or off. The design must be simple and must use a minimum number of parts. The design must employ components, materials, and features to allow for low-friction operation and must last for millions of rotation cycles.

Proposed Design Objectives

Design objectives for a positive-displacement pump are listed in Table 13.13.1.

Proposed Design Solution (Crescent Pump)

One possible design solution is a *crescent pump*, as illustrated in Figure 13.43. The design works by creating fluid pockets that expand to draw fluid in and that collapse to expel fluid out. The pockets are formed in between two gears in mesh, where one is a *ring gear* and the other is a *pinion gear*. Between the ring gear and the pinion gear is a *crescent-shaped feature* that forms a tight seal with the tips of the gear teeth. The pinion gear is attached to the *input shaft* and turns the ring gear, which is free to rotate within the pump housing. Fluid enters the pockets as they expand from the inlet port on the front face, and it exits the pockets as they collapse from the outlet port. A 3D animation of the crescent pump in operation is provided in [**Video 13.43**].

Video 13.43
Crescent Pump

Proposed Design Solution (Radial-Vane Pump)

Another possible design solution is a *radial-vane pump*, as illustrated in Figure 13.44. As in the previous example, this design creates pockets that expand and collapse, drawing fluid in and expelling fluid, respectively. The pockets in this design are formed between a central *rotor* and the pump *housing*. The rotor rotates within the housing and has several spring-loaded *vanes* that radiate outward to press against the housing. As the rotor turns, the expanding pocket draws fluid in from the left, and the collapsing pocket expels fluid on the right. An animation of the radial-vane pump in operation is provided in [**Video 13.44**].

Video 13.44
Radial-Vane Pump

TABLE 13.13.1 DESIGN OBJECTIVES FOR A POSITIVE-DISPLACEMENT PUMP

Design Objective	Description	Value (Units)
Input power	Energy/second supplied to pump via input shaft	1000 (watts)
Flow rate	Volume/second of fluid pumped from input port to output port	500 (ml/sec)
Allowable fluids	Types of fluid handled that can be pumped	Water-based and oil-based liquids
Number of parts	Minimize the number of parts in mechanism	<15 parts
Operating speed	Medium-speed operation	0–300 (rpm)
Operating life	Total number of cycles of operation before repair or replacement	10 million (cycles)
Minimize friction	Employ elements or materials to reduce friction	Plastic and metal parts

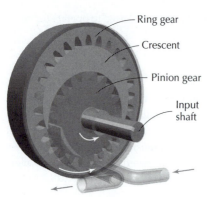

Ring gear

Crescent

Pinion gear

Input shaft

Figure 13.43 Positive-displacement pump incorporating gears and crescent [Video 13.43].

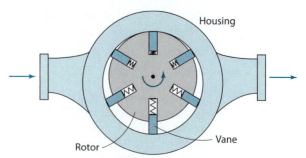

Housing

Rotor

Vane

Figure 13.44 Positive-displacement pump incorporating radial vanes [Video 13.44].

Summary and Suggestions

Two different positive-displacement pump designs are presented, that each meet the design objectives of this problem. They are able to pump fluid from an inlet port to an outlet port with a specific amount of volume per second. As evidenced by their design, fluid can only flow in one direction and cannot leak backwards. Care must be taken when using positive-displacement pumps, since they can generate very high fluid pressure on the outlet side if a downstream valve is closed. The crescent pump is a simpler design requiring very few parts, but the seal between the gear tips and the crescent feature is difficult to maintain and is subject to wear. The radial-vane pump is more complex due to many parts, but forms a better seal with the housing and is less affected by wear. Another form of positive-displacement pump employing cycloidal gear teeth is presented in Section 5.12.

13.13.2 Swash-Plate Compressor

Problem Description

A machine is needed to pump refrigerant fluid from an inlet port to an outlet port. The pumping action should apply a specific amount of pressure to the fluid, and pump a specific amount of fluid per second. All fluid entering the pump should move through it in one direction only. There should be no leakage of fluid backward through the pump when it is on or off. Mechanical input power to the mechanism will be supplied by rotation of an input shaft. The design should be reliable and will need to operate for many cycles, before repair or replacement.

Design Goal Statement

Design a machine to pump refrigerant fluid from an inlet port to an outlet port, making use of rotational power from an input shaft. The machine must move a specific volume of fluid per second, and add a specific amount of pressure to the fluid. The fluid movement must occur in one direction only, and no fluid is permitted to leak backward through the pump, whether it is on or off. The machine must employ components, materials, and features to allow for low-friction operation and must last for many millions of rotation cycles.

Proposed Design Objectives

Design objectives for a swash-plate compressor are listed in Table 13.13.2.

Proposed Design Solution

One possible design solution is an *swash-plate compressor*, as shown in Figure 13.45 and illustrated in Figure 13.46. The design works by using piston-based motion to create fluid pockets within the cylinders. Each of the *pistons* is attached to a *swash plate* (follower) via a spherical joint. The swash plate is tilted with respect to the *housing*, and it rests against an *angled plate*. The angled plate is

TABLE 13.13.2 DESIGN OBJECTIVES FOR A SWASH-PLATE COMPRESSOR

Design Objective	Description	Value (Units)
Input power	Energy/second supplied to pump via input shaft	2000 (watts)
Flow rate	Volume/second of fluid pumped from input port to output port	500 (ml/sec)
Allowable fluids	Types of fluid handled that can be pumped	Refrigerant
Operating speed	High-speed operation	600 (rpm)
Operating life	Total number of cycles of operation before repair or replacement	100 million (cycles)
Minimize friction	Employ elements or materials to reduce friction	Metal parts

attached to the *input shaft*. As the input shaft rotates the angled plate, the swash plate undergoes a cyclic tilting motion, causing the five pistons to sequentially undergo linear reciprocating motion within each of their respective cylinders. As the pistons compress, the fluid is expelled, and when they expand, the fluid is drawn in. *Flapper valves* allow for the expulsion of the compressed fluid and prevent back-flow of fluid. A 3D animation of the compressor in operation is provided in [**Video 13.46**].

Video 13.46
Swash Plate
Compressor

Summary and Suggestions

The compressor mechanism presented here meets the design objectives of this problem. It is able to pump refrigerant fluid from an inlet port to an outlet port with a specific amount of pressure and flow rate. It is typically used in automotive air-conditioning applications and draws its power from an input shaft that is connected to the serpentine belt of an automobile. It can operate with high-speed shaft rotation, and it has a long life on the order of 100 million cycles.

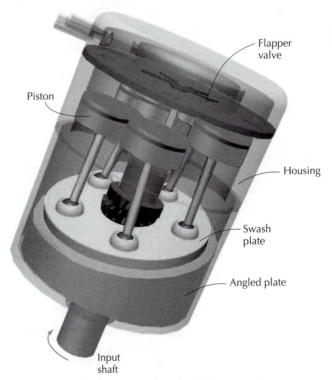

Figure 13.46 Swash-plate compressor [Video 13.46].

Figure 13.45 Swash-plate compressor for air-conditioning application.

An interesting variation of this design employs a *push-plate* mechanism that replaces the angled plate. The push plate is designed to allow for a variable tilt angle. By varying the push-plate angle as it rotates, the amplitude of the swash plate oscillation can be changed. This oscillation amplitude is directly related to the displacement stroke of the pistons, and hence the volume of fluid pumped. When the push plate is perpendicular to the housing, there is zero pumping volume and when the push plate is at its maximum angle, there is maximum pumping volume.

13.13.3 Fluid Dispenser Pump

Problem Description

A device is needed to pump fluid from an inlet port to an outlet port, using human power to supply mechanical energy. The pumping action should allow for movement of specific amounts of fluid per cycle, where fluid is pumped on an intermittent basis. It should accommodate various types of fluid ranging from medium viscosity liquid soap, to high viscosity cream or cosmetic lotion. The design should be simple, and use relatively few parts. Low cost is important, therefore appropriate materials and features should be employed in the design. The pump will need to operate for several hundred cycles in its useful life.

Design Goal Statement

Design a mechanism to pump fluid from an inlet port to an outlet port, making use of human hand power. The mechanism will operate intermittently, only when used by the person. The mechanism must pump fluids with medium to high viscosity, such as liquid soaps or cosmetic lotions. It must pump a specific amount of fluid per cycle. The mechanism must be low cost and use a minimum number of parts. The mechanism only needs to last 1000 cycles during its useful life.

Proposed Design Objectives

Design objectives for a fluid dispenser pump are listed in Table 13.13.3.

Proposed Design Solution

One possible design solution is a *fluid dispenser pump*, as illustrated in Figures 13.47. The design works by using two one-way valves with a small fluid *chamber* between them. Each one-way valve consists of a *ball* and *cone*, and works as follows: When fluid flows in the direction of the narrowing cone, the ball is carried with the fluid and wedges into the cone, thereby blocking further flow of fluid. When fluid flows in the direction of the widening cone, it pushes the ball out of the way and fluid flows around the ball. At the top of the pump is the outlet port within the *push top* structure.

TABLE 13.13.3 DESIGN OBJECTIVES FOR A FLUID DISPENSER PUMP

Design Objective	Description	Value (Units)
Human input power	Energy/second supplied to pump via human power	1.0 (watts)
Intermittent operation	Pump is used for short periods	1–2 (sec)
Flow volume per cycle	Volume/cycle of fluid pumped from input port to output port	5 (ml/cycle)
Allowable fluids	Types of fluid handled that can be pumped	Medium to high viscosity fluids
Number of parts	Minimize the number of parts in mechanism	< 10 parts
Cost	Cost per unit, when mass produced	< 0.50 (s)
Operating life	Total number of operating cycles	1000 (cycles)

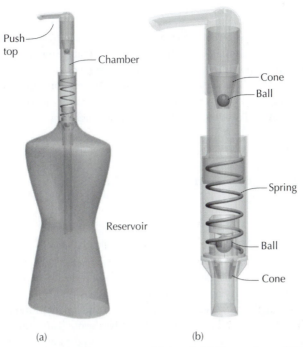

Push top — Chamber

Cone
Ball

Spring

Ball

Cone

(a)   (b)

Figure 13.47 Fluid dispenser pump: (a) Overall, (b) Closeup of pump [Video 13.47].

When the push top is depressed by the human hand, the chamber volume is decreased and fluid flows out from the upper one-way valve. When the push-top is released, an internal compression *spring* expands the chamber volume, and draws fluid in from the *reservoir* via the lower one-way valve. Due to the orientation of the one-way valves, fluid cannot flow from outlet to chamber and cannot flow from chamber to reservoir. An animation of the fluid dispenser in operation is provided in [**Video 13.47**].

Summary and Suggestions

The fluid dispenser pump mechanism presented here meets the design objectives of this problem. It is able to pump medium to high viscosity fluids that reside in a reservoir to an outlet port. It is powered by human action, by pressing down on a pushtop, which contracts a small chamber. The volume of fluid delivered per pump cycle is equal to the contraction volume of the chamber. This design is widely employed in disposable liquid soap dispensers, some shampoo and conditioner dispensers, and other cosmetics dispensers. Because it is a disposable product, it is made with inexpensive plastic materials, where long service life is not needed. As such, it is produced at low cost, to satisfy the applications that it serves.

13.14 VARIABLE OUTPUT MECHANISMS

13.14.1 Zero-Max Drive

Problem Description

Rotational motion and torque must be transmitted from an input shaft to an output shaft. The input rotational speed and torque will be constant, however, there should be a way to adjust the output rotational speed. The output rotational speed should vary between zero and some maximum speed. The adjustment method to vary the speed should be done by a human operator. The arrangement of the shafts is important, where the input and output shafts should be parallel, although they may be offset by some distance. This application needs to have good efficiency to minimize power transmission loss. Also, the transmission of high torques is important.

Design Goal Statement

Design a machine to transmit rotational motion and torque from an input shaft to an output shaft. The machine must have an adjustment mechanism, such that the output shaft speed can be adjusted from zero to a maximum speed. The adjustment is to be done by a human operator, while the machine is operating. The input and output shafts of the machine must be parallel. The machine must employ components, materials, and features to minimize frictional loss to allow for efficient conversion of rotational motion, and ability to transfer high torques.

Proposed Design Objectives

Design objectives for zero-max drive are listed in Table 13.14.10.

Proposed Design Solution

One possible design solution is a *Zero-Max drive*, as illustrated in Figure 13.48(a). It can transmit rotational motion and torque from an *input shaft* to an *output shaft*, where the output speed can be varied. The mechanism is based on the principle of a rocker–rocker four-bar mechanism as described in Section 1.7.1. Figure 13.48(b) shows a skeleton diagram of a single rocker–rocker arrangement, where the rockers correspond to link 4 and link 6. Motion is transmitted from the input shaft through eccentric link 2 and link 3, to drive the oscillatory motion of rocker link 4. Each receiving rocker link 6 is connected to a *one-way clutch* on the output shaft. Each linkage provides intermittent motion to the output shaft in a sequence to ensure the output motion is continuous. Rotational inertia of the output shaft permits the one-way clutches to overrun the driving links and smooth out pulsations imparted by each of the mechanisms. To vary the output speed, the amount of rocking motion can be adjusted by changing the location of the *control lever*. The location of the control lever corresponds to the location of base pivot O_4, which is adjustable. If it is moved along the track illustrated by the dotted line, then link 6 oscillates through a smaller angle for each input rotation, and the speed ratio between the input and output is increased. When the control lever is moved to O_4', the output shaft remains stationary. An animation of this mechanism in operation is provided in [**Video 13.48**].

Video 13.48
Zero-Max Drive

Summary and Suggestions

The Zero-Max mechanism provides for a variable speed drive that meets the design objectives presented. It makes use of a series of rocker–rocker four-bar mechanisms, along with one-way clutches to transfer motion from an input shaft to an output shaft. Speed adjustment can be manually made by moving the control lever, where it can be fixed into position until the next adjustment is required. It is a simple design that can be implemented in a number of applications that require variable speed output, such as sewing machines, conveyor belts, and other production line equipment.

TABLE 13.14.1 DESIGN OBJECTIVES FOR ZERO-MAX DRIVE

Design Objective	Description	Value (Units)
Output torque	Must transmit a specified torque	20 (N-m)
Input speed	Maximum input speed	1800 (rpm)
Variable output speed	Output speed must vary between zero and maximum speed	0 to max speed (rpm)
Shaft alignment	Input and output shafts must be parallel	Parallel
Operating life	Total number of cycles of operation before repair or replacement	100 million (cycles)
Minimize friction	Employ elements or materials to reduce friction	Metal parts

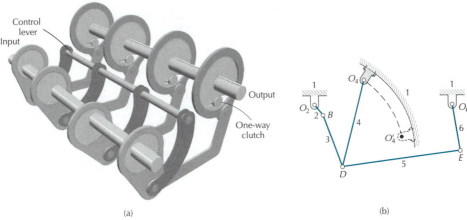

(a)   (b)

Figure 13.48 Zero-max drive [Video 13.48].

13.14.2 Helicopter Rotor Blade Adjustment Mechanism

Problem Description

The rotor blades of a helicopter are needed to provide lift for the aircraft. Also, those rotor blades should provide some amount of horizontal thrust to enable horizontal motion of the helicopter, allowing it to fly along a flight path. Since the flight path and flight conditions are variable, a mechanism is needed to allow the pilot to vary the amount of lift produced by the rotor blades. Also, a mechanism is needed to vary the horizontal thrust produced by the rotor blades. The lift produced by a rotor blade is a function of its aerodynamic shape and angle of attack. The desired mechanism must allow the rotor blades to vary their angle of attack while they spin, to either change the lift or operate in such a way that horizontal thrust can be achieved, or do both simultaneously. This application requires a strong and lightweight mechanical design to achieve the desired operation. It should minimize frictional losses and be able to transmit high forces and torques to the rotor blades.

Design Goal Statement

Design a mechanism to vary the lift produced by helicopter rotor blades by allowing for adjustment of the angle of attack of the rotor blades. The adjustment will be controlled by the pilot at any time while the rotor is spinning. Two types of adjustment are required, one for variable lift and another for horizontal thrust. The mechanism must employ components, materials, and features to produce a strong and lightweight design. Additionally, it must minimize frictional losses and allow for transmission of high forces and torques to the rotor blades.

Proposed Design Objectives

Design objectives for rotor blade adjustment mechanism are listed in Table 13.14.2

Proposed Design Solution

One possible design solution for a *rotor-blade adjustment mechanism* is to use a *swash plate* mechanism to enable a variation in the *angle of attack* of the *main rotor blades*. Figure 13.49(a) illustrates the main rotor blades and *tail rotor blades* of a helicopter. Figure 13.49(b) shows a close-up of the main rotor blades, where the cross section of each blade has the shape of an airfoil, as illustrated in Figure 13.50. Motor torque is supplied to the *driveshaft*, which is connected to the *main rotor hub*, causing it to spin. Four rotor blades are each connected to the rotor hub, and they spin about the axis of the driveshaft to produce the lift forces required to fly the helicopter. The pilot can regulate the amount of lift force by adjusting the angle of attack of the rotor blades. This is done by pivoting the rotor blades with respect to the rotor hub via the *pitch control rods*. Figure 13.49(b) illustrates the components used to change the angle of attack. The mechanism consists of two concentric

TABLE 13.14.2 DESIGN OBJECTIVES FOR ROTOR BLADE ADJUSTMENT MECHANISM

Design Objective	Description	Value (Units)
Rotational speed of rotor	Must transmit a specified torque	0–120 (rpm)
Variable angle of attack	Desired range for angle of attack of rotor blades	0 to 20 (degrees)
Strength	Lift forces exerted on adjustment mechanism	20,000(N)
Weight	Total weight of adjust mechanism	50 (kg)
Operating life	Number of rotor hub rotations before repair or replacement	10 million (cycles)
Minimize friction	Employ elements or materials to reduce friction	Metal parts

Video 13.49
Helicopter
Rotor-Blade
Adjustment
Mechanism

swash plates separated by a set of ball bearings. These components are (a) the *rotating swash plate* that rotates with the rotor hub and (b) the *oscillating swash plate* that does not rotate, but only tilts with respect to the horizontal plane. Four nonrotating *hydraulic actuators* are connected to the oscillating swash plate, and they can move it vertically or tilt it from the horizontal plane in any desired direction. The *scissor links* constrain the rotating swash plate to turn with the driveshaft and blades. The ball bearings allow the rotating swash plate to revolve with respect to the oscillating swash plate, yet transmit vertical forces. Each pitch control rod is pin-jointed to the rotating swash plate and to a blade. Understanding the operation of this mechanism is best done by viewing the animation provided in [**Video 13.49**].

Motion of the oscillating swash plate can be controlled in two basic manners: *collective control* and *cyclic control*. In collective control, as shown in Figure 13.49(c), the oscillating swash plate is driven vertically by all four hydraulic actuators, and in turn the pitch control rods alter the angle of attack of all blades simultaneously by the same amount. The lift forces generated by all blades are equal, driving the helicopter in a vertical direction. In cyclic control, as shown in Figure 13.49(d), the vertical positions of the hydraulic actuators are not equal, which results in a tilting of the oscillating swash plate. Here, the angle of attack of every blade varies periodically during each revolution of the driveshaft. This results in unequal amounts of lift force on opposite sides of the driveshaft, along with thrusting of the helicopter in a horizontal direction. The pilot employs cyclic control to fly the helicopter in a desired horizontal direction by controlling the necessary hydraulic actuators to tilt the swash plate in that corresponding direction.

Summary and Suggestions

The rotor blade adjustment mechanism employs swash plates and hydraulic actuators to vary the angle of attack of helicopter rotor blades. In this way, the amount of lift force and horizontal thrust of a helicopter can be varied, thereby meeting the design objectives presented. The pilot flies the helicopter by manoeuvring the controls, which are translated into linear motion of the actuators to produce collective or cyclic control.

13.15 TOGGLE MECHANISMS

13.15.1 Lever Toggle Mechanisms

Problem Description

A mechanism that can be switched between two different mechanical states is needed. Each mechanical state corresponds to a specific configuration and orientation of the mechanism links. The mechanism should be mechanically stable in either state, even when subjected to disturbances such as small forces or vibration. A specific amount of input force must be applied to switch the

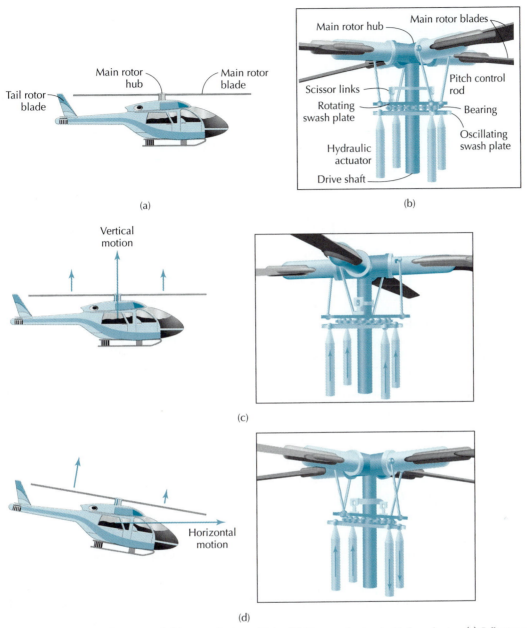

Figure 13.49 Helicopter [Video 13.49]: (a) Main and tail rotor blades, (b) Close-up of main rotor-blade mechanism, (c) Collective control for vertical thrust, (d) Cyclic control for horizontal thrust.

mechanism from one state to the other. The switching process should work equally well in either direction, and it is to be done by manual input. The mechanism should be able to switch back and forth many thousands of times before failure.

Design Goal Statement

Design a mechanism that can be switched between two mechanical states, by the application of a specific amount of human input force. Each mechanical state must be stable and resistant to small force or vibrational disturbances. The mechanism must be able to switch back and forth between states with the same amplitude of input force. The mechanism must function for 100,000 switch cycles before failure.

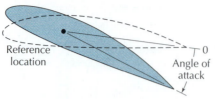

Figure 13.50 Angle of attack of a helicopter blade.

TABLE 13.15.1 DESIGN OBJECTIVES FOR TOGGLE MECHANISMS		
Design Objective	**Description**	**Value (Units)**
Mechanical state A	Defined by angle of lever linkage	60 (degrees)
Mechanical state B	Defined by angle of lever linkage	120 (degrees)
Input force to change states	Amplitude of force needed to change between states A and B, or vice versa	10 (N)
Stability	Amount of force tolerated before significant motion of lever away from a stable position	< 2 (N)
Vibration	Amount of vibration tolerated before significant motion of lever away from a stable position	< 30 (m/sec^2)
Operating life	Total number of cycles of operation before failure	100,000 (cycles)

Proposed Design Objectives

Design objectives for toggle mechanisms are listed in Table 13.15.1.

Proposed Design Solution (Simple Toggle Mechanism)

One possible design solution is a *simple toggle mechanism*, as illustrated in Figure 13.51(a). The design consists of a *lever* linkage that is held in one of two possible stable positions by use of an extension *spring*. The spring is connected to the ground (link 1) at one end and is connected to the lever linkage at the other end. In the position shown, the spring is at a minimum extension with the lever resting against a *stop*. As input force F is applied to the lever, the spring begins to extend as the lever rotates. Eventually, with sufficient amplitude of force F, the lever crosses an inflection point where the spring is at its maximum extension. Further lever rotation will cause the spring to contract again, which will automatically snap the lever onto the other stop. An animation of the simple toggle in operation is provided in [**Video 13.51A**]. A disadvantage of this mechanism is that it may be possible to set the lever in a balanced vertical configuration between the two stops.

Video 13.51A
Simple Toggle
Mechanism

Proposed Design Solution (Double Toggle Mechanism)

Another possible design solution is a *double toggle mechanism*, as illustrated in Figure 13.51(b). As in the previous example, this design has two stable positions for the *lever* linkage, using an extension spring to apply tension to maintain either position. In this mechanism, an additional short lever arm, link 3, has been added. The extension spring is connected to the lever at one end and to link 3 at the other end. This arrangement of links reduces the possibility of balancing the lever between the stops as with a simple toggle mechanism. Here, link 3 can only be in equilibrium when it is in contact with either of its two stops. The movement of link 3 lags behind that of link 2, and therefore link 2 must be moved beyond the vertical position before link 3 will snap through from one stop to the other. An animation of the double toggle in operation is provided in [**Video 13.51B**].

Video 13.51B
Double Toggle
Mechanism

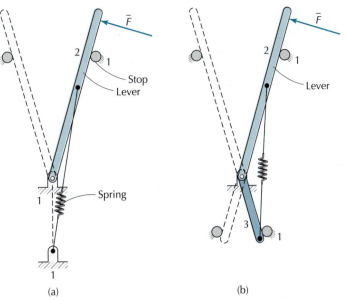

Figure 13.51 Toggle mechanisms: (a) Simple toggle [Video 13.51A], (b) Double toggle [Video 13.51B].

Summary and Suggestions

Two different toggle mechanism designs are presented, that each meet the design objectives of this problem. The designs employ a lever mechanism, where the lever can be switched between two mechanical states, by application of a specific input force. By designing the position of the stops and the selection of extension springs, various toggle mechanisms can be developed with different swing angles and desired force for switching. These types of mechanisms are commonly employed for the mechanical linkages of electric switches. For example, a toggle mechanism used within an electric switch is presented in Section 7.1 and is shown in Figures 7.4 and 7.5. Another example of a toggle mechanism can be found within a door latch design, which is presented Section 2.14 and shown in Figures 2.47 and 2.48.

13.15.2 Ring Binder Mechanism

Problem Description

A device is needed to hold many sheets of paper together via holes in the paper. It should be easy to add paper sheets or to remove them from the device. The device should be able to switch between two different mechanical states. In one mechanical state, it should secure the pages together. In the other mechanical state, it should open in some way so that paper can be added or removed. The device should be mechanically stable in either state, and it should be resistant to force or vibration disturbances. It should be designed so that a user can easily apply finger force to switch the mechanism between the two states, and it should be able to switch back and forth thousands of times before failure.

Design Goal Statement

Design a mechanism to hold many sheets of paper, where sheets can easily be added or removed. The mechanism must switch between two mechanical states, by the application of finger input force. One mechanical state will correspond to adding or removing pages, while the other state will correspond to securing the pages together. It must be stable and resistant to small force or vibrational disturbances. The mechanism must function for 10,000 switch cycles before failure.

Proposed Design Objectives

Design objectives for a ring binder mechanism are listed in Table 13.15.2.

TABLE 13.15.2. DESIGN OBJECTIVES FOR A RING BINDER MECHANISM

Design Objective	Description	Value (Units)
Mechanical state A	Defined by open state of mechanism allowing pages to be added or removed	Open
Mechanical state B	Defined by closed state of mechanism thereby securing pages together	Closed
Input force to change states	Amplitude of force needed to change between states	5 (N)
Stability	Amount of force tolerated before significant motion of lever away from a stable position	< 2 (N)
Operating life	Total number of cycles of operation before failure	10,000 (cycles)

Proposed Design Solution

One possible design solution is a *ring binder mechanism*, as illustrated in Figure 13.52. The design consists of two *ring clips* that form a circular loop. In the design shown, three such loops are employed. Paper sheets to be used with this mechanism must be made with three holes to match the spacing between the three loops. Figure 13.52(b) shows a close-up view of the mechanism in the open position, while Figure 13.52(c) shows a close-up of the mechanism in the closed position. The toggle action is achieved by the relative position of two *leaf springs* at the center of the mechanism. Three ring clips are rigidly attached to one leaf spring on one side, and three are attached to the leaf spring on the other side. In the closed state, the leaf springs press against each other to form a stable v-shape that is concave up. In the open state, the v-shape is concave down. As the mechanism is switched from one state to the other, the leaf springs pass through transition position where both their planes are parallel. A *tab* linkage is used to apply force at the center between the two leaf springs, to push them from one state to the other. An animation of the ring binder mechanism in operation is provided in [**Video 13.52**].

Video 13.52
Ring Binder
Mechanism

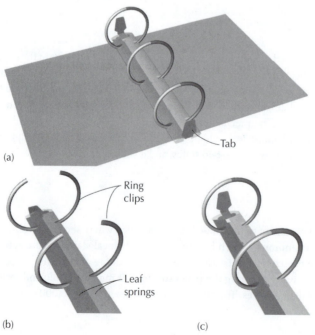

(a)

Tab

Ring clips

Leaf springs

(b)   (c)

Figure 13.52 Ring binder mechanism [Video 13.52]: (a) Overall, (b) Closeup of open position, (c) Closeup of closed position.

Summary and Suggestions

The ring binder mechanism is able to switch between two stable mechanical states, to hold its ring clips in either the open or closed position. This allows paper sheets with appropriate holes to be inserted into, or removed from, the ring clips while it is in the open position. The design employs two leaf springs that press against each other, in either one of two stable v-shaped positions. Transition force is supplied by fingers pressing on the tabs. In this way, the mechanism meets the design objectives of this problem. Binder mechanisms also have wide application in holding sample sheets of fabrics, leather, laminates, or other thin materials, which is convenient for storing sample materials in the trades industries.

13.16 LOCKING MECHANISMS

13.16.1 Key Lock

Problem Description

A mechanism is needed to switch between two different mechanical states. It can only switch between states by the use of an additional mechanical element with a special shape. The specially shaped mechanical element should interface with the main mechanism in some way, which cannot change states without it. The mechanical element should be easily added to, or removed from, the main mechanism and be portable. The two states of the main mechanism should be the engaged or disengaged position of a lever. The lever is to be used for a locking mechanism. The lock should be formed by rotating the lever from one position into another, where the lever enters an opening in a rigid structure, such as a door frame. The mechanism should be designed so that a user can easily apply hand force to add or remove the mechanical element, or to switch the mechanism between the two states.

Design Goal Statement

Design a locking mechanism which can switch between two mechanical states, through the use of a removable mechanical element. The mechanical element must have a special shape that interfaces with the main mechanism, to enable the main mechanism to switch between states. The main mechanism must not switch between states without the removable mechanical element. The mechanism will be used for a lock application, where it must rotate a lever to engage or disengage with an opening in a door frame. The mechanical element should be easily added to, or removed from, the main mechanism with hand force. Also, the main mechanism will be actuated to switch between states with hand force.

Proposed Design Objectives

Design objectives for a key lock mechanism are listed in Table 13.16.1.

Proposed Design Solution

One possible design solution is a *key lock* mechanism, also known as a *pin tumbler lock* [44], as illustrated in Figure 13.53. The design consists of a specially shaped *key* element, which can be inserted into a *keyway* that exists within a *plug*. The plug is rigidly attached to a *lever*, and it rotates within the *housing*. The housing remains stationary. The lever provides the locking action by rotationally engaging with another opening (not shown) in a door frame. The operation of the mechanism is best understood by viewing the animation provided in [**Video 13.53**]. As the animation proceeds, the housing becomes translucent, which reveals a set of *pins* of varying lengths, shown in Figure 13.53(b). These pins are constrained to slide in *pin chambers* within the housing and plug. While in a locked configuration, the pins are arranged in pairs, placed end to end, and are forced downward by compressed *cylinder springs* (shown in Figure 13.53(b) but not part of

TABLE 13.16.1 DESIGN OBJECTIVES FOR A KEY LOCK MECHANISM

Design Objective	Description	Value (Units)
Unlocked state	Open (unlocked) state of mechanism with lever disengaged from opening in door frame	Open
Locked state	Closed (locked) state of mechanism with lever engaged into opening in door frame	Closed
Size of element	Max. length of special shape mechanical element	5 (cm)
Force to add or remove element	Amplitude of force needed to add or remove specially shaped mechanical element	4 (N)
Torque to rotate	Torque needed to rotate lever from engaged to disengaged position.	0.2 (N-m)

the animation). The plug cannot be rotated unless all dividing surfaces between pairs of pins are in alignment with the cylindrical face of the plug. Since the plug is rigidly attached to the lever, the lever cannot rotate and thereby keeps the mechanism locked.

To unlock the mechanism, a key must be inserted into the keyway. As the tip of the key is inserted, the pins are moved in their axial direction. Even when the key is partially inserted (Figure 13.53(b)), the pin ends are not aligned with the cylinder face, thereby preventing the plug from rotating. Only when the key is fully inserted (Figure 13.53(c)), where the shoulder of the key is pushed against the end of the plug, are all dividing surfaces even with the face of the plug. Now the head of the key can then be rotated, along with the plug and lever, as shown in Figure 13.53(d).

Summary and Suggestions

The key lock mechanism is able to switch between locked and unlocked mechanical states, by the use of a key that interfaces with the mechanism. The key has a special shape that allows for

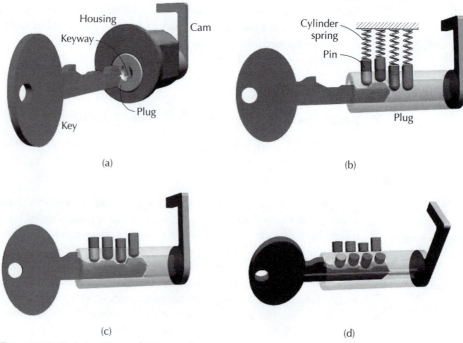

Figure 13.53 Key lock mechanism [Video 13.53].

alignment of internal pins within the mechanism, permitting the plug to rotate within the housing. Without alignment of these internal pins to the plug surface, the plug cannot turn, which keeps it in either the locked or unlocked state. Additionally, the torque to turn the mechanism is transmitted through the key and is supplied by hand force. In this way, the mechanism meets the design objectives of this problem. Key locks have wide application for use in locking doors for homes, automobiles, cabinets, and displays. In this way, only persons who have possession of the key can lock or unlock doors. One disadvantage is that if the key is lost, the mechanism cannot be locked or unlocked, which can cause major inconvenience.

13.16.2 Combination Lock

Problem Description

A mechanism is needed to switch between two different mechanical states and to serve as a lever-based lock. It can only switch between states by the correct manual alignment of its internal linkages. Internal linkage alignment should be achieved by rotational movement, supplied by the human hand. However, the alignment orientations of those internal linkages should be hidden from view for security to prevent unwanted access. To achieve this security, it is desired that a sequence of successive, alternating, mechanical rotations be used to achieve complete internal alignment. Numeric marks on the outer surface of the mechanism should be present to help the user achieve alignment. A specific numeric sequence should correspond to the correct internal alignment, for each successive rotation. The correct sequence of rotational alignment should not be obvious. The two states of the mechanism should correspond to the engaged or disengaged position of a lever, which is to be used for the locking action.

Design Goal Statement

Design a locking mechanism which can switch between two mechanical states, through the alignment of internal linkages. The alignment will be done manually, through a series of successive, alternating rotations. Numeric alignment marks must be provided to assist the user in aligning the mechanism. The internal alignment configuration must not be visible, and the combination of rotations must not be obvious to users who do not know the combination sequence. The mechanism should have a smooth action, since it will be actuated with hand force.

Proposed Design Objectives

Design objectives for a combination lock are listed in Table 13.16.2.

Proposed Design Solution

One possible design solution is a *combination lock*, as illustrated in Figure 13.54. The design consists of a *dial* that is manually rotated, relative to a stationary concentric *ring*. Index marks are inscribed on the periphery of the dial, and are used for aligning numbers (not shown) of the combination, with the correct internal orientation of the *wheels*. The design illustrated here requires that three alternate rotations be made, in proper sequence, for the lock to be opened. The dial is rigidly connected to wheel 1 through a central *spindle* shaft. Wheels 2, 3, and 4 are free to rotate with respect to the spindle. Rotational motion can only be transmitted from one wheel to the next wheel through *extension arms*, which engage tabs located on the back sides of the wheels, as shown in Figure 13.54(b). In this way, rotation can be transmitted from wheel 1 to wheel 2, then from wheel 2 to wheel 3, and finally from wheel 3 to wheel 4, in a cascade fashion. Each wheel has a small cut-out in its periphery, referred to as a *gate*. The lock can be opened only when all of the gates are aligned below the *side bar*. The side bar is rigidly connected to a lever called the *fence*, which in turn is pin-jointed to the *bolt*. It is the bolt that provides locking action by protruding into the recess of an adjacent component (not shown). The lock may be opened by retracting the bolt.

TABLE 13.16.2 DESIGN OBJECTIVES FOR A COMBINATION LOCK

Design Objective	Description	Value (Units)
Unlocked state	Open (unlocked) state of mechanism with lever disengaged with latch	Open
Locked state	Closed (locked) state of mechanism with lever engaged with latch	Closed
Rotational range	Rotational range for the purposes of alignment	360 (degrees)
Number of internal link alignments	Number of internal linkages that need to be aligned in order to lock or unlock mechanism	Three
Torque to rotate linkages	Torque needed to rotate linkages as they are aligned within mechanism	0.1 (N-m)

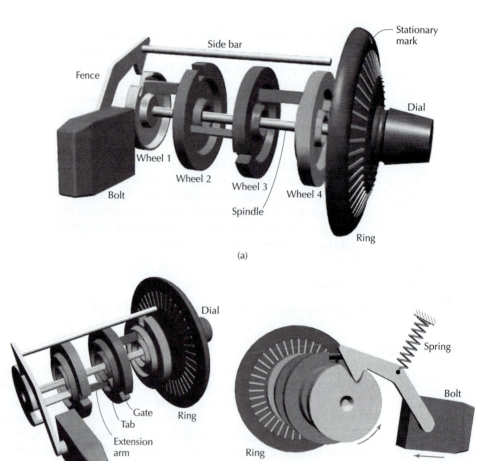

Figure 13.54 Combination lock mechanism [Video 13.54].

The operation of the lock can be better understood by viewing the animation provided in [**Video 13.54**]. The animation begins where the gates of each wheel are at various orientations. If the dial is rotated at least three complete revolutions in one direction, then all extension arms will eventually contact a tab on an adjacent wheel, thereby causing all the wheels to move in unison. The first number of the combination is input by continuing to turn the dial (with all wheels in unison motion) until it is aligned with the *stationary mark* on the ring. This aligns the gate on wheel 4 under the side bar, as illustrated in Figure 13.54(b). Next, the dial is turned in the opposite direction until the extension arms on wheels 1 and 2 contact tabs on their adjacent wheels, resulting in wheel 1, 2, and 3 rotating in unison. This rotation continues until the second number of the combination is aligned with the stationary mark. This brings the gate on wheel 3 into alignment under the side bar. Next, the dial is rotated in the opposite direction until the extension arm on wheel 1 contacts the adjacent wheel. The third number of the combination is then aligned with the stationary mark, and the gate on wheel 2 is aligned under the side bar. Next, the gate on wheel 1 is aligned under the tab on the fence. The side bar and fence then move downward under action of the compression spring (shown in Figure 13.54(c), but not part of the animation). Finally, by providing a small turn of the dial, the bolt is retracted, as shown in Figure 13.54(c).

Summary and Suggestions

The combination lock provides a mechanism that can switch between the locked and unlocked mechanical states. The switch can only take place after a series of internal wheels are manually brought into alignment. When the wheels are correctly aligned, the mechanism is able to swing a lever (fence) and thereby displace a bolt linearly to achieve the unlocked state. In this way, the mechanism meets the design objectives of this problem. Combination locks are commonly employed in stationary applications like cabinets and safes for storage of valuables. Additionally, portable combination locks are frequently used for lockers at schools or for bicycle locks. The major feature of combination locks is that there is no physical key, but rather a numeric combination that can be remembered. Any person who is given the combination would be able to open the lock.

13.17 COMPLIANT MECHANISMS

A majority of the mechanisms presented in this textbook consist of multiple rigid bodies connected together with various joints, such as revolute joints, sliding joints, spherical joints, and so on. Those joints all consist of two or more rigid pieces that rotate, slide, or roll relative to one another. However, there is another category of mechanisms that employ one-piece flexible joints. These are referred to as *compliant mechanisms* [45] and take advantage of the inherent flexibility of materials to create relative motion between stiffer structures at either end of the inherent flexible joint. Compliant mechanisms are usually one-piece structures made of a single piece of material. They are designed to undergo elastic deflections when subjected to applied force. When the forces are removed, the mechanism returns to its original configuration.

A classic example of a compliant mechanism is the simple extension spring, as illustrated in Figure 13.55. An extension spring consists of a single piece of wire wound into a coil, with a loop at each end. By applying a force at each end, one end translates with respect to the other, and when the force is removed, the spring retracts to its original length. A key principle when using extension springs is to ensure that they are not overextended by application of excessive force. Excessive force will lead to plastic (permanent) deformation of the spring, which would typically ruin its performance. This principle also applies to complaint mechanisms in general.

Compliant mechanisms are widely used in many familiar applications. Figure 13.56 shows a snap-lock modular connector used on both ends of an Ethernet patch cord. When the modular connector is inserted into the mating slot of a receptacle (not shown), the *clip* is pushed toward the housing, due to elastic deformation. After fully entering the receptacle, the clip locks the modular connector in place. To remove the connector, the clip must be depressed with finger force. Figure 13.57 shows a DVD (digital video disc) jewel case. A typical case incorporates two compliant mechanisms. The first is the *lid release mechanism*, shown in Figure 13.57(a). In order to open the case, a person must use one hand to depress the lid release mechanism on either side of the

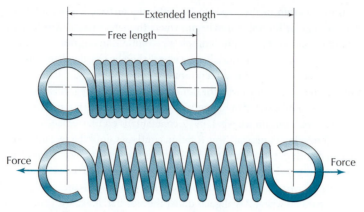

Figure 13.55 Extension spring.

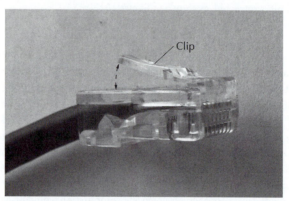

Figure 13.56 Ethernet patch cord with snap-lock modular connector.

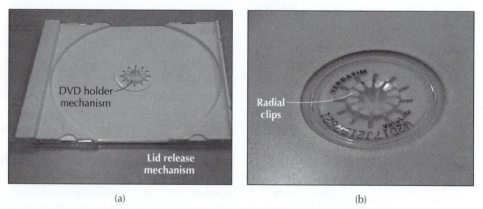

Figure 13.57 DVD (digital video disc) jewel case: (a) Lid release mechanism, (b) DVD holder mechanism.

case, while the other hand is used to lift the lid. The second compliant mechanism, illustrated in Figure 13.57(b), is the *DVD holder mechanism*. The holder is a set of compliant *radial clips* that deflect inward to clear the hole of a DVD as it is inserted and which then expand outward to hold onto the DVD. To release the DVD, the radial clips must be depressed by a finger, while a hand lifts on the DVD.

Some compliant mechanisms employ the toggle switch principle described in Section 13.15.1. In such a case, they do not return to their original position after removal of the applied forces, but rather deflect into one of two stable mechanical states. Figure 13.58 (a) shows a shampoo bottle cap that is made of a single piece of material. It incorporates a number of thin sections that can easily bend and act as *flexible joints*. In order to close the lid, it must be deflected past the transition configuration shown in Figure 13.58(b). If the applied force is removed before the lid reaches the transition configuration, it will return to its open position. If the applied force is removed after the lid passes the transition configuration, the lid will continue on its own to move toward the closed position. Similarly, to open the lid, a force must be applied beyond the transition configuration. Otherwise, it will return toward the closed position.

In some cases, it is possible to design an equivalent compliant mechanism, based on its corresponding classical link-and-joint mechanism. One such example is presented in Figure 13.59. A classical four-bar, rocker–rocker mechanism is shown in Figure 13.59(a), consisting of four rigid links and four turning pairs. The operation of this mechanism is described in Section 1.7.1. The corresponding version of the complaint mechanism is shown in Figure 13.59(b). Here, the turning pairs are replaced with thin sections of material that exhibit less mechanical stiffness than the corresponding thick sections of the links. Hence, these thin sections can easily flex and are labeled as *flexible joints*. The thick sections have much higher stiffness and act essentially as *rigid links*. Through careful design of the flexible joints and their "effective" center of rotation, this compliant mechanism will approximate the motion of the classical mechanism, over a specific range of motion. Beyond the specifically designed range of motion, the compliant mechanism may move irregularly or may be damaged due to excessive deformation.

Compliant mechanisms offer some interesting advantages, and have some disadvantages, in comparison to classical link-and-joint mechanisms. Generally, the benefits are as follows: Compliant mechanisms are usually comprised of a single monolithic piece of material. Therefore, they do not require lubricants at the joints, and no wear occurs at the joints. This allows for a long cycle life, when they are used within their designed elastic deformation limits. Additionally, since

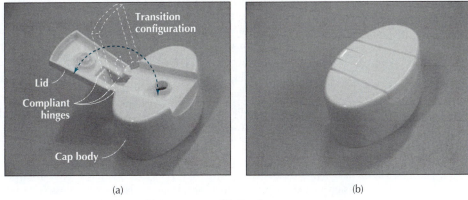

(a)   (b)

Figure 13.58 Shampoo bottle cap: (a) Open position, (b) Closed position.

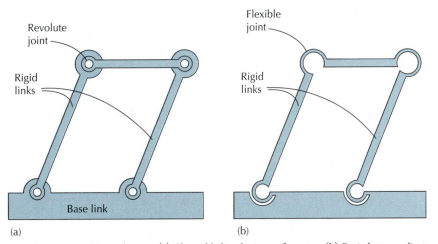

(a)   (b)

Figure 13.59 Four-Bar Mechanism: (a) Classical link-and-joint configuration, (b) Equivalent compliant mechanism configuration.

they are made from a single piece of material, they are easily fabricated with mass-production techniques such as injection molding or modern methods such as 3D printing. This results in low cost to manufacture. Compliant mechanisms have a built-in spring-back effect, where they will always (except for toggle mechanisms) return to their original unloaded configuration. This is useful for many mechanism applications that must reset to their initial position after use. Compliant mechanisms can create highly precise motion with certain materials, since their motion is based exclusively on the material shape, elasticity, and applied force. As such, compliant mechanism translation stages are often used in nanometer-scale motion applications.

Generally, the disadvantages are as follows: The same spring-back effect that is sometimes beneficial can be problematic when it is not desired for a mechanism to return to its initial position, such as maintaining a displaced position without application of force. Also, the spring-back effect requires additional input force to overcome when actuating the mechanism. Compliant mechanisms can be difficult to design, since prediction of the desired motion involves the interplay of material shape, elasticity, and location of applied force, among other aspects. This type of design generally requires application of solid mechanics principles, along with a computer-based computational technique known as Finite Element Analysis (FEA). Additionally, the application of excessive forces can damage compliant mechanisms, hence, they must be designed by considering the internal stresses at the flexible joints, and considerations involving fatigue life. This can be done with the aforementioned FEA technique.

Three design case studies of compliant mechanisms are presented next.

13.17.1 Cable Clip Mechanism

Problem Description

A device is needed to grasp a cable and secure it onto a flat surface such as a wall. The cable has an approximately round cross-section. Operation of the device should be reversible so that cables can be grasped or released. The device should be stable in either the grasp or release configuration and should be resistant to force disturbances. It should be easy and quick to use, where a user can apply finger force to switch between the two configurations. It is anticipated that millions of such devices will be produced; therefore, low cost and ease of mass production are of high importance. For service life, the device should be able to grasp or release the cable hundreds of times before failure.

Design Goal Statement

Design a mechanism to temporarily grasp a cable of round cross-section, onto a flat surface. The mechanism will have two states, where one corresponds to a secured cable state and the other to a released state. It should be mechanically stable in either state, and resistant to force disturbances. The device will be mass-produced with quantities in the millions, and must be low cost. Therefore, a compliant mechanism design approach will be employed. The design must be quick and simple to use with only finger force applied. The service life of the device must be at least 100 grasp and release cycles.

Proposed Design Objectives

Design objectives for a cable clip mechanism are listed in Table 13.17.1.

Proposed Design Solution

One possible design solution is a *cable clip* compliant mechanism, as illustrated in Figure 13.60. The design consists of two *clips*, which are hinged together at the bottom with a flexible *central joint*. Two additional *side joints* hold the clips from either side onto a *flexture link*, and they effectively create a toggle mechanism that has two states. The closed state is shown in Figure 13.60(a), and the open state is shown in Figure 13.60(b). The operation of this compliant mechanism can be better understood by viewing the animation provided in [**Video 13.60**]. In the open state, the two clips have a wide separation at the top, where a cable can be inserted. When a cable is inserted (not shown in the animation), it presses against the central joint, thereby causing both clips to close

Video 13.60
Cable Clip
Mechanism

TABLE 13.17.1 DESIGN OBJECTIVES FOR CABLE CLIP MECHANISM

Design Objective	Description	Value (Units)
Cable size range	Diameter range of cables to be grasped by device	6–12 (mm)
Overall size	Maximum size of device defined by length, width, and height	$25 \times 25 \times 12$ (mm) [L × W × H]
Monolithic design	Employ compliant mechanism design approach to create single piece device	1 piece
Material	Material suitable for compliant mechanism, such as low molecular weight plastic	Nylon or polypropylene
Quantity	Number of units to be manufactured	10,000,000
Cost/Unit	Cost per unit to manufacture including: material, machine time, and labour for production batch	0.25 ($/unit)
Force to actuate	Amount of force needed to open or close device	< 4 (N)
Service life	Total number of cycles of operation before failure	100 (cycles)

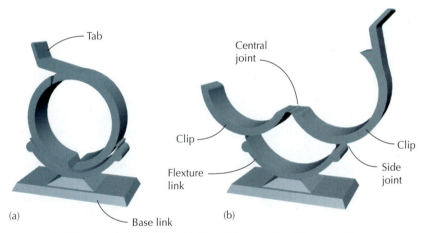

Figure 13.60 Cable clip mechanism [Video 13.60]: (a) Closed position, (b) open position.

around the cable. When the toggle mechanism has passed its transition point, it remains in a stable closed position, thereby securing the cable. To release the cable, a *tab* located on one clip is pushed sideways with finger force. This causes the tops of the clips to separate, and after a transition point, the clips swing open into a stable position. In the process, the cable is ejected from the clip. The entire device is secured to the wall by the *base link*, using an adhesive at the bottom of the base.

Summary and Suggestions

The cable clip compliant mechanism is able to clamp around a cable that is pushed into it. Since it operates like a toggle mechanism, it remains securely closed until such time that it is manually opened. It is fabricated from a single piece of material and employs flexible joints to allow for motion. Since it is one piece, it can be easily fabricated using the injection molding process, and it requires no further assembly steps. This results in an ability to mass produce it in the millions of units, using inexpensive materials such as nylon plastic. Given these features, it satisfies the design objectives of this problem. This mechanism undergoes a large amount of rotation at the flexible joints, in order to achieve the desired range of motion. As such, this will lead to material fatigue, which results in a low cycle life. Regardless, the design will perform effectively, since it will likely be opened and closed only several times during its service.

13.17.2 Compliant Stapler

Problem Description

A mechanism is needed for a medical application, to fasten together two pieces of skin using metal staples. The two pieces of skin are to be brought together, edge to edge, and connected along a straight or curved line. The mechanism should apply the staples from only one side (outer surface of skin) and cannot have any working parts under the skin. The design of the staples should also be considered a part of the project. Due to the medical application, sterility is highly important to prevent infection in patients. The mechanism and the metal staples should be made of nontoxic materials that can be sterilized before use. Additionally, it is desirable to use the mechanism on only one patient and then dispose of it after use. If it is to be used only once, it should be low cost.

Design Goal Statement

Design a medical-grade mechanism to fasten together two pieces of skin, using metal staples. The mechanism must apply the staples from only the outer surface of the skin, and it will have no working parts below the skin. The staples must also be designed to ensure compatibility with the mechanism and the application. The mechanism and staples must be made from materials that are nontoxic and that can be sterilized. The mechanism will only be used once and then disposed, therefore, it must be low cost. Given these requirements, a compliant mechanism approach will be employed.

Proposed Design Objectives

Design objectives for a compliant stapler are listed in Table 13.17.2.

Proposed Design Solution

One possible design solution is a *compliant stapler*, as illustrated in Figure 13.61. The design consists of a compliant mechanism that incorporates a set of metal staples. It is based on the principle of the slider crank mechanism, as described in Section 1.7.2. It can be gripped by one hand, where thumb force acts on the *lever* (crank), to cause it to rotate with respect to the *base link*. The mechanism employs three *flexible joints* and one *sliding joint* to achieve its function. The operation of the compliant stapler can best be understood by viewing the animation provided in [**Video 13.61**]. When the trigger is pressed, force is transferred into the *hammer* (slider), causing it to press the undeformed *staple* against the *crimp stop*, as shown in Figure 13.61(b). As the hammer displaces to its maximum

Video 13.61
Compliant Stapler

TABLE 13.17.2 DESIGN OBJECTIVES FOR COMPLIANT STAPLER

Design Objective	Description	Value (Units)
Staple size	Size of staples to be used, defined by length, width, and height	$12 \times 1 \times 6$ (mm) [$L \times W \times H$]
Staple material	Material for staples must be: strong for binding skin, nontoxic, nonirritating, and sterilizable	Stainless steel or titanium
Compliant design	Employ compliant mechanism design approach	–
Mechanism material	Material suitable for compliant mechanism, such as plastic. It must be nontoxic and sterilizable	Polycarbonate, polyester, or polypropylene
Cost/Unit	Cost per unit to manufacture including: material, machine time, and labor for production batch	< 10 ($/unit)
Force to actuate	Amount of force needed to deploy staple	< 3 (N)
Service life	Total number of cycles of operation before failure	100 (cycles)

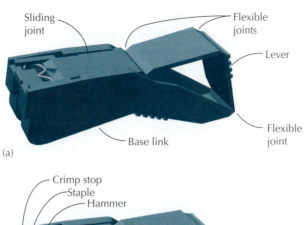

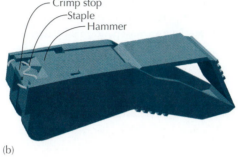

Figure 13.61 Compliant Stapler [Video 13.61]: (a) Initial open position, (b) With undeformed staple, (c) With deformed staple.

extension, it simultaneously forces the staple outward and plastically deforms the staple causing its outer tips to close inward, as shown in Figure 13.61(c). When applied against two adjacent pieces of skin, this closing action connects the two pieces together. When the thumb force is removed, the hammer retracts to its original position, as shown in Figure 13.61(a). This retraction occurs since the three flexible joints have built-in spring force and return to their unloaded position. As the hammer retracts, the next staple is forced upward from the magazine cartridge and is ready to be used.

Summary and Suggestions

The compliant stapler can be used for a medical application to connect two pieces of adjacent skin using metal staples. It presses and deforms metal staples into the skin, from the outside. It is fabricated from a single piece of plastic material and employs flexible joints to allow for motion. When the actuation force is removed, it returns to its initial position. Since it is one piece and made of plastic, it can be fabricated at low cost using the injection molding process. The materials identified are nontoxic and are sterilizable prior to use. Given these features, it satisfies the design objectives of this problem.

13.17.3 Zipper Mechanism

Problem Description

A mechanism is needed to fasten together two sheets of adjacent fabric material. The two fabric sheets are to be brought together and connected along a straight or curved line. It is desirable for the fastening to take place rapidly. Importantly, the method should be reversible so that the two sheets can be rapidly unfastened. It is not desirable to use fasteners such as buttons or rivets. It should be easy to use, where a user can apply hand force to fasten or unfasten the sheets. Low cost and ease of mass production are of high importance, since this unit will be produced in the millions of units. The mechanism should be able to fasten or unfasten the sheets thousands of times before failure.

Design Goal Statement

Design a mechanism to fasten together two adjacent sheets of fabric material, along a straight or curved line. The mechanism must allow for rapid fastening, and it must be reversible so that the two sheets can be rapidly unfastened. The mechanism must be self-contained, where no external fasteners are to be used. It must be easy to use through the application of hand force. Low cost and ease of mass production are of high importance. Given these requirements, a compliant mechanism approach will be employed. The mechanism must perform the fasten and unfasten cycle thousands of times before failure.

Proposed Design Objectives

Design objectives for a zipper mechanism are listed in Table 13.17.3.

Proposed Design Solution

One possible design solution is a *zipper mechanism*, as illustrated in Figure 13.62(a). The design consists of a compliant mechanism that has two rows of *teeth* attached to a flexible *tape*. The tape is in turn connected to the fabric (not shown in figure). The *pull tab* is grasped by the user to move the *slider* up or down the length of the zipper. Depending on the direction of motion of the slider, the teeth that slide within it are either joined together (zipped up) or divided (unzipped). During the zip-up process, the slider passes over the two adjacent rows of teeth, where the slider's internal shape guides the two rows of teeth to mesh together. Each tooth has a *hook* and a *groove*. The zipper *chain* is formed as a hook on a tooth in one row, and it interlocks with a groove in a tooth in the other row, which is repeated back and forth for all teeth in the zipper. Figure 13.62(b) shows a view with part of the slider removed. The *divider* is the name of the internal shape that is used to either (a) separate the teeth while unzipping or (b) guide the teeth together during zipping up. A video of the zipper in operation is provided in [**Video 13.62**].

> **Video 13.62**
> Zipper
> Mechanism

Summary and Suggestions

The zipper mechanism can be used to fasten together two pieces of fabric. It does this by interlocking hundreds of teeth located on two rows of flexible tape, when a slider is passed over them. It is

TABLE 13.17.3 DESIGN OBJECTIVES FOR A ZIPPER MECHANISM

Design Objective	Description	Value (Units)
Fastening strength	Amount of force that can be sustained by fastener before mechanism separation	100 (N)
Compliant design	Employ compliant mechanism design approach	–
Mechanism material	Material suitable for compliant mechanism, such as plastic or fabric	Polyester, or polypropylene
Cost/Unit	Cost per unit to manufacture including: material, machine time, and labor for production batch	0.50 ($/unit)
Force to actuate	Amount of force needed to fasten/unfasten	< 5 (N)
Service life	Total number of cycles of operation before failure	2000 (cycles)

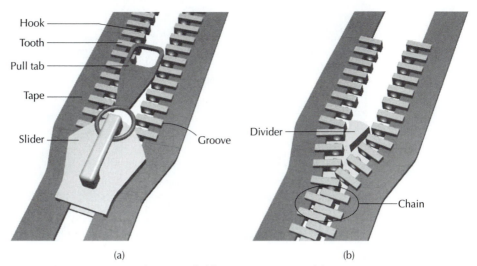

Figure 13.62 Zipper Mechanism [Video 13.62]: (a) With pull tab and slider, (b) Interlocked teeth and divider visible.

classified as a compliant mechanism since the teeth are securely bonded to the tape, yet the tape's flexibility allows the two rows of teeth to mesh or unmesh, by using the slider. When the zipper chain is formed, it has significant strength in a number of directions to keep the two pieces of fabric fastened together. Prior to zippers, buttons were commonly used to connect two pieces of fabric. However, the buttoning process took an excessively long time and a lot of hand dexterity, which led to motivation to develop the zipper. In these ways, the zipper satisfies the design objectives of this problem.

13.18 MICROELECTROMECHANICAL SYSTEMS (MEMS)

A relatively new classification of mechanisms and machines has emerged since the early 1990s, known as *microelectromechanical systems (MEMS)*. MEMS consist of microscopic-sized mechanical and electrical components, which work together to achieve specific functions. MEMS-based mechanisms and machines often have unique performance characteristics that are otherwise not possible to achieve with macro-scale (regular-sized) mechanisms. They are fabricated using the same technology, processes, and materials that are used to fabricate microelectronics, and are made on silicon chips. There are numerous examples of MEMS mechanisms, many of which can be found on the Internet using appropriate keyword searches. Interestingly, many MEMS mechanisms are based on the principles of compliant mechanism design. This is the case, since micro-mechanical mechanisms are greatly affected by a sticky/adhesive effect called stiction [46], which causes high adhesive forces to exist between micro-objects in contact. The stiction problem makes rotational or prismatic joints impractical, and hence compliant flextures are usually used as joints for MEMS mechanisms. Figure 13.63 shows some advanced MEMS mechanisms that were designed and constructed at the University of Toronto and the University of Victoria. Figure 13.63(a) shows a SEM (scanning electron microscope) picture of a 3D rotating inclined micromirror used for switching light signals [47], and Figure 13.63(b) shows a picture of a compliant, thermal-actuator-based microgripper [48]. Figure 13.63(c) shows a SEM picture of a 3D orthogonal test-structure, and Figure 13.63(d) shows a picture of a 3D micro-radiometer. Some examples of commercial MEMS devices are presented next.

A major commercial application of MEMS technology is the micro-accelerometer device [49], originally produced by Analog Devices, Inc. These MEMS devices were originally made for use in the deployment of automobile air bags, whereby the accelerometer detects the unusually high deceleration of a vehicle during a collision and provides a signal to deploy the air bag. A SEM picture of the internal parts of the accelerometer is shown in Figure 13.64. In recent times, the MEMS accelerometer has seen wide use in many other applications, which includes use in game controllers, cell phones, cameras, tablet computers, radio-controlled helicopters, and toys, among others. MEMS accelerometers are compliant mechanisms that operate on the principle of electrostatic

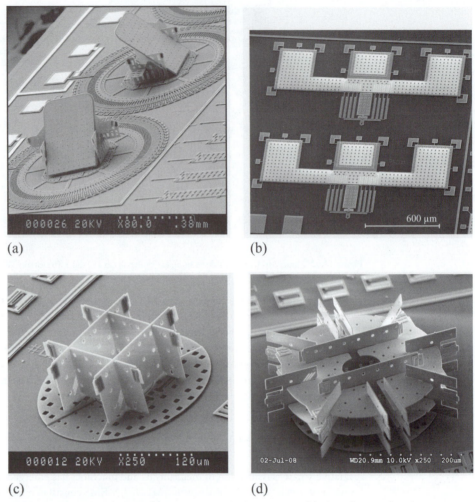

(a)

(b)

(c)

(d)

Figure 13.63 Examples of MEMS: (a) 3D Rotating inclined micromirror, (b) Thermal actuator-based microgrippers, (c) Orthogonal structure, (d) Micro-radiometer.

sensing. Figure 13.65 illustrates a simplified layout of the working components of a MEMS accelerometer. An *inertial mass* is suspended by *four flexible beams*, which are fixed to the chip surface at *anchor points*. Figure 13.65(a) shows the configuration when there is no acceleration. When deceleration occurs, as shown in Figure 13.65(b), there is relative motion between the inertial mass and the rest of the chip. If the deceleration is sufficiently large, this relative motion can be sensed by a *capacitive sensor*, which measures the amount of relative motion as a change in the gap distance, δx. An electric signal proportional to the gap distance is measured by the microelectronics located on the chip (not shown in the figure), to deploy the air bag. This combination of a micro-scaled compliant mechanism, with built-in capacitive sensing, is able to rapidly sense acceleration changes, fits within a 10×5 mm package, and is fabricated and sold for only a few dollars. It matches the performance of traditional inertial sensors, which cost hundreds of dollars and are dozens of times larger in volume. In this sense, it provides performance and value not possible to achieve with past technologies that employ traditional mechanisms and microelectronics.

Another major commercial example of a MEMS device is the digital micromirror device produced by Texas Instruments, referred to as a DLP^{TM} [50], which is used in many video projection systems. A picture of a DLP chip is shown in Figure 13.66. This MEMS device consists of a matrix of micromirrors, where each micromirror corresponds to one pixel of a projected display. Under each micromirror is a microactuator that operates on the principle of electrostatic actuation, to tilt the micromirror back and forth. The tilting is possible since the micromirror is held by a compliant mechanism hinge. To create a projected image, a light source is continuously directed at the matrix of micromirrors, such that light may be reflected toward or away from a projecting lens. Computer control is used to determine which actuators are activated on and off. This effectively allows each pixel of the display to be switched on and off and thereby create a projected image.

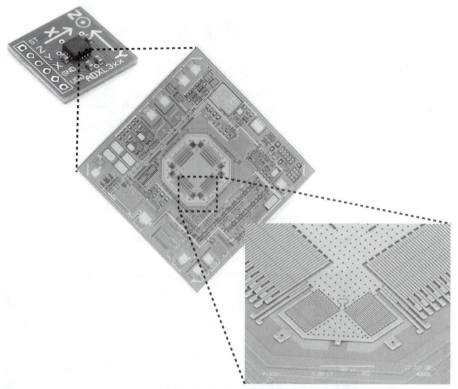

Figure 13.64 SEM image of internal parts of MEMS accelerometer. (Courtesy of Analog Devices, Inc., Norwood, MA.).

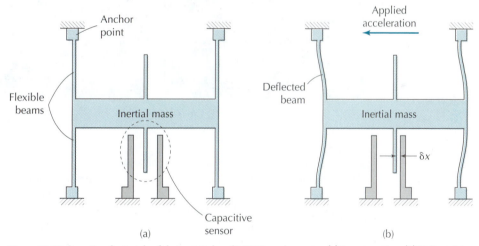

Figure 13.65 Operational principle of electrostatic-based MEMS accelerometer: (a) In rest position, (b) Deflected due to applied acceleration.

MEMS are fabricated by a process that involves the successive deposition, doping, and etching of thin films of material. These materials are typically silicon, polycrystalline silicon, silicon oxide, gold, and other metals and ceramics. This yields micromechanical components that have low aspect ratios (i.e., height versus planar length), usually between 1:10 and 1:1000, and is used to fabricate the aforementioned accelerometer chip and DLP chip technologies.

Research into microassembly at the University of Toronto and the University of Victoria has shown the potential of fabricating and assembling far more complex MEMS. It is now possible to construct 3D microstructures [51] comprised of many parts, such as those shown in Figure 13.63. Figure 13.67(a) shows a 3D microcoil developed at the University of Toronto that can be used to convert electromagnetic radiation into electrical signals, or vice versa. The microassembly process consists of a sequence of grasp, manipulate, join and release operations. It can be described by explaining the process to assemble a microcoil, which is made from a set of microparts that start

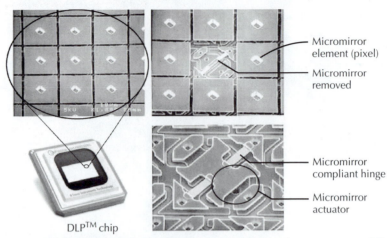

Figure 13.66 DMD chip illustrating array of micro-mirrors (Courtesy of Texas Instruments, Dallas, TX).

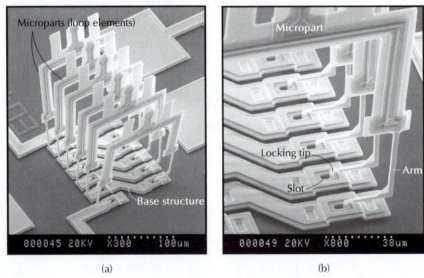

Figure 13.67 (a) Microcoil, (b) Close-up.

out as planar on the surface of a chip, as shown in Figure 13.68(a). A compliant microgripper, adhesive-bonded to the end-effector of a robotic manipulator, is shown in Figure 13.68(b). This microgripper is used to grasp the microparts from the chip, as shown in Figure 13.68(d). After being grasped, the micromanipulator re-orients the microparts in space, and it lines them up with the target joint location. Figure 13.67(b) shows a close-up view of the connection made after the micropart is joined, which is formed by compliant locking tips on the ends of the microparts. This microassembly process allows for the construction of 3D MEMS devices that were not previously possible to build at this scale.

13.19 SUMMARY

A series of case studies in mechanism and machine design have been presented. Each case study began with the provision of loosely structured information in the form of a problem description. Such information is normally provided to the designer by clients, or found by studying the existing situation that requires the new design. The problem description is then translated into a clear design goal statement to guide the design activity. A clear set of design objectives with specific performance targets are established, using information from the problem description as well as information from other sources. The concept generation, design selection method, and preliminary design phases have not been included in these case studies. Instead, possible design solutions

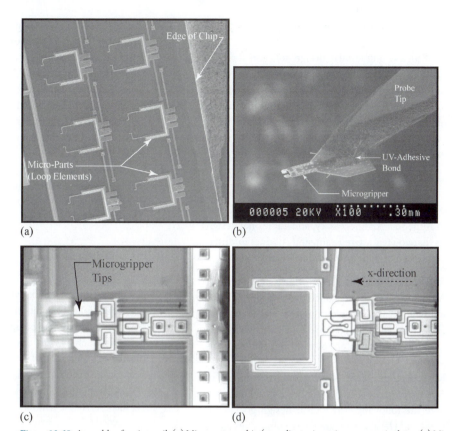

Figure 13.68 Assembly of a microcoil: (a) Microparts on chip (compliant microgripper on manipulator, (c) Microgripper aligned to micropart, (d) Microgripper grasping micropart.

that satisfy the design goal and design objectives have been presented and explained. There are many other possible design solutions for these case studies. Students of mechanism and machine design are encouraged to seek out other possible solutions, or develop their own, using the techniques described in Chapter 12. As well, numerous design exercises are available at the end of Chapter 12 and Chapter 13.

PROBLEMS

Problem Formulation

Problems **P13.1** to **P13.4** are exercises in information gathering. An important step in the design process is to search books, the Internet, patents, or other literature for similar designs that already meet your design goals and design objectives. For each of these problems, review the specified design goal statement and the design objectives and find an existing similar solution that meets those goals and objectives. Ensure the solution is different from the solution described in the case study. Use the information gathering strategies presented in Section 12.3.1, Task 2.

P13.1 Find an existing similar design solution for case study 13.2.2 (constant-velocity coupling) using information gathering strategies. Explain the

solution you found with a paragraph, indicate how the design objectives are met or not met, include a scanned/copied image of the solution, and properly reference all your sources.

P13.2 Find an existing similar design solution for case study 13.4.1 (drill chuck) using information gathering strategies. Explain the solution you found with a paragraph, indicate how the design objectives are met or not met, include a scanned/copied image of the solution, and properly reference all your sources.

P13.3 Find an existing similar design solution for case study 13.8 (variable aperture mechanism) using information gathering strategies. Explain the solution you found with a paragraph, indicate how the design objectives are met or not met, include a scanned/copied image of the solution, and properly reference all your sources.

P13.4 Find an existing similar design solution for case study 13.16.1 (locking mechanism) using information gathering strategies. Explain the solution you found with a paragraph, indicate how the design objectives are met or not met, include a scanned/copied image of the solution, and properly reference all your sources.

Problems **P13.5** to **P13.8** are exercises in problem formulation to develop design goal statements and design objectives. These exercises require reading of the specified section in Appendix A. If necessary, add additional design objectives that you feel are relevant, and conduct information gathering activities to find suitable target values.

P13.5 Read the problem description statement of Appendix A.3 (slicing machine) and prepare the following: (a) a clear design goal statement and (b) at least eight design objectives along with their target values, which are of most relevance to this problem.

P13.6 Read the problem description statement of Appendix A.12 (side-tipping ore railcar) and prepare the following: (a) a clear design goal statement and (b) at least eight design objectives along with their target values, which are of most relevance to this problem.

P13.7 Read the problem description statement of Appendix A.22 (loading machine system) and prepare the following: (a) a clear design goal statement and (b) at least eight design objectives along with their target values, which are of most relevance to this problem.

P13.8 Read the problem description statement of Appendix A.24 (transport mechanism) and prepare the following: (a) a clear design goal statement and (b) at least eight design objectives along with their target values, which are of most relevance to this problem.

Conceptualization and Sketching

Problems **P13.9** to **P13.12**, are exercises in novel concept development and sketching. For each of these problems, create an alternative design concept and sketch, based upon the problem description, design goal statement, and design objectives for the specified case study. Hand drawn sketches with pencil are encouraged, using the format similar to those shown in Example 12.4-B. Ensure the key components of your sketch are labeled, and provide 4–5 sentences describing how the concept works.

P13.9 Create an alternative design concept and sketch for case study 13.3.1 (clutch mechanism). Ensure that your design concept is different from the solution described in the case study.

P13.10 Create an alternative design concept and sketch for case study 13.5 (rotational braking systems). Ensure that your design concept is different from the two solutions described in the case study.

P13.11 Create an alternative design concept and sketch for case study 13.9 (rotational to linear motion systems). Ensure that your design concept is different from three solutions described in the case study.

P13.12 Create an alternative design concept and sketch, for case study 13.13.3 (fluid dispenser pump). Ensure that your design concept is different from the solution described in the case study.

Detailed Design

P13.13 The key lock presented in Section 13.16.1 is designed to open a lock with one key. Using the same principle as the pin tumbler lock presented, design a lock that may be opened with two different keys. Such designs are used in cases where a master key is employed to open the lock in addition to the standard issued key.

(a) Revise the design objectives for this problem, to incorporate the new information, and quantify as needed.

(b) Prepare two alternate concepts for your proposed design. For each alternate concept, provide two diagrams with different views to illustrate your ideas. Ensure the main components of your sketches are clearly labeled.

(c) Write 1–2 paragraphs describing how each of the two alternate concepts work.

(d) Use a design selection strategy, as described in Section 12.3.3.C, to select one of the concepts, and make sure to justify your choice.

P13.14 The synchronizer described in Section 13.3.2 is able to rigidly couple and uncouple the gear and the shaft on which it is mounted. Design a hub and sleeve of a synchronizer that is able to alternately connect one of two gears to the same shaft. This would be suitable for the arrangement illustrated in Figure 6.17, for the dual-clutch transmission. After review of Video 6.17, complete the following tasks:

(a) Revise the design objectives for this problem, to incorporate the new information, and quantify as needed.

(b) Prepare two alternate concepts for your proposed design. For each alternate concept, provide two diagrams with different views to illustrate your ideas. Ensure the main components of your sketches are clearly labeled.

(c) Write 1 or 2 paragraphs describing how each of the two alternate concepts work.

P13.15 The swash-plate compressor illustrated in Figure 13.46 is used to drive the connection rods of the pistons in the pump. The tilt angle of the swash plate is fixed, therefore, for a given rotational speed on the input, the pumping volume/cycle is fixed. Design a variable capacity pump for a given rotational input speed, by employing a swash plate with a variable tilt (*Hint*: Refer to the cyclic control of the helicopter), and complete the following tasks:

(a) Revise the design objectives for this problem, to incorporate the new information, and quantify as needed.

(b) Prepare two alternate concepts for your proposed design. For each alternate concept, provide two or three diagrams with different views to illustrate your ideas. Ensure the main components of your sketches are clearly labeled.

(c) Write 1 or 2 paragraphs describing how each of the two alternate concepts work.

Problems **P13.16** to **P13.18**, are exercises in detailed design and simulation. These exercises require reading of the specified section of Appendix A and the use of Working Model V5.0 simulation software. *Note*: These design exercises are significant variations of the problem described in Appendix A; hence the solutions will be different.

P13.16 Read the problem description statement of Appendix A.7 (windshield wiper) The arrangement described makes use of eight moving linkages, configured in a specific way. Do the following:

(a) Prepare a clear Design Goal Statement.

(b) Prepare at least six Design Objectives along with their target values.

(c) Prepare a new alternate concept to achieve the stated design goal statement, using a different number and arrangement of moving linkages (i.e., not eight moving linkages). *Note:* you may need to revise the design objectives for this problem. Create 1 or 2 sketches of your alternate design concept, and describe your approach.

(d) Use Working Model software to implement your design concept.

(e) Using the evaluation tool set of Working Model, evaluate the quantitative performance of your design concept. Create a table of your design objectives with an additional third column, listing your simulation results for those design objectives (as appropriate).

P13.17 Read the problem description statement of Appendix A.9, overhead garage door. The arrangement described makes use of four moving links, plus the garage door and a spring, configured in a specific way. Do the following:

(a) Prepare a clear Design Goal Statement.

(b) Prepare at least six Design Objectives along with their target values.

(c) Prepare a new alternate concept to achieve the stated design goal statement, using a different number and arrangement of moving linkages (i.e., not four linkages). *Note:* you may need to revise the design objectives for this problem. Create 1 or 2 sketches of your alternate design concept, and describe your approach.

(d) Use Working Model software to implement your design concept.

(e) Using the evaluation tool set of Working Model, evaluate the quantitative performance of your design concept. Create a table of your design objectives with an additional third column, listing your simulation results for those design objectives (as appropriate).

P13.18 Read the problem description statement of Appendix A.20, water pump mechanism. The arrangement described makes use of seven moving linkages, configured in a specific way. Do the following:

(a) Prepare a clear Design Goal Statement.

(b) Prepare at least six Design Objectives along with their target values.

(c) Prepare a new alternate concept to achieve the stated design goal statement, using fewer than seven moving linkages. *Note:* you may need to revise the design objectives for this problem. Create 1 or 2 sketches of your alternate design concept, and describe your approach.

(d) Use Working Model software to implement your design concept.

(e) Using the evaluation tool set of Working Model, evaluate the quantitative performance of your design concept. Create a table of your design objectives with an additional third column, listing your simulation results for those design objectives (as appropriate).

Problems **P13.19** to **P13.20** are design exercises in prototyping and testing, as described in Section 12.3.5 and 12.3.6, respectively. *Note*: These design exercises are significantly challenging and require resources such as tools and materials, a machine shop, or a rapid-prototyping machine, which may or may not be available to the design student. Suggested structural materials include: foam core, cardboard, thin plastic sheets (ABS, acrylic), and thin wood sheets. Suggested fastening materials include: glue, bolts, springs, string.

P13.19 Review and complete problem **P13.13**, as described. Continue on with this problem, by completing the following steps:

(a) Build a "functional prototype" of this design, using appropriate tools and materials. This may include plastic parts, where some (but not all) of those parts may be rapid-prototyped.

(b) Create a suitable test plan for your prototype, as described in Section 12.3.6.

(c) Test your prototype and record the results.

(d) Create a test report and describe how well the prototype satisfies the design objectives.

P13.20 Review Figure 7.3 and [Video 7.3], which describe the inner workings of a toggle switch mechanism for a ball point pen. The mechanism shown in the figure and model is not to scale, is represented in 2D, and runs out of travel. Design a working 3D toggle switch for a ball point pen that does not run out of travel, but can toggle on/off indefinitely, using the principals of the mechanism shown in Figure 7.3. Complete the following steps:

(a) Prepare a clear Design Goal Statement.

(b) List Design Objectives along with their target values.

(c) Draw 1 or 2 sketches that show your design concept of a 3D version of the switch that can toggle indefinitely.

(d) Build a "functional prototype" of your design concept, using appropriate tools and materials. This may include plastic parts, where some (but not all) of those parts may be rapid-prototyped.

(e) Create a suitable test plan for your prototype, as described in Section 12.3.6.

(f) Test your prototype and record the results.

(g) Create a test report and describe how well the prototype satisfies the design objectives.

APPENDIX A

SUPPLEMENTAL DESIGN PROJECTS USING WORKING MODEL 2D

This appendix provides twenty-six design projects in mechanism and machine design. These projects have been employed for several years in an undergraduate engineering course in mechanism and machine design, and they provide students with a hands-on experience to practice mechanism design. The projects are presented with a problem description statement, in a manner that would be typically found in engineering practice. The description provides a mixture of information including: specific performance needs (design objectives), a non-scale sketch illustrating the conceptual idea, and several questions about the design and its simulated performance. Computer-based design and simulation software is in wide-scale use in engineering practice, therefore, these projects are structured to be designed with simulation software specific to kinematics and dynamics. The software tool recommended is Working Model 2D; however, other similar software may be employed. Whereas the examples of Chapter 12 and Chapter 13 focused on the problem formulation, conceptualization, and preliminary aspects of design, the problems of Appendix A focus on the hands-on practice of detailed design, simulation, and testing of mechanisms and machines.

A suggested approach for completing these design projects is as follows: Begin by consolidating the problem description information by writing a clear design goal statement and preparing a list of design objectives, as described in Section 12.3.1. Here, a single concept diagram is already presented that shows the general arrangement of links and kinematic pairs (which are not to scale). Start the detailed design process by creating a Working Model 2D representation of the machine that closely resembles the one shown in the figure. Proceed by adjusting the geometry of the links, their relative position and the location of the joints, to meet the design objectives of the problem. When making these changes, you should attempt to employ the methods of Chapters 1 to 11 to make choices. Your design model must work while employing the given number of links and kinematic pairs. The computer file of the satisfactorily designed mechanism will be part of the material submitted with your design report.

Video A.0
Working Model
Tutorial

A tutorial on the use of Working Model 2D software is provided in [**Video A.0**] and provides instructions on creating models and simulating their performance. A link providing instructions on how to obtain the Working Model 2D software is available on the companion website for this textbook. The simulation of the mechanism should be smooth and move at a reasonable speed, to effectively demonstrate your design solution. (The display rate of the animation may be controlled by using the dialogue box in the World Accuracy menu of the Working Model 2D software.) The entire animation should not exceed 2000 frames and will probably require at least 1000 frames to animate at a reasonable smoothness.

A written report should be prepared for these projects, along with the computer simulation file. The report should not exceed eight double-spaced printed pages, excluding the title page, summary, and appendixes. The following format is suggested:

- **Title page**
- **Summary**
- **Introduction** (Explain the purpose, write a design goal statement, and list the design objectives in table format.)
- **Description of Design** (Include 1 or 2 pictures of the final mechanism design with key components labeled.)
- **Engineering Analysis** (Include any calculations or theory used to assist with determining linkage properties or other aspects in coming up with the design solution.)
- **Simulation and Test Results** (Include a table that lists the simulation results, in comparison to the original design objectives.)
- **Discussion** (Explain the most important parameters that influence the mechanism performance. Also, answer all questions specified for the particular project.)
- **Conclusions**
- **Future Work** (Describe suggestions for further improvement of the mechanism, to aspire to the originally stated design goal statement.)
- **Appendices** (Any additional info that is pertinent to the report.)

A.1 ROCKING CHAIR

Figure A.1(a) shows a traditional rocking chair. For the first part of this project, create the traditional chair using Working Model 2D and perform an analysis of its motion. As shown in Figure A.1(a), θ is the angle of the seat with respect to the horizontal. The ground surface should be created directly on the x axis. Align the center of the chair with the y axis. Create two graphs next to your chair model, with the first showing how x_C (x position of point C) varies with θ, and the second showing how y_C varies with θ, for the range $-30° \leq \theta \leq 30°$.

For the second part of this project, design a new rocking chair using a four-bar mechanism (Figure A.1(b)). Your new design must have motion similar to the traditional rocking chair. In order to design the motion of the new chair, add a plot of x_D versus θ to the first graph and add a plot of y_D versus θ to the second graph. Attempt to match the new motion plots with the traditional chair plots as closely as possible by varying link lengths and joint positions. Employ an appropriate coordinate system for the new design. Specify the link lengths and joint positions.

Employ the same seat dimensions for both chairs, and place both designs within the viewing area of the screen.

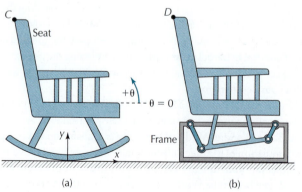

(a) (b)

Figure A.1 Rocking chair.

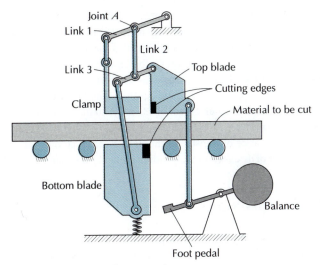

Figure A.2   Clamping and cutting machine.

A.2   CLAMPING AND CUTTING MACHINE

The machine illustrated in Figure A.2 is designed to clamp down on a peice of material to be cut, and then to cut through it with two cutting edges. By applying a downward force on the foot pedal, the clamp first moves down, pressing upon the material to be cut. This action is completed before either of the cutting edges touch the material. Once the clamp is pressed against the material to be cut, the cutting edges continue to move toward the material. The bottom cutting edge and the top cutting edge should contact either side of the material simultaneously, thereby cutting through it. After passing through the material, the cutting edges will continue to move past each other for a short distance, then slow down, and then open again.

Create a computer simulation of this mechanism using Working Model 2D. Use the **Force** option to apply a force to the foot pedal to drive the simulation. Also include a graph showing the contact force between the clamp and the material. Note, the simulation cannot cut the model, but can only simulate the cutting edges moving through it.

Once you have successfully simulated the mechanism, you may try to vary some of the parameters. Referring to the illustration of the clamping and cutting machine, explain the effect of varying the connection point of joint A along link 1. What is the effect of varying the length of link 2?

A.3   SLICING MACHINE

The slicing motion of this machine is obtained from the rotation of two eccentric discs sliding within guides. The two eccentric discs are driven by an input shaft attached to a stand on the ground, as shown in Figure A.3. The two guides, guide A and B, are welded together. Eccentric

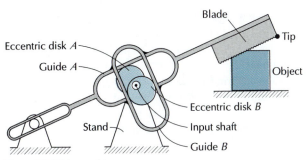

Figure A.3   Slicing machine.

disc A moves only within guide A, and eccentric disc B moves only within guide B. In the position shown, eccentric disc B provides the horizontal cutting movement, and eccentric disc A provides the up-and-down movement. Create a simulation of this mechanism using Working Model 2D. Use the **Torque** button to apply a torque to the input shaft to drive the simulation. In addition, create a point on the tip of the blade. Using the **Tracking** feature, create a track for every fourth frame of motion of the tip, which will trace out the path of the blade in space. This path must be big enough to encompass a square object of 1.0 m by 1.0 m. The overall height of the mechanism must be less than 5.0 m, and the overall length must be less than 10.0 m.

Explain the significance of the size of the discs and the amount of eccentricity. Explain the result if the diameter of disc A (and the width of guide A) were larger than disc B.

A.4 STRAIGHT-LINE MECHANISM

The mechanism illustrated in Figure A.4 lifts objects from the ground, up through a height in a straight line, as traced out by the tip of its fork. The mechanism begins the lift with the tip of the fork positioned at a height below the base of the hydraulic cylinder, as shown. Create a simulation of this described mechanism using Working Model 2D. The tip of the fork must travel at least 7.0 m in the *y* direction, during which it must not vary by more than 20 cm in the *x* direction. The overall height of the mechanism, in the position illustrated, must be less than 4.0 m, and the overall length less than 9.0 m. Use the **Actuator** button to create a simulated hydraulic cylinder to drive the simulation. Create a graph that shows the *x* position of the tip versus the *y* position. Specify the dimensions of links 1, 2, and 3. Also, add a point on the tip of the fork, and using the **Tracking** feature, track this point for every fourth frame of motion. This will trace out the path of the fork in space.

What is the effect of varying the dimensions of link 1? What is the effect of varying link 2?

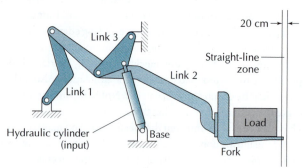

Figure A.4 Straight-line mechanism.

A.5 SINE FUNCTION MACHINE

The machine shown in Figure A.5 consists of two independent four-bar mechanisms, which restrain a block by permitting it to slide along the linkage rods. The block causes the L-shaped link to pivot, which, in turn, drives shaft *M* up and down. Create a simulation of this mechanism, such that the range of motion *Y* varies by no more than 5.0 mm, while angle α varies through the range $45° \le \alpha \le 135°$ and angle β varies through $45° \le \beta \le 135°$. Use the **Torque** button to apply a torque at point *A* to drive link 1 and a torque at point *E* to drive link 4. The base link length of both four-bar mechanisms should be at least 7.0 m. Also, create a graph beside the model showing the motion in the *y* direction versus angle α. (Hint: In order to simplify your model, both four-bar kinematic chains should form parallelograms.)

This machine is called a *sine function machine*, and the motion in the *y* direction can be described by the equation $y = K(\sin(\alpha)/\sin(\beta))$, where *K* is a constant. For the mechanism you have created, what is the value of *K*, and what is the ratio $\alpha{:}\beta$? What is the effect of varying the lengths of link 1 and link 3 together?

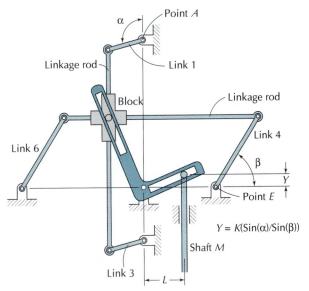

Figure A.5 Sine function machine.

A.6 FLEXIBLE-FINGERED ROBOT GRIPPER

Create the robot gripper mechanism shown in Figure A.6. The gripper has fingers that flex as they close inwards. The starting position for the fingers is shown by the line AB, where links 1 and 2 should be parallel to the line AB. Links 1 and 3 have three joints each, and links 2 and 4 have two joints each. Link 1 should be 50 mm long. Distance $A'A$ should be 80 mm. Both fingers should be able to touch each other tip to tip when they meet.

Once your mechanism is complete, create a round object 30 mm in diameter. By applying a force of 10 N to link 5, as shown in the figure, determine the steady-state contact force exerted upon the round object. Create a graph of this normal force with respect to time. What is the mechanical efficiency of your mechanism (i.e., contact force divided by input force)? How can you increase the mechanical efficiency? What is the effect on finger motion and force, when distance CD on link 5 is made shorter but all other links remain the same?

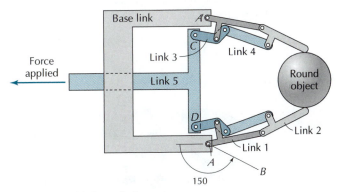

Figure A.6 Flexible-fingered robot gripper.

A.7 WINDSHIELD WIPER

Create the windshield wiper mechanism shown in Figure A.7. Wiper blade 1 and wiper blade 2 must be able to swing through 120° of motion (60° to the left and 60° to the right), as shown in the figure. Both wiper blades should be parallel to each other at all times. The length of the working

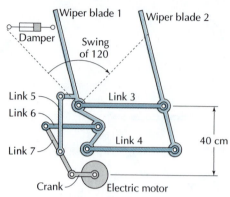

Figure A.7 Windshield wiper.

part of each blade should be 50 cm, and link 3 should be 60 cm long. All links, except both wiper blades, should fit into a space less than 40 cm tall, while moving through the full range of motion.

Create a graph showing the rotation angle of wiper blades 1 and 2 with respect to time. Make the wiper blades oscillate at a frequency of 1.0 Hz. What motor speed is required to do this? Apply a damper with a value of 1.0 N-sec/m at the end of wiper blade 1, as shown in the figure, to simulate water being cleared. Create a graph of motor torque versus time. What is the peak motor torque experienced during a cycle?

A.8 PARALLEL-JAW PLIERS

Create the parallel-jaw pliers shown in Figure A.8. When force is applied to the handles of the pliers, the jaw link will close downward to meet the base link. As this occurs, line segment *AB* on the jaw link must remain parallel to line segment *CD* on the base link. Line segment *CD* should be at an angle of 0° with the horizontal at all times. The overall size of the pliers must not exceed 20 cm in length and 10 cm in height when fully open.

Create a graph showing the rotation angle of line segment *AB* and line segment *CD* with respect to the *y* displacement of the jaw link. A typical adult hand can squeeze with a grip force of 60 N. How much force will a round object 10 mm in diameter experience in the jaws of these pliers, with that force applied? What is the effect of moving the force applied on the handle farther back along the handle (away from the jaws)?

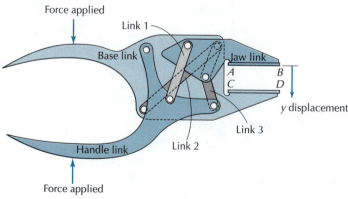

Figure A.8 Parallel-jaw pliers.

A.9 OVERHEAD GARAGE DOOR

An overhead garage door, as illustrated in Figure A.9, must be created such that the clearance height of a vehicle passing below is at least 2.0 m. The garage door is 2.3 m tall. When the door is fully closed, it should be perpendicular to the ground and lined up at the top and bottom with the garage wall. While the door is in motion, no part of it must pass through the garage roof or the garage floor.

For the simulation, the garage door should start in the open position. Apply a downwards vertical force to the garage door at point *B* to enable the motion of the mechanism. Create a trace of point *A* and point *B* on the garage door while it is closing, using the tracking feature of Working Model 2D. What is the purpose of the spring in the mechanism? Create a graph of the spring length and spring tension with respect to the *x* position of point *B*. (Hint: Activate the **Collision** option between the garage roof, door, and floor, to assist you in designing the mechanism.)

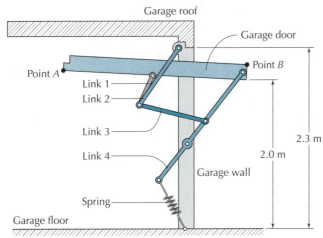

Figure A.9 Overhead garage door.

A.10 OVERHEAD LAMP

Create an overhead lamp, as shown in Figure A.10. It should be capable of moving right to left, while keeping the bottom of the lamp head approximately parallel to the horizontal. Point *A* on the lamp head should be 5.0 cm lower than any other part of the lamp at all times. Design the lamp so that point *A* can extend out (i.e., to the right) as far as 100 cm from the wall and collapse inward as close as 40 cm to the wall. The overall mechanism height should be no more than 1.5 m when fully collapsed, and the overall mechanism length should be no more than 1.5 m when fully extended.

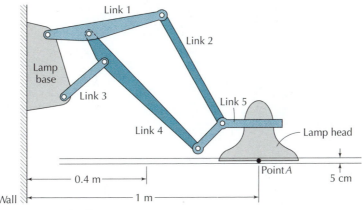

Figure A.10 Overhead lamp.

Start the mechanism in the fully extended position. To move the mechanism, apply a small horizontal force to the lamp head to make the mechanism close. Create a graph showing the y position of point A with respect to the x position of point A. The y position of point A should not deviate more than 5.0 cm throughout the full travel of the lamp head. Also create a graph showing the rotation angle of the lamp head, with respect to the x position of point A.

A.11 STEERING MECHANISM

Create the steering mechanism shown in Figure A.11(a). The tires used are model #P195/75R14. The center-to-center distance between the two tires, when they are both straight, is 142 cm. When the Pitman arm is rotated, about point A, the two tires turn. They must turn in such a way that their motion closely approximates the Ackerman steering condition, which is illustrated in Figure A.11(b). The condition states that lines drawn perpendicular to each of the four tires must have a common point of intersection for any radius of turn. This minimizes lateral sliding action of the tires and reduces tire wear. The center-to-center distance between the front and rear tires is 300 cm.

The steering mechanism can be driven by using a motor at point A, where point A is on the car frame. Create a motor that produces a sinusoidal motion of the Pitman arm. The right front tire should be able to turn to the right by up to 30°. The left front tire should be able to turn to the left by up to 30°.

When your mechanism is completed, create three lines perpendicular to the tires. This can be done by rigidly attaching a thin beam, 1.0 cm wide and 30 m long, to each of the front tires and one through the rear tires. As the motor drives the front tires, the three lines should approximately

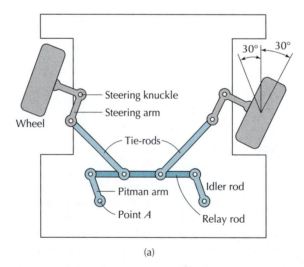

(a)

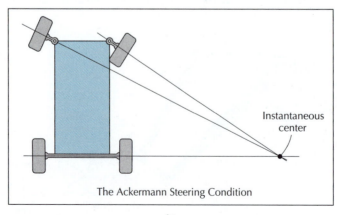

(b)

Figure A.11 Steering mechanism.

have a common point of intersection, as shown in Figure A.11(b), through the full range of motion of the tires. The closer, the better.

Create a graph for each front tire, showing its turning angle with respect to time. What is the purpose of the idler rod? Explain the function of the tie-rods. How does their length and their connection point on the relay rod effect the mechanism?

A.12 SIDE-TIPPING ORE RAILCAR

Create the railcar shown in Figure A.12. The maximum size of the car must be no greater than 2.0 m wide by 1.5 m tall, from the car base to the top door. The tilting bed of the car must be able to pivot about point A, by 20°. As it tilts, three simultaneous motions must occur. (1) The latch must be released as link B rotates up; (2) the top door must rotate upwards to 45° (with respect to the tilting bed); and (3) the side door must rotate downward 90° (with respect to the tilting bed).

In order to drive the mechanism, create the pump jack, using an actuator to exert a force from the car base onto the tilting bed. If necessary, create a stopper mechanism, to ensure the side door stops at 90°, so that it is parallel and in line with the tilting bed.

Create three randomly shaped objects, and place them in the car. When the simulation runs, the three objects should slide across the tilting bed, across the open side door, and fall out of the car. Create a graph showing the angle of the side door and top door with respect to the angle of the tilting bed.

What is the role of links A, B, and C? If the pump jack were required to tilt the car bed when it carries a specific load, a certain force would be required. What is this force if a uniformly distributed load of 20000 N is on the tilting bed?

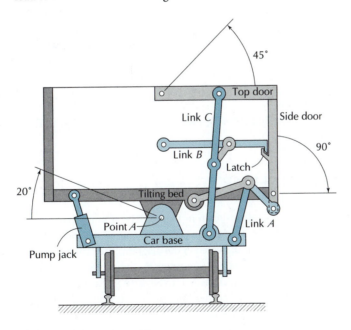

Figure A.12 Side-tipping, ore railcar.

A.13 170° CABINET HINGE

Create the cabinet hinge mechanism shown in Figure A.13. Include the two adjacent cabinet doors and an inner wall, indicated in gray, with the mechanism. All doors are 18 mm thick. The cabinet door is 300 mm long. The initial position of the cabinet door is shown as the dotted rectangle. In the initial position, there is a clearance of 2.0 mm between the door and the right door, and 2.0 mm between the door and the left door. The final position of the door is shown as a dotted rectangle, rotated 170° from its initial position.

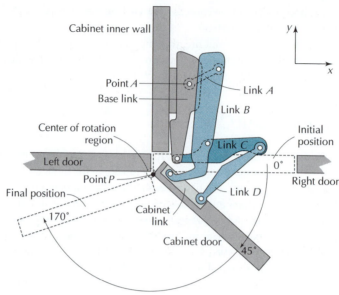

Figure A.13 170° cabinet hinge.

When the hinge mechanism is fully collapsed (when the door is in the initial position), the mechanism should occupy a space of no more than 90 mm deep (depth measured parallel to the inner wall) and 70 mm wide (width measured parallel to the doors). To create the motion of the mechanism, place a motor at point A to drive link A. Create a motor that causes a sinusoidal motion of the door from the 0° position to the 170° position. For the simulation, create a point P on the door as indicated. During the door's rotation, point P should be approximately at the center of rotation. Point P should stay within a circle of diameter 10 mm through the full range of rotation of the door. Create a graph showing the y position of P with respect to the x position. The door must not pass through the right door, left door, or inner wall, throughout its full range of motion.

What is the role of link A? What effect does the initial position and length of link A have on the mechanism?

A.14 ELECTRIC GARDEN SHEAR

Create the shear mechanism shown in Figure A.14. The mechanism must not exceed 40 cm in length. The cutting shears are to be 10 cm deep and 15 cm wide. The rear of the unit must be a maximum of 15 cm wide as indicated in the diagram. The center tip of the gray shear should have a range of motion between point A to point B. The center tip of the gray shear is illustrated at point B in Figure A.14. The mechanism is to be driven by an off-center pin connected to a rotating plate. The rotating plate is connected to a motor at its center. The gray shear has a slot cut into it, which the pin is able to travel within. The motion of the slot, on the gray shear, must stay within the 15 cm band indicated in the diagram, throughout its full range of travel. The motor to be used

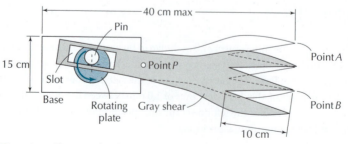

Figure A.14 Electric garden shear.

for this design must deliver 333 mN-m of torque. In order to satisfy the design requirements, the shears must be able to exert a minimum force of 6.0 N on a 2.0 cm-diameter object between any two cutting blades.

Create a graph showing the motor speed and motor torque with respect to time. Create a graph showing the acceleration of the center tip of the gray shear. What effect does the location of point P have on this mechanism? What effect does the location of the 2.0 cm-diameter test object within the shears have on the force exerted on the object by the shears?

A.15 AIRCRAFT LANDING GEAR

Create the aircraft landing gear mechanism as shown in Figure A.15. When the landing gear is fully extended, as illustrated in the figure, the distance from point O to point A must be 0.66 m. The length of the shock strut must be 0.53 m measured from point O to point D. When folded upward, the landing gear mechanism must fit within a space in the aircraft fuselage. The space within the fuselage can be defined as a rectangle, with the top left designated as point C and the bottom right designated as point B. Point O in the figure has x and y coordinates (0, 0). Point C is located at coordinates (−0.83, 0.43), and point B is located at coordinates (0.06, −0.13). All coordinates are measured in meters. The tire diameter is 0.3 m.

Part I: The landing gear mechanism must be driven using an actuator. The approximate location of the actuator is shown in the diagram. Use a sinusoidal function of length as an input for the actuator. The mechanism should move from the fully extended position to the fully folded position in the simulation. No portion of the landing gear mechanism is allowed to penetrate the fuselage during any part of the simulation. The piston should be parallel to the shock strut at all times. Join the piston and the shock strut with a spring damper. Create a graph showing the x and y positions of the tire with respect to time. What is the purpose of the torque arm?

Part II: For the written report, simulate the landing of the landing mechanism. To do this, give the actuator a fixed value for length during the landing simulation. The landing can be simulated by creating a rectangular body (representing the runway) about 1.0 m below the tire. Constrain the runway using the anchor tool in Working Model 2D. Remove the constraint (anchor) from the fuselage, and restrict the motion of the fuselage to pure translation in the y direction only. Apply a force on the fuselage so that it accelerates downward. Measure the maximum contact force between the tire and the runway during landing. Also measure the maximum force at joint O (point O). Answer the following questions in your report: What is the effect of increasing or decreasing the spring stiffness of the shock strut? Should the spring be made rigid? What is the effect of the increasing the damping? Is this an accurate simulation of a landing? Why or why not?

Note: Do not include the landing simulation of Part II in your electronic submission. Submit only the mechanism described in Part I.

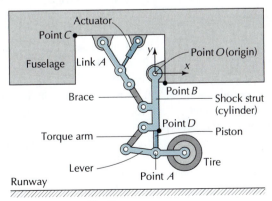

Figure A.15 Aircraft landing gear.

A.16 FOLDOUT SOFA BED MECHANISM

Create the foldout sofa bed mechanism as shown in Figure A.16. The sofa bed is shown in the fully extended position. When fully folded, the sofa bed fits within a rectangular space defined by points A, B, and C. Point O in the figure has x and y coordinates (0, 0). Point A is located at (0.37, –0.12), at the lower right of the rectangular space within the sofa bed. Point B defines the bottom left of the sofa backrest at coordinates (0.04, 0.22). Point C defines the top of the sofa front at coordinates (–0.46, 0.17). The bed frame is 0.66 m long from point L to point H. The bed frame and bed mattress together are 0.15 m thick. The bed mattress can be considered rigidly fixed to the top of the bed frame. The underside of the bed frame is 0.39 m above the floor. The sofa bed mechanism must be driven using an actuator. For the simulation, the sofa bed should begin in the fully extended position and end in the fully folded position. Connect an actuator between point P and point Q in the diagram to create the folding motion. Use a sinusoidal function of length as an input for the actuator. No part of the sofa bed mechanism should penetrate the sofa front, the sofa backrest, the sofa back, or the floor during any part of the simulation. Further, the bed mattress should be horizontal at the fully extended and fully folded positions.

Create a graph showing the angle of the bed mattress with respect to time. Create a graph of the angle of the bed leg with respect to the bed frame. Will the bed frame be able to fold inward without link J? What is the purpose of link J? What is the maximum joint force exerted on link K, and on which joint?

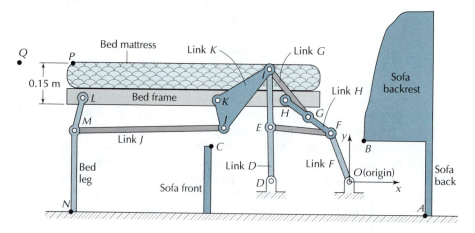

Figure A.16 Foldout sofa bed mechanism.

A.17 WINDOW MECHANISM

Create the window mechanism as shown in Figure A.17. The window mechanism is shown in the partially open position. In the initial position, the window is parallel to the windowsill and fits flush within the windowsill (i.e., point D is on top of point E). In the final position, the window is perpendicular to the windowsill. Point C on the crank slides within the window slot. Point W on the window slides within the windowsill slot. Point O in the figure has x and y coordinates (0, 0). The window is 0.62 m long and 0.04 m wide. The y distance between point O and point W is 0.08 m. Point E is located at coordinates (0.43, 0.11) with respect to point O. The crank is attached to the windowsill via a pin joint at point O. A 10-tooth gear is built into the base of the crank and is in mesh with a 5-tooth gear on the hand crank.

Part I: The window mechanism is to be driven using a motor to rotate the hand crank. Create a geared pair constraint using Working Model 2D between the crank base and the hand crank. (Note, it is not necessary to draw the gear teeth.) Use a sinusoidal function of angular position as an input for the motor. The mechanism simulation should begin in the initial position and end in the final position, as illustrated in the figure. Create a graph showing the angular position of the

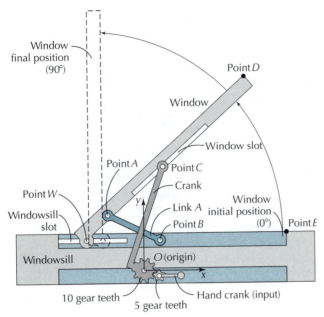

Figure A.17 Window mechanism.

window with respect to the windowsill. What is the minimum length for the window slot, in order for the window mechanism to travel through the full range of motion?

Part II: For the written report, simulate the effects of a strong wind force on the window, while the window is opening. This can be done by applying a horizontal (x direction) force at point D on the window. The amplitude of the wind force should be a function of the angular position of the window, where the wind force is equal to 100 N when the window is in the final position, and 0 N when the window is in the initial position. Create a motor to rotate the hand crank and use a torque to drive the motor. What is the minimum torque required to open the window all the way? Prepare a graph of the torque versus window angular position in your report. Assume that only one-third of this torque could be supplied at the input. What changes could be made to the window mechanism so that it would still be able to open fully, against the wind? Is this an accurate simulation of wind, for the purpose of determining input torque for the mechanism? Why or why not?

Note: Do not include the wind simulation of Part II in your electronic submission. Submit only the mechanism as described in Part I.

A.18 CONVERTIBLE AUTOMOBILE TOP

Create the convertible automobile top mechanism as shown in Figure A.18. The mechanism in the figure is in the fully extended position. The top panel and the rear panel, along with all other links, must be able to fold back and fit within a predefined space. This predefined space is shown in the diagram as a dashed rectangle. Point O has x and y coordinates $(0, 0)$. Point E is located at coordinates $(1.36, -0.44)$. The top panel is 0.84 m long, measured from point T to point B. The rear panel is 0.78 m long, measured from point A to point O. Point T on the top panel is located at coordinates $(-1.275, 0.55)$ in the fully extended position. Also, the top panel must be horizontal in the fully extended position.

The mechanism must be driven by a piston actuator, as shown in the figure. Use a sinusoidal function of length as an input for the actuator. Use the actuator to simulate the mechanism motion from the fully extended position to the fully collapsed position. When the mechanism is fully collapsed, all components, including the piston actuator, must fit within the predefined space. Further, no part of the mechanism must pass through the sides or bottom of the predefined space.

Create a graph showing the angular position of the top panel with respect to time. Using the **Tracking** feature of Working Model 2D, create a track of point T as the mechanism collapses. What is the purpose of link D? What effect does the length of link C have on the mechanism?

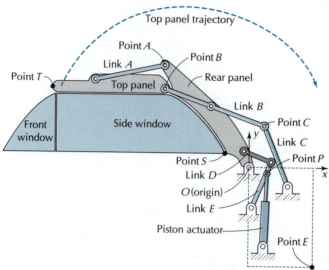

Figure A.18 Convertible automobile top.

A.19 HORIZONTAL-PLATFORM MECHANISM

Create the horizontal-platform mechanism as shown in Figure A.19. The purpose of this mechanism is to provide horizontal motion of the table top, within a specified tolerance. The table top must be able to travel at least 850 mm along the x axis, during which the vertical translation of the table must be less than 4.0 mm. The angular deviation of the table top must be less than 1.0° from the horizontal, throughout the entire motion of the mechanism. In the starting position, as shown in the diagram, all links, including the table top must fit within a space that is 850 mm wide (x direction) by 1050 mm tall (y direction). None of the links may pass through the top of the table top during the motion of the mechanism.

This mechanism must be driven by a motor actuator with a constant velocity. The motor should be located at a joint between two of the links. Which two links should it be placed between? Explain your reasons. Create a graph that displays the table top angle with respect to time. Create two graphs to display (1) the table top vertical travel with respect time and (2) the table top horizontal travel with respect to time. What is the relationship between link 1 and link 3? What kind of application might this mechanism be used for?

The simulation must be able to complete one full cycle.

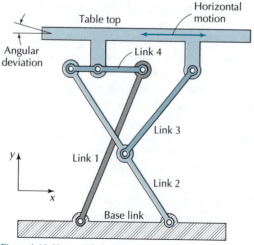

Figure A.19 Horizontal-platform mechanism.

A.20 WATER PUMP MECHANISM

Create the water pump mechanism as shown in Figure A.20. The handle (link 7) is connected to the pump mechanism and body to the left, which is illustrated by dashed lines. The objects that are illustrated by dashed lines should not be included in the simulation. In the initial position, θ must be at $0°$. At maximum rotation, θ must be at least $51°$. Links 1 through 5 are contained in the rectangular mechanism compartment. The inside dimensions of the compartment are 260 mm wide (x direction) by 265 mm tall (y direction). During operation of the mechanism, links 1 through 5 are to remain within the compartment space. No link (except link 6) may penetrate the walls of the compartment. With the handle in the horizontal position, the overall height of the entire mechanism is to be 750 mm or less. The length of the handle is 200 mm, and the length between joint A and joint B is 85 mm.

This mechanism must be driven by a motor actuator with a constant velocity. The motor should be located at a joint between two of the links within the mechanism compartment. Which two links should it be placed between? Is the direction of rotation of the motor important? Explain your answer. Create a graph that shows the angular position of the handle with respect to the angular position of link 2. What function does link 5 serve? What is the effect of changing the distance between joint D and joint E on link 5?

The simulation submission must be able to complete one full cycle.

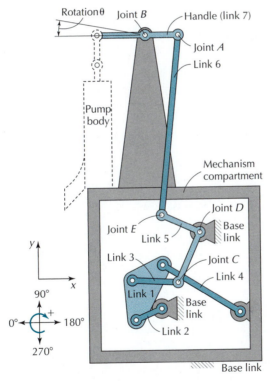

Figure A.20 Water pump mechanism.

A.21 STAIR-CLIMBING MECHANISM

Create the stair-climbing mechanism as shown in Figure A.21. The objective of this mechanism is to climb the staircase and finish with the entire mechanism standing on the upper platform. Create the staircase as shown in the figure. Each step is 254 mm long, and each riser is 178 mm tall. A single leg mechanism (SLM) is illustrated in the upper portion of the diagram. The foot (link 2) has a triangular shape, with two sides 180 mm long, and one side 320 mm long. The complete climbing mechanism consists of four of the SLMs joined together via their respective machine base links, as shown in the lower diagram. To better understand the configuration of the climbing mechanism, think of the climbing mechanism configured as an automobile, where each wheel is replaced by an SLM. Since Working Model 2D software is two-dimensional, the figure shows the two right SLMs drawn with solid lines, while the two left SLMs are drawn with dashed lines. Note that the rotation of link 1 on the SLMs shown in dashed lines, is rotated 180° from that of link 1 on the SLMs shown in solid lines. The entire mechanism at any given position should not be longer than 1150 mm and not taller than 750 mm. None of the links may pass through the staircase during the motion of the mechanism.

Each SLM must be driven by a motor actuator with a constant velocity. The speed of all four actuators must be the same. The motor of each SLM should be located at a joint between two of the links. Which two links should it be placed between? Is the direction of rotation of the motor important? Explain your answer. You will need a counterweight somewhere on the machine base link of the complete mechanism, to keep it from tipping backwards during climbing. Explain the rationale for placement and weight of this counterweight.

The simulation must complete two full cycles (which consist of climbing two steps on the stairs).

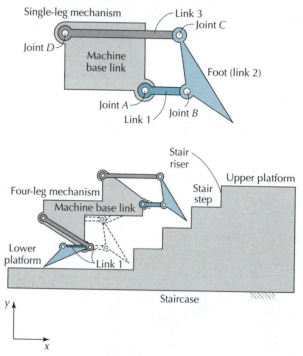

Figure A.21 Stair-climbing mechanism.

A.22 LOADING MACHINE SYSTEM

Create the loading machine system as illustrated in Figure A.22. The objective of this system is to lift the four parcels illustrated and drop them into the bin. The system consists of two separate mechanisms. The first is the bin loader mechanism (BLM) as illustrated at the top of the diagram. The second is the conveyor mechanism (CM) illustrated at the bottom of the diagram. The BLM consists of a fork, two links, and the bin. The BLM must lift parcels, which slide onto the fork, up and over into the bin. The bin is 1.6 m tall and 2.3 m long. The fork is 1.5 m long. The parcels are permitted to pass through the top of the bin but must not pass through the sides or the bottom of the bin. Also, no part of the fork link may penetrate into the bin.

The CM must carry the parcels onto the fork of the BLM. The CM consists of a straight-line Geneva mechanism. The CM will create intermittent *x* translation of the conveyor sidewalls. This intermittent motion occurs when the peg driver, which is mounted on the crank wheel, engages the slot follower located on the follower link. The conveyor sidewalls are attached to the top of the follower link, and the parcels are on top of the conveyor sidewalls. The parcels are spaced 2.0 m apart (center-to-center) and are 400 mm tall and 500 mm long. The fork of the BLM is permitted

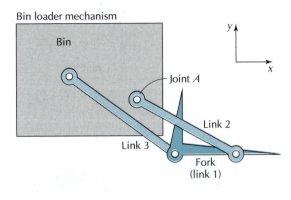

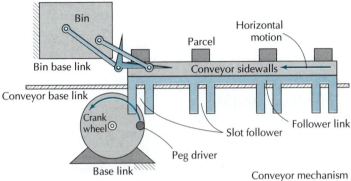

Figure A.22 Loading machine system.

to pass through the conveyor sidewalls during parcel loading, but not through the follower link. The follower link can translate only along the *x* axis.

The BLM must be actuated by a motor using a sine function to control the rotation of link 2. The motor should be located at joint *A* between the bin and link 2. The CM is to be actuated by a motor with a constant velocity, located at the center of the crank wheel. When the peg driver engages the slot follower, the follower link will be advanced by a specific amount along the *x* axis. Create a graph showing the follower link velocity with respect to time. Explain the design strategy needed to create the peg driver and the follower link system. Explain the relation between the parcel motion and the pickup by the fork. Why is the intermittent motion needed? How would it be possible to eliminate the intermittent motion of the conveyor?

The simulation must complete two full cycles (each consists of dropping two parcels into the bin).

A.23   FLIPPER MECHANISM

Create the flipper mechanism as shown in Figure A.23. The purpose of this mechanism is to turn over a flat plate by passing it from the left pad to the right pad. The left pad and right pad rise simultaneously and meet on a line inclined from the vertical axis by angle θ, at which point the flat plate is transferred. The flat plate is 15 mm thick and 240 mm long. In the starting position, as shown in the diagram, all links of the mechanism and the flat plate must fit within a space that is 880 mm wide (*x* direction) by 200 mm tall (*y* direction). The flat plate must start on the left pad. When the flip is complete, it is to rest on the right pad without protruding past the right edge of the right pad.

This mechanism must be driven by a motor with a constant velocity. The motor must be located at point *A* and drive both crank *A* and crank *B*. Create a graph to display the angle of the flat plate with respect to the angle of the left pad. What effect does the angle θ have on the operation of the mechanism? Explain the relevance of the location of the pad edge, for both pads. Replace link 1 with a rod element and create a graph showing the tension of the rod with respect to time. Describe possible applications of this mechanism.

The simulation must complete one full cycle.

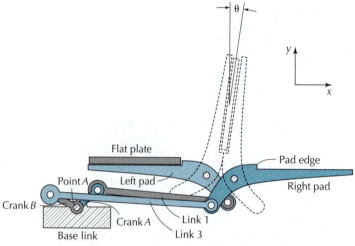

Figure A.23 Flipper mechanism.

A.24 TRANSPORT MECHANISM

Create the transport mechanism as shown in Figure A.24. The objective of this mechanism is to push the parcel intermittently, along the surface of the fixed rail, from the start position to the end position. The parcel is a box 40 mm by 40 mm. The transport link has four posts that are used to push the parcel. The posts are spaced 120 mm apart (pitch distance), and are 40 mm tall. Note the dashed line that represents the approximate path of the transport link while the machine is in operation. With the exception of the posts, no links (including the transport link) are allowed to protrude above the fixed rail surface at any time during operation. The parcel must begin in the start position and after three push cycles must finish in the end position. There are two motion requirements for the transport link. First, the angular deviation of the transport link must be less than 0.1° from the horizontal, throughout the motion cycle. Second, the horizontal path portion, indicated by the blue dashed line, must be at least 120 mm long, during which the transport link may not deviate more than 6.0 mm in the y direction. The entire mechanism, at any given position during its cycle, must fit within a space that is 740 mm in the x direction and 440 mm in the y direction.

This mechanism must be driven by a motor actuator with a constant velocity applied to the crank. Create a graph that shows the x position of the transport link with respect to its y position. Create a graph that shows the angular position of the transport link with respect to time. What

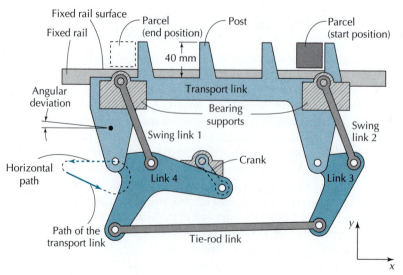

Figure A.24 Transport mechanism.

function does the tie-rod link serve? Can you explain the role of the swing links, link 1 and link 2, on the mechanism and also explain their relation to each other?

The simulation must complete three cycles.

A.25 TYPEWRITER MECHANISM

Create the typewriter mechanism as shown in Figure A.25. The objective of this mechanism is for the impact head to strike the drum by applying a vertical force to the key. It is not necessary to include any of the objects drawn as dashed lines in the simulation. The type bar, which holds the impact head, must start in the horizontal position, as illustrated, and rotate 90° counterclockwise to strike the drum. The drum has a diameter of 160 mm. The impact head must be tangent to the drum during impact and must impact the drum at the midpoint on the right side. The center of the drum must be 157 mm above (positive y direction) point A, and 87 mm to the left (negative x direction) of point A. The entire mechanism (not including the drum) must fit within a space that is 380 mm in the x direction and 150 mm in the y direction, while it is in the starting position.

The mechanism is actuated by applying an intermittent vertical force on the key. The force can be specified as a function of time, such as sin(t). In order for your mechanism to work properly, you will need to add/adjust the following three elements: (1) Adjust the elastic coefficient between the impact head and the drum. Explain the significance of this coefficient and the effects on the mechanism. (2) Convert point A to a rotational damper and adjust the damping value as necessary, for the mechanism to operate properly. Explain the purpose of modeling point A as a rotational damper. (3) Use a spring between link 1 and ground (typewriter body) to have the mechanism return to the starting position when the key force is not present. With regard to the key force, apply a force that you believe is reasonable during normal typing for the average person. Create a graph showing the contact force between the impact head and the drum as a function of time. What is the ratio between the peak contact force and the key input force? What does the ratio tell you about the mechanism?

The simulation must complete three typing cycles.

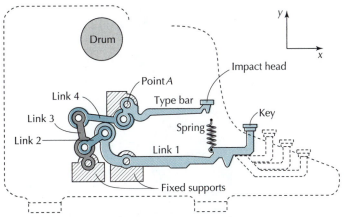

Figure A.25 Typewriter mechanism.

A.26 GRAVITY GRIPPER MECHANISM

Create the gripper mechanism as illustrated in Figure A.26. The objective of the gripper is to lift the circular payload off the ground. There are no actuators to open or close the carrying arms. The gripper should operate as follows: (1) The gripper is lowered onto the payload, while centered above it. (2) As the gripping pads contact the payload, they will not rotate but will push the carrying arms outward. After the gripping pads pass over the maximum width of the payload, the carrying arms will fall inward. (3) As the gripper is raised, the gripping pads will rotate so that they are tangent to the payload. (4) The payload will remain in the gripper's grasp and be raised with the gripper. The gripper should be symmetrical, with the exception of the cross tie link, which connects the two lower links. Set the weight of the counterweights such that the carrying arms remain in a neutral position (as illustrated in the diagram) while at rest. While the gripper is at rest, the space between the gripping pads should be 160 mm. The payload has a diameter of 176 mm. In the starting position, the gripper is centered above the payload and is located at a height such that the pad pivot point is 220 mm above the ground. Use a rod for the cross tie link. When the gripper interacts with the payload, the motion of the carrying arms should be along an inclined line, as illustrated in the diagram.

Create a mechanism to translate the entire gripper up and down onto the payload. Do this by creating a vertical slot above the lift link, and rigidly join the lift link to the vertical slot. Use an actuator to connect the lift link to some fixed point above the lift link. Set **Actuator** property to "length", and enter a function of time in the form of $[(\text{starting length}) + (\text{desired travel})\sin(t)]$ in the input box, to create the necessary up-and-down motion for the lift link, so it can grasp and lift the payload off the ground. Create a graph showing the tension in the cross tie link as a function of time. Adjust the weight of the payload to 10, 100, and 1000 kg. What is the effect on the simulation for each case? If the cross tie link has a cross-sectional area of 1.0 square inch and is made from piano wire grade steel, what is the maximum payload that can be lifted? Will the gripper be able to pick up a 160 mm square object, or an inverted triangle with 160 mm sides? What are the requirements for the payload's shape, for this gripper to work?

The simulation must show the gripper securely grasping the circular payload and lifting it off the ground.

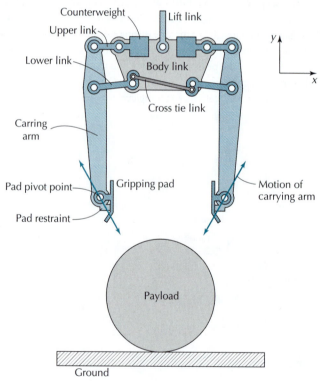

Figure A.26 Gravity gripper mechanism.

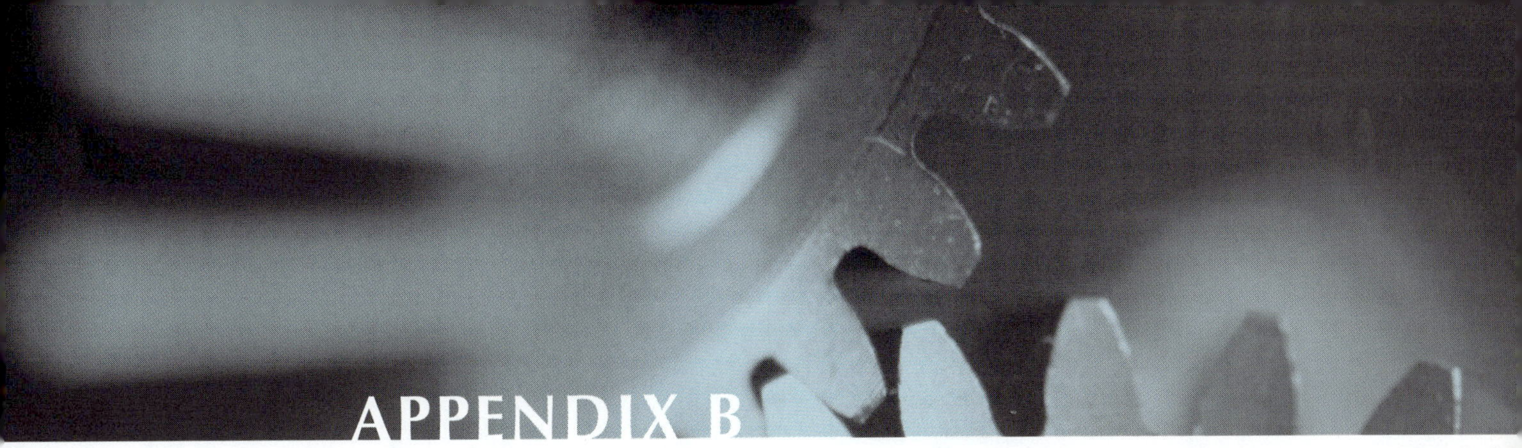

APPENDIX B

SCALARS AND VECTORS

B.1 INTRODUCTION

A *scalar* is a quantity that has magnitude, but no direction associated with it. Mass, energy, and temperature are examples of scalars. Relationships between scalars may be dealt with using algebraic expressions. A *vector,* however, has both direction and magnitude. Examples of vectors are force, displacement, velocity, and acceleration.

Vector quantities can be represented graphically by straight lines with arrowheads indicating directions, such as vectors $\bar{b}$ and $\bar{c}$ shown in Figure B.1. The length of each line is proportional to the magnitude of the vector.

Figure B.1 Representation of vectors

B.2 COMPONENTS OF A VECTOR

A vector may be described by specifying its components. The components are aligned along directions of *unit vectors* defined below.

Two sets of unit vectors are employed in this textbook. They are described in the following subsections.

B.2.1 CARTESIAN COMPONENTS

Consider vector $\bar{b}$, shown in Figure B.2. It may be described by its *Cartesian components* b_x, b_y, and b_z, so that

$$\bar{b} = b_x \bar{i} + b_y \bar{j} + b_z \bar{k}$$

(B.2-1)

where $\bar{i}, \bar{j}$, and $\bar{k}$ are unit vectors in the x, y, and z directions, respectively.

The *magnitude* of the vector may be expressed as

$$b = |\bar{b}| = (b_x^2 + b_y^2 + b_z^2)^{1/2}$$

(B.2-2)

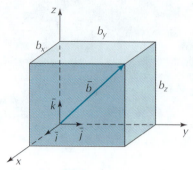

Figure B.2 Rectangular components of a vector.

When $b_z = 0$, as shown in Figure B.3, the direction of the vector may be described by angle θ with respect to the reference shown. The components of this vector are

$$b_x = b \cos \theta; \quad b_y = b \sin \theta$$
(B.2-3)

and therefore

$$\theta = \tan2^{-1}(b_x, b_y), \quad -\pi < \theta < \pi$$
(B.2-4)

B.2.2 Radial-Transverse Component

Vector $\bar{b}$, illustrated in Figure B.3, may be described as

$$\bar{b} = b\bar{i}_r$$
(B.2-5)

where $\bar{i}_r$ is the *unit radial vector*. Also associated with this vector is the *unit transverse vector*, $\bar{i}_\theta$, which is perpendicular to $\bar{i}_r$.

B.3 VECTOR OPERATIONS

B.3.1 Addition and Subtraction

Addition of vectors involves adding the x, y, and z components of the vectors. For instance, if

$$\bar{d} = \bar{b} + \bar{c}$$
(B.3-1)

then the result expressed in terms of the components is

$$\bar{d} = (b_x + c_x)\bar{i} + (b_y + c_y)\bar{j} + (b_z + c_z)\bar{k}$$
(B.3-2)

Graphically, this addition is accomplished by positioning vectors head to tail. Two vectors are shown in Figure B.4(a), and their addition is illustrated in Figure B.4(b).

Subtraction of vectors involves finding the differences of the x, y, and z components individually. If

$$\bar{d} = \bar{b} - \bar{c}$$
(B.3-3)

then the result, expressed in terms of the components, is

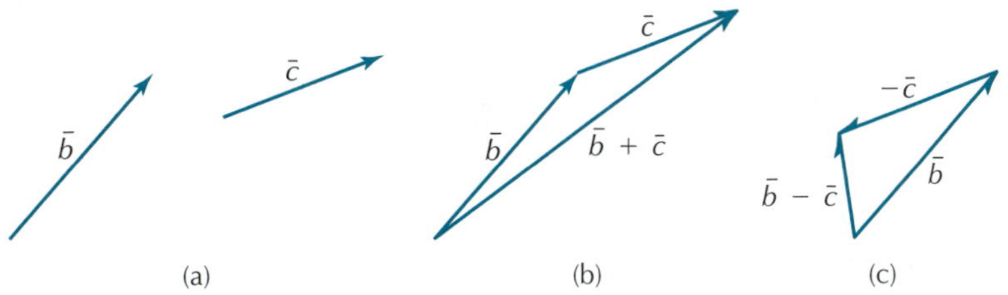

Figure B.4 Addition and subtraction of vectors.

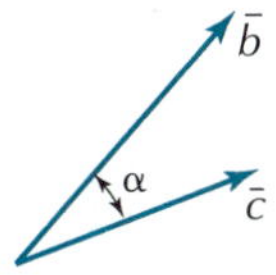

Figure B.5

$$\vec{d} = \left(b_x - c_x\right)\vec{i} + \left(b_y - c_y\right)\vec{j} + \left(b_z - c_z\right)\vec{k} \tag{B.3-4}$$

Subtraction of vectors may be regarded as the addition of the negative value of one vector to the positive value of another, that is,

$$\vec{d} = \vec{b} - \vec{c} = \vec{b} + (-\vec{c}) \tag{B.3-5}$$

Graphically, the difference between two vectors is illustrated in Figure B.4(c).

B.3.2 Dot Product

If we consider two vectors $\vec{b}$ and $\vec{c}$, then their *dot product* or *scalar product* is defined as

$$\boxed{\vec{b} \cdot \vec{c} = |\vec{b}||\vec{c}| \cos\alpha = b_x c_x + b_y c_y + b_z c_z} \tag{B.3-6}$$

where α is the angle between the two vectors as shown in Figure B.5. The dot product of two vectors is a scalar quantity.

B.3.3 Cross Product

If we consider two vectors $\vec{b}$ and $\vec{c}$, then their *cross product* is defined as

$$\vec{d} = \vec{b} \times \vec{c} = \begin{vmatrix} \vec{i} & \vec{j} & \vec{k} \\ b_x & b_y & b_z \\ c_x & c_y & c_z \end{vmatrix}$$
$$= \left(b_y c_z - b_z c_y\right)\vec{i} + \left(b_z c_x - b_x c_z\right)\vec{j} + \left(b_x c_y - b_y c_x\right)\vec{k} \tag{B.3-7}$$

The cross product of two vectors is a vector quantity. The direction of the resultant vector is perpendicular to the two that are employed in the cross product (Figure B.6). For the special case that $b_z = 0$ and $c_z = 0$, then vector $\vec{d}$ is in the direction of vector $\vec{k}$.

B.4 COMPLEX NUMBERS AND VECTORS IN THE COMPLEX PLANE

The *rectangular form* of a complex number is

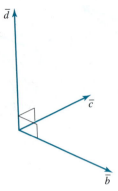

Figure B.6

$$R = x + iy \tag{B.4-1}$$

where x is the *real component* of the complex number, y is the *imaginary component* of the complex number, and

$$i = (-1)^{1/2} \tag{B.4-2}$$

The *polar form* of a complex number is

$$R = re^{i\theta} = r(\cos\theta + i\sin\theta) \tag{B.4-3}$$

where r is the *magnitude* of the complex number and θ is the *complex argument* of the complex number.

A complex number may be represented graphically by a vector in the *complex plane* as shown in Figure B.7, where quantities x, y, and θ are illustrated. For instance, θ is the angle of the vector relative to the horizontal, measured from its tail in the counterclockwise direction.

Using Equations (B.4-1) and (B.4-3), the following expressions relate the rectangular and polar forms of a complex number:

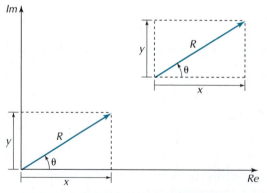

Figure B.7 Graphical representation of a complex number.

$$x = r\cos\theta; \quad y = r\sin\theta$$
$$r(x^2 + y^2)^{1/2}; \quad \theta = \tan 2^{-1}(x,y), \quad -\pi < \theta < \pi \tag{B.4-4}$$

APPENDIX C

MECHANICS: STATICS AND DYNAMICS

C.1 INTRODUCTION

Mechanics is a science that predicts the conditions of either rest or motion of bodies under the action of forces and moments. As indicated in Figure C.1, mechanics may be divided into the topics of *statics*, which deals with bodies at rest, and *dynamics*, which is concerned with bodies in motion. Dynamics may be further subdivided into kinematics and kinetics. *Kinematics* is the description and analysis of the motion of objects without consideration of the applied forces that cause them. *Kinetics* is the study of the effects of forces on the motions of objects.

A *particle* implies that the dimensions of an object are so small that all mass may be assumed to be concentrated at a single point. For a *body*, however, having finite size, different points within the body can have different positions, velocities, and accelerations. It may be considered as an assembly of particles.

Any body will always deform to some extent when acted on by a force. However, a body is considered to be *rigid* when these deformations are small enough to be neglected, and all particles are assumed to remain at fixed distances in relation to one another. In nearly all instances in this textbook, bodies are assumed to be rigid. The valve spring shown in Figure 1.1 is an example of a body that cannot be considered rigid.

This appendix covers the kinematics and kinetics of particles and rigid bodies. The material provides background needed in this textbook dealing with the analysis and design of mechanisms. Analyses will be restricted to planar motion.

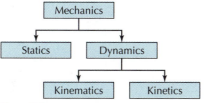

Figure C.1

C.2 KINEMATICS OF A POINT USING CARTESIAN COORDINATES

C.2.1 Position

Consider point B, in Figure C.2, which is constrained to undergo planar motion. The location of the point is depicted by means of a *position vector*, relative to fixed point O.

Employing Cartesian axes incorporating unit vectors $\bar{i}$ and $\bar{j}$, in the x and y directions respectively, the position of point B with respect to point O is

$$\bar{r}_{BO} = (r_{BO})_x\, \bar{i} + (r_{BO})_y\, \bar{j}$$

(C.2-1)

C.2.2 Velocity

The *velocity* of point B is found by taking the time derivative of Equation (C.2-1). If the coordinate system and unit vectors are fixed, then the expression for velocity is

$$\bar{v}_{BO} = \frac{d}{dt}(\bar{r}_{BO}) = \frac{d}{dt}\Big[(r_{BO})_x\, \bar{i} + (r_{BO})_y\, \bar{j}\Big]$$

$$= (\dot{r}_{BO})_x\, \bar{i} + (\dot{r}_{BO})_y\, \bar{j} = (v_{BO})_x\, \bar{i} + (v_{BO})_y\, \bar{j}$$

(C.2-2)

where a dot represents differentiation with respect to time.

In Equation (C.2-2), $\bar{v}_{BO}$ represents the velocity of B relative to point O. However, since point O is fixed, $\bar{v}_{BO}$ is also the absolute velocity of B. The motion of point B is the same with respect to any fixed point, and therefore we may express this velocity as

$$\bar{v}_{BO} = \bar{v}_B$$

(C.2-3)

C.2.3 Acceleration

The *acceleration* of point B is found by taking the time derivative of Equation (C.2-2). The result is

$$\bar{a}_{BO} = \bar{a}_B = \frac{d}{dt}(\bar{v}_{BO}) = \frac{d}{dt}\Big[(v_{BO})_x\, \bar{i} + (v_{BO})_y\, \bar{j}\Big]$$

$$= (\ddot{r}_{BO})_x\, \bar{i} + (\ddot{r}_{BO})_y\, \bar{j}$$

(C.2-4)

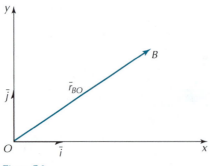

Figure C.2

C.3 KINEMATICS OF A POINT USING RADIAL-TRANSVERSE COORDINATES

C.3.1 Position

Consider point B, in Figure C.3, which is constrained to undergo planar motion. The location of the point is depicted by means of a position vector, relative to fixed point O. Employing the radial-transverse vectors, it is expressed as

$$\overline{r}_{BO} = r_{BO}\,\overline{i}_r \qquad\qquad (C.3\text{-}1)$$

where $\overline{i}_r$ is the *unit radial vector*. Also shown in Figure C.3 is $\overline{i}_\theta$, the *unit transverse vector*, which points 90° counterclockwise from the direction of $\overline{i}_r$

C.3.2 Velocity

Taking the time derivative of Equation (C.3-1), we find the velocity of point B to be

$$\overline{v}_{BO} = \overline{v}_B = \frac{d}{dt}(\overline{r}_{BO}) = \frac{d}{dt}(r_{BO}\,\overline{i}_r) = \dot{r}_{BO}\,\overline{i}_r + r_{BO}\frac{d}{dt}(\overline{i}_r) \qquad\qquad (C.3\text{-}2)$$

Equation (C.3-2) includes a term involving the time derivative of $\overline{i}_r$. This is because, for the radial-transverse coordinate system, the radial direction can be time-dependent.

To determine an expression for the time derivative of $\overline{i}_r$, we consider a small change of the position of point B as it moves along its path. After a short time interval, it has moved to B' as shown in Figure C.4(a).

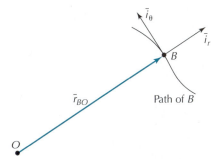

Figure C.3 Position of a point using radial-transverse coordinates

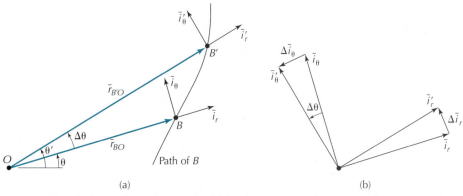

(a)

(b)

Figure C.4 Two closely spaced positions of a point: (a) Radial-transverse coordinate systems, (b) Superimposed coordinate systems

Figure C.4(b) illustrates two sets of unit vectors corresponding to the position vectors for points B and B'. From this figure

$$\Delta \, \overline{i}_r = \overline{i}_r' - \overline{i}_r \; \; ; \; \Delta \, \overline{i}_\theta = \overline{i}_\theta' - \overline{i}_\theta \tag{C.3-3}$$

By definition,

$$\frac{d}{d\theta}(\overline{i}_r) = \lim_{\Delta\theta \to 0} \frac{\Delta \overline{i}_r}{\Delta\theta} \tag{C.3-4}$$

The derivative given in Equation (C.3-4) is a vector quantity and must have both direction and magnitude. With regard to the magnitude, we note that since $\overline{i}_r$ has unit length, then for small values of $\Delta\theta$, we have

$$|\Delta\overline{i}_r| \approx 1 \times \Delta\theta = \Delta\theta \tag{C.3-5}$$

and therefore

$$\left| \frac{d}{d\theta}(\overline{i}_r) \right| \approx \frac{\Delta\theta}{\Delta\theta} = 1 \tag{C.3-6}$$

Furthermore, by inspection of Figure C.4(b), the direction of the vector quantity in Equation (C.3-4) is perpendicular to the radial vector. Thus, it is in the transverse direction.

Combining the information related to the direction and magnitude, we obtain

$$\frac{d}{d\theta}(\overline{i}_r) = \overline{i}_\theta \tag{C.3-7}$$

Applying the chain rule of differentiation, in which we employ Equation (C.3-7), we get

$$\frac{d}{dt}(\overline{i}_r) = \frac{d}{d\theta}(\overline{i}_r)\dot{\theta} = \dot{\theta}\overline{i}_\theta \tag{C.3-8}$$

where parameter $\dot{\theta}$ is the *angular velocity* of the position vector.

Similarly, the time derivative of the unit transverse vector is

$$\frac{d}{dt}(\overline{i}_\theta) = -\dot{\theta}\overline{i}_r \tag{C.3-9}$$

By substituting Equation (C.3-8) in Equation (C.3-2), we obtain the velocity expressed in terms of its radial and transverse components as follows:

$$\boxed{\overline{v}_B = \dot{r}_{BO} \, \overline{i}_r + r_{BO} \, \dot{\theta} \, \overline{i}_\theta} \tag{C.3-10}$$

C.3.3 Acceleration

Returning to Equation (C.3-10), and differentiating this with respect to time, we find the acceleration of point B to be

$$\begin{aligned} \overline{a}_{BO} = \overline{a}_B &= \frac{d}{dt}(\overline{v}_B) = \frac{d}{dt}(\dot{r}_{BO} \, \overline{i}_r + r_{BO} \, \dot{\theta} \, \overline{i}_\theta) \\ &= \ddot{r}_{BO} \, \overline{i}_r + \dot{r}_{BO} \frac{d}{dt}(\overline{i}_r) + \dot{r}_{BO} \, \dot{\theta} \, \overline{i}_\theta + r_{BO} \, \ddot{\theta} \, \overline{i}_\theta + r_{BO} \, \dot{\theta}\frac{d}{dt}(\overline{i}_\theta) \end{aligned} \tag{C.3-11}$$

Substituting Equations (C.3-8) and (C.3-9) and rearranging, we obtain

$$\bar{a}_B = (\ddot{r}_{BO} - r_{BO}\,\dot{\theta}^2)\bar{i}_r + (r_{BO}\,\ddot{\theta} + 2\dot{r}_{BO}\,\dot{\theta})\bar{i}_\theta \qquad\text{(C.3-12)}$$

where $\ddot{\theta}$ is the *angular acceleration* of the position vector.

C.4 DISPLACEMENT MATRICES

Figure C.5 shows a point Q on a rigid body that undergoes planar motion from position 1 to position n ($n = 2, 3, 4,\ldots$). The coordinates of the point before and after this motion may be related by

$$\begin{Bmatrix} x_n \\ y_n \\ 1 \end{Bmatrix} = [D_{1n}]\begin{Bmatrix} x_1 \\ y_1 \\ 1 \end{Bmatrix} \qquad\text{(C.4-1)}$$

where $[D_{1n}]$ is the 3×3 *displacement matrix* [52],

x_1 and y_1 are the coordinates of the point in position 1, and x_n and y_n are the coordinates of the point in position n.

The position vector of the point after the displacement is

$$\bar{r}_n = x_n\,\bar{i} + y_n\,\bar{j} \qquad\text{(C.4-2)}$$

Successive displacements may be described using displacement matrices. For instance, if a point moves from position 1 to position 2, and then from position 2 to position 3, then using displacement matrices, we obtain

$$\begin{Bmatrix} x_2 \\ y_2 \\ 1 \end{Bmatrix} = [D_{12}]\begin{Bmatrix} x_1 \\ y_1 \\ 1 \end{Bmatrix}; \quad \begin{Bmatrix} x_3 \\ y_3 \\ 1 \end{Bmatrix} = [D_{23}]\begin{Bmatrix} x_2 \\ y_2 \\ 1 \end{Bmatrix} \qquad\text{(C.4-3)}$$

Combining the above equations, we get

$$\begin{Bmatrix} x_3 \\ y_3 \\ 1 \end{Bmatrix} = [D_{23}][D_{12}]\begin{Bmatrix} x_1 \\ y_1 \\ 1 \end{Bmatrix} \qquad\text{(C.4-4)}$$

Listed below are forms of the displacement matrix corresponding to typical motions.

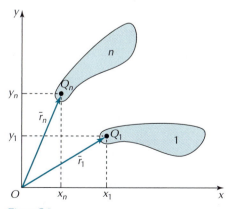

Figure C.5

C.4.1 Translation

Consider the body shown in Figure C.6, which undergoes pure translation (i.e., no rotation). The change in the coordinates of any point on the body is

$$x_n = x_1 + \Delta x; \quad y_n = y_1 + \Delta x \tag{C.4-5}$$

The corresponding displacement matrix is

$$[D_{1n}] = [D_{1n}]_T = \begin{bmatrix} 1 & 0 & \Delta x \\ 0 & 1 & \Delta y \\ 0 & 0 & 1 \end{bmatrix} \tag{C.4-6}$$

C.4.2 Rotation About a Fixed Origin

Consider the body shown in Figure C.7, which undergoes rotation about a fixed origin. In this instance, it is convenient to employ the polar form of a complex number in order to determine the coordinates of a point before and after the motion has taken place. We recognize that

$$|R_1| = |R_n| = r \tag{C.4-7}$$

$$\begin{aligned} R_n = re^{i\theta n} &= re^{i(\theta_1 + \theta_{1n})} = re^{i\theta_1} e^{i\theta_{1n}} \\ &= r\left(\cos\theta_1 + i\sin\theta_1\right)\left(\cos\theta_{1n} + i\sin\theta_{1n}\right) \\ &= \left(x_1\cos\theta_{1n} - y_1\sin\theta_{1n}\right) + i\left(x_1\sin\theta_{1n} + y_1\cos\theta_{1n}\right) \\ &= x_n + iy_n \end{aligned} \tag{C.4-8}$$

where

$$x_1 = r\cos\theta_1; \quad y_1 = r\sin\theta_1 \tag{C.4-9}$$

Using Equation (C.4-8), the displacement matrix is

$$[D_{1n}] = [D_{1n}]_R = \begin{bmatrix} C_{1n} & -S_{1n} & 0 \\ S_{1n} & C_{1n} & 0 \\ 0 & 0 & 1 \end{bmatrix} \tag{C.4-10}$$

where

$$C_{1n} = \cos\theta_{1n}; \quad S_{1n} = \sin\theta_{1n} \tag{C.4-11}$$

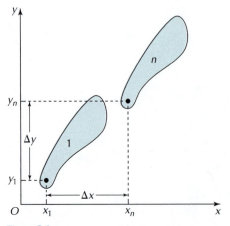

Figure C.6

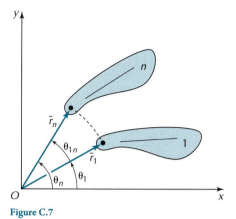

Figure C.7

C.4.3 Rotation About a Fixed Origin Followed by a Translation

Consider the body shown in Figure C.8, which undergoes rotation about a fixed origin (from position 1 to position n'), followed by a translation (from position n' to position n). In this instance, we employ the above results to obtain

$$
\left[D_{1n'}\right]_R = \begin{bmatrix} C_{1n} & -S_{1n} & 0 \\ S_{1n} & C_{1n} & 0 \\ 0 & 0 & 1 \end{bmatrix}; \quad \left[D_{n'n}\right]_T = \begin{bmatrix} 1 & 0 & (x_n - x_{n'}) \\ 0 & 1 & (y_n - y_{n'}) \\ 0 & 0 & 1 \end{bmatrix} \tag{C.4-12}
$$

Combining the above expressions, we obtain

$$
\left[D_{1n}\right] = \left[D_{n'n}\right]_T \left[D_{1n'}\right]_R = \begin{bmatrix} C_{1n} & -S_{1n} & (x_n - x_{n'}) \\ S_{1n} & C_{1n} & (y_n - y_{n'}) \\ 0 & 0 & 1 \end{bmatrix} \tag{C.4-13}
$$

However,

$$
\begin{Bmatrix} x_{n'} \\ y_{n'} \\ 1 \end{Bmatrix} = \left[D_{1n'}\right]_R \begin{Bmatrix} x_1 \\ y_1 \\ 1 \end{Bmatrix} = \begin{bmatrix} x_1 C_{1n} & -y_1 S_{1n} \\ x_1 S_{1n} & +y_1 C_{1n} \\ & 1 \end{bmatrix} \tag{C.4-14}
$$

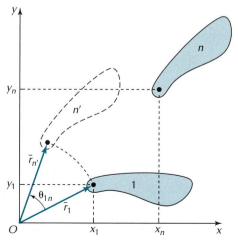

Figure C.8

Combining Equations (C.4-13) and (C.4-14), we get

$$[D_{1n}] = \begin{bmatrix} C_{1n} & -S_{1n} & A_{13n} \\ S_{1n} & C_{1n} & A_{23n} \\ 0 & 0 & 1 \end{bmatrix}$$

(C.4-15)

where

$$A_{13n} = x_n - x_1 C_{1n} + y_1 S_{1n} \, ; \quad A_{23n} = y_n - x_1 S_{1n} - y_1 C_{1n}$$

(C.4-16)

C.5 MOMENT OF A FORCE

An important concept that arises in many problems is the *moment of a force* about some specified point or axis. Referring to Figure C.9, the moment $\overline{M}_o$ of force $\overline{F}$ about point O is defined as

$$\boxed{\overline{M}_o = \overline{r} \times \overline{F}}$$

(C.5-1)

C.6 COUPLE

Figure C.10 illustrates a body subjected to two forces that are equal in magnitude and opposite in direction. The perpendicular distance between the two lines of action is d. The moment produced by these forces is referred to as a *couple*. For instance, the couple about an axis normal to its plane and passing through point O in the plane is

$$M = F(a + d) - Fa \quad \text{or} \quad M = Fd$$

(C.6-1)

The second expression in Equation (C.6-1) is independent of dimension a, which locates the forces with respect to point O. It follows that the moment of a couple has the same value for all points. Figure C.11(a) illustrates a body subjected to a force $\overline{F}$, which passes through point A. This system is equivalent to that shown in Figure C.11(b), in which forces $\overline{F}$ and $-\overline{F}$ are added through point B. It is recognized that force $\overline{F}$ through A and force $-\overline{F}$ through B provide zero net force on the body and a counterclockwise couple of $M = Fd$. The system therefore is in turn equivalent to that shown in Figure C.11(c), where the given force $\overline{F}$ acting at point A is replaced by an equal force acting at point B and a couple. Using similar reasoning, it is also possible to replace a force and a couple acting on a body with a force that has an alternate point of application.

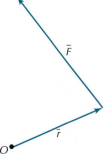

Figure C.9

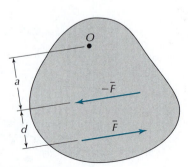

Figure C.10

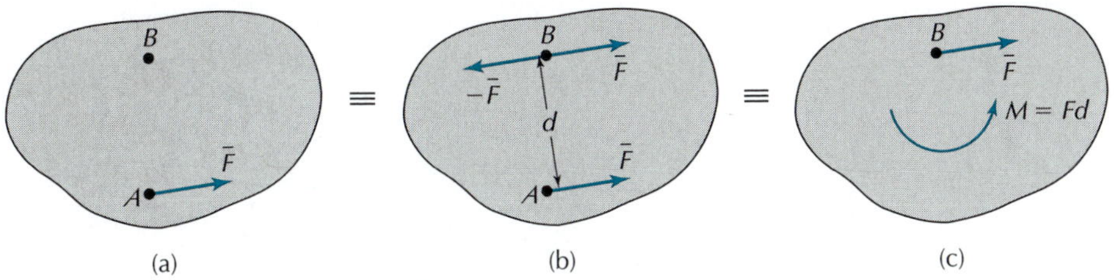

Figure C.11

C.7 NEWTON'S LAWS

Sir Isaac Newton formulated three fundamental laws of mechanics, which were first published in 1687. They provide the basis of studying kinetics.

For purposes of this textbook, *Newton's laws* may be expressed as follows:

1. Every particle remains at rest, or continues to move in a straight line with constant velocity, if there is no resultant force on it.
2. If a resultant force acts on a particle, then the particle will accelerate in the direction of the force, at a rate proportional to that force.
3. For every action between two adjoining bodies, there are actions on each body that are equal, opposite, and simultaneous.

The first law may be regarded as a special case of the second. The second law is expressed mathematically as

$$\overline{F} = m\overline{a} = m(\ddot{x}\overline{i} + \ddot{y}\overline{j}) \tag{C.7-1}$$

where $\overline{F}$ is the resultant force on the particle, m is the mass, and $\overline{a}$ is the acceleration. Vectors $\overline{F}$ and $\overline{a}$ have the same direction. Common sets of units for Equation (C.7-1) are presented in Chapter 2.

Equation (C.7-1) is applied to the analysis of a particle. In the next section, this equation will be employed in the development of the equations of motion for a rigid body.

C.8 KINETICS OF A RIGID BODY UNDERGOING PLANAR MOTION

C.8.1 Linear Motion

Figure C.12(a) illustrates a rigid body subjected to an external force. We consider the body to be an assembly of particles. A typical particle of mass m_k ($k = 1, 2, \ldots$) is shown in Figure C.12(b).

Associated with the body is its *mass center* having coordinates (x_G, y_G) with respect to a fixed coordinate system. They are defined as

$$x_G = \frac{\sum_k m_k x_k}{\sum_k m_k} = \frac{\sum_k m_k x_k}{m}; \quad y_G = \frac{\sum_k m_k y_k}{m} \tag{C.8-1}$$

where the *mass* of the body is

$$m = \sum_k m_k \tag{C.8-2}$$

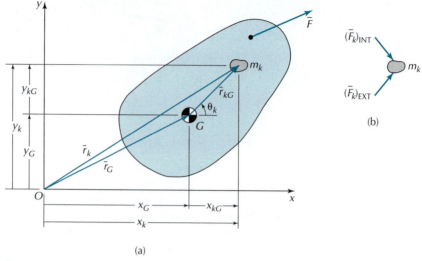

Figure C.12

The position vector to the center of mass is

$$\bar{r}_G = x_G \bar{i} + y_G \bar{j} \tag{C.8-3}$$

Forces on a particle in a body, as shown in Figure C.12(b), may be distinguished by those which are external and those which are internal. External forces are those which are applied to the particle from outside of the body. Examples include gravity and reaction forces from adjacent bodies. Internal forces are those which arise from adjacent particles in the body. Using Equation (C.7-1), the net result is

$$\bar{F}_k = \left(\bar{F}_k\right)_{\mathrm{EXT}} + \left(\bar{F}_k\right)_{\mathrm{INT}} = m_k \bar{a}_k \tag{C.8-4}$$

Now we sum equations for all of the particle masses in the system:

$$\sum_k \bar{F}_k = \sum_k \left(\bar{F}_k\right)_{\mathrm{EXT}} + \sum_k \left(\bar{F}_k\right)_{\mathrm{INT}} = \sum_k m_k \bar{a}_k \tag{C.8-5}$$

However, based on Newton's third law of mechanics, it is necessary that

$$\sum_k \left(\bar{F}_k\right)_{\mathrm{INT}} = \bar{0} \tag{C.8-6}$$

and therefore Equation (C.8-5) reduces to

$$\sum_k \left(\bar{F}_k\right)_{\mathrm{EXT}} = \sum_k m_k \bar{a}_k \tag{C.8-7}$$

Dropping the subscript EXT from Equation (C.8-7) and breaking it into its scalar components, we obtain

$$\sum_k \left(F_k\right)_x = \sum_k m_k \ddot{x}_k; \quad \sum_k \left(F_k\right)_y = \sum_k m_k \ddot{y}_k \tag{C.8-8}$$

Rearranging Equations (C.8-1), and then differentiating twice with respect to time, we get

$$m\,\ddot{x}_G = \sum_k m_k \ddot{x}_k; \quad m\,\ddot{y}_G = \sum_k m_k \ddot{y}_k \tag{C.8-9}$$

Combining Equations (C.8-8) and (C.8-9), we obtain

$$\sum_k \left(F_k\right)_x = m\,\ddot{x}_G; \quad \sum_k \left(F_k\right)_y = m\,\ddot{y}_G \tag{C.8-10}$$

Combining scalar Equations (C.8-10) into vector form, we get

$$\boxed{\overline{F} = \sum_k (F_k)_x \overline{i} + \sum_k (F_k)_y \overline{j} = m\,\overline{a}_G}$$

(C.8-11)

where

$$\boxed{\overline{a}_G = \ddot{x}_G\,\overline{i} + \ddot{y}_G\,\overline{j}}$$

(C.8-12)

Equation (C.8-11) indicates that the net external force on a body equals the mass of the body multiplied by the acceleration of the center of mass. This result is independent of the location at which the force is applied.

C.8.2 Angular Motion

A body of finite size can also have angular motions. For planar motion, we measure moments about a point. For a particle of mass m_k ($k = 1, 2,...$), we can break the moments into internal and external components, that is,

$$(\overline{M}_O)_k = (\overline{M}_O)_{k,\text{EXT}} + (\overline{M}_O)_{k,\text{INT}}$$

(C.8-13)

Using Equation (C.5-1), the moment is

$$(\overline{M}_O)_k = \overline{r}_k \times \overline{F}_k$$

(C.8-14)

where

$$\overline{r}_k = x_k\,\overline{i} + y_k\,\overline{j}$$

(C.8-15)

$$\overline{F}_k = (F_k)_x\,\overline{i} + (F_k)_y\,\overline{j}$$

(C.8-16)

Substituting Equation (C.7-1) for the kth mass in Equation (C.8-16), we obtain

$$\overline{F}_k = m_k\ddot{x}_k\overline{i} + m_k\ddot{y}_k\overline{j}$$

(C.8-17)

Combining Equations (C.8-14), (C.8-15), and (C.8-17), we get

$$(\overline{M}_O)_k = \overline{r}_k \times \overline{F}_k = \begin{vmatrix} \overline{i} & \overline{j} & \overline{k} \\ x_k & y_k & 0 \\ m_k\ddot{x}_k & m_k\ddot{y}_k & 0 \end{vmatrix} = m_k\left(x_k\ddot{y}_k - y_k\ddot{x}_k\right)\overline{k}$$

(C.8-18)

Summing the moments for all particle masses that make up the rigid body, we obtain

$$\overline{M}_O = \sum_k (\overline{M}_O)_{k,\text{ENT}} + \sum_k (\overline{M}_O)_{k,\text{INT}} = \sum_k m_k(x_k\ddot{y}_k - y_k\ddot{x}_k)\overline{k}$$

(C.8-19)

However, based on Newton's third law, we have

$$\sum_k (\overline{M}_O)_{k,\text{INT}} = \overline{0}$$

(C.8-20)

Also, since the moment is in the direction of $\overline{k}$, we can consider Equation (C.8-19) as a scalar equation, that is,

$$M_O = \sum_k m_k(x_k\ddot{y}_k - y_k\ddot{x}_k)$$

(C.8-21)

From Figure C.12(a), we may express

$$x_k = x_G + x_{kG}; \quad \ddot{x}_k = \ddot{x}_G + \ddot{x}_{kG}$$
$$y_k = y_G + y_{kG}; \quad \ddot{y}_k = \ddot{y}_G + \ddot{y}_{kG} \tag{C.8-22}$$

Substituting Equations (C.8-22) in Equation (C.8-21), we get

$$M_O = \sum_k m_k \, x_{kG} \, \ddot{y}_{kG} + x_G \sum_k m_k \, \ddot{y}_{kG} + \ddot{y}_G \sum_k m_k \, x_{kG}$$
$$+ x_G \ddot{y}_G \sum_k m_k - \sum_k m_k \, y_{kG} \, \ddot{x}_{kG} - y_G \sum_k m_k \, \ddot{x}_{kG}$$
$$- \ddot{x}_G \sum_k m_k \, y_{kG} - y_G \ddot{x}_G \sum_k m_k \tag{C.8-23}$$

We recognize that

$$\sum_k m_k \, x_{kG} = 0; \quad \sum_k m_k \, \ddot{x}_{kG} = 0$$
$$\sum_k m_k \, y_{kG} = 0; \quad \sum_k m_k \, \ddot{y}_{kG} = 0 \tag{C.8-24}$$

and

$$x_{kG} = r_{kG} \cos\theta_k; \quad y_{kG} = r_{kG} = \sin\theta_k \tag{C.8-25}$$

Substituting Equations (C.8-24) and (C.8-25) in Equation (C.8-23) and simplifying, we obtain

$$\boxed{M_O = I_G \ddot{\theta} + m(x_G \, \ddot{y}_G - \ddot{x}_G \, y_G)} \tag{C.8-26}$$

where

$$\boxed{I_G = \sum_k m_k r_{kG}^2} \tag{C.8-27}$$

is the *polar mass moment of inertia* of the body with respect to its mass center. Typical values of the polar mass moment of inertia are provided in the following section.

C.9 POLAR MASS MOMENT OF INERTIA

If each particle mass is infinitesimal, Equation (C.8-27) may be expressed as

$$\boxed{I_G = \int_m r_G^2 \, dm} \tag{C.9-1}$$

where r_G is the distance of mass element dm from the mass center.

The polar mass moment of inertia about the mass center may be expressed in terms of its *radius of gyration*, k, using the following relation

$$I_G = mk^2 \tag{C.9-2}$$

where the mass of the body is

$$m = \int_m dm \tag{C.9-3}$$

The following three examples show applications of Equations (C.9-1) and (C.9-2) for geometries often encountered in mechanisms.

EXAMPLE C.1 POLAR MASS MOMENT OF INERTIA OF A SLENDER ROD

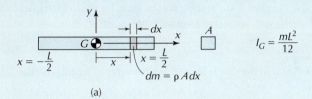

(a)

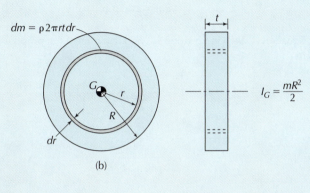

(b)

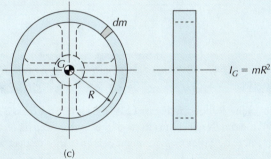

(c)

Figure C.13

For the uniform slender rod of length L shown in Figure C.13(a), determine (a) the polar mass moment of inertia with respect to its mass center and (b) its radius of gyration.

SOLUTION

The center of mass of the rod is located at its midpoint. We thus conveniently set the origin of a coordinate system at the center of mass. Furthermore, we consider an element of mass of

$$dm = \rho\, A\, dx \qquad (C.9\text{-}4)$$

where ρ is the density, and A is the cross-sectional area. Substituting Equation (C.9-4) in Equation (C.9-1) and using the appropriate limits of integration, we obtain

$$I_G = \int_m r_G^2\, dm$$

$$= \int_{x=-L/2}^{x=L/2} x^2 \rho\, A\, dx = \rho A \frac{x^3}{3}\Big|_{x=-L/2}^{x=L/2}\ \rho A\frac{L^3}{12} = \frac{mL^2}{12} \qquad (C.9\text{-}5)$$

where

$$m = \rho A L \qquad (C.9\text{-}6)$$

Using Equations (C.9-2) and (C.9-5), the radius of gyration is

$$k = \left(\frac{I_G}{m}\right)^{1/2} = \left(\frac{mL^2/12}{m}\right)^{1/2} = \frac{L}{\sqrt{12}} \qquad (C.9\text{-}7)$$

EXAMPLE C.2 POLAR MASS MOMENT OF INERTIA OF A SOLID DISC

For a solid disc of radius R and thickness t shown in Figure C.13(b), determine the polar mass moment of inertia with respect to its mass center.

SOLUTION

The center of mass of the disc is located at the center of its cross section. We consider an element of mass of

$$dm = 2\pi r t \rho \, dr \qquad \text{(C.9-8)}$$

Substituting Equation (C.9-8) in Equation (C.9-1) and using the appropriate limits of integration, we obtain

$$I_G = \int_m r_G^2 dm = \int_0^R r^2 2\pi r t \rho \, dr = 2\pi t \rho \frac{r^4}{4} \bigg|_{r=0}^{r=R} = \frac{mR^2}{2} \qquad \text{(C.9-9)}$$

where

$$m = \rho \pi R^2 t \qquad \text{(C.9-10)}$$

EXAMPLE C.3 POLAR MASS MOMENT OF INERTIA OF A RIM DISC

For a rim disc of radius R shown in Figure C.13(c), assume the mass is uniformly distributed at the periphery of the rim. Neglect the effects of the spokes, which are shown as dashed lines in the figure. Determine the polar mass moment of inertia with respect to its mass center.

SOLUTION

The center of mass of the disc is located at the center of the circular rim. The polar mass moment of inertia is

$$I_G = \int_m r_G^2 dm = R^2 \int dm = mR^2 \qquad \text{(C.9-11)}$$

C.10 WORK, ENERGY, AND POWER

C.10.1 Kinetic Energy

The *kinetic energy* due to the motion of a particle is

$$T_k = \frac{1}{2}m_k v_k^2 \qquad \text{(C.10-1)}$$

Considering a rigid body to be a composition of a series of particles, it can be shown that the kinetic energy of the body is

$$T = \sum_k T_k = \frac{1}{2}mv_G^2 + \frac{1}{2}I_G\dot{\theta}^2 \qquad \text{(C.10-2)}$$

where v_G is the speed of the mass center.

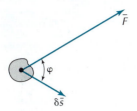

Figure C.14

C.10.2 Work Done by a Force

When applying Newton's second law of motion, the forces and accelerations are those at an instant and thus give instantaneous solutions; however, in many cases it is necessary to determine the change in motion over a finite time interval. The *work-energy* method is one way of doing this, where the interval is one of displacement, and the change of motion is defined in terms of velocity.

Consider a force $\overline{F}$ applied to a particle or rigid body, as shown in Figure C.14, and suppose that the point of application of $\overline{F}$ moves a small distance $\delta\overline{s}$, during which the force is assumed to remain constant. The work done by $\overline{F}$ during this displacement is

$$\delta W = \overline{F} \cdot \delta\overline{s} = F\delta s \cos\varphi \tag{C.10-3}$$

where φ is the angle between $\overline{F}$ and $\delta\overline{s}$. By summing all such small displacements over a finite distance, the work is

$$W = \int_s F\cos\varphi\, ds \tag{C.10-4}$$

and is measured in Newton-meters (N-m) or joules (J). For angular motions, the infinitesimal amount of work done may be expressed as

$$dW = M\, d\theta \tag{C.10-5}$$

C.10.3 Power

Power is the time rate of change of doing work. For a body undergoing linear translation we have

$$\frac{dW}{dt} = Fv\cos\varphi \tag{C.10-6}$$

where v is the speed of the body.

For rotating systems, using Equation (C.10-3), we obtain

$$\frac{dW}{dt} = M\frac{d\theta}{dt} = M\omega \tag{C.10-7}$$

where ω is the rotational speed of the body.

For the SI system, the unit of power is

$$1 \text{ watt} = 1\frac{\text{N-}m}{\text{sec}} \tag{C.10-8}$$

APPENDIX D

INDEX OF COMPANION WEBSITE

This appendix contains a listing of all the digital files included on the companion website that accompanies this textbook. Also indicated is the type of software on which each file is based.

The video animation files have been stored in mp4 format. These video files can be viewed with software such as Windows Media Player, QuickTime Player, or other similar video player software. Other files employ *Working Model 2D* and require either a copy of the software or the file player version of the software.

Also included on the companion website are video clips of real mechanical systems in operation.

CHAPTER 1 INTRODUCTION

File Name	Video player software	Working Model 2D player
Video 1.1: Single-cylinder piston engine	•	
Video 1.4: Engine assembly	•	
Video 1.5: Slider crank mechanism	•	
Video 1.7: Slider crank mechanism with offset	•	
Video 1.8: Four-bar mechanism	•	
Video 1.10: Equivalent four-bar mechanisms	•	
Video 1.12: Washing machine mechanism	•	
Video 1.14A: Four-bar with coupler point	•	•
Video 1.14B: Four-bar with coupler point	•	•
Video 1.15A: Chebyshev straight-line mechanism	•	
Video 1.15B: Roberts straight-line mechanism	•	
Video 1.15C: Watt straight-line mechanism	•	
Video 1.16: Film transport mechanism	•	
Video 1.20: Chain drive	•	
Video 1.22: Toothed gears	•	
Video 1.23: Internal gear	•	
Video 1.24: Rack and pinion	•	
Video 1.25: Rack and pinion steering	•	
Video 1.26: Disc cam mechanism	•	
Video 1.27: Disc cams	•	
Video 1.28A: Wedge cam mechanism	•	

(*Continued*)

CHAPTER 1 (CONTINUED)

File Name	Video player software	Working Model 2D player
Video 1.28B: Cylindrical cam mechanism	•	
Video 1.28C: End cam mechanism	•	
Video 1.29: Fishing reel	•	
Video 1.30: Prosthetic hand	•	
Video 1.31: Chalmers suspension	•	
Video 1.35: Linear bearing sliding stage	•	
Video 1.42: Front-end loader mechanism	•	•
Video 1.43: Slider crank mechanism and its three inversions	•	
Video 1.45: Four-bar mechanism and its three inversions	•	
Video 1.48: Speed governor	•	
Video 1.51: Parallelogram four-bar mechanism	•	
Video 1.53: Variable base link four-bar mechanism	•	•
Video 1.54: Variable-offset slider crank mechanism	•	•
Video 1.57: Cognates of a four-bar mechanism	•	
Video 1.58: Cayley's triangles	•	
Video 1.61: Cognate of a slider crank mechanism	•	

CHAPTER 2 MECHANICS OF RIGID BODIES

File Name	Video player software	Working Model 2D Player
Video 2.16: Coriolis acceleration	•	•
Video 2.23: Shaping machine	•	
Video 2.36: Instantaneous centers of a four-bar mechanism	•	
Video 2.47: Door latch mechanism	•	

CHAPTER 3 GRAPHICAL KINEMATIC ANALYSIS OF PLANAR MECHANISMS

File Name	Video player software	Working Model 2D player
Video 3.4: Mechanism incorporating a two degree of freedom pair	•	
Video 3.7: Wheeled system	•	
Video 3.8: Mechanism and velocity polygon	•	
Video 3.9: Illustration of relative velocities	•	
Video 3.15: Mechanism and acceleration polygon	•	

CHAPTER 4 ANALYTICAL KINEMATIC ANALYSIS OF PLANAR MECHANISMS

File Name	Video player software	Working Model 2D player
Video 4.20: Parallel-motion mechanism	•	
Video 4.21: Umbrella	•	

- *Mathcad program: fourbarkin, kinematic analysis of a four-bar mechanism*
- *Mathcad program: slidercrankkin, kinematic analysis of a slider crank mechanism*
- *Mathcad program: fourbarcpkin, kinematic analysis of a four-bar mechanism with a coupler point*
- *Mathcad program: slidercrankcpkin, kinematic analysis of a slider crank mechanism with a coupler point*

CHAPTER 5 GEARS

File Name	Video player software	Working Model 2D player
Video 5.2: Continuously variable traction drive	•	
Video 5.4: Continuously variable belt drive	•	
Video 5.5A: Straight spur gears	•	
Video 5.5B: Helical spur gears	•	
Video 5.7: Miter gears	•	
Video 5.8A: Herringbone gears, narrow	•	
Video 5.8B: Herringbone gears, wide	•	
Video 5.9A: Plain bevel gears	•	
Video 5.9B: Spiral bevel gears	•	
Video 5.10: Hypoid gear set	•	
Video 5.11A: Worm and wheel gears, right-hand worm	•	
Video 5.11B: Worm and wheel gears, left-hand worm	•	
Video 5.12A: Worm and wheel, double-start worm, right-hand worm	•	
Video 5.12B: Worm and wheel, double-start worm, left-hand worm	•	
Video 5.15: Generation of an involute profile	•	
Video 5.18: Straight spur gears	•	
Video 5.19: External–internal meshing gears	•	
Video 5.20: Rack and pinion gears	•	
Video 5.27: Antibacklash gear	•	
Video 5.32: Milling of a straight spur gear	•	
Video 5.34: Straight spur gear and hob	•	
Video 5.37: Hobbing of a straight spur gear	•	
Video 5.39: Helical spur gear and hob	•	
Video 5.40: Hobbing of a helical spur gear	•	
Video 5.42A: Shaping of a straight spur gear, external gear	•	
Video 5.42B: Shaping of a straight spur gear, internal gear	•	
Video 5.43A: Shaping of a straight spur gear	•	
Video 5.43B: Shaping of an internal gear	•	
Video 5.45: Planing of a plain bevel gear	•	
Video 5.47: Rotary broaching operation	•	

(Continued)

CHAPTER 5 (CONTINUED)

File Name	Video player software	Working Model 2D player
Video 5.52: Two spur gears in mesh	•	
Video 5.62: Meshing of gears requiring undercutting	•	
Video 5.63: Generation of a cycloid	•	
Video 5.64: Cycloidal gerotor pump	•	
Video 5.66: Cycloidal gerotor pump	•	
Video 5.67: Supercharger blower	•	
Video 5.68: Cycloidal spur gears	•	

CHAPTER 6. GEAR TRAINS

File Name	Video player software	Working Model 2D player
Video 6.3: Mechanical clock	•	
Video 6.6A: Simple gear train	•	
Video 6.8A: Reverted gear train	•	
Video 6.9: Gear train used in a winch	•	
Video 6.14: Manual transmission	•	
Video 6.16: Manual transmission	•	
Video 6.17: Dual-clutch transmission	•	
Video 6.21: Planetary gear train	•	
Video 6.25: Pencil sharpener	•	
Video 6.28: Two-speed automatic transmission	•	
Video 6.33B: Two-stage planetary gear train	•	
Video 6.34: Differential gear train	•	
Video 6.38: Front wheel drive gear train	•	
Video 6.39: Differential incorporating spur gears	•	
Video 6.41: Torsen differential	•	
Video 6.45: Harmonic drive	•	

CHAPTER 7. CAMS

File Name	Video player software	Working Model 2D player
Video 7.2: Ballpoint pen	•	
Video 7.3: Ballpoint pen mechanism—planar representation	•	
Video 7.5: Toggle mechanism electric switch	•	
Video 7.6: Heart switch	•	
Video 7.7: Door knob mechanism	•	
Video 7.8: Caster wheel	•	
Video 7.9: Disc cam mechanisms	•	

(Continued)

CHAPTER 7 (CONTINUED)

File Name	Video player software	Working Model 2D player
Video 7.30: Positive-motion disc cam mechanism	•	
Video 7.31: Constant-breadth cam Mechanism	•	
Video 7.33: Movie projector mechanism	•	
Video 7.57: Moving-headstock milling machine	•	
Video 7.58A: Single knife-edge follower, disc cam mechanism	•	
Video 7.58B: Multiple knife-edge followers, disc cam mechanism	•	
Video 7.58C: Roller follower, disc cam mechanism	•	
Video 7.59: Cylindrical cam mechanism	•	
Video 7.60: Face cam mechanism	•	
Video 7.61A1: Manufacture of a disc cam—scribing radial lines	•	
Video 7.61A2: Manufacture of a disc cam—scribing circular arcs	•	
Video 7.61B: Manufacture of a disc cam—scribing rises and falls	•	
Video 7.61C: Manufacture of a disc cam—rough cut	•	
Video 7.61D: Manufacture of a disc cam—fine cut	•	

• Cam Design, kinematic analysis of disc cam mechanisms

CHAPTER 8. GRAPHICAL FORCE ANALYSIS OF PLANAR MECHANISMS

File Name	Video player software	Working Model 2D player
Video 8.7: Force polygon	•	
Video 8.11: Scissor jack	•	
Video 8.15: Front-end loader mechanism	•	
Video 8.17: Parking brake mechanism	•	
Video 8.23: Dynamic analysis of a four-bar mechanism	•	•

CHAPTER 9. ANALYTICAL FORCE ANALYSIS AND BALANCING OF PLANAR MECHANISMS

File Name	Video player software	Working Model 2D player
Video 9.26: Shaking force analysis	•	

- *Mathcad program: fourbarforce, dynamic analysis of a four-bar mechanism*
- *Mathcad program: slidercrankforce, dynamic analysis of a slider mechanism*
- *Mathcad program: fourbarbalance, balancing of a four-bar mechanism*
- *Mathcad program: slidercrankbalance, balancing of a slider crank mechanism*

CHAPTER 10 FLYWHEELS

File Name	Video player software	Working Model 2D player
Video 10.3: Flywheel analysis	•	
Video 10.4: Engine flywheel	•	•

CHAPTER 11 SYNTHESIS OF MECHANISMS

File Name	Video player software	Working Model 2D player
Video 11.2: Corkscrew mechanisms		
Video 11.21A: Headlight cover, undesirable design	•	
Video 11.21B: Headlight cover, desirable design	•	

- *Mathcad program: fourbarfuncsyn, function synthesis of a four-bar mechanism*
- *Mathcad program: slidercrankfuncsyn, function synthesis of a slider crank mechanism*
- *Mathcad program: fourbarrbg, rigid-body guidance synthesis of a four-bar mechanism*

CHAPTER 13 DESIGN CASE STUDIES FOR MECHANISMS AND MACHINES

File Name	Video player software	Working Model 2D player
Video 13.3: Hooke's coupling	•	
Video 13.4: Double Hooke's coupling	•	
Video 13.5: Constant-velocity coupling	•	
Video 13.8: Oldham coupling	•	
Video 13.10: Offset drive	•	
Video 13.11: Clutch mechanism	•	
Video 13.13: Synchronizer	•	
Video 13.14: Cross section of a synchronizer	•	
Video 13.15: Drill chuck mechanism	•	
Video 13.18: Manual lathe chuck	•	
Video 13.19: Power lathe chuck	•	
Video 13.20: Disc brake	•	
Video 13.21: Drum brake	•	
Video 13.22: Ratchet mechanism	•	
Video 13.23: Ratchet wrench	•	
Video 13.24: Sprag clutch	•	
Video 13.25: Escapement mechanism	•	
Video 13.26: Iris mechanism	•	
Video 13.27: Flow regulator	•	
Video 13.28: Lead screw and nut	•	
Video 13.29: Recirculating ball screw	•	

(Continued)

CHAPTER 13 (CONTINUED)

File Name	Video player software	Working Model 2D player
Video 13.30: Straight-line dwell mechanism	•	
Video 13.31: Circular-arc dwell mechanism	•	
Video 13.32: Geneva mechanism	•	
Video 13.33: Counting mechanism	•	
Video 13.34: Scotch yoke mechanism	•	
Video 13.35: Application of scotch yoke mechanism	•	
Video 13.36: Overhead valve piston engine cross-section	•	
Video 13.37: Sleeve valve piston engine	•	
Video 13.38: V-2 piston engine	•	
Video 13.39A: Radial piston engine	•	
Video 13.39B: Radial piston engine with propeller	•	
Video 13.40: Airplane radial piston engine	•	
Video 13.42: Wankel engine	•	
Video 13.43: Cresent pump	•	
Video 13.44: Radial-Vane pump	•	
Video 13.46: Swash-plate compressor	•	
Video 13.47: Fluid dispenser pump	•	
Video 13.48: Zero-max drive	•	
Video 13.49: Helicopter rotor-blade adjustment mechanism	•	
Video 13.51A: Simple toggle mechanism	•	
Video 13.51B: Double toggle mechanism	•	
Video 13.52: Ring binder mechanism	•	
Video 13.53: Key lock mechanism	•	
Video 13.54: Combination lock mechanism	•	
Video 13.60: Cable clip mechanism	•	
Video 13.61: Compliant stapler	•	
Video 13.62: Zipper mechanism	•	

APPENDIX A DESIGN PROJECTS USING WORKING MODEL 2D

File Name	Video player software	Working Model 2D player
Video A.0 Working Model tutorial	•	

ANSWERS TO SELECTED PROBLEMS

Chapter 1

P1.2:   (a) $r_2 < 0.5$ cm; (b) $1.5 < r_2 < 3.5$ cm; (c) $r_2 = 0.5$ cm or $r_2 = 1.5$ cm or $r_2 = 3.5$ cm;
        (d) $0.5 < r_2 < 1.5$ cm or $3.5 < r_2 < 5.5$ cm
P1.4:   (a) $r_2 \leq 1.5$ cm; (b) $r_2 \geq 0$ cm
P1.6:   (a) crank rocker; (b) change point; (c) rocker–rocker
P1.10: (a) 1; (b) 1
P1.12: (a) 1; (b) 1
P1.18: (a) 1; (b) 1

Chapter 2

P2.4:   (a) 50 rpm; (b) 31.4 cm/sec
P2.6:   (a) 13.0 cm; (b) 210°, 330°; (c) 1; (d) 1; (e) 1
P2.8:   (a) 124 cm/sec; (b) 85.6 cm/sec; (c) 1.45
P2.10: (a) 53.1°, 306.9°; (b) 0, 180°; (c) 0, 53.1°, 180°, 306.9°
P2.12: 18.4°, 85.2°, 198.4°, 311.7°
P2.14: 80.4°, 120°
P2.26: (a) $m = 1$
P2.28: (a) $m = 1$
P2.30: (a) $m = 1$
P2.32: 1193 kN-m
P2.34: 40 kN
P2.36: 0.36 m
P2.38: 319 mm
P2.40: (a) 13.1 N ; (b) 6.94 N
P2.42: 6000 N
P2.44: (a) 9990 N ; (b) 740 N
P2.48: 20.5 N
P2.50: 91 N-cm
P2.52: 39 N

Chapter 3

P3.2:   47.1 in/sec @ 80°
P3.4:   (a) 21.9 rad/sec CW; (b) 19 in/sec @ 180°
P3.6:   3.7 rad/sec CW
P3.8:   7.2 in/sec @ 345°
P3.10: (a) 3.5 rad/sec CW; (b) 1.9 rad/sec CW; (c) 43 cm/sec @ 0°
P3.12: (a) 5.42 rad/sec CW; (b) 493 cm/sec^2 @ 65°; (c) 21.7 cm/sec @ 280°
P3.14: (a) 2.23 rad/sec CCW; (b) 2.12 rad/sec CCW; (c) 25.0 cm/sec @ 242°
P3.16: (a) 10 rad/sec CW; (b) 20 rad/sec CCW

P3.18: 4.5 rad/sec CW, 38.2 in/sec @ 75°

P3.20: 0.28 rad/sec CW, 0.94 in/sec @ 140°

P3.22: (a) 18.4 cm/sec @ 300°; (b) 18.0 cm/sec @ 339°; (c) 11.2 cm/sec @ 50°; (d) 1.7

P3.24: (b) 47.1 in/sec @ 330°, 17.6 in/sec @ 18°, 27.9 in/sec @ 10°, 10.9 rad/sec CW, 11.2 rad/sec CW

P3.26: (b) 43 cm/sec @ 210°, 627 cm/sec^2 @ 260°

P3.28: (b) 4.4 rad/sec CCW, 6.4 rad/sec CCW, 1060 cm/sec^2 @ 72°

P3.30: (a) 152 cm/sec^2 @ 65°

P3.32: (b) 4.2 rad/sec CW, 6.4 rad/sec CW, 109 cm/sec @ 330°

P3.34: (c) 120 cm/sec^2 @ 220°, 27.1 rad/sec^2 CW, 13.9 rad/sec^2 CCW

P3.36: (c) 80 rad/sec^2 CW

P3.38: (c) 33 rad/sec^2 CCW

P3.40: (c) 634 rad/sec^2 CW, 650 cm/sec^2 @ 90°

P3.42: (d) 177 rad/sec^2 CCW, 353 rad/sec^2 CCW

P3.46: (c) 41 rad/sec^2 CCW

P3.48: (a) 6.0 rad/sec CW, (b) 48 cm/sec @ 270°, (c) 576 cm/sec^2 @ 0°

Chapter 4

P4.2:   12.5 rad/sec CW

P4.4:   (d) 13.6 in/sec

P4.6:   (e) 2030 cm/sec^2 @ 0° and 180°

P4.8:   (a) 7.29 rad/sec CW; (b) 718 cm/sec^2 @ 258°

P4.12: 1.69 rad/sec CCW

P4.16: (c) 439 cm/sec^2 @ 253°

P4.18: (a) 8.5 rad/sec CCW; (b) 178 cm/sec @ 353°

P4.22: (a) 0.606 rad/sec CCW; (b) 44.9 cm/sec @ 292°

P4.24: (a) 2.21 rad/sec CCW; (b) 42.5 cm/sec @ 123°

P4.26: (c) 8032 cm/sec^2 @ 240°

P4.28: 64.4°, 309 cm/sec @ 180°, 260 cm/sec^2 @ 0°

Chapter 5

P5.2:   450 rpm

P5.4:   (a) 1.33; (b) 30

P5.6:   15, 25, 2356 mm/sec

P5.8:   (a) 12.56 mm; (d) 282 mm

P5.10: 16, 64

P5.12: 26, 65, 18°

P5.14: (a) 45.105 mm, 43 mm, 48 mm, 52 mm

P5.16: (a) 0.785 in, 3.0 in, 180 in/sec, 1.79

P5.18: (a) 0.433 in, 3.308 in

P5.20: (a) 0.577 in, 2.205 in, 7.351 in, 5.146 in, 18 rad/sec, 132.3 in/sec

P5.24: (a) 0.546 in; (b) 1.48

P5.28: (a) 0.785 in, 3.0 in, 10.0 in, 7.0 in, 18 rad/sec, 180 in/sec, 1.91

P5.30: (a) 166.7 rpm CW; (b) 1; (c) 131 cm/sec; (e) 58.9 liters/min

Chapter 6

P6.2: $N_2 = 120, N_3 = 39, N_4 = 150, N_5 = 115$ (other solutions possible)

P6.6:   (a) 1.496 in; (b) 0.477 in; (c) 1.64; (d) 39, 19 33, 27; (e) 0.517 in; (f) 14.8°; (g) 0; (h) 3.127 in; (i) 3.445 in

P6.8:   (d) 37, 23, 34, 31

P6.10: (e) 35.6°

P6.12: (a) 7.48 in; (c) 1.86; (e) 35.7°; (f) 7.79 in

P6.14: (c) 1.93; (e) 24.6°

P6.20: 3

P6.22: 6328 rpm CW
P6.24: 28.6 rpm CW
P6.26: 17,542 rpm (opposite to output)
P6.28: 28.6 rpm CCW
P6.30: (a) 1.75 rev CCW; (b) 7.15 rev CCW; (c) 2.75 rev CW
P6.32: (b) (i) 0.25 rev CCW
P6.34: (a) 55 rpm CCW; (b) 400 rpm CW; (c) 8
P6.36: 19.61 rad/sec
P6.40: (a) 800 rpm CW; (b) 200 rpm CCW
P6.42: 19.9 rad/sec (opposite to input)
P6.46: 1.4 rpm CCW
P6.48: 546.8 N-m (same direction as output)

Chapter 7

P7.2: 60 rpm
P7.8: (b) φ @ $\theta = 30° = 23°$
P7.14: increase base circle to 10 cm to eliminate undercutting
P7.16: dwell from 162° to 198°

Chapter 8

P8.2: (a) 14 lb @ 90°; (b) 129 lb @ 270°
P8.4: 158 lb @ 0°; 17.5 lb @ 90°
P8.6: 6.1 in-lb CW
P8.8: 900 N-m CW
P8.10: 2.49 N-m CW
P8.12: 30 N
P8.14: 116 N-m CCW
P8.16: 12,300 N, 7300 N
P8.18: 0.5 N-m CW
P8.20: (a) 9300 in-lb CW; (b) 55.5 in-lb CCW; (c) 89.7 in-lb CCW
P8.22: 4.16 N-m CW
P8.24: (a) 0.34 N-m CCW; (b) 0.66 N-m CCW; (c) 1.85 N-m CCW
P8.26: 0.011 N-m CCW
P8.28: (a) 0.64 N-m CCW; (b) 1.21 N-m CCW; (c) 1.00 N-m CCW

Chapter 9

P9.4: 387 gr @ −175.8°; 448 gr @170°
P9.6(a): 3.6 gr

Chapter 10

P10.2: (a) 0.1053; (b) 490 kg
P10.4: (a) 641 N-m; (b) 0.0488
P10.6: (a) 136 N-m; (b) 0.975 m
P10.8: 0.274
P10.10: (a) 387.2 rpm, 412.8 rpm; (b) 51.6 rad/sec^2, 35.9 rad/sec^2; (c) 41.2 cm
P10.12: 84 kJ
P10.14: (a) 209 M-watt; (b) 168 kg; (c) 61.7 rad/sec^2
P10.16: 47.4 cm

Chapter 11

P11.2: $r_2 = 1.70$ cm, $r_3 = 2.33$ cm, $r_4 = 4.05$ cm
P11.4: $r_2 = 1.759$ cm, $r_3 = 0.724$ cm, $r_4 = 2.066$ cm

REFERENCES

1. Dechev, N., Cleghorn, W. L., and Naumann, S., Multiple Finger, Passive Grasp Prosthetic Hand, *Mechanism and Machine Theory*, Vol. 36, No. 10, pp. 1157–1173, 2001.

2. Uicker, J. J., Pennock, G. R., and Shigley, J. E., *Theory of Machines and Mechanisms*, fourth edition, Oxford University Press, New York, 2010.

3. Mabie, H. H., and Reinholtz, C. F., *Mechanisms and Dynamics of Machinery*, fourth edition, Wiley, New York, 1987.

4. Beer, F. P., and Johnson, E. R., *Vector Mechanics for Engineers*, McGraw-Hill, New York, 1988.

5. Zarkandi, S., Application of Mechanical Advantage and Instant Centers on Singularity Analysis of Single-DOF Planar Mechanisms, *Journal of Mechanical Engineering*, Vol. ME 91, No. 1, pp. 50–57, 2010.

6. Lévai, Z., Theory of Epicycle Gears and Epicycle Change-Speed Gearing, Budapest Technical University of Building, Civil Transport Engineering, Professorate for Motor Vehicles, Doctorate dissertation, No. 29, 1966.

7. Cleghorn, W. L., and Tyc, G., "Kinematic Analysis of Multiple Stage Planetary Gear Trains Using a Microcomputer," *International Journal of Mechanical Engineering Education*, Vol. 15, No. 1, pp. 57–69, 1987.

8. Chocholek, S. E., The Development of a Differential for the Improvement of Traction Control, In *Traction Control and Anti-Wheel-Spin Systems for Road Vehicles*, IMechE, Automobile Division, London, England, pp. 75–82, 1988.

9. Cleghorn, W. L., and Podhorodeski, R. P., Disc Cam Design Using a Microcomputer, *International Journal of Mechanical Engineering Education*, Vol. 16, No. 4, pp. 235–250, 1988.

10. Berkof, R. S., and Loewen, G. G., A New Method for Completely Force Balancing Simple Linkages, *ASME J. Engineering for Industry*, Vol. 91, No. 1, pp. 21–26, 1969.

11. Kraus, H., and Babbidge, H. D., Symptoms of Withdrawal: An Assessment of Mechanical Corkscrews, *Chartered Mechanical Engineer*, Vol. 24, No. 12, pp. 70–75, 1977.

12. Hartenberg, R. S., and Denavit, J., *Kinematic Synthesis of Mechanisms*, McGraw-Hill, New York, 1964.

13. Hrones, J. A., and Nelson, G. L., *Analysis of the Four Bar Linkage*, Wiley, New York, 1951.

14. Erdman, A. G., Sandor, G. N., and S. Kota, *Mechanism Design: Analysis and Synthesis*, fourth edition, Prentice-Hall, Englewood Cliffs, NJ, 2001.

15. Hyman, B., *Fundamentals of Engineering Design*, second edition, Prentice Hall, Upper Saddle River, NJ, 2003.

16. Eggert, R. J., *Engineering Design*, second edition, Prentice Hall, Upper Saddle River, NJ, 2010.

17. Ulrich, K. T., and Eppinger, S. D., *Product Design and Development*, fifth edition, McGraw-Hill, New York, 2012.

18. Parmley, R. O., *Illustrated Sourcebook of Mechanical Components*, McGraw-Hill Professional, New York, 2000.

19. Sclater, N., and Chironis, N., *Mechanisms and Mechanical Devices Sourcebook*, McGraw-Hill, New York, 1996.

20. Otto, K. N., and Wood, K. L., *Product Design: Techniques in Reverse Engineering and New Product Development*, Prentice-Hall, Engleword Cliff, NJ, Chapter 5, 2001.

21. Dixon J., and Poli, C., *Engineering Design and Design for Manufacturing: A Structured Approach*, Fieldstone Publications, Nashville, TN, 1995.

22. Akiyama, N. K., *Function Analysis: Systematic Improvement of Quality and Performance*, Productivity Press, New York, 1991.

23. Michalko, M., *Thinkertoys: A Handbook of Creative-Thinking Techniques*, second edition, Ten Speed Press, Berkeley, CA, 2006.

24. Linsey, J. S., and Becker, B., Effectiveness of Brainwriting Techniques: Comparing Nominal Groups to Real Terms in *Design Creativity*, Springer, London, 2011.

25. Spotts, M. F., *Design of Machine Elements*, Prentice-Hall, sixth edition, 1985.

26. Hyman, B., *Fundamentals of Engineering Design*, Prentice-Hall, second edition, Chapter 9, Upper Saddle River, NJ, 2003.

27. Ugural, A. C., and Fenster, S. K., *Advanced Mechanics of Materials and Applied Elasticity*, fifth edition, Prentice Hall, Upper Saddle River, NJ, 2012.

28. Solecki, R., and Conant, R. J., *Advanced Mechanics of Materials*, Oxford University Press, New York, 2003.

29. Logan, D. L., *A First Course in the Finite Element Method*, fifth edition, Cengage Learning, 2011.

30. Arora, J. S., *Introduction to Optimum Design*, third edition, Elsevier Academic Press, Amsterdam, 2012.

31. Verzuh, E., *The Fast Forward MBA in Project Management*, fourth edition, Wiley, Hoboken, NJ, 2011.

32. McMaster Carr (July 1, 2014). Home Page [Online]. Available: http://www.mcmaster.com/

33. Stock Drive Products (July 1, 2014). Home Page SDP/SI [Online]. Available: http://www.sdp-si.com/

34. Digikey (July 1, 2014). Home Page [Online]. Available: http://www.digikey.com/

35. Metal Supermarkets (July 1, 2014). Home Page [Online]. Available: http://www.metalsupermarkets.com/

36. Hess, E., *The CNC Cookbook: An Introduction to the Creation and Operation of Computer Controlled Mills, Router Tables, Lathes, and More*, Scitech Publications, Nagur, Pakistan, 2009.

37. Steen, W. M., and Mazumder, J., *Laser Material Processing*, fourth edition, Springer, Berlin, 2010.

38. Guitrau, E. B., *The EDM Handbook*, Hanser Publications, Cincinnati, 2009.

39. Gibson, I., and Rosen, D. W., *Additive Manufacturing Technologies: Rapid Prototyping to Direct Digital Manufacturing*, Springer, Berlin, 2010.

40. Hicks, C. R., and Turner, K. V., *Fundamental Concepts in the Design of Experiments*, fifth edition, Oxford University Press, New York, 1999.

41. Antony, J., *Design of Experiments for Engineers and Scientists*, Elsevier, Amsterdam, 2003.

42. Sclater, N., and Chironis, N. P., *Mechanisms and Mechanical Devices Sourcebook*, fourth edition, McGraw-Hill Professional, New York, 2001.

43. Stone, R., *Introduction to Internal Combustion Engines,*" third edition, Society of Automotive Engineers Inc., Washington, D.C., 1999.

44. Alth, M., *All About Locks and Locksmithing*, Hawthorn Books, New York, 1972.

45. Howell, L. L., *Compliant Mechanisms*, Wiley-Interscience, Hoboken, NJ, 2001.

46. Spengen, W. M., Puers, R., and De Wolf, I., A Physical Model to Predict Stiction in MEMS, *Journal of Micromechanics and Microengineering*, Vol. 12, pp. 702–713, 2002.

47. Basha, M. A., Dechev, N., Safavi-Naeini, S., and Chadhuri, S., A Scalable $1 \times N$ Optical MEMS Switch Architecture Utilizing A Microassembled Rotating Micromirror, *IEEE Journal of Selected Topics in Quantum Electronics*, Vol. 13, No. 2, pp. 336–347, March/April 2007.

48. Dechev, N., Cleghorn, W. L., and Mills, J. K., Microassembly of 3-D MEMS Structures Utilizing a MEMS Microgripper with a Robotic Manipulator, *Proceedings of the IEEE International Conference on Robotics and Automation (ICRA 2003)*, Taipei, Taiwan, September 14–19, 2003.

49. Analog Devices (July 1, 2014). MEMS Acelerometers, [Online]. Available: http://www.analog.com/en/mems-sensors/mems-accelerometers/products/index.html

50. Texas Instruments (July 1, 2014). DLP technology product [Online]. Available: http://www.dlp.com/

51. Dechev, N., Cleghorn, W. L., and Mills, J. K., Microassembly of 3D Microstructures Using a Compliant, Passive Microgripper, *Journal of Microelectromechanical Systems*, Vol. 13, No. 2, pp. 176–189, 2004.

52. Suh, C. H., and Radcliffe, C. W., Synthesis of Plane Linkages with Use of the Displacement Matrix, *ASME J. Engineering for Industry*, Vol. 84, No. 2, pp. 206–214, 1967.

TRIGONOMETRIC
IDENTITIES

$$\cos\left(u \pm \frac{\pi}{2}\right) = \mp \sin u; \qquad \sin\left(u \pm \frac{\pi}{2}\right) = \pm \cos u$$

$$\sin(u + \pi) = -\sin u; \qquad \sin(-u) = -\sin u$$

$$\cos(u + \pi) = -\cos u; \qquad \cos(-u) = \cos u$$

$$\tan\left(u + \frac{\pi}{2}\right) = -\tan u; \qquad \tan(-u) = -\tan u$$

$$\sin^2 u + \cos^2 u = 1$$

$$\tan^2 u + 1 = \sec^2 u$$

$$\cot^2 u + 1 = \csc^2 u$$

$$\sin^2 u = \tfrac{1}{2}\left(1 - \cos 2u\right)$$

$$\cos^2 u = \tfrac{1}{2}\left(1 + \cos 2u\right)$$

$$\sin(u \pm v) = \sin u \cos v \pm \cos u \sin v$$

$$\cos(u \pm v) = \cos u \cos v \mp \sin u \sin v$$

$$\sin^{-1} u = \cos^{-1}\left(1 - u^2\right)^{\frac{1}{2}}$$

$$-\frac{\pi}{2} < \tan^{-1}\left(\frac{y}{x}\right) < \frac{\pi}{2}$$

$\tan2^{-1}(x, y)$ considers the signs of both x and y, $-\pi < \tan2^{-1}(x, y) < \pi$

Note: $\tan2^{-1}\left(|x|, |y|\right) = \tan^{-1}\left(\dfrac{|y|}{|x|}\right)$

$$\frac{d}{dx}\sin u = \cos u \frac{du}{dx}; \quad \frac{d}{dx}\cos u = -\sin u \frac{du}{dx}$$

$$\frac{d}{dx}\sin^{-1}u = \frac{1}{\sqrt{1-u^2}}\frac{du}{dx}; \qquad -\frac{\pi}{2} \le \sin^{-1}u \le \frac{\pi}{2}$$

$$\frac{d}{dx}\cos^{-1}u = -\frac{1}{\sqrt{1-u^2}}\frac{du}{dx}; \quad 0 \le \cos^{-1}u \le \pi$$

$$\frac{d}{dx}\tan^{-1}u = -\frac{1}{1+u^2}\frac{du}{dx}; \qquad -\frac{\pi}{2} \le \tan^{-1}u \le \frac{\pi}{2}$$

$$\frac{d}{dx}\cot^{-1}u = -\frac{1}{1+u^2}\frac{du}{dx}; \qquad -\frac{\pi}{2} \le \cot^{-1}u \le \frac{\pi}{2}$$

COSINE LAW

$$a^2 = b^2 + c^2 - 2b\,c\,\cos\alpha$$

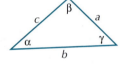

SINE LAW

$$\frac{a}{\sin\alpha} = \frac{b}{\sin\beta} = \frac{c}{\sin\gamma}$$

INDEX